D1570475

COMPUTER AIDED CHEMICAL THERMODYNAMICS OF GASES AND LIQUIDS

COMPUTER AIDED CHEMICAL THERMODYNAMICS OF GASES AND LIQUIDS

Theory, Models, and Programs

PAUL BENEDEK
AND
FRANCIS OLTI

EÖTVÖS UNIVERSITY
BUDAPEST, HUNGARY

A Wiley-Interscience Publication

JOHN WILEY & SONS

New York · Chichester · Brisbane · Toronto · Singapore

***Library of Congress Cataloging in Publication Data*:**

Benedek, Pál.
Computer aided chemical thermodynamics of gases and liquids.

"A Wiley-Interscience publication."
Includes index.
1. Thermodynamics—Data processing. I. Olti, Francis. II. Title.

QD504.B45 1985 541.3′95′02854 84-29168
ISBN 0-471-87825-1

Printed in the United States of America

10 9 8 7 6 5 4 3 2 1

To two Catherines
and Dávid, Kálmán, and Zoltán

FOREWORD

The three cornerstones of present science and technology are energy, entropy, and equilibrium, and they are in the focus of this book on chemical thermodynamics. Although there are many excellent texts dealing with the subject, our rapidly changing world demands, and offers at the same time, new ways and approaches to the treatment and use of this branch of applied thermodynamics. The wide proliferation of personal computers in the "post-slide-rule" era permits easy access to computers having sufficient capacity to handle most calculations required in chemical thermodynamics. The goal of this book is to exploit this opportunity and to provide an easy-to-follow guide to the calculation of physicochemical properties related to energy, entropy, and equilibria. Yet, the book goes much deeper than merely giving a tool for such calculations needed in chemical engineering practice. Beginning with the basic concepts of physics it develops the fundamental framework of thermodynamics. Upon introducing elementary quantum chemistry and statistical thermodynamics the authors provide a definite treatment of chemical thermodynamics in a lucid and well-organized fashion. Thus the book is a self-contained treatise on chemical thermodynamics profusely supplemented with computer programs. On that account the programs serve as aids for applying the theory to particular situations and studying the effect of changes in composition, temperature, and pressure.

Classical books on chemical thermodynamics did not go beyond examples and problems which could not have been solved with the use of a slide rule. Recently, "cookbooks" with computer programs have become available to chemical engineers for use in calculation of problems; however, they do not provide information on the underlying fundamental science. The present book superbly fills the gap by combining a rigorous theoretical treatment with easy-to-use, highly practical computer programs. It is likely to serve as a model for scientific and engineering texts of the future to be used in the classroom and by practicing chemical engineers. I was very impressed with the success by one of the authors, P. B., in using this text in both undergraduate and graduate courses on chemical engineering thermodynamics while teaching at Yale in the current academic year.

The authors are to be complimented for this much needed compound text on chemical thermodynamics which carries on the legacy of J. W. Gibbs and B. F. Dodge, who pioneered this field at Yale, and brings it befittingly in line with the opportunities and necessities brought about by the computer revolution.

CSABA HORVÁTH

Department of Chemical Engineering
Yale University
New Haven, Connecticut

PREFACE

This book deals with the theory of thermodynamics, while making use of structural models for both pure substances and mixtures. Theory and structural models are used together with an algorithm to calculate meaningful thermodynamic properties.

The book is informally divided into four parts. The first part consists of Chapter 1, which deals with the elements of both classical and quantum mechanics. The reason for incorporating these subjects into the book is very simple: we use the basic ideas of classical mechanics (coordinates, force, pressure, energy, principles of conservation, and virial theorem) in developing thermodynamics in the second part of the book.

Three chapters belong to the second part. Chapter 2 introduces the first new physical variable that can not be treated within classical mechanics: temperature. The Zeroth Law, which leads to this idea, exists in the common frontier of mechanics and thermodynamics. The First Law is the subject of Chapter 4. Unlike most textbooks in declaring this Law, we do not use the notion of heat. In a conventional treatment the First Law is nothing more than the definition of internal energy. We do not need this because it was introduced in Chapter 1 by the apparatus of classical mechanics. So internal energy, volume, and amount of substance, being defined independently of thermodynamics, are well-known variables used in the formulation of the first law, which in this context states that these variables determine all other variables of a body as well.

One of these other variables is the entropy. We encounter entropy in Chapter 5 where it is immediately used to formulate the general criteria of equilibrium. Properties of the entropy function are described by the Second Law. This approach excludes both the ideas of heat and the Carnot cycle. In Chapter 6 we introduce a few new variables, such as enthalpy and free energy, to aid in the formulation of criteria in special types of equilibria. This chapter elaborates on all the relationships among the thermodynamic variables. By the end of the Chapter the complete structure of thermodynamics has been depicted. The characteristic common to all these relationships is that they are valid at any temperature and pressure and for any substance and composition, because, in developing thermodynamics, we never used information about the molecular structure of matter. Direct, practical application is possible only if one is able to substitute in the appropriate thermodynamic relationship an explicit expression of a model taking into account the specific molecular structure of the substance. This may be described by intermolecular and intramolecular forces and energies.

In Chapter 1, the fundamentals of both intermolecular forces and quantum mechanics, which was invented to describe intramolecular forces and energy, were introduced, however, only in the third part of this book do we systematically develop molecular structural models of substances. We first discuss the models of the one-component body (Chapter 3) and later those of mixtures (Chapter 11). Chapter 7, on statistical thermodynamics, is presented here not as an alternative to classical thermodynamics, but as a method to develop a family of models for the calculation of intramolecular energy of molecules with a given structure. The appropriate thermodynamic relationships together with a specific model are necessary for the calculation of thermodynamic properties of substances.

We show how to calculate thermodynamic properties for both one-component and multicomponent systems (Chapters 8–10), while Chapters 12 and 13 present calculations for phase or chemical equilibrium. These five chapters comprise the fourth part of the book and they are really the application of thermodynamics, together with specific models, to chemical systems, that is, chemical thermodynamics.

The appropriate general relationships, together with a specific model, taking pressure, temperature, and composition of a given substance into account, are sufficient for any state and/or equilibrium calculation. At this point, however, a new question arises: How do we connect the general relationship with the specific model. This question is not a fundamental one; it belongs to the practical side of the application. We demonstrate the answer to this question by means of algorithms and computer programs presented in each of the appropriate chapters.

Theory, models, and algorithms are inseparable.

PAUL BENEDEK
FRANCIS OLTI

Budapest, Hungary
May 1985

ACKNOWLEDGMENTS

The authors express their thanks to Professor Pál Fejes, Professor Iván Pallai, and Dr. Miklós Preisich for critical comments and advice during the preparation of the manuscript.

The authors also wish to thank Dr. Gabriella Bogyó and Dr. Lajos Timár for the English translation, and also our editors at John Wiley & Sons for polishing up numerous fine points in the text.

I would like to convey my additional gratitude to Yale University's Department of Chemical Engineering for giving me the opportunity to teach chemical thermodynamics to both graduate and undergraduate students during the 1984–85 academic year. And I also wish to thank these students for their collaboration in this educational adventure.

P. B.

CONTENTS

ALGORITHMS AND PROGRAMS

Classical Mechanics and Quantum Mechanics

Equations of State

Elements of Statistical Thermodynamics

Enthalpy

Calculation of the Thermodynamic Properties of the One-Component Homogeneous Phase

Mixtures

Vapor–Liquid Equilibria

Chemical Equilibrium

LIST OF SYMBOLS

a	Activity	—
a	Amplitude	m
a_{nm}	Group interaction parameter	K
a_{RK}	Dimensionless parameter of the Redlich–Kwong equation of state	
$\mathbf{a}[t]$	Acceleration	$\mathrm{m\ s^{-2}}$
A_{12}, A_{21}	Parameters of excess-free-enthalpy equations	
$A_{\mathrm{w}i}$	Van der Waals surface of group i in a molecule	$\mathrm{cm^2\ molecule^{-1}}$
$\mathbf{A}$	Atom matrix	
b	Covolumen	$\mathrm{m^3\ mole^{-1}}$
b_{RK}	Dimensionless parameter of the Redlich–Kwong equation of state	—
$B[T]$	Second virial coefficient	$\mathrm{m^3\ mole^{-1}}$
B_{e}	Rotational constant	$\mathrm{s^{-1}}$
$B_{\mathrm{r}}[T_{\mathrm{r}}]$	Reduced second virial coefficient	—
$B_{\mathrm{r}}^{(0)}$	Reduced second virial coefficient of simple fluid	—
$B_{\mathrm{r}}^{(1)}$	Reduced second virial coefficient, specific size and shape contribution	—
$B_{\mathrm{r}}^{(\mathrm{p})}$	Reduced second virial coefficient, specific polarity contribution	—
c	Cohesion energy	$\mathrm{J\ m^{-3}}$
c_P	Molar heat capacity at constant pressure	$\mathrm{J\ K^{-1}\ mole^{-1}}$
c_v	Molar heat capacity at constant volume	$\mathrm{J\ K^{-1}\ mole^{-1}}$

$\exp(-\beta E_i)$	Boltzmann factor	—
E	Total energy (of a molecule)	J
E	Superscript indicating excess property	—
$E_0, E_1, E_2, \ldots$	Eigenvalues of the Schrödinger equation	
E_{trans}	Energy of translation	J
E_{k}	Kinetic energy	J
E_{p}	Potential energy	J
E_{rot}	Energy of the rotator	J
E_{vibr}	Energy of the harmonic oscillator	J
f	Molar free energy	J mole^{-1}
f	Fugacity	Pa
$\bar{f}_i$	Partial fugacity of component i	Pa
F	Free energy	J
F_{d}	Dispersion force	N
F_{G}	Degrees of freedom	—
$\mathbf{F}$	Force	N
$\mathbf{F}_{ik}$	Force acting on the ith particle from the kth particle	N
$\mathbf{F}_{i0}$	Force acting on the ith particle from without	N
g	Molar free enthalpy	J mole^{-1}
g	Coordination factor	—
g	Combinatorial or statistical weight	—
g^{E}	Molar excess free enthalpy of mixing	J mole^{-1}
g_{ii}	Energy needed for the removal of one mole of molecules i from a surrounding liquid of molecules i	J mole^{-1}
g_{ij}	Energy needed for the removal of one mole of molecules i from a surrounding liquid of molecules j	J mole^{-1}
G	Free enthalpy	J
ΔG°	Standard free enthalpy of reaction	J mole^{-1}
$\Delta G_{\text{f}}^\circ$	Standard free enthalpy of formation	J mole^{-1}
h	Molar enthalpy	J mole^{-1}

Δh	Isothermal enthalpy correction	J mole^{-1}
Δh_{vap}	Molar enthalpy of vaporization	J mole^{-1}
H	Enthalpy	J
ΔH°	Standard enthalpy of reaction	J mole^{-1}
ΔH_f°	Standard enthalpy of formation	J mole^{-1}
$\mathbf{h}$	Planck constant	6.626176×10^{-34} J s
i	Subscript for component index	—
id	Superscript indicating ideal mixture	—
I	Moment of inertia	kg m^2
I	First ionization energy	J
$\mathbf{I}$	Momentum	kg m s^{-1}
j	Subscript for component index	—
J	Jacobi determinant	—
J	Rotational quantum number	—
k	Number of component in a system	—
k	Force constant	kg s^{-2}
$\mathbf{k}$	Boltzmann constant	1.381×10^{-23} J K^{-1}
k_i	Henry constant of component i	Pa
K	Chemical-reaction equilibrium constant	—
K_i	Vapor–liquid equilibrium ratio or volatility of component i	—
l	Largest diameter of the molecule	m
l	Superscript indicating liquid	—
$\bar{l}$	Mean free path	m
L	Dimension of length	—
L	Operator of Legendre transformation	—
L	Liquid-to-feed ratio	—
m	Parameter in Soave and PR equations of state	—
m	Mass	kg
M	Actual value of an arbitrary extensive property of one mole of mixture at a given temperature, pressure, and composition	—
M_i	Molar value of property M of com-	—

	ponent i in pure state at temperature and pressure of mixture	
$\overline{M}_i$	Partial molar value of property M of component i	—
M_i^E	Excess M of component i	—
ΔM	Molar value of mixing of property M	—
ΔM_i	Partial molar value of mixing of property M of component i	—
M	Dimension of mass	—
n	Translational quantum number	—
n	Number of molecules	—
N	Dimension of amount of substance	—
N_j	Amount of substance j	mole
$\boldsymbol{N}$	Avogadro's number	6.022×10^{23} mole
P	Pressure	Pa
P_c	Critical pressure	Pa
P_i	Probability of quantum state i	—
$\overline{P}_i$	Partial pressure of component i	Pa
P_r	Reduced pressure	—
P_u	Standard pressure	$101{,}325 \text{ N m}^{-2}$
$P°$	Equilibrium vapor pressure	Pa
ps	Subscript indicating "pseudo-one-component"	
q	Molecular partition function	—
q_i	Relative surface (lattice parameter) of group i	—
Q	Canonical partition function	—
r	Radius of rotator	m
r	displacement	m
(r)	Superscript indicating reference molecule	—
r_i	Number of lattice points occupied by molecule i	—
r_i	Relative volume (lattice parameter) of group i	—
r_{ij}	Distance between two points i, j	m
$\mathbf{r}[t]$	Radius vector	m

$\boldsymbol{R}$	Universal gas constant	8.31441 J K^{-1} $mole^{-1}$
$\mathbf{R}$	Reaction matrix	—
s	Molar entropy	J K^{-1} $mole^{-1}$
Δs	Isothermal entropy correction	J K^{-1} $mole^{-1}$
S	Entropy	J K^{-1}
t	Time	s
t	Empirical temperature	°C
T	Thermodynamic temperature	K
T_b	Bubble point temperature	K
T_{bp}	Boiling point	K
T_B	Boyle temperature	K
T_c	Critical temperature	K
T_d	Dew-point temperature	K
T_{hbp}	Normal boiling point	K
T_r	Reduced temperature	—
T	Dimension of temperature	—
u	Molar internal energy	J $mole^{-1}$
Δu	Isothermal internal-energy correction	J $mole^{-1}$
U	Internal energy	J
v	Vibrational quantum number	—
v	Molar volume	m^3 $mole^{-1}$
v_c	Critical volume	m^3 $mole^{-3}$
v_c	Volume of one cell in liquid phase	m^3 $molecule^{-1}$
v_f	Free volume of a molecule of diameter σ	m^3 $molecule^{-1}$
v_r	Reduced volume	—
V	Volume	m^3
V_{wi}	Van der Waals volume of segment i	cm^3 $mole^{-1}$
$\mathbf{v}[t]$	Velocity	m s^{-1}
$\bar{v}$	Mean velocity	m s^{-1}
w	Exchange energy	J/molecule
W	Molar exchange energy	J/mole
W	Work	J
x_i	Mole fraction of component i	—

$x^{(a)}$	Mole fraction at azeotropic composition	—
X	Extent of reaction	—
X_m	Group fraction of group *m*	—
y_i	Mole fraction of component *i* (in gas phase)	—
z	Coordination number	—
z	Compressibility factor	—
z_c	Critical compressibility factor	—
Z	Collision number	s^{-1}
α	Inhomogeneity parameter in NRTL equation	—
α	Polarizability	m^3
α	Relative volatility or separation factor	—
α_v°	Ratio of two equilibrium vapor pressures	—
γ_i	Activity coefficient of component *i*	—
Γ_k	Residual activity coefficient of group *k* in the mixture	—
$\Gamma_k^{(i)}$	Residual activity coefficient of group *k* in pure *i*	—
δ	Solubility parameter	—
δ_{ij}	Binary cross effect of second virial coefficient	—
ε	Vapor ratio	—
ϵ	Energy quantum	J
$\varepsilon[r_{ij}]$	Pair potential	J
ε_r	Eigenvalue of rotator energy	J
ε_t	Eigenvalue of translational energy	J
ε_v	Eigenvalue of harmonic-oscillator energy	J
ε_0	Depth of potential well or minimum of pair potential	J
ε_k	Kinetic energy of a single molecule per degree of freedom	J
Θ	Dimension of time	—
Θ	Absolute temperature	K
Θ_m	Surface fraction of group *m*	—

Θ_r	Rotational characteristic temperature	K
Θ_v	Vibrational characteristic temperature	K
ι	Superscript for phase index	—
μ	Joule–Thomson coefficient	—
μ	Electric dipole moment	C m
μ	Reduced mass	kg
μ_j	Chemical potential of component j	J mole^{-1}
ν	Frequency	s^{-1}
ν_j	Stoichiometric coefficient of component j in equation of a chemical reaction	—
$\nu_m^{(i)}$	Stoichiometric index of group m in molecule i	—
π	Vector of parameters in equation of state	—
ρ	Density	kg m^{-3}
ρ	Molar density	mole m^{-3}
ρ_n	Particle density	m^{-3}
σ	Collision diameter	m
ϕ	Phi factor for characterization of polarity	—
ϕ	Phase angle	—
$\phi[x, y, z]$	Potential	J
ϕ_j	Volume fraction of component j	—
ϕ_r	Ideal reduced volume	—
φ	Number of phases at equilibrium	—
ψ	Legendre transform	—
ψ	Wave function	—
$\psi_{01}, \psi_{11}, \psi_2, \ldots$	Eigenfunction of the Schrödinger equation	s^{-1}
ω	Omega factor or acentric factor	—
$\langle x \rangle$	Expected value of x	
$\circ$	Reference state	
$^{(0)}$	Simple molecule	
*	Ideal gas	
$\bar{}$	Accent for "partial molar"	

COMPUTER AIDED CHEMICAL THERMODYNAMICS OF GASES AND LIQUIDS

INTRODUCTION

In this book various methods are used for the explication of the philosophy of chemical thermodynamics. Some are in style; others are based on the use of computers.

All the themes are introduced in a classical way. The subject is explained by the description of an actual experiment or a thought experiment; a mathematical model of the phenomenon is constructed and treated in the usual way with mathematical tools; and finally, the physical content of the results is discussed.

In certain details we depart from custom:

1. Equations are not numbered. If we need a previous equation within the same deduction, we state the equation unequivocally. If we need an equation which has figured earlier, then we write it again, to spare the reader the trouble of turning the pages.
2. In the discussion of the relationships of some physical property with other physical properties, or in mathematical language, of multivariable functions, italic letters are used to denote both variables and functions. For example

$$U = U[S, V, N_j]$$

at the LHS means the value of the function U to the independent variables S, V, N_j. We use the convention of placing the independent variables of the function in brackets.

3. We nest parentheses for grouping in the style of programming languages, for example

$$y = ((dx + c)x + b)x + a$$

We do not use braces.

4. The symbol of natural logarithm is ln in the text, LOG in the BASIC programs.
5. There are no true-to-scale graphs, such as were used for manual calculations, and for observing trends. Instead, there are programmed calculation algorithms, with which you can generate on your personal computer the values needed.
6. There are no examples, exercises, or illustrations to be performed by manual calculation. Instead, programs are supplied, which can be stored in the personal computer and read at any time, but they are *not substance-specific*, and can be used also
   (a) for other components,
   (b) for other compositions,
   (c) in other pressure–temperature regions.

   We suggest for each such program a few examples for calculation, but recommend that you extend these cases on your own initiative. In this way you can also acquire the thermodynamic knowledge of materials that you as a chemist will need.

As concerns the program given, an elementary knowledge of the programming language BASIC is assumed. The programs were written for the Apple II, because these professional personal computers are rather widely used. However, there should be no difficulty in the adaption of the programs to any other computer which has a periphery suitable for the storage of the source program and the data files. As data files have to be frequently read, it is convenient if the computer has a floppy-disk unit.

For the sake of convenient trial runs, a small property data file called PHDAT was compiled, containing 14 properties of 23 substances. With the aid of the service programs, this data file can be easily extended, or exchanged, with data selected from those on 232 substances given in the appendix at the end of the book. Similarly, data files are used for the storage of measuring data, which are used for checking the results calculated with the programs in Chapter 10.

The computer programs are intentionally not too well polished. Only the most necessary error checking has been performed; the graphics could be far more elegant, and the output graphs more attractive. Our aim with these programs was not an exercise in computation techniques, but the clear, concise imparting of algorithms.

Those who are using this book for serious study are advised to perform for themselves the programming work specified under the subtitle "Algorithm and Program," after considering the essential steps of algorithm given under "Algorithm," and to run their program for the model examples given under "Example," comparing their results with the results given there. The program listing should be used only as suggestive source material, or as a check to help the student when he or she is stuck. We think that by this means the subject matter of a chapter read once can be memorized without toil, by pleasant, interesting programming work, free of logical gaps, imparting to the reader the continual reward of success.

Those who turn the pages of our book to facilitate their daily engineering work can seek the program they need for their task—we hope they will usually find it—and run it with as few alterations as possible. For this purpose it suffices to compile with the service programs already mentioned a property data file for the material of interest to the reader.

1

CLASSICAL MECHANICS, QUANTUM MECHANICS

Let us consider a single particle (molecule) moving in a fixed (x, y, z) coordinate system. The location of this particle in the coordinate system at time t is given by the radius vector $\mathbf{r}(t)$. The derivative of the radius vector $\mathbf{r}$ with respect to time is the velocity $\mathbf{v}(t)$ of the particle:

$$\mathbf{v}(t) = \dot{\mathbf{r}}(t) = \frac{d\mathbf{r}(t)}{dt}$$

The acceleration of the particle is the second derivative of the radius vector with respect to time:

$$\mathbf{a}(t) = \ddot{\mathbf{r}}(t) = \frac{d^2\mathbf{r}(t)}{dt^2}$$

We are now acquainted with three vectors. Each has three components in the rectangular coordinate system:

$$\mathbf{r} = x + y + z$$

$$\mathbf{v} = v_x + v_y + v_z$$

$$\mathbf{a} = a_x + a_y + a_z$$

If the acceleration of the particle is zero, then the particle performs a uniform straight-line motion.

THE THREE LAWS OF MECHANICS

Let us consider now two particles (molecules) in the fixed (x, y, z) coordinate system. When the distance r_{ij} between the particles is sufficiently large, each preserves its constant velocity $\mathbf{v}_i$ or $\mathbf{v}_j$, respectively:

$$\left.\frac{d\mathbf{v}_i}{dt}\right|_{r_{ij}\gg\sigma} = 0 \qquad \left.\frac{d\mathbf{v}_j}{dt}\right|_{r_{ij}\gg\sigma} = 0$$

This is Newton's first law. It is practically applicable when the distance between the two particles (molecules) is more than four to five times the diameter σ of either particle (considered as spherical).

If the two particles approach each other, interaction occurs, as a result of which both are accelerated, and their independent straight-line uniform motion ceases. The change in motion arises from the force acting on each particle. This vector is by definition

$$\mathbf{F}_i \equiv \frac{d\mathbf{I}_i}{dt} \qquad \mathbf{F}_j \equiv \frac{d\mathbf{I}_j}{dt}$$

On the right side is the derivative of momentum with respect to time. This is Newton's second law.

The momentum vector is defined by the third law. The interacting two particles acquire accelerations $\mathbf{a}_{ij}$ and $\mathbf{a}_{ji}$, respectively. The two accelerations are opposite in direction, but not necessarily equal in magnitude when molecules i and j are not identical.

According to Newton's third law, the force by which particle j acts on particle i is equal and opposite to the force by which particle i acts on particle j:

$$\mathbf{F}_{ij} = -\mathbf{F}_{ji}$$

Consequently there exist characteristic scalar parameters m_i and m_j satisfying

$$m_i\mathbf{a}_{ij} \equiv -m_j\mathbf{a}_{ji}$$

These are the masses of the particles.

THE EQUATIONS OF MOTION

The momentum of the particle is the product $m\mathbf{v}$. According to the second law,

$$\mathbf{F}_i = \frac{d\mathbf{I}_i}{dt}$$

while on the other hand,

$$\mathbf{F}_i = m_i \mathbf{a}_i = m_i \frac{d\mathbf{v}_i}{dt} = \frac{d(m_i \mathbf{v}_i)}{dt}$$

Thus the momentum is equal to the product of the mass and the velocity of the particle:

$$\mathbf{I}_i = m_i \mathbf{v}_i$$

Force is a vector quantity; its components are

$$\mathbf{F} = F_x + F_y + F_z$$

The components of the force acting on particle m can be written also in the following form:

$$F_x = X = m\ddot{x}$$

$$F_y = Y = m\ddot{y}$$

$$F_z = Z = m\ddot{z}$$

In the coordinate system (x, y, z) the force $m\mathbf{a}$ acting on particle of mass m is a function of radius vector $\mathbf{r}$ and velocity $\dot{\mathbf{r}}$ of the particle:

$$m\mathbf{a} = F[\mathbf{r}, \dot{\mathbf{r}}, t]$$

Separating into components, we have three differential equations of second order:

$$m\ddot{x} = F_x[x, y, z, \dot{x}, \dot{y}, \dot{z}, t]$$

$$m\ddot{y} = F_y[x, y, z, \dot{x}, \dot{y}, \dot{z}, t]$$

$$m\ddot{z} = F_z[x, y, z, \dot{x}, \dot{y}, \dot{z}, t]$$

These differential equations are called the equations of motion of the particle. If the space–time dependences of F_x, F_y, F_z are explicitly known, then the equations of motion can be solved as initial-value problems in terms of the three initial space coordinates and the three initial velocity components. In other words, the history of the motion of the particle is determined by the six parameters (initial values) fixed at the point of time $t = t_0$. It is usual to say that the particle has six degrees of freedom.

Free Fall

Let us consider a resting particle of mass m at the origin of the coordinate system at a time $t_0 = 0$, on which a constant force mg acts in direction z. No force is acting in directions x and y. Let us describe the motion of the particle in space and time.

The six initial values are the following:

$$x_0 = 0 \qquad v_x|_0 = 0$$

$$y_0 = 0 \qquad v_y|_0 = 0$$

$$z_0 = 0 \qquad v_z|_0 = 0$$

Moreover,

$$F_x = 0$$

$$F_y = 0$$

$$F_z = mg = \text{const}$$

Hence, the three equations of motion are

$$m\ddot{x} = 0$$

$$m\ddot{y} = 0$$

$$m\ddot{z} = mg$$

Double integration of the equations of motion yields velocity and space coordinates:

$$\dot{x} = v_x|_0 \qquad x = v_x|_0 \cdot t + x_0$$

$$\dot{y} = v_y|_0 \qquad y = v_y|_0 \cdot t + y_0$$

$$\dot{z} = gt + v_z|_0 \qquad z = \tfrac{1}{2}gt^2 + v_z|_0 \cdot t + z_0$$

Substituting the initial values, we have

$$x[t] = 0$$

$$y[t] = 0$$

$$z[t] = \tfrac{1}{2}gt^2$$

There is no displacement in the x and y directions; the particle falls in the z

direction. The distance covered is proportional to the square of the time elapsed since t_0.

Reflection of a Particle

It is important to be able to express the effect of force in two ways: in terms of the time over which it is exerted, and in terms of the distance. The first is given by the *momentum theorem*, according to which the integral of force with respect to time is equal to the change in momentum. This is rather evident, since

$$\int_{t_0}^{t} \mathbf{F}[t]\,dt = \int_{t_0}^{t} \frac{d\mathbf{I}}{dt}\,dt$$

$$= \mathbf{I}[t] - \mathbf{I}[t_0]$$

Consider a particle (molecule) of mass m, which moves between two parallel walls at unit distance from one another on a path perpendicular to the walls, and turns back at the walls by elastic impact. The velocity of the particle is v_x. What is the change in momentum during a single impact? What is the change in momentum per unit time at one of the walls?

The momentum of the particle moving towards the wall is mv_x, and after impact, moving in the opposite direction, $-mv_x$. The change in momentum during a single impact is

$$\Delta I_x = mv_x - (-mv_x) = 2mv_x$$

To answer the second question, we have to consider that the particle hits one of the walls during unit time on $v_x/2$ occasions. Accordingly, the change of momentum per unit time is

$$(2mv_x)\left(\frac{v_x}{2}\right) = mv_x^2$$

This is by definition the force:

$$F_x \equiv \frac{dI_x}{dt} = \frac{\Delta I_x}{\Delta t} = mv_x^2$$

WORK, KINETIC ENERGY, AND POTENTIAL

We now consider the effect of force in terms of distance along the path of the particle in motion. Here the integral of force along the path covered must be taken into consideration. The elementary work performed by a force $\mathbf{F}$ over a displacement $d\mathbf{r}$ is

$$dW = \mathbf{F} \cdot d\mathbf{r}$$

$$= X\,dx + Y\,dy + Z\,dz$$

Thus, the work is the product of the displacement and the component of force pointing in the direction of the displacement.

The work done along the path between points A and B is

$$W_{AB} = \int_A^B \mathbf{F} \cdot d\mathbf{r}$$

$$= \int_A^B (X\,dx + Y\,dy + Z\,dz)$$

The components of an elementary displacement are functions of time t, and thus

$$W_{AB} = \int_{t_A}^{t_B} \left(X\frac{dx}{dt} + Y\frac{dy}{dt} + Z\frac{dz}{dt} \right) dt$$

Expanding the components of force, we have

$$W_{AB} = m\int_{t_A}^{t_B} (\ddot{x}\dot{x} + \ddot{y}\dot{y} + \ddot{z}\dot{z})\, dt$$

Here, however,

$$\int \ddot{x}\dot{x}\, dt = \int \frac{1}{2}\frac{d}{dt}(\dot{x})^2\, dt$$

and therefore the work done along the path between points A and B is

$$W_{AB} = \tfrac{1}{2}m\int_{t_A}^{t_B} \frac{d}{dt}(\dot{x}^2 + \dot{y}^2 + \dot{z}^2)\, dt$$

$$= \tfrac{1}{2}m\int_{t_A}^{t_B} \frac{d}{dt}\mathbf{v}^2$$

$$= \tfrac{1}{2}m\mathbf{v}_B^2 - \tfrac{1}{2}m\mathbf{v}_A^2$$

Thus we obtain that the work done along the path between points A and B is equal to the change of the quantity $\frac{1}{2}m\mathbf{v}^2$, that is, to the difference in the values of $\frac{1}{2}m\mathbf{v}^2$ at points A and B. The quantity $\frac{1}{2}m\mathbf{v}^2$ is called the kinetic energy E_k

of the particle of mass m, and we have

$$W_{AB} = E_k[B] - E_k[A]$$

The work done is equal to the difference in the values of kinetic energy at the initial and end points of the path. A force for which this work is independent of the path is called a conservative force.

If there exists a function $\phi[x, y, z]$ of which the negative partial derivatives with respect to coordinates x, y, z give the force components X, Y, Z,

$$X = -\frac{\partial \phi}{\partial x}$$

$$Y = -\frac{\partial \phi}{\partial y}$$

$$Z = -\frac{\partial \phi}{\partial z}$$

then this function is called a potential. The force is then the negative gradient of the potential:

$$\begin{aligned} \mathbf{F} &= X + Y + Z \\ &= -\operatorname{grad} \phi \end{aligned}$$

A number of conclusions and useful concepts follow from the definition of potential.

A potential is a scalar quantity with the dimensions of energy. Points at which potential ϕ is equal to a constant C form a surface defined by the following equation:

$$\phi[x, y, z] = C$$

If all possible values are given to C, a set of surfaces is obtained. These are the equipotential surfaces. The force acts perpendicularly to an equipotential surface.

If a potential exists, then:

1. The elementary work done is equal to the total differential of potential, taken with negative sign.
2. The work performed along the path AB is equal to the negative potential difference; thus it is independent of the path.
3. Along a closed path the work done is zero.

Necessary and sufficient conditions for the existence of potential are the following:

$$\frac{\partial Z}{\partial y} - \frac{\partial Y}{\partial z} = 0$$

$$\frac{\partial X}{\partial z} - \frac{\partial Z}{\partial x} = 0$$

$$\frac{\partial Y}{\partial x} - \frac{\partial X}{\partial y} = 0$$

The Work Performed by a Gravitational Force

What work is performed by the gravitational force on particle of mass m, when moving it from point A to point B?

If a potential exists, then the work of gravitational force is equal to the negative potential difference. The particle of mass m is located at the origin of a coordinate system (x, y, z). This is point A. If the z axis is directed vertically downward, the components of the force acting on the particle are

$$X = 0 \qquad Y = 0 \qquad Z = mg$$

The potential exists, because the conditions for its existence are fulfilled:

$$\frac{\partial Z}{\partial y} - \frac{\partial Y}{\partial z} = \frac{\partial (mg)}{\partial y} - \frac{\partial 0}{\partial z} = 0$$

$$\frac{\partial X}{\partial z} - \frac{\partial Z}{\partial x} = \frac{\partial 0}{\partial z} - \frac{\partial (mg)}{\partial x} = 0$$

$$\frac{\partial Y}{\partial x} - \frac{\partial X}{\partial y} = \frac{\partial 0}{\partial x} - \frac{\partial (mg)}{\partial y} = 0$$

The partial derivatives of the potential are

$$-\frac{\partial \phi}{\partial x} = X = 0 \qquad -\frac{\partial \phi}{\partial y} = Y = 0 \qquad -\frac{\partial \phi}{\partial z} = Z = mg$$

Accordingly, the potential at the origin A is

$$\phi_A = mgz = mg0 = 0$$

and at point B, at a depth z,

$$\phi_B = -mgz$$

Hence, the work done by the gravitational force is

$$W_{AB} = -(\phi_B - \phi_A)$$
$$= mgz$$

In the gravitational field the equipotential surfaces are perpendicular to the z axis. This means that a displacement of the particle in the x and y directions does not change the work done, which is determined solely by the distance z. Observe that the mass m gets from the place of higher potential to a place of lower potential through the action of the gravitational force.

CONSERVATION OF ENERGY

It has been mentioned already that potential is a quantity with the dimensions of energy. It is therefore also called potential energy (E_p). In a conservative field of force the sum of the potential energy and the kinetic energy of the particle is constant. Indeed, this is evident, as the work done along the path between points A and B is equal to the change in kinetic energy,

$$W_{AB} = E_\mathrm{k}[B] - E_\mathrm{k}[A]$$

while on the other hand this same work is equal to the potential difference:

$$W_{AB} = \phi[A] - \phi[B] = E_p[A] - E_p[B]$$

The left sides are identical; thus

$$E_\mathrm{p}[A] + E_\mathrm{k}[A] = E_\mathrm{p}[B] + E_\mathrm{k}[B]$$

At any point of the path the sum of the potential and kinetic energies of the particle is the same:

$$E_\mathrm{p} + E_\mathrm{k} = \mathrm{const}$$

LINEAR HARMONIC OSCILLATOR

The law of energy conservation is valid also when elastic forces are acting. To see this, first of all, we solve the equation of motion for a mass m, moving under the action of an elastic force. The simplest form of elastic forces is

$$\mathbf{F} = -k\mathbf{r},$$

where $\mathbf{r}$ is the displacement reckoned from the position of rest, and k, the force constant, a positive number. The force is directed towards the position of rest and is proportional to the displacement. The components of the elastic force are

$$X = -kx$$

$$Y = -ky$$

$$Z = -kz$$

A potential corresponding to the elastic force exists, because the necessary and sufficient conditions for its existence are fulfilled:

$$\frac{\partial Z}{\partial y} - \frac{\partial Y}{\partial z} = 0$$

$$\frac{\partial X}{\partial z} - \frac{\partial Z}{\partial x} = 0$$

$$\frac{\partial Y}{\partial x} - \frac{\partial X}{\partial y} = 0$$

The partial derivatives of the potential are in turn

$$-\frac{\partial \phi}{\partial x} = X = -kx$$

$$-\frac{\partial \phi}{\partial y} = Y = -ky$$

$$-\frac{\partial \phi}{\partial z} = Z = -kz$$

Let us first consider the linear case ($Y = Z = 0$), when the potential (i.e., the potential energy) is

$$\phi = \frac{k}{2}x^2$$

The constant of integration is zero, because the potential at the point $x = 0$ is by convention zero. The equation of motion of a mass m under the action of the elastic force will then be

$$m\ddot{x} = -kx$$

or

$$\ddot{x} = -\frac{k}{m}x$$

Let us introduce the following notation:

$$\omega = \sqrt{\frac{k}{m}}$$

Then the equation of motion can be written in the following form:

$$\ddot{x} + \omega^2 x = 0$$

This is an initial-value problem for a differential equation of second order. Let the two initial values be

$$x[0] = 0$$

$$v_x[0] = a\sqrt{\frac{k}{m}}$$

where on the right-hand side the amplitude a can be freely chosen under the given physical conditions (m, k).

The solution of the equation of motion is

$$x = a \sin \omega t$$

The velocity of the mass is obtained by differentiating:

$$v_x = \dot{x} = \omega a \cos \omega t$$

and the acceleration by differentiating the velocity:

$$\ddot{x} = -\omega^2 a \sin \omega t$$

Let us substitute back in the equation of motion:

$$-\omega^2 a \sin \omega t + \omega^2 (a \sin \omega t) = 0$$

This shows that x is actually the solution of the equation of motion, as the acceleration, $\ddot{x}$, is $-\omega^2$ times the displacement.

Now the statement can be proved that the sum of potential energy and kinetic energy is constant:

$$E = E_{\mathrm{k}} + E_{\mathrm{p}}$$

where

$$E_{\mathrm{k}} = \tfrac{1}{2} m \dot{x}^2 = \tfrac{1}{2} m \omega^2 a^2 \cos^2 \omega t$$

$$E_{\mathrm{p}} = \tfrac{1}{2} k x^2 = \tfrac{1}{2} m \omega^2 a^2 \sin^2 \omega t$$

Thus, the total energy is

$$E = \tfrac{1}{2} m\omega^2 a^2$$

This, however, is constant, because the mass m is constant, ω depends only on it and the physical nature of the elastic force (on the force constant), and the amplitude a is determined by the initial conditions and is independent of t.

The mass m is caused to vibrate (oscillate) by the action of the elastic force. The farther the mass moves away from its position $x = 0$, the greater is the elastic restoring force; as the mass moves out, the potential energy increases, while the kinetic energy diminishes. The potential energy reaches its maximum value when the velocity of the mass (and its kinetic energy) decreases to zero. Now a motion in the opposite direction begins; the velocity steadily increases to attain its maximum at $x = 0$, where the potential energy is zero. Since the kinetic energy of the mass has its maximal value at this point, this produces a displacement in the opposite direction. Thus the elastic force gives rise to simple harmonic oscillation, in which the displacement x is a periodic function of time t.

The period or vibration time, t_0, is the time after which the vibrating mass returns to the identical phase:

$$x[t + t_0] = x[t]$$

The sine function takes on identical x values when its argument changes by 2π, that is, $t_0\omega = 2\pi$. Thus the period is

$$t_0 = \frac{2\pi}{\omega} = 2\pi\sqrt{\frac{m}{k}}$$

and the frequency of vibration

$$\nu = \frac{1}{t_0} = \frac{1}{2\pi}\sqrt{\frac{k}{m}}$$

This means that the body performs ν vibrations per unit time. The frequency is determined by the physical characteristics of the oscillator and is not affected by the amplitude, which, on the other hand, determines the total energy of the oscillator.

The Energy of the Linear Harmonic Oscillator

What are the potential energy and the kinetic energy of the linear harmonic oscillator as a function of displacement?

From the preceding expression,

$$2\pi\nu = \sqrt{\frac{k}{m}} = \omega$$

The potential energy can also be written in the form

$$E_p = \tfrac{1}{2}m(2\pi\nu)^2 a^2 \sin^2\omega t$$

The displacement is

$$x = a \sin \omega t$$

Thus

$$x^2 = a^2 \sin \omega t$$

Substituting this into the expression for the potential energy, we have

$$E_p = \tfrac{1}{2}m(2\pi\nu)^2 x^2$$

The expression for the total energy is known:

$$E = \tfrac{1}{2}m(2\pi\nu)^2$$

The kinetic energy is the difference between total energy and potential energy:

$$\begin{aligned} E_k &= E - E_p \\ &= \tfrac{1}{2}m(2\pi\nu)^2 a^2 - \tfrac{1}{2}m(2\pi\nu)^2 x^2 \\ &= \tfrac{1}{2}m(2\pi\nu)^2 (a^2 - x^2) \end{aligned}$$

This relationship gives the kinetic energy as a function of distance measured from the origin. It will be needed later.

The Reduced Mass

Consider now in the (x, y, z) coordinate system two particles of mass m_1 and m_2 with radius vectors $\mathbf{r}_1$ and $\mathbf{r}_2$, respectively. When there are no external forces and the internal forces between the particles depend only on their mutual distance, the equations of motion according to the third law are

$$m_1\ddot{\mathbf{r}}_1 = \mathbf{F}[|\mathbf{r}_1 - \mathbf{r}_2|]$$

$$m_2\ddot{\mathbf{r}}_2 = -\mathbf{F}[|\mathbf{r}_1 - \mathbf{r}_2|]$$

Thus

$$m_1\ddot{\mathbf{r}}_1 + m_2\ddot{\mathbf{r}}_2 = 0$$

Let us take instead of $\mathbf{r}_1$ and $\mathbf{r}_2$ a single position vector $\mathbf{r}_0 = (m_1\mathbf{r}_1 + m_2\mathbf{r}_2)/(m_1 + m_2)$. Then

$$m_1\ddot{\mathbf{r}}_1 + m_2\ddot{\mathbf{r}}_2 = (m_1 + m_2)\ddot{\mathbf{r}}_0$$

and hence

$$\ddot{\mathbf{r}}_0 = \frac{m_1\ddot{\mathbf{r}}_1 + m_2\ddot{\mathbf{r}}_2}{m_1 + m_2} = 0$$

The vector $\mathbf{r}_0$ is the mean of the vectors $\mathbf{r}_1$ and $\mathbf{r}_2$, weighted with the respective masses, directed to the center of mass of the composite body consisting of the two particles. Since the acceleration of the center of mass is zero, this means that the center of mass performs uniform straight-line motion. On the other hand, the two particles may move away from the center of mass or approach it; they can be brought into vibration by an elastic interaction force. The equation of motion of the center of mass is

$$\mu\ddot{\mathbf{r}} = \mathbf{F}[\mathbf{r}]$$

where

$$\mu = \frac{m_1 m_2}{m_1 + m_2}$$

is the reduced mass. Thus, the equation of motion containing the reduced mass is the equation of motion of a single particle of mass μ.

Here an essential technique is shown, namely, how a two-body problem can be reduced to a single-body problem.

Algorithm and Program: Calculation of Reduced Mass

Write a program to calculate the reduced mass of two-particle systems.

Algorithm.

1. Read the masses of the particles.
2. Calculate the reduced mass.
3. Print the result.

Program M7. A possible realization:

```
]LOAD M7
]LIST

 10  REM :CALCULATION OF REDUCED MASS
 20  INPUT " INPUT M1(KG): "; M1
 25  PRINT
 30  INPUT " INPUT M2(KG): "; M2
 40  M1 = M1 * 1E25
 50  M2 = M2 * 1E25
 60  MU = M1 * M2 / (M1 + M2)
 70  MU = MU * 1E - 25
 90  M1 = M1 * 1E - 25
```

```
100  M2 = M2 * 1E - 25
120  PRINT : PRINT " REDUCED MASS
      = "; MU; " KG"
130  PRINT : PRINT : PRINT
140  INPUT "DO YOU NEED ANOTHER C
     ALCULATION (Y / N) "; A$
150  PRINT : PRINT
160  IF A$ = "Y" THEN GOTO 20
170  END
```

Example. Run the program for:

1. A hydrogen atom (consisting of one proton and one electron).
2. A hydrogen molecule (consisting of two hydrogen atoms).
3. A chlorine molecule (consisting of two chlorine atoms).
4. A hydrochloric acid molecule (consisting of one hydrogen and one chlorine atom).

```
]RUN
 INPUT M1(KG): 9.105534E - 31
 INPUT M2(KG): 1.672648E - 27
 REDUCED MASS = 9.10057986E - 31 KG

DO YOU NEED ANOTHER CALCULATION (Y / N) Y

 INPUT M1(KG): 1.627AA
?REENTER
 INPUT M1(KG): 1.672648E - 27
 INPUT M2(KG): 1.672648E - 27
 REDUCED MASS = 8.36324002E - 28 KG

DO YOU NEED ANOTHER CALCULATION (Y / N) Y

 INPUT M1(KG): 1.672648E - 27
 INPUT M2(KG): 5.854268E - 26
 REDUCED MASS = 1.62618556E - 27 KG

DO YOU NEED ANOTHER CALCULATION (Y / N) Y

 INPUT M1(KG): 5.854268E - 26
 INPUT M2(KG): 5.854268E - 26
 REDUCED MASS = 2.927134E - 26 KG

DO YOU NEED ANOTHER CALCULATION (Y / N) N
```

Algorithm and Program: Calculation of the Frequency of Vibration

Write a program for the calculation of the frequency of vibration. The reduced mass and the force constant k are known.

Algorithm.

1. Read the force constant and the reduced mass.
2. Calculate the frequency of vibration.
3. Print the result.

Program VIBR. A possible realization:

```
]LOAD VIBR
]LIST

 10  REM : CALCULATION OF VIBRATION
      FREQUENCY
 20  PRINT : PRINT : INPUT "ENTER
      K "; K
 30  PRINT
 40  INPUT "ENTER REDUCED MASS "; M
      U
 50  PRINT : PRINT
 60  NU = SQR (K / MU) / 2 / 3.141
      5
 70  PRINT "VIBRATION FREQUENCY: "
      ; NU
 80  PRINT : PRINT
 90  INPUT "DO YOU NEED ANOTHER CA
      LCULATION ?(Y / N) "; A$
100  IF A$ = "Y" THEN GOTO 20
110  END
```

Example. Run the program for a hydrogen molecule and a hydrochloric acid molecule:

```
]RUN

ENTER K 162.559
ENTER REDUCED MASS 8.36324002E - 28

VIBRATION FREQUENCY: 7.01699505E + 13

DO YOU NEED ANOTHER CALCULATION ?(Y / N) Y

ENTER K 483
ENTER REDUCED MASS 1.62618556E - 27

VIBRATION FREQUENCY: 8.67404071E + 13
DO YOU NEED ANOTHER CALCULATION ?(Y / N) N
```

MOTION OF A BODY CONSISTING OF PARTICLES

When a body consists of particles but is treated as a unified whole, the forces on its particles are classified as internal and external. Internal forces act between the particles of the body; external forces originate outside the body, but act on the particles of the body. By writing for each particle of the body its own equation of motion and solving the resulting system of differential equations, the description of the behavior of the body as a whole can be obtained. However, this procedure is rather cumbersome. Instead, we proceed as follows: The equation of motion of the ith particle of an n-particle body is written in vector form:

$$m_i \ddot{\mathbf{r}}_i = \mathbf{F}_{io} + \sum \mathbf{F}_{ik}$$

where $\mathbf{F}_{io}$ means the force acting on the ith particle from without, and $\mathbf{F}_{ik}$ the force acting on the ith particle due to the kth particle. Summation of the equations of motion of all the particles of the body moving as a whole gives

$$\sum_{i=1}^{n} m_i \ddot{\mathbf{r}}_i = \sum_{i=1}^{n} \mathbf{F}_{io} + \sum_{i=1}^{n} \sum_{k=1}^{n} \mathbf{F}_{ik}$$

The double sum is zero according to the third law,

$$\mathbf{F}_{ik} = -\mathbf{F}_{ki}$$

so that we can write for the body as a whole

$$\sum_{i=1}^{n} m_i \ddot{\mathbf{r}}_i = \sum^{n} \mathbf{F}_{io}$$

Let us define the vector $\mathbf{r}_0$ in the following way:

$$\mathbf{r}_0 = \frac{\sum\limits_{i=1} m_i \mathbf{r}_i}{\sum\limits_{i=1}^{n} m_i}$$

It is directed toward the center of mass of the body. The velocity and acceleration of the center of mass are

$$\dot{\mathbf{r}}_0 = \frac{\sum\limits_{i=1}^{n} m_i \dot{\mathbf{r}}_i}{M} \quad \text{and} \quad \ddot{\mathbf{r}}_0 = \frac{\sum\limits_{i=1}^{n} m_i \ddot{\mathbf{r}}_i}{M}$$

respectively. Hence, the total external force acting on the body is equal to

$$M\ddot{\mathbf{r}}_0 = \sum_{i=1}^{n} \mathbf{F}_{io}$$

This means that the center of mass of the body moves as if the total mass M of the n particles were concentrated there and subjected to a force equal to the sum of external force vectors acting on the individual particles.

If no external forces act, then

$$M\ddot{\mathbf{r}} = 0$$

so the acceleration of the center of mass is zero, and the center of mass performs uniform straight-line motion. However, this does not mean that all the particles perform uniform straight-line motion; quite the contrary, these particles are able to carry out a variety of motions.

A typical example of particles moving as a body is a polyatomic molecule. Atomic nuclei vibrate within the molecule around their equilibrium positions, while the molecule as an entirety may also translate and rotate.

Another typical example of particles moving as a body is a phase consisting of polyatomic molecules. The total energy of such a phase consists of two parts:

$$E = \tfrac{1}{2}M\dot{\mathbf{r}}^2 + E_u$$

The first term on the right-hand side is the kinetic energy of the phase of mass M, moving as a body at velocity $\dot{\mathbf{r}}$. The second term on the right-hand side comprises the kinetic energy arising from the relative (translational, rotational, vibrational) motions of the molecules within the phase and the potential energy arising from their interaction. This is called the internal energy of the phase:

$$E_u = \frac{1}{2}\sum_i m_i\dot{\mathbf{r}}_i^2 + \frac{1}{2}\sum_i\sum_k \varepsilon[r_{ik}]$$

where $\varepsilon[r_{ik}]$ is the potential between molecules i and k, depending on the distance r_{ik}.

Internal Energy

Prove that the total energy of the phase, moving as a whole at velocity $\mathbf{v}_0 = \dot{\mathbf{r}}_0$, is

$$E = \tfrac{1}{2}M\dot{\mathbf{r}}_0^2 + E_u$$

Consider two coordinate systems, one at rest, and the other moving at a constant velocity $\mathbf{v}_0$ with respect to it. This means that the body is at rest in the second coordinate system. The velocity of the molecules of the phase in the fixed coordinate system is

$$\mathbf{v}_i = \mathbf{v}_i^{\square} + \mathbf{v}_0$$

The relation between the respective momenta is

$$\mathbf{I} = \sum_i m_i \mathbf{v}_i = \sum_i m_i \mathbf{v}_i^{\square} + \mathbf{v}_0 \sum_i m_i$$

$$= \mathbf{I}^{\square} + \mathbf{v}_0 \sum_i m_i$$

and between the energies,

$$E = \frac{1}{2} \sum_i m_i \mathbf{v}_i^2 = \frac{1}{2} \sum_i m_i (\mathbf{v}_i^{\square} + \mathbf{v}_0)^2$$

$$= \frac{1}{2} \sum_i m_i \mathbf{v}_i^{\square 2} + \mathbf{v}_0 \sum_i m_i \mathbf{v}_i^{\square} + \tfrac{1}{2} M \mathbf{v}_0^2$$

$$= E^{\square} + \mathbf{v}_0 \mathbf{I}^{\square} + \tfrac{1}{2} M \mathbf{v}_0^2$$

Since the body is at rest in the coordinate system moving at velocity $\mathbf{v}_0$,

$$\mathbf{I}^{\square} = 0.$$

This means that $E^{\square}$ is the energy of the body when its center of mass is at rest, that is, its internal energy

$$E^{\square} = E_u$$

Thus

$$E = E_u + \tfrac{1}{2} M \mathbf{v}_0^2$$

which was the very thing to be proved.

THE KINETIC ENERGY OF TRANSLATIONAL MOTION

Let us now consider the case when the particles form a body, but individually do not possess either external or internal potential energy. Let the particles be monatomic molecules:

1. Each molecule is of identical mass m.
2. The molecules are in random linear uniform motion at an average velocity $\bar{v}$, in the root-mean-square sense to be defined below.
3. During motion elastic impact occurs between the molecules, and a total volume V is accessible to the molecules.
4. The molecules can be treated as point masses.

The state of the body consisting of n molecules is determined exclusively by the spatial positions and velocities of the molecules. Consider the molecules in a cube with edges of unit length. The walls of the cube are not rigid and are under an external pressure P. Particles in the cube and hitting the walls of the cube are in equilibrium with this pressure. Let us separate the velocity of the molecules into components according to the space coordinates x, y, z:

$$\mathbf{v} = v_x + v_y + v_z$$

The velocity of the molecules can be only a positive quantity, but not so that of the components. To avoid negative velocity components, let us consider the squares of the velocity components. These are equal to one another on average, because although the molecules of the gas are in random motion, macroscopically the gas does not move in any direction:

$$\bar{v}_x^2 = \bar{v}_y^2 = \bar{v}_z^2$$

where we define the average (root-mean-square) velocities $\bar{v}_x = \sqrt{\overline{v_x^2}}$ and so on. It follows from this that a third of the mean square of the velocity is equal to the mean square of each velocity component:

$$\frac{\bar{\mathbf{v}}^2}{3} = \bar{v}_x^2 = \bar{v}_y^2 = \bar{v}_z^2$$

The corresponding components of the kinetic energy are

$$\tfrac{1}{2}m\bar{v}_x^2 = \tfrac{1}{2}m\bar{v}_y^2 = \tfrac{1}{2}m\bar{v}_z^2$$

This means that mean energy is uniformly partitioned between the three degrees of freedom of translational motion. This fact is an instance of the *equipartition of energy*.

Consider now a gas containing two types of molecules of masses m_1, m_2 and rms velocities $\bar{v}_1$, $\bar{v}_2$, respectively, under the assumptions made above. Then not only molecules of identical type, but also those of different type will collide. In this case equipartition of energy is extended to molecules of both types:

$$\tfrac{1}{2}m_1\bar{\mathbf{v}}_1^2 = \tfrac{1}{2}m_2\bar{\mathbf{v}}_2^2$$

That is, the mean kinetic energy of the molecule is independent of the type of molecule. Thus the mean velocity of the molecule of smaller mass is higher than that of the molecule of larger mass.

So far monatomic molecules have been discussed. Diatomic and polyatomic molecules have, besides the three spatial directions of translation, further kinds of motion, namely, intramolecular motion, thus possessing further degrees of freedom. The equipartition of energy extends also to these degrees of freedom.

Now let us investigate the x component of the motion of a single selected particle in a cube with edges of unit length. Applying the momentum theorem, we have that the average change in momentum in a single collision is $2m\bar{v}_x$, so that per unit time it is

$$(2m\bar{v}_x)\left(\frac{\bar{v}_x}{2}\right) = m\bar{v}_x^2$$

This change in momentum per unit time is the force. Summed over all the molecules, the force acting on the wall is

$$\sum_{i=1}^{n} \frac{dI_i}{dt} = F_x = (nm)\bar{v}_x^2$$

The area of the wall is unity, and the force acting on a unit surface is the pressure. Thus the pressure of the n molecules, taken as a whole, in equilibrium with the external pressure P is

$$P = (nm)\bar{v}_x^2 = (nm)\frac{\bar{\mathbf{v}}^2}{3}$$

Now, let us consider separately two types of molecule, but in vessels of identical volume. It has been shown that the mean kinetic energy of translation is identical in the two gases:

$$\tfrac{1}{2}m_1\bar{\mathbf{v}}_1^2 = \tfrac{1}{2}m_2\bar{\mathbf{v}}_2^2$$

Let the pressure of the two gases be identical. In this case

$$P = (n_1m_1)\frac{\bar{\mathbf{v}}_1^2}{3} = (n_2m_2)\frac{\bar{\mathbf{v}}_2^2}{3}$$

The second equality is possible only if the number of molecules is equal in the two equal volumes:

$$n_1 = n_2$$

This statement is called Avogadro's hypothesis.

In the above deduction nm is the total mass of molecules contained in 1 cm^3 of volume, that is, the mass density:

$$\rho = nm$$

The density is also the ratio of molar mass to molar volume:

$$\rho = \frac{M}{v}$$

Thus for the pressure we can write

$$P = \frac{M}{v}\frac{\bar{\mathbf{v}}^2}{3}$$

Let us multiply both sides of the equation by $\frac{3}{2}v$:

$$\frac{3}{2}Pv = \frac{1}{2}M\bar{\mathbf{v}}^2$$
$$= \bar{E}_k$$

The right-hand side of the equation is the total kinetic energy of a mole. One mole consists of **N** molecules; therefore

$$\bar{E}_k = \mathbf{N}\left(\tfrac{3}{2}\varepsilon_k\right)$$

where ε_k is the kinetic energy of a single molecule per degree of freedom. The quantity on the left side is $\frac{3}{2}Pv$, called by Clausius the external virial. The virial law states that the total kinetic energy of a system consisting of N noninteracting particles is equal to the external virial of the system.

From the preceding equation we have for the pressure

$$Pv = \tfrac{2}{3}\bar{E}_k$$

The product Pv is constant at a given value of $\bar{E}_k$. This is the Boyle–Mariotte law.

INTERMOLECULAR FORCES

The specific expression for the kinetic energy has not caused any difficulty. However, the specific form of the potential energy needs further consideration.

Let us first consider for this purpose two spherically symmetric molecules of the same kind, such as argon atoms. If the two molecules are effectively at infinite distance from one another, which in practice means more than five times the molecular diameter, then the attractive force between them becomes zero, and the mutual potential energy of the two molecules is also zero. However, as attractive forces bring the molecules to a finite distance r from one another, they perform work between infinity and r: this is the potential energy of the molecules. Conversely, the work to be performed by external forces to remove molecules at distance r to an infinite distance from one another is equal to the potential energy of the pair of molecules.

The nature of the attractive forces must be briefly discussed. How can attraction be interpreted, say in the case of a noble gas, whose molecules are monatomic, spherical, and electrically neutral? The spherical symmetry actually exists on a time average, but there are moments when the electron cloud "slips," and the centers of gravity of positive and negative charge of the molecule do not coincide. This means that the molecule changes has a dipole moment, and the resulting field induces a dipole moment in the neighboring molecule. The resulting electrostatic force between the dipoles is attractive. The magnitude of the induced dipole moment depends on the polarizability of the atom, that is, on the facility with which "slip" is produced. The resulting *van der Waals force* called also the London or dispersive force) acting between two identical molecules is

$$F_d = \frac{9I\alpha^2}{2r^7}$$

where I is the first ionization energy,‡ α the polarizability, and r the distance between the two molecules. Hence, the corresponding potential energy is

$$\varepsilon = -\frac{3I\alpha^2}{4r^6}$$

The polarizability depends on the number of electrons in the molecule, so the *van der Waals force* increases with molecular weight.

If the molecule is a permanent dipole, it will attract another permanent dipole. There will also be attraction between a permanent dipole and an induced dipole. In these two cases the potential energy is, respectively,

$$\varepsilon = -\frac{2}{3kT}\frac{\mu}{r^6}$$

and

$$\varepsilon = -\frac{\mu^2\alpha}{r^6}$$

where μ is the dipole moment.

The three forces decrease as the seventh power of the distance; the three potential energies, as the sixth power of the distance.

The three potential energies of different origin are additive. Table 1.1 shows the relative magnitudes of the coefficients of $1/r^6$. It can be noted that even in the case of permanent dipoles the *van der Waals force* has the greatest magnitude.

It is seen in Table 1.1 that molecules can be divided into two groups, polar and nonpolar, depending on whether or not, besides the *van der Waals forces*, attractive forces involving permanent dipoles play a role. At first, only substances composed of nonpolar molecules will be discussed. The substances at the top of the table, from helium to methane represent this class. A further characteristic of these substances with the exception of chlorine is their

‡This first ionization energy is the energy required for the endothermic reaction in which the neutral molecule is decomposed into a positive ion and an electron: $M \rightarrow I^+ + e$

TABLE 1.1
Coefficients of $1/r^6$ for the Three Different Additive Potential Energies

Substance	$\frac{2\mu}{3kT}$	$\mu^2\alpha$	$\frac{3}{4}I\alpha$
	Monatomic Molecules		
He			1.23
Ne			4.67
Ar			55.4
Kr			108
Xe			233
	Nonpolar Polyatomic Molecules		
Cl_2			321
CH_4	0	0	112
	Polar Molecules		
CO	0.0034	0.057	67.5
HI	0.35	1.68	382
HBr	6.2	4.05	176
HCl	18.6	5.4	105
	Molecules Forming Hydrogen Bonds		
NH_3	84	10	93
H_2O	190	10	47

rotational symmetry. Thus the size and shape of the molecule are further aspects of classification. Methane has tetrahedral rotational symmetry, while, for example, propane and acetone have no point symmetry.

An exceptionally high dipole–dipole potential energy, comparable in order of magnitude with the energy of chemical bonding, is to be observed in NH_3 and H_2O. This indicates a weak chemical interaction: the presence of a hydrogen bond.

So far only attractive forces have been discussed. There exists also a force of repulsion, which plays a role when molecules get very near to one another. Owing to repulsion, the distance between the molecules cannot approach zero. The attractive and repulsive forces act independently of one another and are additive; the work done by the resultant force determines the potential energy of a pair of molecules. The function relating the potential energy of a pair of molecules to the distance between them is called the pair potential.

THE PAIR POTENTIAL

The form of the pair potential depends on the properties of the individual molecules in a given pair. A mathematical form suitable for its description will

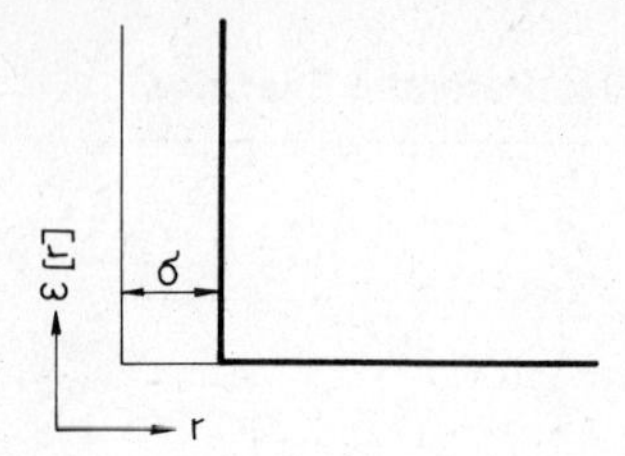

FIGURE 1.1. Pair potential between rigid spheres: no attractive force.

now be given. Here naturally a compromise must be made between realism and tractability of the mathematical model.

1. If to a good approximation the molecules are rigid spheres, which are not deformed on collision, and no attractive force acts between the molecules, then the pair potential has the following form (Figure 1.1):

$$\varepsilon[r] = \begin{cases} \infty & r < \sigma \\ 0 & r \geq \sigma \end{cases}$$

This model can be applied with satisfactory results, if bodies are to be described in which the molecules are located at a relatively large distance from one another.

2. The simplest model that takes attraction into consideration is one that preserves the assumption of rigid spheres with respect to repulsion. Thus the pair potential is infinite if the distance is smaller than the collision diameter. At a greater distance there is no repulsion, but only attraction, which gives rise to a constant pair potential. Then beyond a certain distance attraction ceases too, and the pair potential is zero. This model is a mathematical simplification (discretization) of a continuous pair potential, taking into consideration both attraction and repulsion (Figure 1.2):

$$\varepsilon[r] = \begin{cases} \infty & r < \sigma_1 \\ -\varepsilon_0 & \sigma_1 \leq r \leq \sigma_2 \\ 0 & r > \sigma_2 \end{cases}$$

3. Discussing above the nature of the attractive potentials, it was found that these are inversely proportional to the sixth power of the molecular distance. The repulsion, increasing more rapidly at a shorter distance, is suitably approximated as the inverse twelfth power of the molecular distance (Figure 1.3):

$$\varepsilon[r] = A_1 r^{-12} - A_2 r^{-6}$$

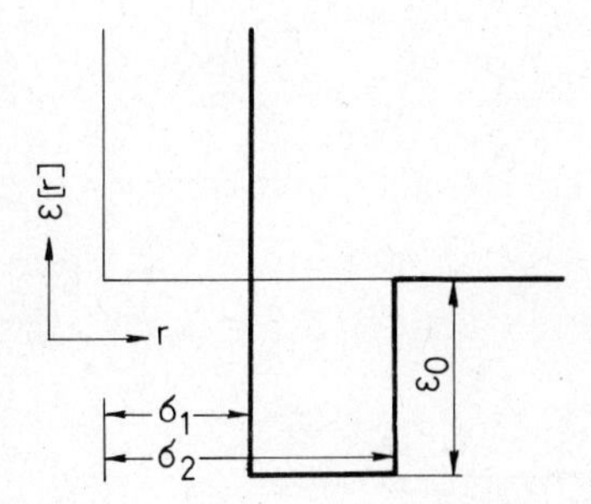

FIGURE 1.2. Pair potential between rigid spheres: discretized attractive force.

This is the Lennard-Jones pair potential. In this relationship A_1 and A_2 are parameters depending on the nature of the molecule.

There is the following relationship between intermolecular force and pair potential:

$$F = -\frac{d\varepsilon}{dr}$$

The pair potential is zero if the two molecules are at infinite distance from one another. It passes through a minimum at that distance r_m where the attractive and repulsive forces cancel (Figure 1.4). At a distance greater than this, $d\varepsilon/dr > 0$, from which it follows that the resultant of intermolecular forces is negative: the two molecules attract each other. At a distance smaller than that belonging to the minimum of the pair potential, $d\varepsilon/dr < 0$, so the resultant of

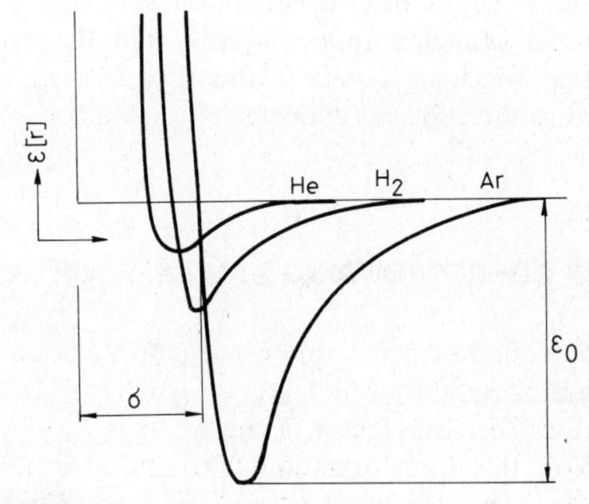

FIGURE 1.3. Lennard-Jones pair potential for three substances.

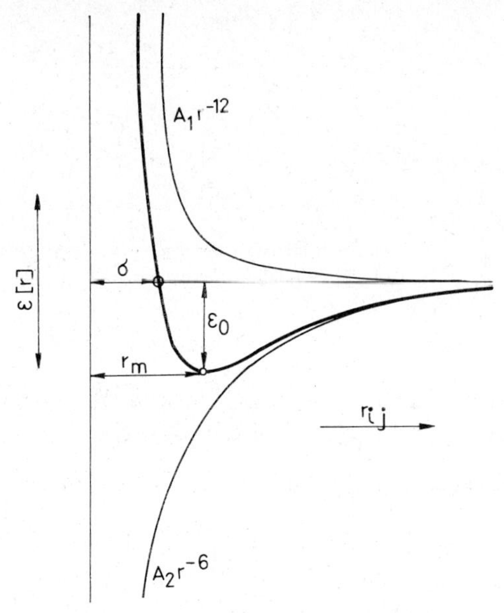

FIGURE 1.4. Repulsive and attractive parts of the Lennard-Jones pair potential.

intermolecular forces is positive: the two molecules repel each other. Accordingly the two molecules are in equilibrium when their potential energy is minimal. It is worth noting that if the potential energy of the two molecules is not minimal, then the attractive or repulsive force produces a displacement and intermolecular forces do work. Owing to the work of intermolecular forces the potential energy of the two molecules decreases. The work done is equal to the negative of the change in potential.

The curve of the potential intersects the abscissa at point σ. At this point the pair potential is changing rather steeply, and the centers of the two molecules cannot approach each other further. For this reason, this distance is usually considered as the collision diameter.

THE LAW OF CORRESPONDING STATES: FOUR HYPOTHESES

1. The pair potentials of two or more substances can be made to coincide by the introduction of suitable scale factors (Figure 1.3). On the r axis σ is a suitable scale factor. The scale factor of the ε axis is ε_0, the minimum of the pair potential. With this transformation the reduced distance r/σ and the reduced potential $\varepsilon/\varepsilon_0$ have been introduced. With these reduced variables we

can write

$$f\left[\frac{r}{\sigma}, \frac{\varepsilon}{\varepsilon_0}\right] = 0$$

This means that the pair potential of simple spherically symmetric molecules, expressed with the reduced variables, is a universal function. In other words, two molecule pairs of different kinds are in corresponding states if their reduced potentials are equal and if their reduced distances are also equal. This statement, in two different wordings of equivalent content, is the *law of corresponding states*. Using the Lennard-Jones expression, we have

$$\frac{\varepsilon}{\varepsilon_0} = 4\left(\left(\frac{\sigma}{r}\right)^{12} - \left(\frac{\sigma}{r}\right)^{6}\right)$$

2. The use of the pair potential makes possible the calculation of the total potential energy of n particles, as the sum of the potential energies (pair potentials) of all the possible molecule pairs formed from n particles. The average potential energy of a molecule is then

$$\bar{\varepsilon} = \frac{\varepsilon_0\left(\sum_i \sum_k f\left[\frac{\sigma}{r_{ik}}\right]\right)}{2n}$$

Evidently, this summation can be carried out only if the geometrical arrangement of the particles (i.e., the distances of all the possible molecule pairs, r_{ik}) is known.

3. The purpose of this summation is to calculate the macro properties of a body consisting of n particles. This can be actually done with the apparatus of classical statistical mechanics.

4. If the pair potentials and geometrical arrangement are known and the macro variables can be calculated, then the law of corresponding states can be expressed also in terms of reduced macro variables. Molecular parameters and macro parameters can be expressed in terms of one another. This is the point of view from which states of aggregation and critical states must be discussed.

Algorithm and Program: Lennard-Jones Potential

Write a program for the calculation of the Lennard-Jones potential in tabular form.

Algorithm.

1. Reading the collision diameter and the pair-potential minimum.
2. Reading the tabulation parameters.
3. Calculate the Lennard-Jones potential.
4. Print the results.

Program LJ. A possible realization:

```
]PR#0
]LOAD LJ
]LIST

 10  REM :CALCULATION OF
 20  REM :LENNARD-JONES
 30  REM :PAIR POTENTIAL
 40  REM :IN TABULATED FORM
 50  PRINT : PRINT
 60  INPUT "ENTER COLLISION-DIAMETER
     (NM) "; SI
 70  PRINT
 80  PRINT "ENTER THE PAIR POTENTIAL"
 90  INPUT "MINIMUM (JOULE)
     ";E0
100  PRINT
110  PRINT "ENTER BEGINNING DISTANCE"
120  INPUT " (NM)
     ";R1
130  IF R1 < = 0 THEN GOTO 110
140  PRINT
150  INPUT "ENTER STEPSIZE (NM)
     ";DR
160  PRINT
170  INPUT "ENTER THE NUMBER OF POINTS ";N
180  PRINT : PRINT : PRINT
190  PRINT " EPSILON R "
200  PRINT "------------------------------"
210  FOR I = 1 TO N
220  R = R1 + (I - 1) * DR
230  E = 4 * E0 * ((SI / R) ∧ 12 - (SI / R) ∧ 6)
240  PRINT TAB( 4);E; TAB( 22);R
250  NEXT I
260  PRINT : PRINT
270  INPUT "NEXT RANGE ? (Y / N) ";
     A$
280  IF A$ = "Y" THEN GOTO 110
290  PRINT : PRINT
```

```
300   INPUT "NEXT SUBSTANCE ? (Y / N
      ) "; A$
310   IF A$ = "Y" THEN GOTO 40
320   END
```

Example. Run the program for argon, helium, and hydrogen.

```
]RUN

ENTER COLLISION-DIAMETER (NM) 0.3418
ENTER THE PAIR POTENTIAL
MINIMUM (JOULE) 171.244E - 23

ENTER BEGINNING DISTANCE
 (NM) 0.02

ENTER STEPSIZE (NM) 0.03

ENTER THE NUMBER OF POINTS 20
```

EPSILON	R
4.25187113E - 06	.02
7.13338635E - 11	.05
2.53389614E - 13	.08
5.54300224E - 15	.11
3.05736991E - 16	.14
2.94393686E - 17	.17
4.08121283E - 18	.2
7.20924797E - 19	.23
1.47142379E - 19	.26
3.0860558E - 20	.29
4.93364654E - 21	.32
-7.87793587E - 22	.35
-1.70644702E - 21	.38
-1.52749572E - 21	.41
-1.17443496E - 21	.44
-8.63371586E - 22	.47
-6.27682345E - 22	.5
-4.57328244E - 22	.53
-3.35833914E - 22	.56
-2.49149414E - 22	.59

```
NEXT RANGE ? (Y / N) N

NEXT SUBSTANCE? (Y / N) Y
```

ENTER COLLISION-DIAMETER (NM) 0.257

ENTER THE PAIR POTENTIAL
MINIMUM (JOULE) 1.40862E − 22

ENTER BEGINNING DISTANCE
(NM) 0.02

ENTER STEPSIZE (NM) 0.03

ENTER THE NUMBER OF POINTS 20

EPSILON	R
1.1420662E − 08	.02
1.91596566E − 13	.05
6.80105337E − 16	.08
1.48136019E − 17	.11
8.03554012E − 19	.14
7.35644845E − 20	.17
8.88394412E − 21	.2
1.0379096E − 21	.23
−3.53505665E − 23	.26
−1.40725009E − 22	.29
−1.1062597E − 22	.32
−7.44738557E − 23	.35
−4.8760157E − 23	.38
−3.21050017E − 23	.41
−2.14852118E − 23	.44
−1.46588001E − 23	.47
−1.01987999E − 23	.5
−7.22960409E − 24	.53
−5.21492276E − 24	.56
−3.82264222E − 24	.59

NEXT RANGE? (Y / N) N

NEXT SUBSTANCE ? (Y / N) Y

ENTER COLLISION-DIAMETER (NM) 0.2915

ENTER THE PAIR POTENTIAL
MINIMUM (JOULE) 5.2478E − 22

ENTER BEGINNING DISTANCE
(NM) 0.02

ENTER STEPSIZE (NM) 0.03

ENTER THE NUMBER OF POINTS 20

EPSILON	R
1.92903446E − 07	.02
3.23630066E − 12	.05
1.14930297E − 14	.08
2.51033631E − 16	.11
1.37657744E − 17	.14
1.30281122E − 18	.17
1.72780653E − 19	.2
2.73553905E − 20	.23
4.11083141E − 21	.26
6.80680639E − 23	.29
−5.14082222E − 22	.32
−4.66763265E − 22	.35
−3.40571465E − 22	.38
−2.36104261E − 22	.41
−1.62475184E − 22	.44
−1.12675957E − 22	.47
−7.91866508E − 23	.5
−5.64966033E − 23	.53
−4.09273752E − 23	.56
−3.00879964E − 23	.59

NEXT RANGE ? (Y / N) N

NEXT SUBSTANCE ? (Y / N) N

The Lennard – Jones Potential

What is the relationship between the collision diameter σ and the local coordinate r_m of potential minimum?

The minimum of the pair potential is located where

$$\frac{d\varepsilon}{dr} = 0$$

Let us differentiate the Lennard–Jones expression for the pair potential written with reduced coordinates:

$$\frac{d\varepsilon}{dr} = 4\varepsilon_0\left(-\frac{12\sigma^{12}}{r^{13}} + \frac{6\sigma^6}{r^7}\right)$$

The expression in parentheses is the algebraic sum of repulsive and attractive forces. This sum is zero if

$$r_m = 2^{1/6}\sigma$$

This is the relationship sought.

The Critical State

The kinetic energy of the molecules acts against the mean potential energy. At not too small mean distances, the potential energy holds the molecules together, while the kinetic energy tends to separate them. Depending on whether kinetic or potential energy dominates, molecules can exist in two arrangements:

1. They are sufficiently far from one another so that the whole geometric volume is available to them (this is the gas configuration, Figure 1.5*a*).
2. They are sufficiently near to one another so that each molecule can move only in a volume limited by the neighboring molecules, in a cell (this is the liquid configuration, Figure 1.5*b*).

If, however, the mean kinetic energy is equal to or higher than the maximum value of the mean potential energy, then the latter configuration cannot exist. At any mean molecular distance, however small, that is larger than the collision diameter, molecules will then be able to move in the whole available volume. The state of molecule n in which the mean kinetic energy is just equal to the maximum of the mean potential energy is called the *critical state*. In the critical state the mean distance of the molecules is r_c. This is a characteristic individual molecular parameter. In the critical state the molar volume v_c is

$$v_c = Nr_c^3$$

It is known from the pair potential that at the closest packing the volume of an elementary cell is σ^3, and the molar volume calculated from it $N\sigma^3$. According to the law *of corresponding states*, the ratio r_c/σ is independent of the nature of the substance, as long as we deal with simple molecules. This means that a bijective mapping exists between the molecular parameter σ and the macro

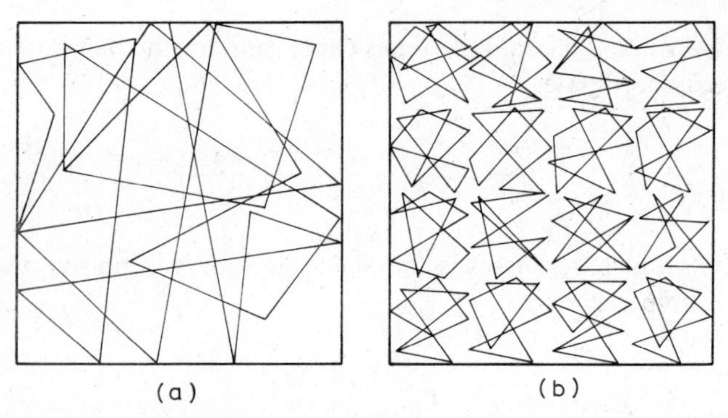

FIGURE 1.5. Translational motion of molecules (*a*) in a gas, (*b*) in the cells of a liquid.

parameter v_c. It should suffice to say here that a similar mapping can be elaborated between other molecular and macro parameters.

Extension of the Law of Corresponding States to Nonsymmetric Nonpolar Substances

It is immediately evident to chemists that the types of molecules that are rotationally symmetric or treatable as such (noble gasses, oxygen, nitrogen, carbon monoxide) are few in number, while the overwhelming majority of molecules met with in practice is of the asymmetric type. Asymmetry of the molecule is inconsistent with the statement made above, that the pair potential depends only on the distance between the two molecules. Indeed, if the molecule cannot be considered as rotationally symmetric (as, e.g., in the case of propane, Figure 1.6), then it is not even obvious what one should call the distance between two molecules. Be that as it may, the pair potential curve intersects the r axis, and the distance of the intersection measured from the origin (distance parameter) remains a suitable scale factor, while the scale factor of the ordinate is the greatest depth of the potential well. It is essential that the form of the pair potential function does not change, and the Lennard–Jones exponents can be maintained.

The displacement of the two scale factors depends on a single third molecular parameter, the *shape factor* of the molecule. The shape factor is the ratio of the largest diameter of the molecule to the distance parameter. Thus, the extension of corresponding states to nonpolar nonsymmetric molecules can be written in the following form:

$$f\left[\frac{\varepsilon}{\varepsilon_0}, \frac{r}{\sigma}, \frac{l}{\sigma}\right] = 0$$

The pair potential of nonpolar, nonsymmetric molecules, expressed in terms of three reduced variables, is a universal function. In other words, two asymmetric molecule pairs are in corresponding states if their reduced distances,

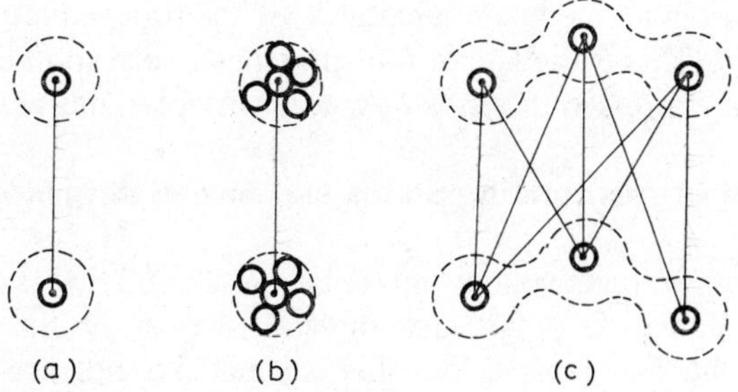

FIGURE 1.6. Distances between two molecules: (*a*) argon, (*b*) methane, (*c*) propane.

```
    H   H    H
    |   |    |
H - C - C  - C - H
    |   |    |
    H   H    H
       (a)

    H   H    H
    |   |    |
H - C - C  - C - H
    |   ||   |
    H   O    H
       (b)
```

FIGURE 1.7. Homomorphous molecules: (*a*) propane, (*b*) acetone.

reduced potentials, and shape factors are identical. Of the three reduced variables two are independent. This is the law of three-parameter corresponding states.

The law of three-parameter corresponding states can also be expressed in terms of macroscopic parameters.

Extension of the Law of Corresponding States to Polar Substances

This extension does not cause difficulties if it is assumed that the difference between the potentials of nonpolar and polar molecule pairs with identical shape factors (Figure 1.7) depends at identical reduced distances only on the reduced dipole moment, $\mu/\varepsilon_0\sigma$. Here μ, the dipole moment, is the fourth molecular parameter. Then for polar substances the law of four-parameter corresponding states is valid:

$$f\left[\frac{\varepsilon}{\varepsilon_0}, \frac{r}{\sigma}, \frac{l}{\sigma}, \frac{\mu}{\varepsilon_0\sigma}\right] = 0$$

The pair potential of polar molecules, expressed with four reduced variables, is a universal function. In other words, two polar-molecule pairs are in corresponding states if their reduced distances, reduced potentials, shape factors, and reduced dipole moments are identical. Of the four reduced variables three are independent. This is the law of four-parameter corresponding states, which also has a form expressed in terms of macro variables and parameters.

Law of Corresponding States and Associating Molecules

The concept of homomorphous molecules was used to extend the law of corresponding states to polar substances. However, in the case of certain molecules (alcohols, organic acids) this concept has not proved satisfactory. Detailed investigations showed that certain polar molecules dimerize, less so in

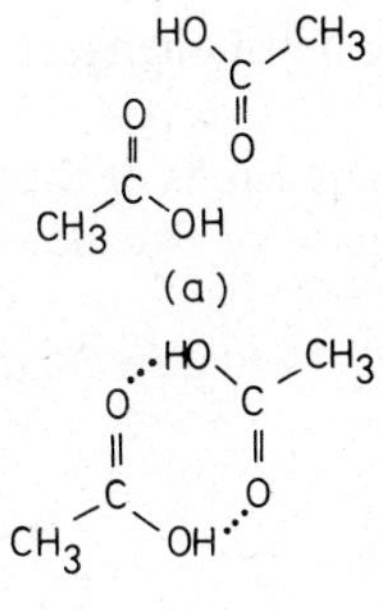

(a)

(b)

FIGURE 1.8. Acetic acid: (*a*) two independent monomers, (*b*) dimer/.

the gas phase, and more so in the liquid phase. This means that the substance is not single-component, but is a mixture of monomeric and dimeric molecules; moreover, the dimer is not homomorphous with the same nonpolar molecule as the monomer (Figure 1.8). In this case a new approach is needed.

TRANSITION TO QUANTUM MECHANICS

In the foregoing, the mechanics of multiparticle bodies and, in conjunction with it, the potential energy of attraction and repulsion between the particles were discussed. Let us now consider the case in which the pair potential is zero, that is, the energetic behavior of the molecules is not disturbed by the other molecules in the vicinity. This focuses our attention on the intramolecular energy of the molecules.

If the molecule is monatomic and in its ground state, its energy consists entirely of the kinetic energy of translational motion. A diatomic molecule can rotate during translation about its center of mass in space, and the two atoms can oscillate along the straight line passing through their centers of mass.

The basic assumption of classical mechanics is that the possible values of energy E form a continuum and energy can change continuously. A clear case of this is the harmonic oscillator discussed already, for which it was shown that its total energy is constant: the sum of potential and kinetic energies. In contrast, a fundamental consequence of quantum mechanics for bound systems is that the energy E can take on only discrete values and therefore can change only from one discrete value to another. The difference between the discrete energy values is negligible in the case of sufficiently large masses and distances, but not in the case of atoms and molecules, and decidedly not in the case of electrons.

The constraining of energy to discrete values is called quantization. Quantum theory deduces the rules of quantization from a few basic principles, which will not be discussed here. However, we will show the application of quantization rules to cases which will be needed in the building up of chemical thermodynamics.

Quantization rules involve Planck's constant $\boldsymbol{h}$ which has dimensions of the product of linear momentum and distance. In physics this quantity is called action. The value of Planck's constant is

$$\mathbf{h} = 6.626\,176 \times 10^{-34} \text{ J s.}$$

Intuitive Quantization Rules

Let us consider as the first and simplest example a particle of mass m, moving to and fro between two plane walls at a distance L from each other, perpendicular to the walls. The quantization rule is expressed in terms of the momentum:

$$|I|2L = n\mathbf{h}$$

where $|I|$ = absolute value of the momentum of the particle,
L = distance between the two walls,
n = quantum number, an integer.

The factor 2 arises from the fact that the particle covers in a complete cycle (i.e., from a wall back to the same wall) a distance of $2L$. The quantization rule is: the product of the momentum of the particle and the path length of a total cycle is equal to an integer multiple of Planck's constant.

The particle possesses only kinetic energy, given by

$$E_k = \tfrac{1}{2}m\dot{x}^2 = \frac{I^2}{2m} = n^2\frac{\mathbf{h}^2}{8mL^2}$$

This means that the kinetic energy of particles can have only discrete values.

Algorithm and Program: Number of Quantum States

Write a program to calculate the square of the number of quantum states if one particle is present in a cube.

Algorithm.

1. Read the mass of the particle, the size of the cube, and the temperature corresponding to the kinetic energy.
2. Calculate the kinetic energy of the particle from the temperature.

3. Calculate the square of the quantum number.

Program DEG. A possible realization:

```
]LOAD DEG
]LIST

 10  REM :NUMBER OF QUANTUM STATES

 20  REM : INPUT THE MASS OF THE PARTICLE
 30  PRINT : PRINT : PRINT
 40  INPUT "ENTER THE MASS (KG) ";
      M
 50  PRINT
 60  INPUT "ENTER THE SIZE OF THE
      CUBE (M) ";L
 70  REM :CALCULATION OF THE ENERGY
 80  REM : OF ONE PARTICLE
 90  PRINT
100  INPUT "THE TEMPERATURE (K) ?
      ";T
110  EK = 3 / 2 * 1.381E - 23 * T
120  H = 6.626176E - 34
130  N2 = 8 * EK / H * M / H * L *
      L
140  PRINT : PRINT
150  PRINT " SQUARE OF "
160  PRINT "QUANTUM NUMBER: "; INT
     (N2)
170  PRINT : PRINT
180  INPUT "NEXT TEMPERATURE ? (Y /
     N) ";A$
190  PRINT : PRINT
200  IF A$ = "Y" THEN GOTO 100
210  PRINT : PRINT
220  INPUT "NEXT CUBE SIZE ? (Y /
     N) ";A$
230  PRINT : PRINT
240  IF A$ = "Y" THEN GOTO 60
250  INPUT "NEXT MASS ? (Y / N) ";A
     $
260  PRINT : PRINT
270  IF A$ = "Y" THEN GOTO 30
280  END
```

Example. Run the program for hydrogen and hydrochloric acid molecules at cube sizes 10^{-6}, 10^{-7}, 10^{-8} m and at temperatures 0.1, 1, and 10 K, respectively.

```
]RUN

ENTER THE MASS (KG) 1.6261855E - 27

ENTER THE SIZE OF THE CUBE (M) 1E - 6

THE TEMPERATURE (K) ? 0.1

SQUARE OF
QUANTUM NUMBER: 61378

NEXT TEMPERATURE ? (Y / N) Y

THE TEMPERATURE (K) ? 1

SQUARE OF
QUANTUM NUMBER: 613788

NEXT TEMPERATURE ? (Y / N) Y

THE TEMPERATURE (K) ? 10

SQUARE OF
QUANTUM NUMBER: 6137889

NEXT TEMPERATURE ? (Y / N) N

NEXT CUBE SIZE ? (Y / N) Y

ENTER THE SIZE OF THE CUBE (M) 1E - 7

THE TEMPERATURE (K) ? 0.1

SQUARE OF
QUANTUM NUMBER: 613

NEXT TEMPERATURE ? (Y / N) Y

THE TEMPERATURE (K) ? 1

SQUARE OF
QUANTUM NUMBER: 6137

NEXT TEMPERATURE ? (Y / N) Y

THE TEMPERATURE (K) ? 10

SQUARE OF
QUANTUM NUMBER: 61378

NEXT TEMPERATURE ? (Y / N) N

NEXT CUBE SIZE ? (Y / N) Y

ENTER THE SIZE OF THE CUBE (M) 1E - 8

THE TEMPERATURE (K) ? 0.1

SQUARE OF
QUANTUM NUMBER: 6

NEXT TEMPERATURE ? (Y / N) Y
```

```
THE TEMPERATURE (K) ? 1

SQUARE OF
QUANTUM NUMBER: 61

NEXT TEMPERATURE ? (Y / N) Y

THE TEMPERATURE (K) ? 0

SQUARE OF
QUANTUM NUMBER: 0

NEXT TEMPERATURE ? (Y / N) Y

THE TEMPERATURE (K) ?
?REENTER
THE TEMPERATURE (K) ? 10

SQUARE OF
QUANTUM NUMBER: 613

NEXT TEMPERATURE ? (Y / N) N

NEXT CUBE SIZE ? (Y / N) N

NEXT MASS ? (Y / N) Y

ENTER THE MASS (KG) 8.36324E - 28

ENTER THE SIZE OF THE CUBE (M) 1E - 6

THE TEMPERATURE (K) ? 0.1

SQUARE OF
QUANTUM NUMBER: 31566

NEXT TEMPERATURE ? (Y / N) Y

THE TEMPERATURE (K) ? 1

SQUARE OF
QUANTUM NUMBER: 315662

NEXT TEMPERATURE ? (Y / N) Y

THE TEMPERATURE (K) ? 10

SQUARE OF
QUANTUM NUMBER: 3156629

NEXT TEMPERATURE ? (Y / N) N

NEXT CUBE SIZE ? (Y / N) Y

ENTER THE SIZE OF THE CUBE (M) 1E - 7

THE TEMPERATURE (K) ? 0.1

SQUARE OF
QUANTUM NUMBER: 315
```

```
NEXT TEMPERATURE ? (Y / N) 1

NEXT CUBE SIZE ? (Y / N) Y

ENTER THE SIZE OF THE CUBE (M) 1E - 7

THE TEMPERATURE (K) ? 1

SQUARE OF
QUANTUM NUMBER: 3156

NEXT TEMPERATURE ? (Y / N) Y

THE TEMPERATURE (K) ? 10

SQUARE OF
QUANTUM NUMBER: 31566

NEXT TEMPERATURE ? (Y / N) N

NEXT CUBE SIZE ? (Y / N) Y

ENTER THE SIZE OF THE CUBE (M) 1E - 8

THE TEMPERATURE (K) ? 0.1

SQUARE OF
QUANTUM NUMBER: 3

NEXT TEMPERATURE ? (Y / N) Y

THE TEMPERATURE (K) ? 1

SQUARE OF
QUANTUM NUMBER: 31

NEXT TEMPERATURE ? (Y / N) Y

THE TEMPERATURE (K) ? 10

SQUARE OF
QUANTUM NUMBER: 315

NEXT TEMPERATURE ? (Y / N) N

NEXT CUBE SIZE ? (Y / N) N

NEXT MASS ? (Y / N) N
```

Remark. The formula used for the calculation of the kinetic energy from temperature in this program will be discussed in the next chapter.

Translational Motion. Let us now consider a freely moving particle of mass m, enclosed in a box. The particle has three degrees of freedom. According to quantum theory, each quantum state is characterized by as many quantum numbers as there are degrees of freedom. If we have a rectangular box with edges a, b, and c, and the position of the particle is suitably expressed in an

(x, y, z) coordinate system parallel to the edges of the box, as was done in classical mechanics, then the motion of the particle can be separated according to coordinates x, y, z. This means that an independent equation of motion can be written for any of the three directions defined. Using the quantization rule already applied, we have in turn according to the three degrees of freedom:

$$|I_x|2a = n_x\mathbf{h}$$

$$|I_y|2b = n_y\mathbf{h}$$

$$|I_z|2c = n_2\mathbf{h}$$

where n_x, n_y, n_z are three independent quantum numbers.

The energy of the particle (here too, only kinetic energy is involved) is given by

$$E_\mathrm{k} = \frac{I_x^2 + {}_y^2 + I_z^2}{2m} = \frac{\mathbf{h}^2}{8m}\left(\frac{n_x^2}{a^2} + \frac{n_y^2}{b^2} + \frac{n_z^2}{c^t}\right)$$

If the box is a cube (i.e., $a = b = c = L$), then the energy is

$$E_\mathrm{k} = \left(n_x^2 + n_y^2 + n_z^2\right)\frac{\mathbf{h}^2}{8mL^2}$$

The same energy belongs to each quantum state (n_x, n_y, n_z) for which the sum $n_x^2 + n_y^2 + n_z^2$ is the same.

Algorithm and Program: Quantum Numbers

Write a program for the determination of each combination of quantum numbers having a given sum of squares.

Algorithm.

1. Reading the sum of squares of the quantum numbers.
2. Determine the largest possible quantum number.
3. Calculate with nested loops the sum of squares of the quantum numbers for each combination. If it is equal to the given one, then print.

Program KVANT. A possible realization:

```
]LOAD KVANT
]LIST

 10  REM :CALCULATION OF QUANTUM
      NUMBERS
 20  PRINT : PRINT
 30  INPUT "INPUT THE SUM OF SQUAR
      ES ";S
 40  PRINT : PRINT : PRINT
```

```
50  PRINT " NX NY NZ": PRINT
     "-------------------"
60  N = INT ( SQR (S))
70  FOR I = 1 TO N
80  FOR J = 1 TO N
90  FOR K = 1 TO N
100 IF I * I + J * J + K * K = S
     THEN GOTO 120
110 GOTO 130
120 PRINT TAB( 5);I; TAB( 12);J
     ; TAB( 19);K
130 NEXT K
140 NEXT J
150 NEXT I
160 PRINT : PRINT : PRINT
170 INPUT "DO YOU WANT ANOTHER RUN
     ?(Y / N) ";A$
180 IF A$ = "Y" THEN GOTO 20
190 END
```

Example. Run the program for sums of squares 66, 70, 81.

```
]RUN

INPUT THE SUM OF SQUARES 66
```

NX	NY	NZ
1	1	8
1	4	7
1	7	4
1	8	1
4	1	7
4	5	5
4	7	1
5	4	5
5	5	4
7	1	4
7	4	1
8	1	1

```
DO YOU WANT ANOTHER RUN ?(Y / N) Y

INPUT THE SUM OF SQUARES 70
```

NX	NY	NZ
3	5	6
3	6	5
5	3	6
5	6	3
6	3	5
6	5	3

```
DO YOU WANT ANOTHER RUN ?(Y / N) Y
INPUT THE SUM OF SQUARES 81
```

NX	NY	NZ
1	4	8
1	8	4
3	6	6
4	1	8
4	4	7
4	7	4
4	8	1
6	3	6
6	6	3
7	4	4
8	1	4
8	4	1

```
DO YOU WANT ANOTHER RUN ?(Y / N) N
```

It is seen in the last run of the preceding example that the energy of the particle is the same in twelve different quantum states. We say in this case that the energy level is twelvefold degenerate.

Let us consider now **the harmonic oscillator**. With an extension according to the quantization rule used above, we can write

$$\int I\,dx = v\mathbf{h}$$

where the integration is performed over a whole cycle and the integer v is the vibrational quantum number. Rewriting the left hand-hand side of the equation, we have

$$\int I\,dx = \int I\dot{x}\,dt = \int_0^{1/\nu} 2E_{\mathrm{k}}\,dt = \frac{2\overline{E}_{\mathrm{k}}}{\nu}$$

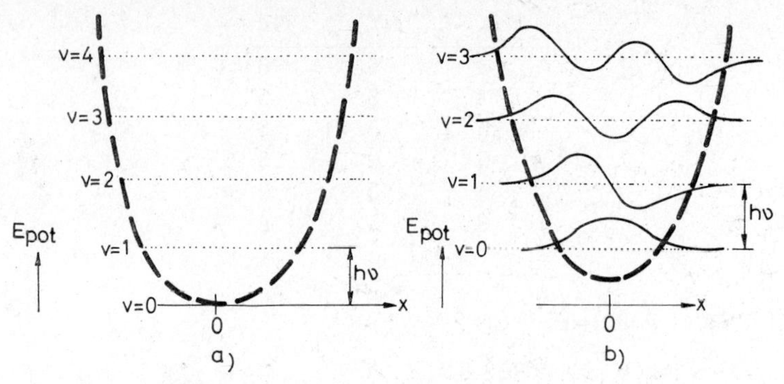

FIGURE 1.9. The energy levels of a linear harmonic oscillator: (*a*) solution obtained by naive quantization rule, (*b*) solution of Schrödinger equation.

where E_k is the mean kinetic energy over a whole cycle, that is, for the period $1/\nu$. The mean kinetic energy of the harmonic oscillator is equal to its mean potential energy:

$$\overline{E}_k = \overline{E}_p$$

and according to convention $E_p(0) = 0$. Hence, the total energy of the harmonic oscillator is

$$E = 2\overline{E}_k$$

Substituting this into the equation of the quantization rule, we have

$$\int I\,dx = \frac{E}{\nu} = v\mathbf{h}$$

Accordingly the energy of the harmonic oscillator is

$$E = v\mathbf{h}\nu$$

This relationship was obtained by the extension of the quantization rule valid for translational motion (Figure 1.9*a*). This extension of the quantization rule describes correctly the separation of the energy levels, but is erroneous by a constant term as concerns the absolute magnitude of energy. The correct result (Figure 1.9*b*) is

$$E = \left(v + \tfrac{1}{2}\right)\mathbf{h}\nu$$

The smallest possible value of the energy of the oscillator is $E_0 = \frac{1}{2}\mathbf{h}\nu$. This is called the zero-point energy.

THE SCHRÖDINGER EQUATION

To get the above result, we need a more exact treatment involving the Schrödinger equation of the actual system. The Schrödinger equation is

$$-\frac{\mathbf{h}^2}{8\pi^2 m}\left(\frac{\partial^2\psi[x,y,z]}{\partial x^2}+\frac{\partial^2\psi[x,y,z]}{\partial y^2}+\frac{\partial^2\psi[x,y,z]}{\partial z^2}\right)$$

$$= E_{\mathrm{p}}[x,y,z]\psi[x,y,z] = E\psi$$

or, written in a condensed form,

$$-\frac{\mathbf{h}^2}{8\pi^2 m}\nabla^2\psi + E_{\mathrm{p}}\psi = E\psi$$

$$\left(-\frac{\mathbf{h}^2}{8\pi^2 m}\nabla^2 + E_{\mathrm{p}}\right)\psi = E\psi$$

where ψ = wave function,
$\mathbf{h}$ = Planck's constant,
m = mass of the particle,
E = total energy of the particle,
E_{p} = potential energy of the particle,
$\nabla^2 = \frac{\partial^2}{\partial x^2} + \frac{\partial^2}{\partial y^2} + \frac{\partial^2}{\partial z^2}$

The Schrödinger equation is a postulate. ψ is the wave function, whose physical significance we need not discuss. Here we only mention that:

1. de Broglie's relationship gives the wavelength which can be assigned to a momentum **I**.
2. The wave motion is periodic, and quantization is introduced into the model through this periodicity.

The Schrödinger equation has solutions only at discrete energy values $E_0, E_1, E_2, \ldots$. These are the eigenvalues of the Schrödinger equation, which belong to eigenfunctions $\psi_0, \psi_1, \psi_2, \ldots$.

Schrödinger Equation for Translational Motion

Schrödinger's wave equation can be written for one particle in the following form:

$$\nabla^2\psi + \frac{8\pi^2 m}{\mathbf{h}^2}(E - E_{\mathrm{p}})\psi = 0$$

If the case is that of an ideal gas, in which the potential energy of the particle is zero, then

$$\nabla^2\psi + \frac{8\pi^2 m}{\mathbf{h}^2} E_\mathrm{k}\psi = 0$$

Owing to the equipartition of energy, the equation can be separated:

$$E_\mathrm{k} = E_x + E_y + E_z$$

and

$$E_x = E_y = E_z$$

Let us therefore deal with the one-dimensional Schrödinger equation:

$$\frac{d^2\psi}{dx^2} + \frac{8\pi^2 m}{\mathbf{h}^2} E_x\psi = 0$$

This is an ordinary linear, homogeneous differential equation of second order, with a solution known from calculus:

$$\psi[x] = C\sin\left(\sqrt{\frac{8\pi^2 mE_x}{\mathbf{h}^2}}\, x - \phi\right)$$

where the constant ϕ is the so-called phase angle. The constants ϕ and C can be determined from the boundary conditions. In the foregoing we have investigated by an intuitive quantization rule the energy of the particle moving to and fro along length L. This particle is replaced now by a standing wave. Under any circumstances the amplitude is zero at the two walls:

$$\psi[0] = 0$$

$$\psi[L] = 0$$

It follows from the first boundary condition that the phase angle ϕ is zero, and from the second boundary condition that

$$\sin\left(\sqrt{\frac{8\pi^2 mE_x}{\mathbf{h}^2}}\, L\right) = 0$$

In view of the properties of the sine function, this is possible if

$$\frac{8\pi^2 mE_x}{\mathbf{h}^2} L^2 = n^2\pi^2$$

where n is an integer. The eigenvalues of the energy are

$$E = n^2 \frac{\mathbf{h}^2}{8mL^2} \qquad n = 0, 1, 2, 3, \ldots$$

Thus, the same result is obtained as by the application of the intuitive quantization rule.

Now we can pass over to the discussion of the harmonic oscillator and show the application of the Schrödinger equation to this case.

Schrödinger Equation of the Linear Harmonic Oscillator

We know already that one-dimensional Schrödinger equation

$$\frac{d^2\psi}{dx^2} + \frac{8\pi^2 m}{\mathbf{h}^2}(E - E_\mathrm{p})\psi = 0$$

The known expression

$$(E - E_\mathrm{p}) = \tfrac{1}{2}m(2\pi\nu)^2(a^2 - x^2)$$

substituted into the preceding equation, gives

$$\frac{d^2\psi}{dx^2} + \frac{16\pi^2\nu^2m^2}{\mathbf{h}^2}(a^2 - x^2)\psi = 0$$

Let us introduce the following notation:

$$\alpha = \frac{4\pi\nu m}{\mathbf{h}}$$

and with it a new variable instead of x:

$$\xi = x\sqrt{\alpha}$$

From this,

$$x = \frac{\xi}{\sqrt{\alpha}}$$

and in this notation the second derivative of ψ is given by

$$\frac{d^2\psi}{dx^2} = \alpha\frac{d^2\psi}{d\xi}$$

Rewriting the Schrödinger equation with the new variable, we have

$$\alpha \frac{d^2\psi}{d\xi^2} + \alpha^2\left(a^2 - \frac{\xi^2}{\alpha}\right)\psi = 0$$

and dividing by α, we have

$$\frac{d^2\psi}{d\xi^2} + (\alpha a^2 - \xi^2)\psi = 0$$

It has been shown that the total energy of the oscillator is

$$E = \tfrac{1}{2}m(2\pi\nu a)^2 = 2\pi^2\nu^2 a^2 m$$

Let us solve this for a^2:

$$a^2 = \frac{E}{2\pi^2\nu^2 m}$$

From the Schrödinger equation written with the new variable the parameters α and a^2 are eliminated:

$$\frac{d^2\psi}{d\xi^2} + \left(\frac{4\pi^2\nu m}{\mathbf{h}} \cdot \frac{E}{2\pi^2\nu^2 m} - \xi^2\right)\psi = 0$$

After reduction we have

$$\frac{d^2\psi}{d\xi^2} + \left(\frac{2E}{\mathbf{h}\nu} - \xi^2\right)\psi = 0$$

This is the differential equation to be solved.

Let us look for the solution in the following form:

$$\psi = C_0 e^{-\xi^2/2}$$

Its first derivative with respect to ξ will then be

$$\frac{d\psi}{d\xi} = -\xi\psi$$

and its second derivative

$$\frac{d^2\psi}{d\xi^2} = -\psi - \frac{d\psi}{d\xi}\xi = -\psi + \xi^2\psi = (\xi^2 - 1)\psi$$

From this the differential equation is

$$\frac{d^2\psi}{d\xi^2} + (1 - \xi^2)\psi = 0$$

By comparison with the Schrödinger equation we have

$$\frac{d^2\psi}{d\xi^2} + \left(\frac{2E}{\mathbf{h}\nu} - \xi^2\right)\psi = \frac{d^2\psi}{d\xi^2} + (1 + \xi^2)\psi$$

which is possible only if

$$\frac{2E}{\mathbf{h}\nu} = 1$$

whence the energy of the harmonic oscillator is

$$E_0 = \tfrac{1}{2}\mathbf{h}\nu$$

We started above from the presumption that ψ is a linear function of $\exp(-\xi^2/2)$. Let us now see the solution in the following form:

$$\psi = C_1 2\xi c^{-\xi^2/2}$$

The first and second derivatives of the wave function are

$$\frac{d\psi}{d\xi} = \left(\frac{1}{\xi} - \xi\right)\psi$$

$$\frac{d^2\psi}{d\xi^2} = \left(\frac{1}{\xi} - \xi^2\right)\psi + \left(\frac{1}{\xi^2} - 1\right)\psi$$

$$= (\xi^2 - 3)\psi$$

That is to say, we have again the differential equation

$$\frac{d^2\psi}{d\xi^2} + (3 - \xi^2)\psi = 0$$

which, compared with the differential equation to be solved, shows that

$$\frac{2E}{\mathbf{h}\nu} = 3$$

that is,

$$E_1 = \tfrac{3}{2}\mathbf{h}\nu$$

It can be said in general that the energy eigenvalues of the Schrödinger equation of the harmonic oscillator are

$$E_{\text{vibr}} = \left(v + \tfrac{1}{2}\right)\mathbf{h}\nu \qquad v = 0, 1, 2, \ldots$$

where v is the vibrational quantum number. Here the noteworthy result is obtained that the energy levels of the linear harmonic oscillator are at a distance $\mathbf{h}\nu$ from one another (Figure 1.9*b*), and that even at the lowest energy level the energy is $\mathbf{h}\nu/2$. This is called the zero-point energy and means that, contrary to the classical oscillator, the quantum oscillator has some energy even at absolute zero, and is not at complete rest.

Schrödinger Equation of the Rotator

If a point mass m revolves in a circular orbit of radius r in a space free of forces, then its potential energy is zero, and the two-dimensional Schrödinger equation can be written in the following form:

$$\frac{\partial^2\psi}{\partial x^2} + \frac{\partial^2\psi}{\partial y^2} + \frac{8\pi^2 m}{h^2}E\psi = 0$$

We introduce polar coordinates:

$$x = r\cos\varphi$$

$$y = r\sin\varphi$$

Evidently (because of the circular orbital)

$$x^2 + y^2 = r^2$$

Now we can write Schrödinger's equation in the following form:

$$\frac{\partial^2\psi}{\partial\varphi^2} + \frac{8\pi^2 mr^2}{h^2}E\psi = 0$$

The following notation is introduced:

$$\alpha = \frac{2\pi}{\mathbf{h}}\sqrt{2mr^2E}$$

to obtain the equation in the following form:

$$\frac{\partial^2\psi}{\partial\varphi^2} + \alpha^2\psi = 0$$

Let us seek the solution in the following form:

$$\psi = A\sin(\alpha\varphi + B)$$

ψ must be a periodic function with period 2π, and so

$$\alpha = \frac{2\pi}{\mathbf{h}}\sqrt{2mr^2E} = J \qquad J = 0, 1, 2, \ldots$$

whence we have for the eigenvalues of energy

$$E = J^2\frac{\mathbf{h}^2}{8\pi^2mr^2} = J^2\frac{\mathbf{h}^2}{8\pi^2I}$$

where $I = mr^2$ is the moment of inertia,
$J =$ the rotational quantum number.

However, experimental facts show that our initial assumption, namely, that the plane rotator is an adequate model of the rotating molecule, does not hold true. If the calculations are made for a three-dimensional rotator, then we obtain by a more circuitous calculation the following energy eigenvalues:

$$E = E_{\text{rot}} = J(J + 1)\frac{\mathbf{h}^2}{8\pi^2I}$$

which differs from the plane rotator only in that the variable J^2 is replaced by the quantum-mechanical quadratic $J(J + 1)$.

It is customary to use *rotational constant*

$$B_e = \frac{\mathbf{h}}{8\pi^2I}$$

which has the dimensions of frequency. Its unit is s^{-1}. In the case of diatomic molecules

$$I = \mu r^2,$$

where $\mu =$ reduced mass,
$r =$ distance between the two particles.

Algorithm and Program: Moment of Inertia

Write a program for the calculation of the moment of inertia of a two-particle system. The reduced mass and the distance between the two particles are known.

Algorithm.

1. Read the reduced mass and the distance between the two particles.
2. Calculate the moment of inertia.
3. Print the result.

Program NYOM. A possible realization:

```
]PR#0
]LOAD NYOM
]LIST

 10  REM :CALCULATION OF MOMENT
      OF INERTIA
 20  PRINT : PRINT : PRINT
 30  INPUT "ENTER REDUCED MASS (KG)
      ";MU
 40  PRINT
 50  INPUT "ENTER DISTANCE OF THE
      TWO PARTICLES (M) ";R
 60  R = R * 1E9
 70  I = MU * R * R * 1E2
 80  PRINT : PRINT
 90  PRINT "MOMENT OF INERTIA :"
100  PRINT
110  PRINT TAB( 5);I;"*10^- 20 (K
      G*M^2)"
120  PRINT : PRINT
130  INPUT "NEXT CALCULATION ? (Y /
      N) ";A$
140  IF A$ ≈ "Y" THEN GOTO 20
150  END
```

Example. Run the program for the hydrogen atom, hydrogen molecule, chlorine molecule, and hydrochloric acid molecule.

```
]RUN

ENTER REDUCED MASS (KG) 9.1005798E - 31
ENTER DISTANCE OF THE TWO PARTICLES (M) 5.29177E - 11

MOMENT OF INERTIA :
    2.54841987E - 31*10^ - 20 (KG*M^2)
```

```
NEXT CALCULATION ? (Y / N) Y

ENTER REDUCED MASS (KG) 8.36324E - 28
ENTER DISTANCE OF THE TWO PARTICLES (M) 7.416E - 11

MOMENT OF INERTIA :
   4.59953579E - 28*10^ - 20 (KG*M^2)

NEXT CALCULATION ? (Y / N) Y

ENTER REDUCED MASS (KG) 2.927134E - 26
ENTER DISTANCE OF THE TWO PARTICLES (M) 1.988E - 10

MOMENT OF INERTIA :
   1.15684551E - 25*10^ - 20 (KG*M^2)

NEXT CALCULATION ? (Y / N) Y

ENTER REDUCED MASS (KG) 1.62618556E - 27
ENTER DISTANCE OF THE TWO PARTICLES (M) 1.2746E - 9

MOMENT OF INERTIA :
   2.64190945E - 25*10^ - 20 (KG*M^2)

NEXT CALCULATION ? (Y / N) N
```

Algorithm and Program: Rotational Energy

Write a program for the calculation of rotational energy. The rotation constant and the quantum numbers are known.

Algorithm.

1. Read the rotation constant and the largest rotational quantum number.
2. Calculate the rotational energy for the different quantum numbers:

$$E_{\text{rot}} = J(J+1)B_e\mathbf{h}$$

3. Print the results.

Program EROT. A possible realization:

```
]LOAD EROT
]LIST

 10  REM :CALCULATION OF ROTATION
      ENERGY
 20  PRINT : PRINT : PRINT
 30  PRINT "ENTER THE ROTATION CONSTANT"
 40  INPUT " S^- 1
      ";B
 50  PRINT
 60  PRINT "ENTER THE MAXIMAL ROTATION"
```

```
70  PRINT
80  INPUT "QUANTUM NUMBER ";J
90  PRINT " J*(J + 1)
     E ROT"
100 PRINT"---------------------------------"
110 FOR I = 1 TO J
120 JJ = I * (I + 1)
130 ER = JJ * B * 6.626176E - 34
140 PRINT TAB( 3);I; TAB( 13);J
     J;TAB( 25);ER
150 PRINT
160 NEXT I
170 PRINT : PRINT : PRINT
180 INPUT "NEXT QUANTUM NUMBER ?
     (Y / N) ";A$
190 PRINT : PRINT
200 IF A$ = "Y" THEN GOTO 60
210 INPUT "NEXT MOLECULE ? (Y / N)
     ";A$
220 IF A$ = "Y" THEN GOTO 20
230 END
```

Example. Calculate the rotational energy of the hydrogen molecule, chlorine molecule, and $H^{35}Cl$ molecule for quantum numbers 1 to 5.

```
]RUN

ENTER THE ROTATION CONSTANT
   S^- 1 1.824672E12

ENTER THE MAXIMAL ROTATION
QUANTUM NUMBER 5
```

J	J*(J + 1)	E ROT
1	2	2.41811956E - 21
2	6	7.2543587E - 21
3	12	1.45087174E - 20
4	20	2.41811956E - 20
5	30	3.62717935E - 20

```
NUMBER QUANTUM NUMBER ? (Y / N) Y

ENTER THE MAXIMAL ROTATION
QUANTUM NUMBER 7
```

J	J*(J + 1)	E ROT
1	2	2.41811956E − 21
2	6	7.2543587E − 21
3	12	1.45087174E − 20
4	20	2.41811956E − 20
5	30	3.62717935E − 20
6	42	5.07805108E − 20
7	56	6.77073478E − 20

NEXT QUANTUM NUMBER ? (Y / N) N

NEXT MOLECULE ? (Y / N) Y

ENTER THE ROTATION CONSTANT
S^− 1 7.254767E9

ENTER THE MAXIMAL ROTATION
QUANTUM NUMBER 5

J	J*(J + 1)	E ROT
1	2	9.61427261E − 24
2	6	2.88428178E − 23
3	12	5.76856356E − 23
4	20	9.61427261E − 23
5	30	1.44214089E − 22

NEXT QUANTUM NUMBER ? (Y / N) N

NEXT MOLECULE ? (Y / N) Y

ENTER THE ROTATION CONSTANT
S^ − 1 3.176734E11

ENTER THE MAXIMAL ROTATION
QUANTUM NUMBER 6

J	J*(J + 1)	E ROT
1	2	4.20991972E − 22
2	6	1.26297592E − 21
3	12	2.52595183E − 21
4	20	4.20991972E − 21
5	30	6.31487958E − 21
6	42	8.8083142E − 21

NEXT QUANTUM NUMBER ? (Y / N) N

NEXT MOLECULE ? (Y / N) N

2

PRESSURE AND TEMPERATURE

Let us pass on now to the characterization of the macroscopic body, the continuum. Macroscopic investigations concern precisely delimited bodies of finite extent. Precise delimitation means the characterization of the continuous surface—the *wall*—separating the body from its environment, the part of the material world outside the body.

The Adiabatic and the Diathermic Wall

The constraints we use with respect to the properties of the wall are extreme abstractions of the cases occurring in practice. In particular, when we say that the wall of the body investigated is *rigid*, this means that no kind of change within or outside the body produces a change in volume of the body:

$$dV = 0$$

This cannot be realized in practice. Indeed, the wall is not treated as a physical, but as a mathematical boundary of the body: the wall does not form a physical part of the body.

Now two types of wall with special properties must be described. (We do not formally explain what is meant by the volume and the pressure of the system, because these concepts are well known from geometry and mechanics.) If no chemical substance can penetrate the wall, so that the amount of substance in the body is constant, and if the pressure of the body is a uniquely defined function of volume,

$$P = f[V]$$

then we say that the wall of the body is of the *first kind*, or that it is *adiabatic*.

Thus, the pressure of an adiabatic body does not change when the body is transferred to another environment.

However, there are also rigid walls, impermeable to chemical substances, for which the aforesaid relationship is not valid, that is, the pressure is not a unique function of volume:

$$P \neq f[V]$$

In this case we say that the wall of the body is of the second kind, or *diathermic*. The pressure of a body of diathermic and rigid wall changes when the body is transferred into another ("colder" or "hotter") environment. The expressions "cold" or "hot" are of physiological rather than physical origin, but a mode of expression with precise physical meaning will be sought.

Let us start from the following considerations: Experience based on physiological sensation shows that different bodies A, B, and C can be arranged according to their "hotness" into a well-defined order. This means that if

A is hotter than B and
B is hotter than C, then

A is hotter than C.

The case is the same as with real numbers. If

$$\frac{a > b \text{ and } \quad b > c, \text{ then}}{a > c}$$

Therefore, it is feasible to arrange the bodies in order of their "hotness" in a series, and assign higher numbers to hotter bodies. These numbers can be then called the *temperatures* of the bodies. There are many scales consistent with the ordering and of these a suitable one must be chosen.

MECHANICAL EQUILIBRIUM AND THE MEASUREMENT OF PRESSURE

Before we do this, let us discuss the measurement of pressure, a property known already from mechanics. Consider for this purpose a body containing a given amount of substance, constrained by an adiabatic, nonrigid wall (Figure 2.1). It was said of this body that its pressure is a function of its volume. Thus, a single datum, the volume, is sufficient for the description of the body. When two adiabatic bodies are put in contact, this contact has no effect on the pressure of either of the bodies.

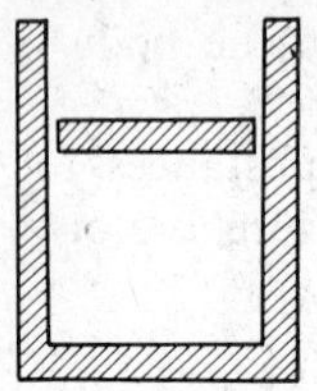

FIGURE 2.1. Adiabatic, movable wall.

However, the situation is different if the two bodies are joined through a common, adiabatic, movable wall (Figure 2.2). In this case a change begins, during which pressure diminishes in one of the bodies and increases in the other, and consequently the volume of one body increases, while that of the other is reduced. This change continues until the pressures are equalized:

$$P_1 = P_2$$

Now the two bodies are in *mechanical equilibrium*. Thus, both bodies have a property, the pressure, the magnitude of which is the same in mechanical equilibrium.

Measurement of pressure is based on just this fact. For this purpose the body shown in Figure 2.3 is connected through a displaceable wall (in this case the surface of mercury in the U tube) with another body: the manometer. The manometer is a body with a displaceable wall, to some property of which (in our case the height difference of the two mercury levels) an arbitrary pressure scale can be assigned. The measurement of pressure in this way is subject to following criteria:

1. Mechanical equilibrium shall exist between the body to be measured and the manometer.
2. The clearance over the manometer shall be negligible in comparison with the volume of the body to be measured.

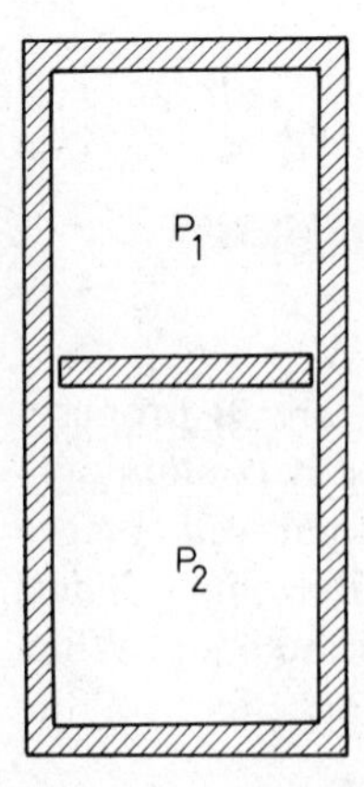

FIGURE 2.2. Two bodies connected by an adiabatic, movable wall.

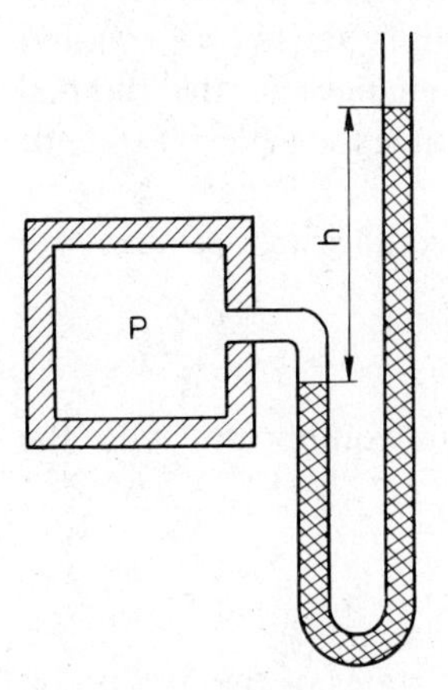

FIGURE 2.3. Measurement of pressure.

THERMAL EQUILIBRIUM AND THE ZEROTH LAW OF THERMODYNAMICS

As mentioned already, contact between two adiabatic bodies has no effect on the pressure of either body. However, things are different if a diathermic aperture is made in the common wall of the two bodies.

As has already been said, a change starts in both bodies, the pressure decreasing in one of them and increasing in the other. We say that the states of both bodies change. Thus, the state of either body is not a well-defined function of its own volume, but also depends on the state of the other body. When the macroscopically observable change in state ceases, we say that the two bodies are in *thermal equilibrium*.

Let us denote the pressure and the volume of the two bodies in thermal equilibrium by p', v' and P', V', respectively (Figure 2.4a). Now when the

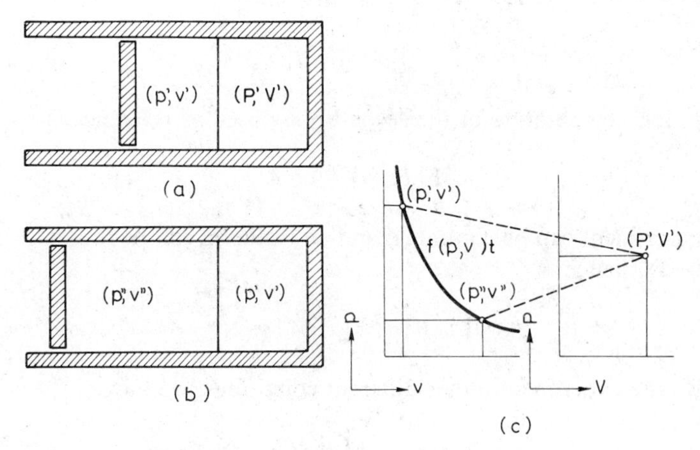

FIGURE 2.4. Two bodies in thermal equilibrium.

state of the first body is changed by external intervention, states (p'', v'') and (p''', v''') can be achieved, which do not produce changes in the thermal equilibrium of the two bodies, that is to say, do not change the (P', V') state of the second body (Figure 2.4*b*).

Thus, if the first and second bodies are in thermal equilibrium, so that

$$(p', v') \Leftrightarrow (P', V')$$

and if the first body is also in the state (p'', v'') in equilibrium with the unchanged second body, that is,

$$(p'', v'') \Leftrightarrow (P', V')$$

then there is also thermal equilibrium between the two states of the first body:

$$(p', v') \Leftrightarrow (P', V')$$

This statement is usually called the *zeroth law of thermodynamics.*

EMPIRICAL TEMPERATURE

Along the curve shown in Figure 2.4*c* an infinity of states can be realized according to the zeroth law, for each of which thermal equilibrium with the unchanged second body exists, so that these states are also in thermal equilibrium among themselves. Thus in these states the first body possesses a property, which has the same value for all the states; this property is called *empirical temperature*. A curve in the (p, v) coordinate system along which states having the same temperature are realized is called an *isotherm* (Figure 2.4*c*). This concept of temperature is independent of physiological sensations.

The equation of the isotherm for the first body is

$$\mathrm{f}[p, v] = t_1^1$$

The empirical temperature of the second body used as reference is

$$\mathrm{F}[P, V] = \text{const}$$

The experiment can be continued and an isotherm can be found also for the second body; hence

$$\mathrm{f}[p, v] = \mathrm{F}[P, V] = t_1^2$$

Repeating the experiment under different conditions, we have

$$\mathrm{f}[p, v] = \mathrm{F}[P, V] = t_2^2$$

No conditions were imposed on the material nature of the reference body; any homogeneous substance (gas, mercury) can be used. A reference phase of this kind is called a *thermometer*. Temperature as established with this thermometer depends on the properties of the reference substance and on the function $F(P, V)$ used to assign an empirical temperature to the reference substance of the thermometer.

Measure of Temperature

Furthermore, it follows from our reasoning that P and V are not the only independent variables which can be chosen for the interpretation of temperature; any two properties of the thermometer will do. Let these two properties be X and Y. The equation of the isotherm, expressed in terms of X and Y, is

$$t = \mathrm{f}[X, Y]$$

Of the two properties let X be that used for the measurement of temperature, and with that object let $Y = Y_0$ be constant (Figure 2.5). X values corresponding to the intersections of the isotherms are used for the characterization of temperature; that is to say, with the parameter Y_0 fixed, the temperature will be a function of X only. On this basis five important types of thermometer are described in Table 2.1.

Let us now choose the function f. Let the temperature be a linear function of X:

$$t[X, Y] = aX + b \qquad Y_0 = \text{const}$$

It follows from this convention that equal changes in temperature correspond to equal changes of X. Let us introduce the following two standard temperature points:

1. Temperature of ice in equilibrium with its liquid phase:

$$t[X_i] = aX_i + b = 0°\text{C} \qquad Y_0 = \text{const}$$

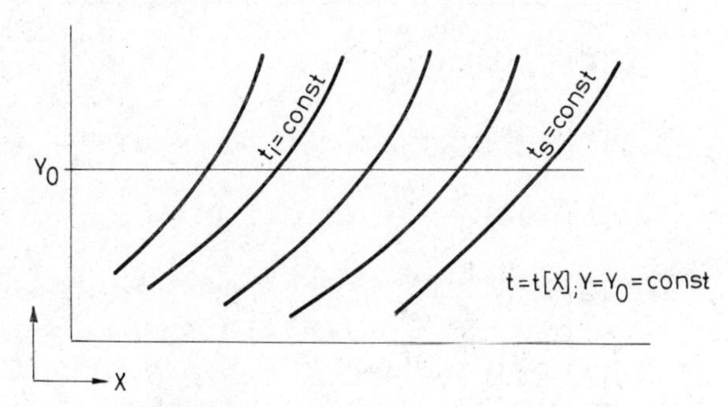

FIGURE 2.5. Measurement of temperature.

TABLE 2.1
Five Types of Thermometer

Type	X	Symbol
Mercury	Length	L
Gas (constant volume)	Pressure	P
Gas (constant pressure)	Volume	V
Resistance	Electric resistance	R
Thermocouple	Thermoelectric force	$\mathscr{E}$

2. Temperature of water in equilibrium with its vapor at standard atmospheric pressure (this is the normal boiling point)

$$t[X_s] = aX_s + b = 100°\text{C} \qquad Y_0 = \text{const}$$

It follows from this that the temperature corresponding to the value X establishing itself in the thermometer at thermal equilibrium is

$$t[X] = \frac{X - X_i}{X_s - X_i} \times 100°\text{C} \qquad Y_0 = \text{const}$$

On this basis temperature as measured by the five kinds of thermometers mentioned in Table 2.1 is defined by the equations in Table 2.2. By bringing the same body into thermal equilibrium with five different thermometers and inserting the readings into these equations, five different values will be obtained for the temperature, as shown in Table 2.3.

TABLE 2.2
Temperature Measured by Five Different Types of Thermometers

$t[P] = 100\dfrac{P - P_i}{P_s - P_i}\,°\text{C}$	$V = \text{const}$
$t[V] = 100\dfrac{V - V_i}{V_s - V_i}\,°\text{C}$	$P = \text{const}$
$t[R] = 100\dfrac{R - R_i}{R_s - R_i}\,°\text{C}$	
$t[\mathscr{E}] = 100\dfrac{\mathscr{E} - \mathscr{E}_i}{\mathscr{E}_s - \mathscr{E}_i}\,°\text{C}$	
$t[L] = 100\dfrac{L - L_i}{L_s - L_i}\,°\text{C}$	

TABLE 2.3
Temperature Measured by Five Different Types of Thermometers

Type		$t[X]$ (°C)
Hydrogen thermometer of constant volume	P	40,000
Air thermometer of constant volume	P	40,001
Platinum resistance thermometer	R	40,360
Platinum–platinum-rhodium thermocouple	$\mathscr{E}$	40,297
Mercury thermometer	L	40,111

Note that the temperature measured depends on the properties of the thermometric substance. This means that there is no quantity definable as "the" temperature of which the property X is a linear function.

The finding that values obtained with different gas thermometers agree rather well suggests a new idea.

ABSOLUTE TEMPERATURE SCALE

It seems convenient to describe temperature with a reference substance whose properties are independent of the individual properties of any chemical substance, and use for it a function characterizing by numbers the "hotness" of bodies, so that a larger number corresponds to a hotter body. The first demand leads to a reference substance in the gaseous state, since the thermal behavior of gases is very similar. It is found in most cases that for a given quantity of gas the products PV are equal at constant temperature:

$$P'V' = P''V'' = \text{const}$$

Actually the product of pressure and volume is only roughly constant, and the changes depend on various factors. However, experience shows that the value of the product Pv of one mole of any gas, measured at the same temperature and extrapolated to pressure $P = 0$, is the same (Figure 2.6). For example, independently of the nature of the gas, the volume of 1 mole of gas at the temperature of melting ice is $v^*_{0°\text{C}} = 0.0224138\ \text{m}^3$ at atmospheric pressure.

Therefore, for the reference substance a gas was chosen for which the Boyle–Mariotte law is strictly valid:

$$(Pv)_t^* = \text{const}$$

This theoretically defined substance is called an ideal gas.

Thus, if the ideal gas is the reference substance of the thermometer, then we can write for the empirical temperature in the case of a thermometer of

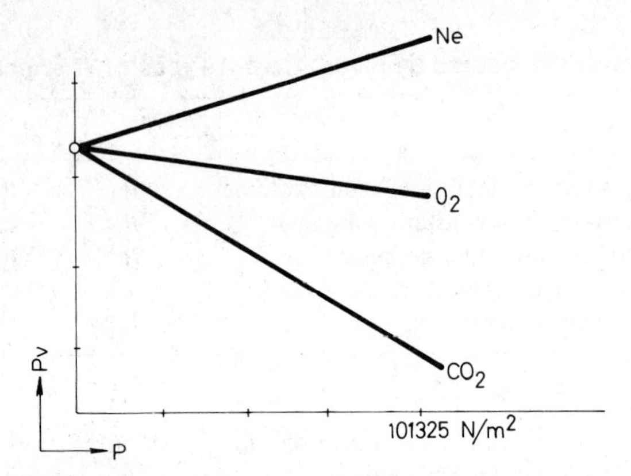

FIGURE 2.6. Extrapolation of the product Pv to zero pressure.

constant volume

$$t = \lim_{P \to 0} t[P] = \lim_{P \to 0} 100 \frac{P - P_i}{P_s - P_i} \qquad V = \text{const}$$

and in the case of a thermometer of constant pressure

$$t = \lim_{V \to \infty} t[V] = \lim_{P \to 0} 100 \frac{V - V_i}{V_s - V_i} \qquad P = \text{const}$$

This is the Celsius temperature scale, which is completely independent of the individual material properties of the thermometer.

Moreover, it follows from the above that the temperature of melting ice is

$$\Theta_i = \lim_{P \to 0} 100 \frac{P_i}{P_s - P_i} \qquad V = \text{const}$$

and

$$\Theta_i = \lim_{P \to 0} 100 \frac{V_i}{V_s - V_i} \qquad P = \text{const}$$

Θ_i is a universal constant, the temperature of melting ice on a temperature scale on which the unit temperature difference is a hundreth part of $\Theta_s - \Theta_i$, as on the Celsius scale, but the zero point is shifted down by just Θ_i (Figure 2.7). On this scale the temperature is

$$\Theta = \lim_{P \to 0} 100 \frac{P}{P_s - P_i} \qquad V = \text{const}$$

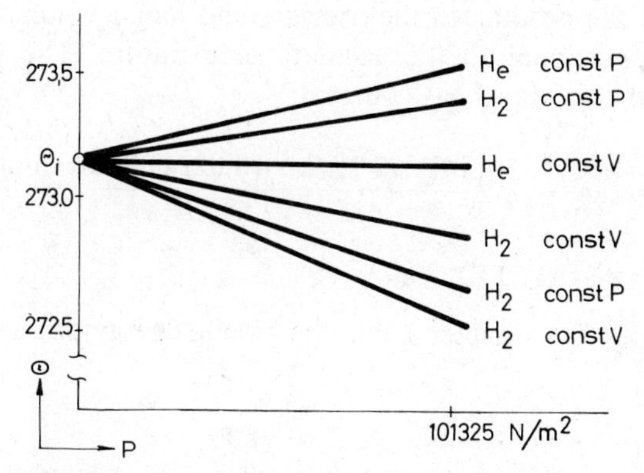

FIGURE 2.7. Ice point on the absolute temperature scale.

and

$$\Theta = \lim_{P\to 0} 100 \frac{V}{V_s - V_i} \qquad P = \text{const}$$

This temperature is called the absolute temperature:

$$\Theta = \lim_{P\to 0} \mathrm{F}[P, V]$$

The unit of absolute temperature is one kelvin (1 K), a hundreth part of the temperature difference between the boiling point of water and the temperature of melting ice. The absolute temperature scale is shifted with respect to the Celsius scale by the value Θ_i:

$$\Theta = \Theta_i + t$$

where

$$\Theta_i = 273.165 \pm 0.015 \text{ K}$$

resulting from the fact that at constant volume the pressure of an ideal gas becomes zero at a temperature of $-(273.165 \pm 0.15)$°C.

Equation of State of the Ideal Gas

Our equation $\Theta = \lim_{P\to 0}\mathrm{F}(P, V)$ can be rewritten now, given the definition of absolute temperature, in the following more suitable form:

$$\lim_{P\to 0} (Pv)_\Theta = (Pv)^*_\Theta = \mathrm{f}[\Theta] = \mathbf{R}\Theta$$

This means that the product of the pressure and molar volume, $(Pv)^*$, of the ideal gas is proportional to the absolute temperature. The proportionality factor is called the universal gas constant:

$$\mathbf{R} = \frac{(Pv)^*}{\Theta} = \frac{(101{,}325\ \mathrm{N/m^2})(0.0224138\ \mathrm{m^3/mole})}{273.15\ \mathrm{K}}$$

$$= 8.31441\ \mathrm{J\,K^{-1}\,mole^{-1}}$$

It was mentioned in Chapter 1 that the kinetic energy of a gas consisting of **N** particles is

$$\overline{E}_{\mathrm{k}} = \tfrac{3}{2}Pv$$

Since this relationship applied to a gas in which no forces act between the molecules, the product Pv written for this gas is just equal to the product $(Pv)^*$ in the definition of the ideal gas:

$$Pv = \lim_{P \to 0} (Pv)_{\Theta} = (Pv)^*_{\Theta}$$

Thus we have

$$(Pv)^* = \mathbf{R}\Theta$$

and

$$(Pv)^* = \tfrac{2}{3}\overline{E}_{\mathrm{k}}$$

The left-hand sides of the last two equations are equal, so that the right-hand sides are also equal:

$$\tfrac{2}{3}\overline{E}_{\mathrm{k}} = \mathbf{R}\Theta$$

With this equality a relationship has been established between the body consisting of **N** particles, discussed in Chapter 1, and the ideal gas discussed as a continuum in the present chapter:

$$\tfrac{2}{3}\overline{E}_{\mathrm{k}} = \mathbf{Nk}\Theta$$

As a result of this relationship the universal gas constant is proportional to Avogadro's number:

$$\mathbf{R} = \mathbf{Nk}$$

where the proportionality constant (Boltzmann's constant) can be interpreted as the gas constant for a single molecule:

$$\mathbf{k} = \frac{\mathbf{R}}{\mathbf{N}}$$

$$= 1.381 \times 10^{-23}\ \mathrm{J\,K^{-1}}$$

The kinetic energy of a single molecule, per degree of freedom, is

$$\varepsilon_{\mathrm{k}} = \tfrac{1}{2}\mathbf{k}\Theta$$

Immediately the following conclusion can be drawn: What we call temperature is in the case of gases a quantitative characteristic of the mean *translational* energy of the molecules.

THE CARNAP CRITERIA

This chapter deals with the concept and measurement of temperature. The question may arise whether this discussion has covered everything needed for the description of a physical quantity such as temperature. What is needed in general was formulated by Carnap as the following five criteria:

1. Unit.
2. Equality.
3. Lower–higher relation.
4. Zero point.
5. Scale rule.

1. The unit has been established: the kelvin.

2. Equality was elucidated by saying that the temperatures of two bodies are equal if they are in direct or indirect thermal equilibrium with one another. For this purpose the zeroth law was stated.

3. The lower–higher relation was given by the convention specifying on the thermometer the direction of increasing temperature.

4. The zero point of temperature follows from the property of the ideal gas (as thermometer) that for a gas thermometer at constant volume the relationship between pressure and temperature is linear. Zero temperature is that temperature at which the pressure becomes zero by linear extrapolation.

5. The scale rule is linear, that is, the number that measures the temperature is the multiplier of the temperature unit.

Thus, the five Carnap criteria needed for the description of a physical quantity are met in the case of temperature.

Temperature is a property of bodies. In Chapter 5 the thermodynamic temperature will be introduced and denoted by T. It will be shown that absolute temperature Θ and thermodynamic temperature T are identical. Anticipating this, temperature will be denoted in what follows by T.

3

EQUATIONS OF STATE

The equation of state is the relationship for a given homogeneous one-component phase:

$$P = \mathrm{f}[v, T]$$

where P = pressure of the phase,
v = molar volume,
T = temperature.

The equation of state assigns to each temperature and molar volume of a homogeneous one-component thermodynamical body a pressure. The specific form of the function f between the macro variables depends on the shape, size, and electrical properties of the molecules forming the body.

Equation of State of the Ideal Gas

The equation of state has been introduced in the preceding chapter:

$$P = \frac{\mathbf{R}T}{v}$$

It was shown that the molecular model of the ideal gas can be discussed on the basis of classical mechanics, as was done in Chapter 1.

Molecular Size and Repulsive Forces

Molecules are considered in the naive kinetic theory as point masses without extent. Actually molecules are masses with finite size. These, similar to point masses, perform uniform straight-line motion, during which they collide with one another. The collisions are all assumed to be elastic, but are modeled in one of two ways. In the simpler model, the molecules behave as spheres of definite diameter with rigid walls; in the more sophisticated model they compress, their centers of mass approach one another within a molecular diameter, and they are then separated by repulsion and recover their original form. In the first case the pair potential function has the form shown in Figure 1.1; in the second it decays very rapidly, but without discontinuity:

$$\varepsilon[r] = \gamma r^{-\delta}$$

where γ and δ are parameters and the exponent is such that $9 < \delta < 15$. Since the volume of the molecule is determined by the extent of the electron cloud surrounding the molecular core, the compression mentioned above is the deformation of the electron cloud.

In certain cases a combination of the electron clouds can also occur. This is accompanied by the formation of a new molecule; that is, a collision of this kind is a chemical reaction.

In the following the molecule will be treated as a rigid sphere.

Collision Cross-section and Mean Free Path

A molecule of diameter σ exposes a nominal target surface (collision cross-section) of $(\sigma/2)^2\pi$ to a molecule moving toward it. If the number density of these molecules is ρ_n and their mean velocity perpendicular to the target surface is $\bar{v}_x$, then the centers of mass of the molecules penetrating the target surface in unit time are located in a cylinder of volume $(\sigma/2)^2\pi\bar{v}_x$ (Figure 3.1). Their number is

$$Z = \left(\frac{\sigma}{2}\right)^2 \pi\bar{v}_x\rho_n$$

This target model is not actually correct, since there is an impenetrable molecule at the position of the "target," which deflects molecules approaching it from their original direction. This is demonstrated in Figure 3.2. The collision number calculated on the basis of this model is

$$Z = \sqrt{2}\left(\frac{\sigma}{2}\right)^2 \pi\bar{v}_x\rho_n$$

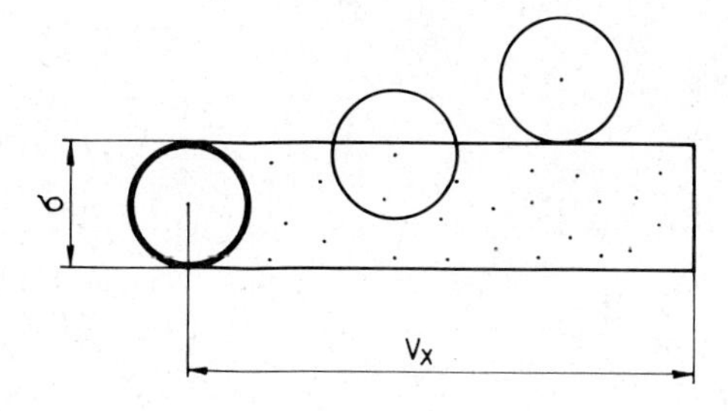

FIGURE 3.1. Collision cross-section and collision number.

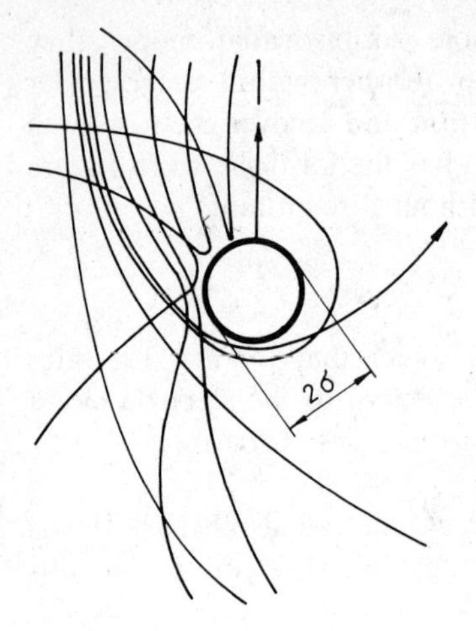

FIGURE 3.2. Deflection of the approaching molecule by the target molecule.

This can be interpreted also by saying that an average molecule, while covering during unit time a distance $\bar{v}_x$, collides on Z occasions with other molecules. The distance covered between two successive collisions is

$$\bar{l} = \frac{\bar{v}_x}{\sqrt{2}\,(\sigma/2)^2 \pi \bar{v}_x \rho_n} = \frac{1}{\sqrt{2}\,(\sigma/2)^2 \pi \rho_n}$$

This distance $\bar{l}$ is called the *mean free path*. The mean free path is inversely proportional to the density and the collision cross-section.

Algorithm and Program: Collision Number and Velocity

Write a program for the calculation of the collision number and x component of the mean velocity. The mass and size of the particle and the temperature and pressure are known.

Algorithm.

1. Read the reduced mass and diameter of the particle.
2. Suppose that

$$v_x = v_y = v_z$$

$$\bar{v}^2 = v_x^2 + v_y^2 + v_z^2 = 3v_x^2$$

$$v_x^2 = \frac{\bar{v}^2}{3}$$

$$\tfrac{1}{2}m\frac{\bar{v}^2}{3} = \tfrac{3}{2}\mathbf{k}T = E_\mathrm{k}$$

$$\frac{\bar{v}^2}{3} = v_x^2 = \frac{3\mathbf{k}T}{m}$$

and thus

$$v_x = \sqrt{\frac{3\mathbf{k}T}{m}}$$

Calculate v_x.

3. Calculate the density ρ = (number of particles in a unit cube) × (mass of one particle):

$$p = \rho v_x = \rho_n m v_x^2 \qquad \rho = m\rho_n$$

$$\rho_n m = \frac{p}{v_x^2} = \frac{pm}{3\mathbf{k}T} = \rho$$

$$\rho_n = \frac{p}{3\mathbf{k}T}$$

4. Calculate the collision number

$$Z = \sqrt{2}\left(\frac{\sigma}{2}\right)^2 \pi v_x \rho_n$$

5. Calculate the mean free path

$$\bar{l} = \frac{1}{\sqrt{2}\,(\sigma/2)^2 \pi \rho_n}$$

6. Print the results.

Program UTK. A possible realization:

```
]PR#0
]LOAD UTK
]LIST

 10  REM :CALCULATION OF NUMBER OF
      COLLISIONS
 20  REM :CALCULATE THE X COMPONENT
 30  REM :OF AVERAGE SPEED FROM
 40  REM :TEMPERATURE
 50  PRINT : PRINT : PRINT
 60  INPUT "ENTER THE MASS OF THE
      PARTICLE (KG) ";MU
 70  PRINT
 80  INPUT "ENTER THE DIAMETER (M)
      ";D
 90  PRINT
```

```
100  INPUT "ENTER THE TEMPERATURE
      (K) ";T
110  K = 1.381E - 23
120  VX = SQR (3*K*T / MU)
130  REM :CALCULATION OF DENSITY
      FROM
140  REM :PRESSURE
150  PRINT : PRINT
160  INPUT "ENTER THE PRESSURE (P
      A) ";P
170  RO = P / 3 / K / T
180  PI = 3.1415
190  PRINT
200  Z = SQR(2)*D*D / 4*PI
      *VX*RO
210  L = 4 / SQR(2) / D / D / PI
      /RO
220  PRINT : PRINT
230  PRINT "NUMBER OF COLLISIONS
      ";Z
240  PRINT : PRINT
250  PRINT "MEAN FREE PATH ";L;"
      M"
260  PRINT : PRINT
270  INPUT "NEXT PRESSURE ? (Y / N)
      ";A$
280  IF A$ = "Y" THEN GOTO 150
290  PRINT : PRINT
300  INPUT "NEXT TEMPERATURE ? (Y
      /N) ";A$
310  IF A$ = "Y" THEN GOTO 90
320  PRINT :
330  INPUT "NEXT PARTICLE ? (Y / N
      ) ";A$
340  IF A$ = "Y" THEN GOTO 50
350  END
```

Example. Run the program for methane at pressures 10^{-5}, 5×10^{-5}, 10^{-4}, 10 Pa and at temperatures 1, 10, 100 K.

```
]RUN

ENTER THE MASS OF THE PARTICLE (KG) 8.363E - 28
ENTER THE DIAMETER (M) 3.418E - 9
ENTER THE TEMPERATURE (K) 1
ENTER THE PRESSURE (PA) 1E - 5

NUMBER OF COLLISIONS 697.104157
MEAN FREE PATH .319285169 M
```

```
NEXT PRESSURE ? (Y / N) Y

ENTER THE PRESSURE (PA) 5E - 5

NUMBER OF COLLISIONS 3485.52078
MEAN FREE PATH .0638570339 M

NEXT PRESSURE ? (Y / N) Y

ENTER THE PRESSURE (PA) 1E - 4

NUMBER OF COLLISIONS 6971.04157
MEAN FREE PATH .0319285169 M

NEXT PRESSURE ? (Y / N) N

NEXT TEMPERATURE ? (Y / N) Y

ENTER THE TEMPERATURE (K) 10
ENTER THE PRESSURE (PA) 1E - 5

NUMBER OF COLLISIONS 220.44369
MEAN FREE PATH 3.1928517 M

NEXT PRESSURE ? (Y / N) Y

ENTER THE PRESSURE (PA) 5E - 5

NUMBER OF COLLISIONS 1102.21845
MEAN FREE PATH .638570339 M

NEXT PRESSURE ? (Y / N) Y

ENTER THE PRESSURE (PA) 1E - 4

NUMBER OF COLLISIONS 2204.4369
MEAN FREE PATH .319285169 M

NEXT PRESSURE ? (Y / N) N

NEXT TEMPERATURE ? (Y / N) Y

ENTER THE TEMPERATURE (K) 100
ENTER THE PRESSURE (PA) 1E - 5

NUMBER OF COLLISIONS 69.7104158
MEAN FREE PATH 31.9285169 M

NEXT PRESSURE ? (Y / N) Y

ENTER THE PRESSURE (PA) 5E - 5

NUMBER OF COLLISIONS 348.552079
MEAN FREE PATH 6.38570339 M

NEXT PRESSURE ? (Y / N) Y

ENTER THE PRESSURE (PA) 1E - 4

NUMBER OF COLLISIONS 697.104158
MEAN FREE PATH 3.1928517 M
```

NEXT PRESSURE ? (Y / N) N

NEXT TEMPERATURE ? (Y / N) N

NEXT PARTICLE ? (Y / N) N

SIZE OF THE MOLECULES AND THE COVOLUMEN

If the diameter of the molecules is σ, then the centers of the two molecules can approach one another to a distance σ. Alternatively it can be said that for the center of molecule A there is around the center of molecule B a spherical forbidden volume of radius 2σ (Figure 3.3), within which it cannot get. The volume of the sphere is $\frac{4}{3}\pi\sigma^3$. The forbidden volume for one mole of molecules, that is, for $\mathbf{N}/2$ molecule pairs, is

$$\tfrac{1}{2}(\mathbf{N} - 1)\left(\tfrac{4}{3}\pi\sigma^3\right) \approx \tfrac{2}{3}\mathbf{N}\pi\sigma^3 \equiv b$$

This forbidden molar volume b was called by van der Waals the *covolumen*. Thus the covolumen is the forbidden volume and not the geometrical volume of a mole of molecules. The latter is actually

$$\tfrac{4}{3}\mathbf{N}\pi\left(\frac{\sigma}{2}\right)^3 = \tfrac{1}{6}\mathbf{N}\pi\sigma^3 = \frac{b}{4}$$

Hence, the geometrical volume of the molecules is one-fourth of the covolumen.

At a given temperature the mean free path is inversely proportional to the pressure. With increasing pressure, molecules get nearer to one another and the mean free path decreases in the gas space.

Let us assume that a sufficiently high pressure is applied, and the molecules (of finite extent) get sufficiently near to one another. If the potential energy of attraction is higher than the mean kinetic energy at this mean distance of the molecules, then neighboring molecules remain in the vicinity of one another; some geometrical arrangement is reached, which is sufficiently characterized by the number of the immediate neighbors of any selected molecule. The number of immediate neighbors is called the *coordination number*. The coordination of molecules means that molecules move only in a part of the space limited by the

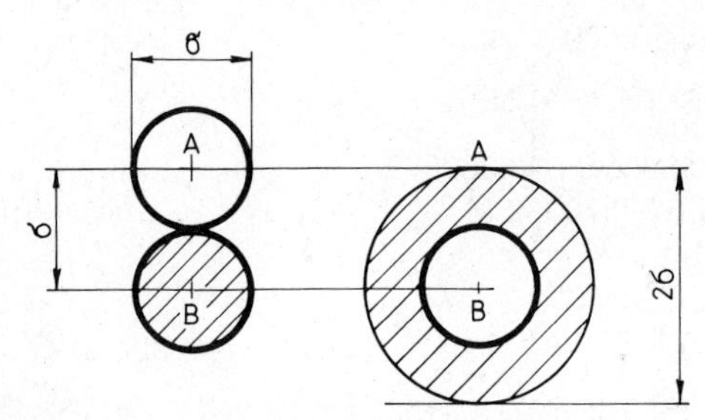

FIGURE 3.3. Forbidden volume in collision of rigid molecules.

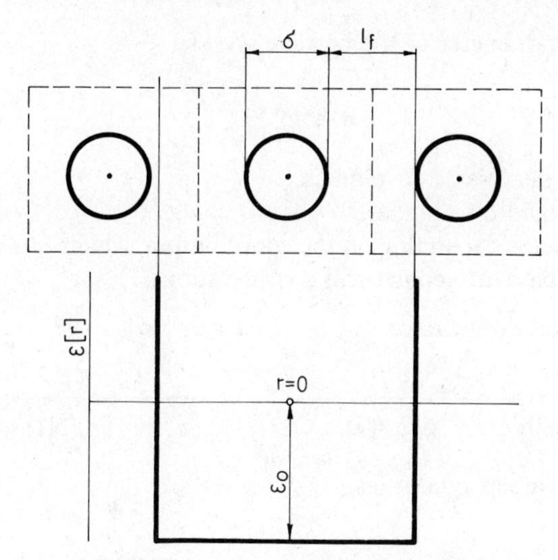

FIGURE 3.4. Molecule in a potential well.

neighboring molecules (in a cell) and the total volume is not available for them. This model was qualitatively discussed in Chapter 1. Thus, the pair potential of the molecules moving in the cell and that of the neighboring molecules (Figure 3.4) will be

$$\varepsilon[r] = \begin{cases} \infty & r \leq \dfrac{\sigma}{2} - \bar{l} \\ \varepsilon_0 & \dfrac{\sigma}{2} - l < r < \dfrac{\sigma}{2} + \bar{l} \\ 0 & r \geq \dfrac{\sigma}{2} + \bar{l} \end{cases}$$

A molecule of diameter σ can move along the r axis between particles fixed at the centers of two neigboring cells. The path which can be freely covered is, in the notation of the figure,

$$2\bar{l} = 2\left(v_{\mathrm{c}}^{1/3} - \sigma\right) = v_{\mathrm{f}}^{1/3}$$

where v_{c} is the volume of one cell, and v_{f} the free volume belonging to the particle of diameter σ, in which this particle can move.

The mean linear velocity of the particle in the x direction is $\bar{v}_x$; its momentum, $m\bar{v}_x$; the change in momentum in a single impact at one of the walls of the cell, $2m\bar{v}_x$. In unit time the particle collides with this cell wall, of area $v_{\mathrm{c}}^{2/3}$, on $\bar{v}_x/2l$ occasions, so that the pressure is

$$P = \frac{(2m\bar{v}_x)\,\bar{v}_x}{2\bar{l}v_{\mathrm{c}}^{2/3}} = \frac{2\mathbf{k}T}{v_{\mathrm{f}}^{1/3}v_{\mathrm{c}}^{2/3}} = \frac{2\mathbf{k}T}{2\left(v_{\mathrm{c}}^{1/3} - \sigma\right)v_{\mathrm{c}}^{2/3}}$$

where $\mathbf{k}$ is Boltzmann's constant.

The collision diameter of the particle is

$$\sigma = g\left(\frac{b}{\mathbf{N}}\right)^{1/3}$$

where b = van der Waals covolumen,
$\mathbf{N}$ = Avogadro's number,
g = a factor depending on the coordination, which can be calculated on the basis of geometrical considerations.

After substitution we have

$$P = \frac{2\mathbf{k}T}{2\left(v_c^{1/3} - g(b/\mathbf{N})^{1/3}\right)v_c^{2/3}} = \frac{\mathbf{k}T}{\left(v_c - g(b/\mathbf{N})\right)^{1/3} v_c^{2/3}}$$

Passing over to molar quantities,

$$v = \mathbf{N}\bar{v}_c$$

$$\mathbf{R} = \mathbf{N}\mathbf{k}$$

so that the equation of state is

$$P = \frac{\mathbf{R}T}{v - gb^{1/3}v^{2/3}}$$

This equation of state takes into consideration the covolumen of the particles and their coordination in the liquid phase.

The Virial Theorem and Equations of State

It was shown in the discussion of the molecular model of ideal gases that the mean kinetic energy $\bar{E}$ of a body consisting of $\mathbf{N}$ particles is equal to its external virial:

$$E = \tfrac{3}{2}Pv$$

This is valid for ideal gases, where there is no interaction between the particles. However, if such interaction exists, and external ($\mathbf{F}_{io}$) and internal ($\sum_{i=1}^{\mathbf{N}}\mathbf{F}_{ik}$) forces act on a particle in position (x, y, z), the the equation of motion of the ith particle is

$$m_i\frac{d^2\mathbf{r}_i}{dt^2} = \mathbf{F}_{io} + \sum_{k=1}^{\mathbf{N}}\mathbf{F}_{ik}$$

Scalar multiplication of the equation by $d\mathbf{r}_i/dt$ and summation over i gives

$$\sum_{i=1}^{\mathbf{N}} m_i\frac{d^2\mathbf{r}_i}{dt^2}\cdot\frac{d\mathbf{r}_i}{dt} = \frac{d}{dt}\left(\frac{1}{2}\sum_{i=1}^{\mathbf{N}} m_i\left(\frac{d^2\mathbf{r}_i}{dt^2}\right)^2\right)$$

$$= \sum_{i=1}^{\mathbf{N}}\mathbf{F}_{io}\frac{d\mathbf{r}_i}{dt} + \sum_{i=1}^{\mathbf{N}}\sum_{k=1}^{\mathbf{N}}\mathbf{F}_{ik}\frac{d\mathbf{r}_i}{dt}$$

After integration we have

$$\frac{1}{2}\sum_{i=1}^{N} m_i\left(\frac{d^2\mathbf{r}_i}{dt}\right)^2 = \sum_{i=1}^{N} \mathbf{F}_{io}\,d\mathbf{r}_i + \sum_{i=1}^{N}\sum_{k=1}^{N} \mathbf{F}_{ik}\,d\mathbf{r}_i$$

On the left-hand side of this equation we have the kinetic energy of one mole of the substance, which for ideal gases is

$$\tfrac{3}{2}\mathbf{R}T$$

This is equal to the so-called external virial, which is identical with the first summation on the right-hand side of the equation: $\frac{3}{2}Pv$. If the internal forces have a potential, then $\mathbf{F}_{ik}$ is a function of the distance r_{ik}, and the second (double) summation on the right-hand side of the equation (the internal virial) can be written in the following form:

$$-\frac{1}{2}\sum_{i=1}^{N}\sum_{k=1}^{N} \varepsilon[r_{ik}]$$

where ε_{ik} is the pair potential. The factor $\frac{1}{2}$ follows from the fact that the potential between particles i and k must occur only once in the summation. Substitution gives

$$E = \tfrac{3}{2}\mathbf{R}T = \tfrac{3}{2}Pv - \frac{1}{2}\sum_{i=1}^{N}\sum_{k=1}^{N} \varepsilon[r_{ik}]$$

This means that if the form of the function $\varepsilon[r_{ik}]$ and the geometry of the arrangement of the particles are known, then the relationship between temperature, pressure, and volume—that is, the equation of state—can be deduced from the specific form of the virial theorem. Thus we cannot expect to obtain a single universal equation of state, since the pair potential and coordination are specific properties of the substance.

If, for example, the specific form of the virial theorem is based on the consideration that the internal virial is proportional to the density,

$$\sum_{i=1}^{N}\sum_{k=1}^{N} \varepsilon[r_{ik}] = \frac{a'}{v}$$

then the specific form of the virial theorem is

$$Pv + \frac{a'}{v} = \mathbf{R}T$$

or, after rearrangement,

$$P + \frac{a'}{v^2} = \frac{\mathbf{R}T}{v}$$

This is interpreted as the equality of the sum of *external* and *internal pressures* to $\mathbf{R}T/v$. When the volume is corrected for the van der Waals covolumen, the van der Waals equation is obtained:

$$P + \frac{a}{v^2} = \frac{\mathbf{R}T}{v - b}$$

Correcting the volume in the way already described, with a term depending on both volume and coordination, yields the following equation of state:

$$P + \frac{a}{v^2} = \frac{\mathbf{R}T}{v - gb^{1/3}v^{2/3}}$$

Expanding the right-hand side in powers of b/v and substituting $g = 0.7163$, corresponding to coordination in the body-centered cubic cell, the following equation of state is obtained:

$$P + \frac{a}{v^2} = \frac{\mathbf{R}T}{v}\left(1 + \frac{b}{v} + 0.625\left(\frac{b}{v}\right)^2 + 0.2869\left(\frac{b}{v}\right)^3 + 0.1928\left(\frac{b}{v}\right)^4\right)$$

This is the equation of state of Hirschfelder, Stevenson, and Eyring.

THE VIRIAL EQUATION AND THE COMPRESSIBILITY FACTOR

Kammerlingh Onnes suggested writing the equation of state corresponding to any specific model of the virial theorem in the following form:

$$P = \frac{\mathbf{R}T}{v}\left(1 + \frac{\mathrm{B}[T]}{v} + \frac{\mathrm{C}[T]}{v^2} + \frac{\mathrm{D}[T]}{v^3} + \frac{\mathrm{E}[T]}{v^4} + \cdots\right)$$

This equation of state is called the virial equation, and $\mathrm{B}[T]$, $\mathrm{C}[T]$, $\mathrm{D}[T]$, $\mathrm{E}[T]$, ... are the second, third, fourth, fifth, ... virial coefficients. All the virial coefficients are substance-specific, and besides this depend only on temperature.

The sum in parentheses on the right-hand side is the compressibility factor:

$$z \equiv \frac{Pv}{\mathbf{R}T} = \left(1 + \frac{\mathrm{B}[T]}{v} + \frac{\mathrm{C}[T]}{v^2} + \cdots\right)$$

The compressibility factor is the ratio of the molar volume of a body at temperature T and pressure P to the molar volume of the ideal gas at the same temperature and pressure. We have

$$Pv = z\mathbf{R}T$$

$$Pv^* = \mathbf{R}T$$

and hence

$$z = \frac{v}{v^*}$$

Some plots of the compressibility factor of nitrogen at various temperatures are shown in Figure 3.5.

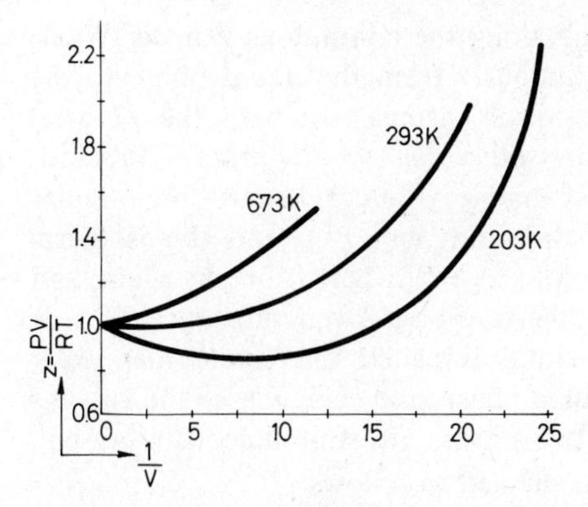

FIGURE 3.5. Compressibility factor of nitrogen.

Phase Transition and Critical Point

Of the aforementioned equations of state, the van der Waals equation is one with rather simple mathematical structure that describes qualitatively the (P, v, T) relationship of the same substance in both the liquid and the gas phases. The van der Waals isotherm is of the third degree in v (Figure 3.6):

$$v^3 - v^2\left(b - \frac{\mathbf{R}T}{P}\right) + v\frac{a}{P} - \frac{ab}{P} = 0$$

Along one section of the isotherm, expansion in volume leads to an increase in pressure:

$$\frac{dp}{dv} > 0$$

This is inconsistent with physical experience. This section of the van der Waals isotherm, which is physically meaningless, is broader than the section satisfying

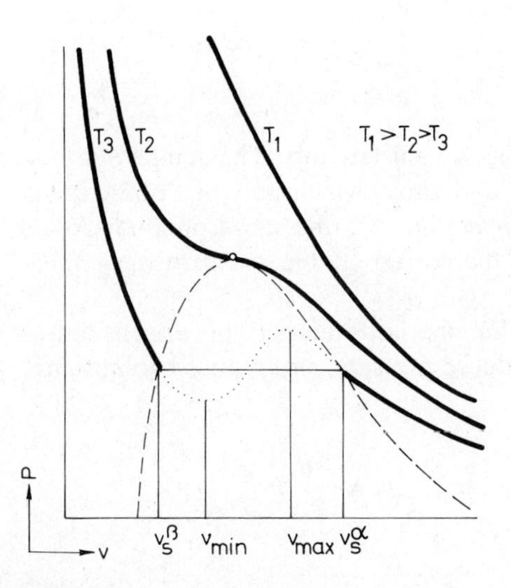

FIGURE 3.6. Three van der Waals isotherms.

$dp/dv < 0$. Furthermore, by proceeding along the continuous van der Waals isotherm, the substance could go continuously from the liquid phase to the gaseous phase (and vice versa), but this is inconsistent with the physical experience that the phase transition is an isothermal, isobaric process. Proceeding along a given isotherm from low molar volumes toward high molar volumes, one encounters a break at the point (v_s^β, P°), and the isotherm continues as a straight line up to the point (v_s^α, P°), where it breaks again, and continues along the van der Waals isotherm. For a given substance P° is a single-valued function of temperature and is called the *equilibrium vapor pressure*. Each $v < v_s^\beta$ represents the liquid phase, and each $v > v_s^\alpha$ the gaseous phase. In the range $v_s^\beta < v < v_s^\alpha$ the two phases are simultaneously present, and the vapor ratio at volume v can be defined as follows:

$$\varepsilon = \frac{v - v_s^\beta}{v_s^\alpha - v_s^\beta}$$

The pressure at which the isobar of the two-phase region intersects the van der Waals isotherm satisfies the following condition:

$$P^\circ\left(v_s^\alpha - v_s^\beta\right) = \int_{v_s^\beta}^{v_s^\alpha} P[v]\, dv$$

So far a single isotherm has been studied. Let us consider now a set of isotherms. On each isotherm, $\Delta v = v_s^\alpha - v_s^\beta$ for pressure P° is the difference of the molar volumes of the gas phase and the liquid phase. This difference decreases with increasing temperature and becomes zero at a definite T value. Above this temperature the function $P[v]$ is monotonic.

When the difference of the molar volumes of the two equilibrium phases decreases to zero—that is, the molar volumes of the two equilibrium phases are identical,

$$v^\alpha = v^\beta$$

—the difference between the two phases vanishes too. The temperature at which the two phases cease to coexist, and above which only one phase exists at any pressure, is called *critical temperature* T_c; the corresponding molar volume is the *critical volume* v_c, and the corresponding pressure the *critical pressure* P_c.

These are suitable scale factors for the introduction of dimensionless variables. The reduced temperature, reduced pressure, and reduced volume are, respectively,

$$T_r = \frac{T}{T_c} \qquad v_r = \frac{v}{v_c} \qquad P_r = \frac{P}{P_c}$$

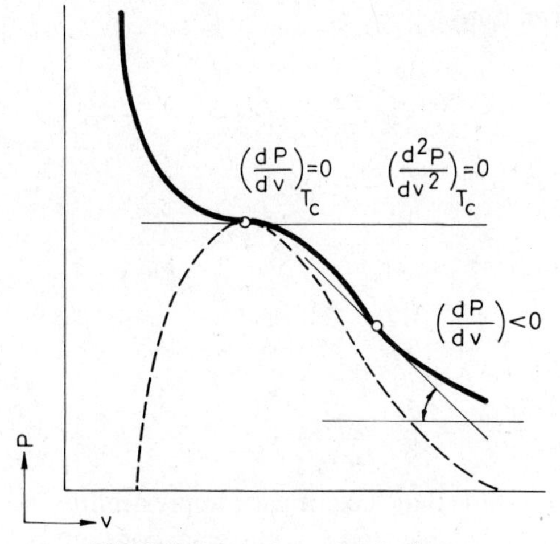

FIGURE 3.7. The critical isotherm.

It will be shown that different substances having identical variables are in corresponding states.

Law of Corresponding States with Macro Variables

Any equation of state to describe the (P, v, T) conditions for the vapor and liquid phases must satisfy the condition that the critical isotherm has an inflection at the critical point (Figure 3.7):

$$\left(\frac{\partial P}{\partial v}\right)_{T_c} = 0$$

$$\left(\frac{\partial^2 P}{\partial v^2}\right)_{T_c} = 0$$

Let us now apply these conditions to the van der Waals equation:

$$P_c = \frac{\mathbf{R}T_c}{v_c - b} - \frac{a}{v_c^2}$$

$$\left(\frac{\partial P}{\partial v}\right)_{T_c} = -\frac{\mathbf{R}T_c}{(v_c - b)^2} + \frac{2a}{v_c^3} = 0$$

$$\left(\frac{\partial^2 P}{\partial v^2}\right)_{T_c} = \frac{2\mathbf{R}T_c}{(v_c - b)^3} - \frac{6a}{v_c^4} = 0$$

The following relationships are obtained from this system of equations by algebraic transformations:

$$v_c = 3b$$

$$a = \frac{9\mathbf{R}T_c v_c}{8} = \frac{27\mathbf{R}T_c b}{8} = \frac{27\mathbf{R}^2 T_c^2}{64 P_c}$$

$$P_c = \frac{3\mathbf{R}T_c}{2v_c} - \frac{9\mathbf{R}T_c}{8v_c} = \frac{3\mathbf{R}T_c}{8v_c}$$

$$b = \frac{\mathbf{R}T_c}{8P_c}$$

$$\frac{P_c v_c}{\mathbf{R}T_c} = \frac{3}{8} \equiv z_c$$

The last relationship says that the critical compressibility factor is independent of the nature of the substance and is a universal constant of the van der Waals gas.

Rewriting the van der Waals equation with reduced variables, we have

$$P = \frac{\mathbf{R}T}{v - b} - \frac{a}{v^2}$$

$$P_r P_c = \frac{\mathbf{R}T_r T_c}{v_r v_c - b} - \frac{a}{(v_r v_c)^2}$$

Substituting the two relationships obtained for a and b, and taking into consideration that

$$\frac{\mathbf{R}T_c}{P_c v_c} = \frac{8}{3}$$

we obtain the reduced form of the van der Waals equation of state:

$$P_r = \frac{8}{3} \frac{T_r}{v_r - \frac{1}{3}} - \frac{3}{v_r^2}$$

By the introduction of the reduced state variables an equation of state can be deduced from the van der Waals equation, which is valid if the pressure, temperature, and molar volume are measured on a suitably reduced dimensionless scale. Accordingly, substances for which the three reduced state variables are equal are in corresponding states. The law of corresponding states is the statement that there exists a function

$$\mathrm{f}[P_r, T_r, v_r; P_c, T_c] = 0$$

of the reduced state variables with macro parameters P_c, T_c which is valid for all the substances belonging to its range of validity, that is, for substances consisting of rotationally symmetric simple molecules. It is not necessary for this function to be identical with the reduced equation of state deduced from the van der Waals equation; indeed, in practice they are not identical.

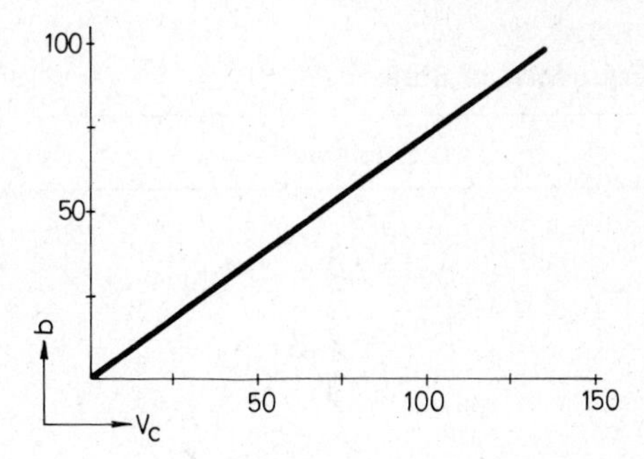

FIGURE 3.8. Covolumen and critical volume.

MOLECULAR AND MACRO PARAMETERS OF THE LAW OF CORRESPONDING STATES

One of the relationships sought between molecular and macro parameters was shown by way of example in Chapter 1. It was stated that for simple molecules the ratio of the molecular parameter σ to the macro parameter v_c is constant. In this class of substances the ratio of the covolumen to the critical volume (Figure 3.8) is 0.75, that is,

$$\frac{b}{v_c} = \frac{\frac{2}{3}\pi \mathbf{N}\sigma^3}{v_c} = 0.75$$

The collision diameter as a molecular parameter is concealed in the covolumen.

A relationship of this kind exists between the molecular energy parameter ε_0 and a macroscopic energy parameter: the critical temperature. The relationship between these molecular and macro parameters is linear (Figure 3.9*a*):

$$\frac{\varepsilon_0}{\mathbf{k}} = 0.77T_c$$

where **k** is Boltzmann's constant.

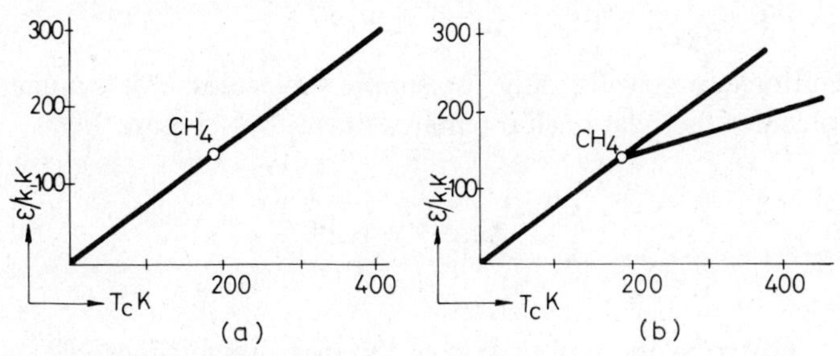

FIGURE 3.9. Energy parameter and critical temperature: (*a*) simple molecule, (*b*) polyatomic molecule.

TABLE 3.1
Classification of Equations of State

	Single-Parameter	Two-Parameter
Linear	$P = \frac{\mathbf{R}T}{v}$	
	$P = \frac{\mathbf{R}T}{v - b}$	
Quadratic	$P = \frac{\mathbf{R}T}{v}\left(1 + \frac{\mathrm{B}[T]}{v}\right)$	
Cubic		$P = \frac{\mathbf{R}T}{v}\left(1 + \frac{\mathrm{B}[T]}{v} + \frac{\mathrm{C}[T]}{v^2}\right)$
		$P = \frac{\mathbf{R}T}{v - b} - \frac{a}{v^2}$
		$P = \frac{\mathbf{R}T}{v - b} - \frac{aT}{v(v - b)}$
		$P = \frac{\mathbf{R}T}{v - b} - \frac{a}{T^{0.5}v(v + b)}$
		$P = \frac{\mathbf{R}T}{v - b} - \frac{a}{Tv^2}$
		$P = \frac{\mathbf{R}T}{v - b} - \frac{a}{v(v + b) + b(v - b)}$
Fourth-degree		
Fifth-degree	$P = \frac{\mathbf{R}T}{v}\left(1 + \frac{b}{v} + 0.625\left(\frac{b}{v}\right)^2 + 0.2869\left(\frac{b}{v}\right)^3 + 0.1928\left(\frac{b}{v}\right)^4\right)$	
Sixth-degree		

This relationship is valid only for simple molecules. For symmetric polyatomic molecules the relationship remains linear, but (Figure 3.9*b*)

$$\frac{\varepsilon_0}{\mathbf{k}} = 82.789 + 0.282T_c$$

Methane is located at the intersection of the two straight lines.

Three-Parameter	Six-Parameter
$P = \frac{\mathbf{R}T}{v-b} - \frac{a}{v(v+b)+c(v-b)}$ $P = \frac{\mathbf{R}T}{v-b} - \frac{a}{v(v+c)}$	
$P = \frac{\mathbf{R}T}{v}\left(1 + \frac{\mathrm{B}[T]}{v} + \frac{\mathrm{C}[T]}{v^2} + \frac{\mathrm{D}[T]}{v^3}\right)$	
	$P = \frac{\mathbf{R}T}{v}\left(1 + \frac{\mathrm{B}[T]}{v} + \frac{\mathrm{C}[T]}{v^2} + \frac{\mathrm{D}[T]}{v^5}\right) + \frac{\mathrm{E}[T]}{v^2}\left(1 + \frac{\mathrm{F}[T]}{v^2}\right)\exp\left(-\frac{\mathrm{G}[T]}{v^2}\right)$

Selection of a Suitable Equation of State

In the foregoing different equations of state were discussed. These and several dozen other equations of state reported in the literature can be classed according to the following aspects (Table 3.1):

1. The power to which the macro variable v is raised.
2. The number of parameters contained in the equation of state. (For this purpose the universal gas constant is not considered as a parameter.)

Of these the quadratic virial equation, the cubic Redlich–Kwong equation, and the Lee–Kesler equation of sixth degree will be used in this book. But why just these?

In general, we may ask how to select an equation of state when a certain behavior is required.

There are two approaches. According to one of these, using the apparatus of kinetic gas theory or statistical mechanics, a model is developed that takes into account specific intermolecular forces. The other approach does not start from first principles; instead a mathematical construction, fitting the experimental data well, is formulated with suitable parameters. Owing to its great practical importance the second approach will now be discussed.

Before fitting, three preliminary decisions must be made:

1. Specification of the desired range of validity with respect to density.
2. Choice of experimental data.
3. Specification of accuracy.

The requirements are conflicting. The same accuracy that can be achieved with a relatively simple few-parameter equation in the low-density range will require a multiparameter equation of complicated structure at higher density. If the density is at most one-fifth the critical density, we can get on well with a two-parameter equation of state, but four or five parameters may be needed if the density reaches half the critical density. In the vicinity of critical density the required number of parameters is at least six, and it increases to a dozen if the density is one and a half times the critical density.

THE VIRIAL EQUATION OF STATE AND THE SECOND VIRIAL COEFFICIENT

The explicit form for the compressibility coefficient of the virial equation, truncated after the second term, is

$$z = \frac{Pv}{\mathbf{R}T} = 1 + \mathrm{B}[T]\frac{1}{v}$$

In general the slope of the curve of the compressibility coefficient at constant temperature is a function of pressure, but in a gas at very low pressure it becomes

$$\lim_{1/v \to 0} \left(\frac{\partial z}{\partial (1/v)} \right)_T = \mathrm{B}[T]$$

The second virial coefficient has the dimension of volume and depends only on the temperature. At low temperatures B is negative, at sufficiently high temperatures positive. There exists a temperature at which the second virial coefficient is zero:

$$\mathrm{B}[T_{\mathrm{B}}] = 0$$

This is the Boyle temperature T_{B}.

The second virial coefficient is a measure of the deviation from the equation of state of ideal gases, due to the work done by intermolecular forces—more exactly, of the ratio of potential and kinetic energies. When the pair potential is known, the second virial coefficient can be calculated with the apparatus of classical statistical mechanics:

$$\mathrm{B}[T] = 2\pi \mathbf{N} \int_0^{\infty} (1 - e^{-\varepsilon[r]/\mathrm{k}T}) r^2 \, dr$$

For the pair potential $\varepsilon[r]$ it is assumed, after Lennard-Jones, that

$$\frac{\varepsilon}{\varepsilon_0} = 4\left(\left(\frac{\sigma}{r}\right)^6 - \left(\frac{\sigma}{r}\right)^{12}\right)$$

where σ and ε_0 are parameters characteristic of the molecule.

Thus, knowing the molecular characteristics, the second virial coefficient can be calculated.

The Reduced Second Virial Coefficient

The theory of corresponding states is particularly suitable for calculations where macro parameters are used instead of σ and ε_0 in the normalization. In the case of the second virial coefficient this can be achieved as follows. Since B has the dimensions of volume, $\mathrm{B}/v_{\mathrm{c}}$ is dimensionless. Our aim is now to express $\mathrm{B}/v_{\mathrm{c}}$ as a universal function of another dimensionless quantity: T/T_{c}.

The second virial coefficient is the partial derivative of the isotherm as $1/v \to 0$:

$$\mathrm{B}[T] = \lim_{1/v \to 0} \left(\frac{\partial (Pv/\mathbf{R}T)}{\partial (1/v)}\right)_T$$

Multiplying both numerator and denominator by the constant $\mathbf{R}T_{\mathrm{c}}/P_{\mathrm{c}}v_{\mathrm{c}}$, we have

$$\mathrm{B}[T] = \lim_{1/v \to 0} \left(\frac{P_{\mathrm{c}}v_{\mathrm{c}}^2}{\mathbf{R}T_{\mathrm{c}}}\right)\left(\frac{\partial (Pv/\mathbf{R}T)(\mathbf{R}T_{\mathrm{c}}/P_{\mathrm{c}}v_{\mathrm{c}})}{\partial (1/v)}\right)_{T/T_{\mathrm{c}}}$$

and from this the reduced second virial coefficient is

$$\mathrm{B}_{\mathrm{r}}[T_{\mathrm{r}}] = \frac{\mathrm{B}[T]}{v_{\mathrm{c}}} = \lim_{1/v_{\mathrm{r}} \to 0} \left(\frac{P_{\mathrm{c}}v_{\mathrm{c}}}{\mathbf{R}T_{\mathrm{c}}}\right)\left(\frac{\partial (P_{\mathrm{r}}v_{\mathrm{r}}/T_{\mathrm{r}})}{\partial (1/v_{\mathrm{r}})}\right)_{T_{\mathrm{r}}}$$

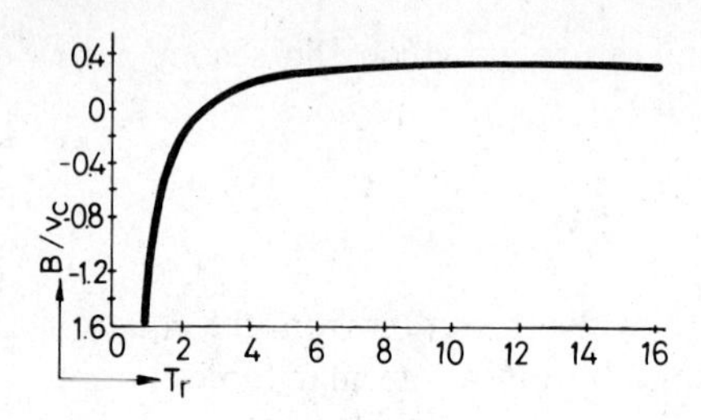

FIGURE 3.10. Reduced second virial coefficient.

The product of the critical state variables is a universal constant. The partial derivative is only a function of T_r, since it is interpreted in the limit $1/v_r \to 0$, and hence is a function of neither v_r nor P_r. Consequently the reduced second virial coefficient is a universal function of the reduced temperature (Figure 3.10).

The dependence of the dimensionless second virial coefficient on temperature can be well described with the following polynomial (Table 3.3):

$$\mathrm{B}_r[T_r] = b_1 - \frac{b_2}{T_r} - \frac{b_3}{T_r^2} - \frac{b_4}{T_r^3}$$

Algorithm and Program: Second Virial Coefficient for Simple Fluids

Write a program for the calculation of the reduced second virial coefficient of a simple fluid. The critical temperature is known.

Algorithm.

1. Read the critical temperature.
2. Read the actual temperature.
3. Calculate the reduced second virial coefficient.
4. Print the result.

Program VIRBNORMAL. A possible realization:

```
]PR#0
]LOAD VIRBNORMAL,D2
]LIST

 10  REM :CALCULATION OF DIMENSION
      LESS
 20  REM :SECOND VIRIAL COEFFICIENT
 30  REM :FOR SIMPLE COMPOUND
```

```
 40  READ B1,B2,B3,B4
 50  DATA 0.1181193,0.265728,0.15
      479,0.0030323
 60   PRINT : PRINT
 70   INPUT "ENTER CRITICAL TEMPERATURE
      (K) ";TC
 80   PRINT : PRINT
 90  MR = 0.3978
100   INPUT "ENTER TEMPERATURE (K)
      ";T
110   PRINT
120  TR = T / TC
130  B = B1 - B2 / TR - B3 / TR /
      TR - B4 / TR / TR / TR
140   PRINT : PRINT
150   PRINT "REDUCED TEMPERATURE"
      ;TR
160   PRINT
170   PRINT "REDUCED SECOND VIRIAL
      COEFF. ";B
180   PRINT : PRINT
190   PRINT
200   INPUT "NEXT TEMPERATURE ? (Y
      /N) ";A$
210   IF A$ = "Y" THEN GOTO 100
220   PRINT
230   INPUT "NEXT COMPOUND ? (Y / N)
      ";A$
240   IF A$ = "Y" THEN GOTO 60
250   END
```

Example. Run the program for methane in the temperature range 200–700 K.

```
]RUN

ENTER CRITICAL TEMPERATURE (K) 190.1

ENTER TEMPERATURE (K) 200

REDUCED TEMPERATURE 1.05207785
REDUCED SECOND VIRIAL COEFF. -.276904154

NEXT TEMPERATURE ? (Y / N) Y
ENTER TEMPERATURE (K) 300

REDUCED TEMPERATURE 1.57811678
REDUCED SECOND VIRIAL COEFF. -.113188571

NEXT TEMPERATURE ? (Y / N) Y
ENTER TEMPERATURE (K) 400
```

```
REDUCED TEMPERATURE 2.10415571
REDUCED SECOND VIRIAL COEFF. -.0434546888

NEXT TEMPERATURE ? (Y / N) Y
ENTER TEMPERATURE (K) 500

REDUCED TEMPERATURE 2.63019463
REDUCED SECOND VIRIAL COEFF. -5.45234711E - 03

NEXT TEMPERATURE ? (Y / N) Y
ENTER TEMPERATURE (K) 600

REDUCED TEMPERATURE 3.15623356
REDUCED SECOND VIRIAL COEFF. .0182930298

NEXT TEMPERATURE ? (Y / N) Y
ENTER TEMPERATURE (K) 700

REDUCED TEMPERATURE 3.68227249
REDUCED SECOND VIRIAL COEFF. .0344785108

NEXT TEMPERATURE ? (Y / N) N
NEXT COMPOUND ? (Y / N) N
```

Algorithm and Program: Virial Equation of State for Simple Fluids

Write a program for the calculation of the compressibility factor, molar volume, and molar density of a simple fluid in the gaseous state. The critical temperature and critical pressure of the substance are known.

Algorithm.

1. Read the critical temperature and critical pressure.
2. Read the existing temperature and pressure.
3. Calculate the reduced temperature and reduced pressure.
4. Calculate the reduced second virial coefficient (see program VIRB-NORMAL).
5. Solve the reduced form of the virial equation of state

$$V_r^2 - \frac{T_r}{P_r} V_r - \frac{T_r}{P_r} B = 0$$

   for V_r.
6. Calculate the compressibility factor $Z = P_r V_r / T_r$.
7. Calculate the molar volume $V = \dfrac{Z\mathbf{R}T}{P}$ and the molar density $\rho = 1/v$.
8. Print the results.

Program ESVIRNORMAL. A possible realization:

```
]PR # 0
]LOAD ESVIRNORMAL,D2
]LIST

 10  REM :CALCULATION OF COMPRESSIBILITY
 20  REM :FACTOR BY VIRIAL EQUATION
      OF
 30  REM :STATE OF REDUCED FORM
 40  REM FOR NORMAL COMPOUND
 50  READ B1,B2,B3,B4
 60  DATA 0.1181193,0.265728,0.15
      479,0.0030323
 70  PRINT : PRINT
 80  INPUT "ENTER CRITICAL TEMPERATURE
      (K) ";TC
 90  INPUT "ENTER CRITICAL PRESSURE
      (PA) ";PC
100  PRINT : PRINT
110  INPUT "ENTER PRESSURE (PA) "
      ;P
120  INPUT "ENTER TEMPERATURE (K)
      ";T
130  PRINT
140  TR = T / TC
150  PR = P / PC
160  B = B1 - B2 / TR - B3 / TR /
      TR - B4 / TR / TR / TR
170  BA = TR / PR
180  IF BA ^ 2 + 4 * BA * < 0 THEN
      PRINT "NEGATIVE SQR ARGUMEN
     T ": GOTO 260
190  VR = (BA + SQR (BA ^ 2 + 4 *
      BA * B)) / 2
200  Z = PR * VR / TR
210  PRINT : PRINT
220  PRINT "COMPRESSIBILITY FACTOR
      : ";Z
230  V = Z * 8.3144 * T / P
240  PRINT "MOLAR VOLUME (M3 / MOL)
      ";V
250  PRINT "DENSITY (MOL / M3) ";1
      / V
260  PRINT : PRINT : PRINT
270  PRINT
280  INPUT "NEXT TEMPERATURE ? (Y
      /N) ";A$
290  IF A$ = "Y" THEN GOTO 120
```

```
300   INPUT "NEXT PRESSURE ? (Y / N)
      ";A$
310   IF A$ = "Y" THEN GOTO 110
320   PRINT
330   INPUT "NEXT COMPOUND ? (Y / N)
      ";A$
340   IF A$ = "Y" THEN GOTO 70
350   END
```

Example. Run the program for methane at pressures 0.5×10^5, 10^5, 10×10^5, 50×10^5, 100×10^5, 1000×10^5 Pa and at temperatures 200, 400, and 1000 K.

```
]PR # 0
]RUN

ENTER CRITICAL TEMPERATURE (K) 190.56
ENTER CRITICAL PRESSURE (PA) 4.6002E6

ENTER PRESSURE (PA) 0.5E5
ENTER TEMPERATURE (K) 200

COMPRESSIBILITY FACTOR : .997110469
MOLAR VOLUME (M3 / MOL) .0331615011
DENSITY (MOL / M3) 30.1554503

NEXT TEMPERATURE ? (Y / N) N
NEXT PRESSURE ? (Y / N) Y
ENTER PRESSURE (PA) 1E5
ENTER TEMPERATURE (K) 200

COMPRESSIBILITY FACTOR : .994204043
MOLAR VOLUME (M3 / MOL) .0165324202
DENSITY (MOL / M3) 60.4872117

NEXT TEMPERATURE ? (Y / N) N
NEXT PRESSURE ? (Y / N) Y
ENTER PRESSURE (PA) 10E5
ENTER TEMPERATURE (K) 200

COMPRESSIBILITY FACTOR : .938607293
MOLAR VOLUME (M3 / MOL) 1.56079129E - 03
DENSITY (MOL / M3) 640.700652

NEXT TEMPERATURE ? (Y / N) Y
ENTER TEMPERATURE (K) 400

COMPRESSIBILITY FACTOR : .995429476
MOLAR VOLUME (M3 / MOL) 3.31055954E - 03
DENSITY (MOL / M3) 302.063742

NEXT TEMPERATURE ? (Y / N) N
NEXT PRESSURE ? (Y / N) Y
```

```
ENTER PRESSURE (PA) 50E5
ENTER TEMPERATURE (K) 200

NEGATIVE SQR ARGUMENT

NEXT TEMPERATURE ? (Y / N) Y
ENTER TEMPERATURE (K) 400

COMPRESSIBILITY FACTOR : .976709377
MOLAR VOLUME (M3 / MOL) 6.49660195E - 04
DENSITY (MOL / M3) 1539.26623

NEXT TEMPERATURE ? (Y / N) N
NEXT PRESSURE ? (Y / N) Y
ENTER PRESSURE (PA) 100E5
ENTER TEMPERATURE (K) 400

COMPRESSIBILITY FACTOR : .952220808
MOLAR VOLUME (M3 / MOL) 3.16685788E - 04
DENSITY (MOL / M3) 3157.70407

NEXT TEMPERATURE ? (Y / N) N
NEXT PRESSURE ? (Y / N) Y
ENTER PRESSURE (PA) 1000E5
ENTER TEMPERATURE (K) 200

NEGATIVE SQR ARGUMENT

NEXT TEMPERATURE ? (Y / N) Y
ENTER TEMPERATURE (K) 500

COMPRESSIBILITY FACTOR : .949326628
MOLAR VOLUME (M3 / MOL) 3.94654065E - 05
DENSITY (MOL / M3) 25338.6469

NEXT TEMPERATURE ? (Y / N) Y
ENTER TEMPERATURE (K) 1000

COMPRESSIBILITY FACTOR : 1.21145552
MOLAR VOLUME (M3 / MOL) 1.00725258E - 04
DENSITY (MOL / M3) 9927.99646

NEXT TEMPERATURE ? (Y / N) N
NEXT PRESSURE ? (Y / N) N
NEXT COMPOUND ? (Y / N) N
```

TWO-PARAMETER CUBIC EQUATIONS OF STATE

It is required of a cubic equation of state to follow the general tendency of experimental evidence, that is, to behave as shown in Figures 3.5 and 3.6. This is done by the van der Waals equation, so let us start from this equation:

$$P = \frac{\mathbf{R}T}{v - b} - \frac{a}{v^2}$$

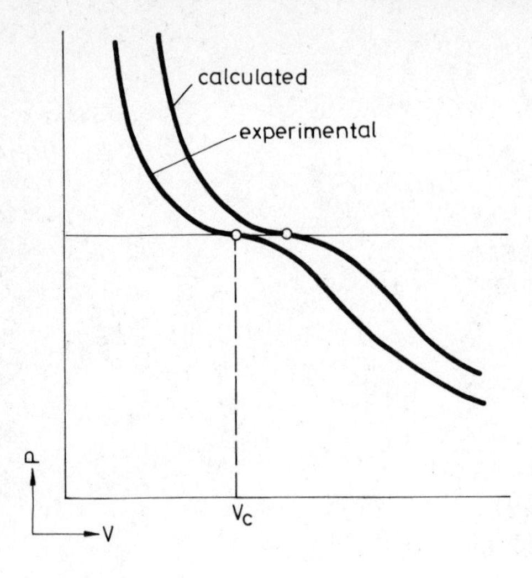

FIGURE 3.11. Deviation between calculated and experimental critical isotherms.

The most difficult task of the equation of state is to describe the critical isotherm well. The mathematical requirement is that both the first and second derivatives of pressure shall be zero at the critical point. After the elimination of the parameters a and b the van der Waals equation can be written in the following form:

$$P[v, T; P_c, T_c] = \frac{\mathbf{R}T}{v - \mathbf{R}T_c/8P_c} - \frac{27\mathbf{R}^2T_c^2}{64P_c v^2}$$

If we plot the isotherm calculated in this way (Figure 3.11) and compare it with the isotherm obtained from data measured for argon, we see that the critical volume is higher than the true value v_c by a factor of 1.29. We conclude that the van der Waals equation needs some modification.

On inspection it appears that the measured and the calculated isotherms can be superimposed by shifting the calculated isotherm parallel to the v axis toward the critical point. This is a linear coordinate transformation without any effect on the first and second derivatives. All we have to do is increase the volume variable in the van der Waals equation by an additive constant:

$$P = \frac{\mathbf{R}T}{v - b + c} - \frac{a}{(v + c)^2}$$

After the elimination of a and b, we have

$$P = \frac{\mathbf{R}T}{v - \dfrac{\mathbf{R}T_c}{8P_c} + c} - \frac{27\mathbf{R}^2T_c}{64P_c(v + c)^2}$$

Taking into consideration the definition of the reduced state variables and the fact that the critical compressibility coefficient is a constant,

$$z_c = \frac{P_c v_c}{\mathbf{R}T_c} = \frac{3}{8}$$

the equation of state can be written in the following reduced form:

$$P_r = \frac{T_r}{z_c\left(v_r - \dfrac{0.125}{z_c} + \dfrac{c}{v_c}\right)} - \frac{27}{64z_c^2\left(v_r + \dfrac{c}{v_c}\right)^2}$$

In this equation z_c and c are parameters. One of them can be eliminated by assuming some convenient relationship between the two. A good fit is obtained with

$$\frac{c}{v_c} = \frac{1}{25z_c} = \frac{0.04}{z_c}$$

By the elimination of the parameter c our reduced equation of state takes the following form:

$$P_r = \frac{T_r}{z_c(v_r - 0.085/z_c)} - \frac{27}{64z_c(v_r + 0.04/z_c)^2}$$

Let us remember that so far only the critical isotherm has been described, and we know nothing about the behavior of this corrected equation of state along any other isotherm. The critical isostere beyond the critical point is linear, and is tangent to the vapor-pressure curve at the critical point. The slope at the critical point as calculated from our reduced equation of state, is

$$\left(\frac{\partial P_r}{\partial T_r}\right)_{v_r=1} = \frac{1}{z_c(v_r - 0.085/z_c)} = 4.85$$

whereas the value measured for argon is

$$\left(\frac{\partial P_r}{\partial T_r}\right)_{v_r=1} = 6$$

The difference is significant and can be explained by the absence of a temperature-dependent term in the second member of our reduced equation of state.

Now we have several possibilities for including the temperature dependence:

1. One is to introduce a multiplier which is a linear function of temperature:

$$P_r = \frac{T_r}{z_c(v_r - 0.085/z_c)} - \frac{27(A + BT_r)}{64z_c(v_r + 0.04/z_c)^2}$$

Differentiation with respect to temperature gives

$$\left(\frac{\partial P_r}{\partial T_r}\right)_{v_r} = \frac{1}{z_c(v_r - 0.085/z_c)} - \frac{27B}{64z_c(v_r + 0.04/z_c)^2}$$

Let us apply these two equations to the critical point, where

$$T_r = v_r = 1$$

$$\left(\frac{\partial P_r}{\partial T_r}\right)_{v_r=1} = 6$$

The solution of the system of equations is

$$A = 1.3$$

$$B = -0.3$$

The final form of the equation of state with these parameters is

$$P_r = \frac{T_r}{z_c(v_r - 0.085/z_c)} - \frac{9(4 - T_r)}{64z_c^2(v_r + 0.04/z_c)^2}$$

This equation of state is due to Martin. It represents two modifications of the van der Waals equation: a linear transformation in the volume variable, and substitution of a linear function of the temperature for the parameter a.

2. Maintaining the above transformation, but substituting for the parameter a a value proportional to the square root of the temperature, the following reduced equation of state is obtained:

$$P_r = \frac{T_r}{z_c(v_r - b_{RK}/z_c)} - \frac{a_{RK}}{T_r^{0.5} z_c v_r(v_r + b_{RK}/z_c)}$$

This is the reduced form of the Redlich–Kwong equation of state. The two RK parameters arise from the condition at the critical point:

$$a_{RK} = 0.42748$$

$$b_{RK} = 0.08664$$

It is easy to see the similarity between the Martin and the RK equations of state. The first term on the right-hand side is identical for the two equations, disregarding the small difference in parameter values:

$$\frac{T_r}{z_c(v_r - 0.085/z_c)} \approx \frac{T_r}{z_c(v_r - 0.08664/z_c)}$$

The second term on the right-hand side of the two equations can be compared as follows:

$$\frac{9(4 - T_r)}{64z_c^2(v_r + 0.04/z_c)^2} \leftrightarrow \frac{0.42748}{T_r^{0.5} z_c^2 v_r(v_r + 0.08664/z_c)}$$

Reducing in the denominator by z_c^2, we can see that

$$\left(v_r + \frac{0.04}{z_c}\right)^2 \leftrightarrow v_r\left(v_r + \frac{0.08664}{z_c}\right)$$

Thus, the difference is represented by the following factors:

Martin	Redlich–Kwong
$\dfrac{9(4 - T_r)}{64}$	$\dfrac{0.42748}{T_r^{0.5}}$

The Martin equation of state can be used more advantageously in the vicinity of the critical point, where the isosteres are linear, while the RK equation of state is better in the region of higher densities, where the isosteres are bent.

THE REDLICH–KWONG EQUATION OF STATE

The RK equation of state plays a prominent role among two-parameter equations of state—not because it is the "best," but because it is most widely used, and its application has been worked out in many fields. The RK equation of state is used in five forms:

1. In terms of the pressure:

$$P = \frac{RT}{(v - b)} - \frac{a}{T^{0.5}v(v + b)}$$

where the two parameters with dimension can be calculated from the critical constants:

$$a = 0.42748 \frac{R^2 T_c^{2.5}}{P_c}$$

$$b = 0.08664 \frac{RT_c}{P_c}$$

2. In terms of the compressibility factor:

$$z = \frac{1}{1-h} - \left(\frac{A^2}{B}\right)\left(\frac{h}{1+h}\right)$$

where

$$h = \frac{BP}{z}$$

$$A = \left(\frac{a}{\mathbf{R}^2 T^{2.5}}\right)^{0.5}$$

$$B = \frac{b}{\mathbf{R}T}$$

(here B is not to be mistaken for the second virial coefficient). This equation can be solved for the compressibility factor by iteration.

3. As a cubic equation in z, with the parameters A and B used before:

$$z^3 - z^2 + (AP - BP - B^2P^2)z - (ABP^2) = 0$$

This equation can be solved with the Cardano formula.

4. The reduced form of the RK equation of state has been discussed:

$$P_r = \frac{T_r}{z_c(v_r - 0.08664/z_c)} - \frac{0.42748}{T_r^{0.5} z_c v_r (v_r + 0.08664/z_c)}$$

Let us eliminate from the equation the critical compressibility factor, and introduce instead of the reduced volume the *ideal* reduced volume as a new variable. The critical compressibility factor is

$$z_c = \frac{P_c v_c}{\mathbf{R}T_c}$$

The ideal reduced volume is by definition

$$\phi_r \equiv \frac{vP_c}{\mathbf{R}T_c}$$

TABLE 3.2
Parameters for Redlich – Kwong Equation of State

Compound	ω	i	α_i	β_i
Methane	0	0	0.40967470	$0.88023049 \times 10^{-1}$
		1	$0.22458091 \times 10^{-1}$	$-0.52911710 \times 10^{-2}$
		2	$-0.91956459 \times 10^{-2}$	$0.19177828 \times 10^{-2}$
n-Octane	0.3978	0	-0.19457486	$0.64435102 \times 10^{-1}$
		1	0.11008389×10	$0.18707489 \times 10^{-1}$
		2	-0.42560414	$0.84749789 \times 10^{-3}$

The reduced form of the RK equation with this reduced variable is

$$P_r = \frac{T_r}{\phi_r - 0.08664} - \frac{0.42748}{T_r^{0.5}\phi_r(\phi_r + 0.08664)}$$

5. The performance of the RK equation of state can be improved by describing the parameters a and b with polynomials of second degree in reduced temperature:

$$a = \alpha_0 + \alpha_1 T_r^{-1} + \alpha_2 T_r^{-2}$$

$$b = \beta_0 + \beta_1 T_r^{-1} + \beta_2 T_r^{-2}$$

The coefficients of the polynomial are contained in Table 3.2.

Algorithm and Program: Redlich – Kwong Equation of State

Write a program for the calculation of the compressibility factor, molar volume, and density of gases using the compressibility-factor form of the RK equation of state.

Algorithm.

1. Read the critical temperature and pressure.
2. Read the actual temperature and pressure.
3. Calculate a, b, A, and B.
4. Suppose that $z = 1$, and calculate h.
5. Calculate a new z from h.
6. Stop the iteration if the change of z in two consecutive steps is less than 10^{-5} or the number of iterations exceeds 50.
7. Calculate the molar volume and density.
8. Print the results.

Program ESRK. A possible realization:

```
]PR#0
]LORD ESRK
]LIST

 10  REM :CALCULATION OF COMPRESSI
      BILITY
 20  REM :FACTOR BY REDLICH-KWONG
 30  REM :EQUATION OF STATE
 40  R = 8.3144
 50  CA = 0.4278
 60  CB = 0.0867
 70  PRINT : PRINT
 80  INPUT "ENTER CRITICAL TEMPERA
      TURE (K) ";TC
 90  PRINT
100  INPUT "ENTER CRITICAL PRESSU
      RE (PA) ";PC
110  PRINT : PRINT
120  INPUT "ENTER PRESSURE (PA) "
      ;P
130  PRINT
140  INPUT "ENTER TEMPERATURE (K)
      ";T
150  AL = CA * R * R * TC ∧ 2.5 /
      PC
160  BL = CB * R * TC / PC
170  B = BL / R / T
180  A2 = AL / R / R / T ∧ 2.5
190  Z1 = 1
200  I = 0
210  H2 = B * P / Z1
220  I = I + 1
230  IF I > 50 THEN PRINT : PRINT
      "ITERATION NOT CONVERGENT " :
      GOTO 330
240  Z2 = 1 / (1 - H2) - A2 * H2 /
      B / (1 + H2)
250  IF ABS (Z2 - Z1) < 1E - 5 THEN
      GOTO 280
260  Z1 = Z2
270  GOTO 210
280  PRINT : PRINT
290  PRINT "COMPRESSIBILITY FACTO
      R ";Z2
300  PRINT
310  PRINT "MOLAR VOLUME (M3 / MOL)
      ";Z2 * R * T / P
320  PRINT "DENSITY (MOL / M3) ";P /
      Z2 / R / T
```

```
330  PRINT : PRINT
340  INPUT "NEXT TEMPERATURE ? (Y
      /N); "A$
350  IF A$ = "Y" THEN GOTO 130
360  PRINT
370  INPUT "NEXT PRESSURE ? (Y / N)
      ";A$
380  IF A$ = "Y" THEN GOTO 120
390  PRINT
400  INPUT "NEXT COMPOUND ? (Y / N)
      ";A$
410  IF A$ = "Y" THEN GOTO 40
420  END
```

Example. Run the program for propane in the pressure range 0.5×10^5–500×10^5 Pa and temperature range 250–500 K.

```
]RUN

ENTER CRITICAL TEMPERATURE (K) 369.86
ENTER CRITICAL PRESSURE (PA) 4.2455E6

ENTER PRESSURE (PA) 0.5E5
ENTER TEMPERATURE (K) 300

COMPRESSIBILITY FACTOR .992715157
MOLAR VOLUME (M3 / MOL) .0495229854
DENSITY (MOL / M3) 20.1926437

NEXT TEMPERATURE ? (Y / N) Y
ENTER TEMPERATURE (K) 400

COMPRESSIBILITY FACTOR .99679655
MOLAR VOLUME (M3 / MOL) .0663021218
DENSITY (MOL / M3) 15.0824736

NEXT TEMPERATURE ? (Y / N) Y
ENTER TEMPERATURE (K) 500

COMPRESSIBILITY FACTOR .998383969
MOLAR VOLUME (M3 / MOL) .0830096367
DENSITY (MOL / M3) 12.0467941

NEXT TEMPERATURE ? (Y / N) N
NEXT PRESSURE ? (Y / N) Y
ENTER PRESSURE (PA) 2E5
ENTER TEMPERATURE (K) 250

COMPRESSIBILITY FACTOR .950298295
MOLAR VOLUME (M3 / MOL) 9.87645018E - 03
DENSITY (MOL / M3) 101.250954

NEXT TEMPERATURE ? (Y / N) y
ENTER TEMPERATURE (K) 300
```

```
COMPRESSIBILITY FACTOR .970345939
MOLAR VOLUME (M3 / MOL) .0121017664
DENSITY (MOL / M3) 82.632565

NEXT TEMPERATURE ? (Y / N) Y
ENTER TEMPERATURE (K) 400

COMPRESSIBILITY FACTOR .987119766
MOLAR VOLUME (M3 / MOL) .0164146172
DENSITY (MOL / M3) 60.9213112

NEXT TEMPERATURE ? (Y / N) N
NEXT PRESSURE ? (Y / N) Y
ENTER PRESSURE (PA) 5E5
ENTER TEMPERATURE (K) 270

COMPRESSIBILITY FACTOR .894031247
MOLAR VOLUME (M3 / MOL) 4.01400004E - 03
DENSITY (MOL / M3) 249.12805

NEXT TEMPERATURE ? (Y / N) Y
ENTER TEMPERATURE (K) 300

COMPRESSIBILITY FACTOR .922938572
MOLAR VOLUME (M3 / MOL) 4.60420828E - 03
DENSITY (MOL / M3) 217.192607

NEXT TEMPERATURE ? (Y / N) Y
ENTER TEMPERATURE (K) 400

COMPRESSIBILITY FACTOR .967454147
MOLAR VOLUME (M3 / MOL) 6.43504061E - 03
DENSITY (MOL / M3) 155.399175

NEXT TEMPERATURE ? (Y / N) N
NEXT PRESSURE ? (Y / N) Y
ENTER PRESSURE (PA) 10E5
ENTER TEMPERATURE (K) 270

COMPRESSIBILITY FACTOR .756200867
MOLAR VOLUME (M3 / MOL) 1.69758625E - 03
DENSITY (MOL / M3) 589.071688

NEXT TEMPERATURE ? (Y / N) Y
ENTER TEMPERATURE (K) 300

COMPRESSIBILITY FACTOR .833004314
MOLAR VOLUME (M3 / MOL) 2.07777932E - 03
DENSITY (MOL / M3) 481.283065

NEXT TEMPERATURE ? (Y / N) Y
ENTER TEMPERATURE (K) 400

COMPRESSIBILITY FACTOR .933672712
MOLAR VOLUME (M3 / MOL) 3.10517136E - 03
DENSITY (MOL / M3) 322.043419
```

```
NEXT TEMPERATURE ? (Y / N) N
NEXT PRESSURE ? (Y / N) Y
ENTER PRESSURE (PA) 30E5
ENTER TEMPERATURE (K) 250

ITERATION NOT CONVERGENT

NEXT TEMPERATURE ? (Y / N) Y
ENTER TEMPERATURE (K) 340

ITERATION NOT CONVERGENT

NEXT TEMPERATURE ? (Y / N) Y
ENTER TEMPERATURE (K) 400

COMPRESSIBILITY FACTOR .781675723
MOLAR VOLUME (M3 / MOL) 8.66555284E - 04
DENSITY (MOL / M3) 1153.99446

NEXT TEMPERATURE ? (Y / N) N
NEXT PRESSURE ? (Y / N) Y
ENTER PRESSURE (PA) 100E5
ENTER TEMPERATURE (K) 400

ITERATION NOT CONVERGENT

NEXT TEMPERATURE ? (Y / N) Y
ENTER TEMPERATURE (K) 500

COMPRESSIBILITY FACTOR .721309762
MOLAR VOLUME (M3 / MOL) 2.99862894E - 04
DENSITY (MOL / M3) 3334.85743

NEXT TEMPERATURE ? (Y / N) N
NEXT PRESSURE ? (Y / N) Y
ENTER PRESSURE (PA) 500E5
ENTER TEMPERATURE (K) 500

ITERATION NOT CONVERGENT

NEXT TEMPERATURE ? (Y / N) N
NEXT PRESSURE ? (Y / N) N
NEXT COMPOUND ? (Y / N) N
```

Remark. It can be seen that the method is not convergent at high pressures and low temperatures. To avoid iteration problems use the polynomial form of the equation and find the roots with the Cardano equation.

THE PENG – ROBINSON EQUATION OF STATE

The Peng–Robinson equation of state belongs to the family of two-parameter cubic equations of state:

$$P = \frac{\mathbf{R}T}{v - b} - \frac{a[T]}{v(v + b) + b(v - b)}$$

This equation is based on the recognition that the performance of a two-parameter equation of state can be improved by making the "parameter" a temperature-dependent, and by improving the form of the function with the additive term $b(v - b)$ in the denominator of the second term.

The resulting expressions for a and b, calculated from experimental data, are

$$a = 0.45724 \frac{\mathbf{R}^2 T_c^2}{P_c} \alpha^2[T]$$

$$b = 0.0778 \frac{\mathbf{R} T_c}{P_c}$$

The correction factor in α for temperature dependence is

$$\alpha = 1 + 0.37464 \left(1 - T_r^{0.5}\right)$$

The Peng–Robinson equation of state also can be written as a cubic equation in z:

$$z^3 - (1 - B)z^2 + (A - 3B^2 - 2B)z - (AB - B^2 - B^3) = 0$$

where

$$A = \frac{aP}{\mathbf{R}^2 T^2}$$

$$B = \frac{bP}{\mathbf{R}T}$$

The equation can be solved with Newton's iteration formula.

Algorithm and Program: Graphical Representation of the Peng – Robinson Equation of State

Write a program for the plotting of the function

$$z^3 - (1 - B)z^2 + (A - 3B^2 - 2B)z - (AB - B^2 - B^3) = \mathrm{f}[z]$$

where A and B are the Peng–Robinson constants.

Algorithm.

1. Read the critical temperature and pressure.
2. Read the actual temperature and pressure.

3. Calculate a, b, A, B.
4. Calculate and plot values of the function $f[z]$ in the range $-5 \leq z \leq 6$.

Program GRPR. A possible realization:

```
]PR#0
]LOAD GRPR
]LIST

 10  REM :GRAPHICAL REPRESENTATION
      OF
 20  REM :THE PENG-ROBINSON EQUATI
      ON OF STATE
 30  HPLOT 0,0
 40  HGR
 50  HPLOT 10,95 TO 279,95
 60  HPLOT 125,0 TO 125,191
 70  R = 8.3144
 80  INPUT "ENTER CRITICAL TEMPERA
      TURE (K) ";TC
 90  INPUT "ENTER CRITICAL PRESSUR
      E (PA) ";PC
100  PRINT
110  INPUT "ENTER TEMPERATURE (K)
      ";T
120  INPUT "ENTER PRESSURE (PA)"
      ;P
130  TR =  T / TC
140  AL = 0.45724 * R * R * TC * T
      C / PC * (1 + 0.37464 * (1 -
      SQR (TR))) ^ 2
150  BL  =  0.0778 * R * TC / PC
160  A  =  AL * P / R / R / T / T
170  B  =  BL * P / R / T
180  HPLOT 0,0
190  FOR Z  =   - 5 TO 6 STEP 0.1
200  X  =  Z * 25 + 125
210  F  =  Z ^ 3 - (1 - B) * Z ^ 2 +
      (A - 3 * B ^ 2 - 2 * B) * Z -
      (A * B - B ^ 2 - B ^ 3)
220  Y  =  95 - 0.30 * F
230   IF Z  =   - 5 THEN HPLOT X,Y
240   HPLOT TO X,Y
250   NEXT Z
260   HPLOT 0,0
270   PRINT
280   INPUT "NEXT PRESSURE ? (Y / N)
      ";A$
290  IF A$ = "Y" THEN GOTO 120
```

```
300  INPUT "NEXT TEMPERATURE ? (Y
       /N) ";A$
310  IF A$ = "Y" THEN GOTO 110
320  INPUT "NEXT COMPOUND ?(Y / N)
       ";A$
330  IF A$ = "Y" THEN GOTO 80
340  TEXT
350  END
```

Example. Run the program for propane at temperatures 300 and 500 K, and pressures 0.5×10^5, 30×10^5, and 500×10^5 Pa.

```
]RUN
ENTER CRITICAL TEMPERATURE (K) 369.86
ENTER CRITICAL PRESSURE (PA) 4.2455E6

ENTER TEMPERATURE (K) 300
ENTER PRESSURE (PA) 0.5E5

NEXT PRESSURE ? (Y / N) Y
ENTER PRESSURE (PA) 30E5

NEXT PRESSURE ? (Y / N) Y
ENTER PRESSURE (PA) 500E5

NEXT PRESSURE ? (Y / N) N
NEXT TEMPERATURE ? (Y / N) Y
ENTER TEMPERATURE (K) 500
ENTER PRESSURE (PA) 0.5E5

NEXT PRESSURE ? (Y / N) Y
ENTER PRESSURE (PA) 500E5

NEXT PRESSURE ? (Y / N) N
NEXT TEMPERATURE ? (Y / N) N
NEXT COMPOUND ?(Y / N) N
```

Remarks. You can see on the screen that after the largest root there is no inflection on the curve of f(z), and hence the equation of state can be solved by Newton's method. The recommended initial guess is $z = 6$ because there is no root larger than 6.

Algorithm and Program: Peng – Robinson Equation of State for Simple Fluids

Write a program for the calculation of the compressibility factor, the molar volume, and the molar density of a simple fluid in the gaseous state. Solve the equation by Newton's method.

Algorithm.

1. Read the critical temperature and critical pressure of the compound.
2. Read the actual temperature and pressure.
3. Calculate of a, b, A, B.
4. Make initial guess $z = 6$.
5. Perform a Newtonian iteration step

$$z_{new} = z_{old} - \frac{f[z_{old}]}{f'[z_{old}]}$$

where

$$f[z] = z^3 - (1 - B)z^2 + (A - 3B^2 - 2B)z - (AB - B^2 - B^3)$$

6. Stop the iteration and go to step 7 if the number of steps exceeds 50 or the change of z in two consecutive steps is less than 10^{-5}; otherwise return to step 5.
7. Calculate the molar volume and density, or print the iteration error.
8. Print the results.

Program ESPRNORMAL. A possible realization:

```
]PR#0
]LOAD ESPRNORMAL,D2
]LIST

 10  REM :CALCULATION OF COMPRESSI
      BILITY
 20  REM :FACTOR BY PENG-ROBINSON
 30  REM :EQUATION OF STATE
 40  REM :FOR SINGLE COMPOUNDS
 50  DEF FN F(Z) = Z ^ 3 - (1 - B
      ) * Z ^ 2 + (A - 3 * B ^ 2 -
      2 * B) * Z - (A * B - B ^ 2 -
      B ^ 3)
 60  DEF FN FD(Z) = 3 * Z ^ 2 - 2
      * (1 - B) * Z + (A - 3 * B ^
      2 - 2 * B)
 70  R = 8.3144
 80  PRINT : PRINT
 90  INPUT "ENTER CRITICAL TEMPERA
      TURE (K) ";TC
100  PRINT
110  INPUT "ENTER CRITICAL PRESSU
      RE (PA) ";PC
```

```
120  PRINT : PRINT
130  INPUT "ENTER PRESSURE (PA) "
      ;P
140  INPUT "ENTER TEMPERATURE (K)
      ";T
150  TR = T / TC
160  M = 0.37464
170  AL = 0.45724 * R * R * TC * T
      C / PC * (1 + M * (1 - SQR
      (TR))) ∧ 2
180  BL = 0.0778 * R * TC / PC
190  A = AL * P / R / R / T / T
200  B = BL * P / R / T
210  Z1 = 6
220  I = 0
230  H = FN F (Z1) / FN FD(Z1)
240  I = I + 1
250  IF I > 50 THEN PRINT : PRINT
      "ITERATION NOT CONVERGENT": GOTO
      350
260  Z2 = Z1 - H
270  IF ABS (Z2 - Z1) < 1E - 5 THEN
      GOTO 300
280  Z1 = Z2
290  GOTO 230
300  PRINT : PRINT
310  PRINT "COMPRESSIBILITY FACTO
      R ";Z2
320  PRINT
330  PRINT "MOLAR VOLUME (M3 / MOL)
      ";Z2 * R * T / P
340  PRINT "DENSITY (MOL / M3) ";P
      / Z2 / R / T
350  PRINT
360  INPUT "NEXT TEMPERATURE ? (Y
      /N) ";A$
370  IF A$ = "Y" THEN GOTO 140
380  INPUT "NEXT PRESSURE ? (Y / N)
      ";A$
390  IF A$ = "Y" THEN GOTO 130
400  INPUT "NEXT COMPOUND ? (Y / N)
      ";A$
410  IF A$ = "Y" THEN GOTO 80
420  END
```

Example. Run the program for methane at pressures 0.5×10^5, 10^5, 10×10^5, 50×10^5, and 100×10^5 Pa and at temperatures 200, 400, and 1000 K.

```
]RUN
ENTER CRITICAL TEMPERATURE (K) 190.56
ENTER CRITICAL PRESSURE (PA) 4.6002E6
ENTER PRESSURE (PA) 0.5E5
ENTER TEMPERATURE (K) 200

COMPRESSIBILITY FACTOR .996370966
MOLAR VOLUME (M3 / MOL) .033136907
DENSITY (MOL / M3) 30.1778316

NEXT TEMPERATURE ? (Y / N) N
NEXT PRESSURE ? (Y / N) Y
ENTER PRESSURE (PA) 1E5
ENTER TEMPERATURE (K) 200

COMPRESSIBILITY FACTOR .992731097
MOLAR VOLUME (M3 / MOL) .0165079269
DENSITY (MOL / M3) 60.5769585

NEXT TEMPERATURE ? (Y / N) N
NEXT PRESSURE ? (Y / N) Y
ENTER PRESSURE (PA) 10E5
ENTER TEMPERATURE (K) 200

COMPRESSIBILITY FACTOR .925169722
MOLAR VOLUME (M3 / MOL) 1.53844623E - 03
DENSITY (MOL / M3) 650.006469

NEXT TEMPERATURE ? (Y / N) Y
ENTER TEMPERATURE (K) 400

COMPRESSIBILITY FACTOR .992708024
MOLAR VOLUME (M3 / MOL) 3.30150864E - 03
DENSITY (MOL / M3) 302.891832

NEXT TEMPERATURE ? (Y / N) N
NEXT PRESSURE ? (Y / N) Y
ENTER PRESSURE (PA) 50E5
ENTER TEMPERATURE (K) 200

COMPRESSIBILITY FACTOR .522286048
MOLAR VOLUME (M3 / MOL) 1.73699805E - 04
DENSITY (MOL / M3) 5757.05886

NEXT TEMPERATURE ? (Y / N) Y
ENTER TEMPERATURE (K) 400

COMPRESSIBILITY FACTOR .96889151
MOLAR VOLUME (M3 / MOL) 6.44460125E - 04
DENSITY (MOL / M3) 1551.68638

NEXT TEMPERATURE ? (Y / N) N
NEXT PRESSURE ? (Y / N) Y
ENTER PRESSURE (PA) 100E5
ENTER TEMPERATURE (K) 400
```

```
COMPRESSIBILITY FACTOR .951343729
MOLAR VOLUME (M3 / MOL) 3.16394092E - 04
DENSITY (MOL / M3) 3160.61527

NEXT TEMPERATURE ? (Y / N) N
NEXT PRESSURE ? (Y / N) Y
ENTER PRESSURE (PA) 1000E5
ENTER TEMPERATURE (K) 200

COMPRESSIBILITY FACTOR 2.10440014
MOLAR VOLUME (M3 / MOL) 3.49936489E - 05
DENSITY (MOL / M3) 28576.6141

NEXT TEMPERATURE ? (Y / N) Y
ENTER TEMPERATURE (K) 500

COMPRESSIBILITY FACTOR 1.45178315
MOLAR VOLUME (M3 / MOL) 6.03535291E - 05
DENSITY (MOL / M3) 16569.0394

NEXT TEMPERATURE ? (Y / N) Y
ENTER TEMPERATURE (K) 1000

COMPRESSIBILITY FACTOR 1.28325088
MOLAR VOLUME (M3 / MOL) 1.06694611E - 04
DENSITY (MOL / M3) 9372.54461

NEXT TEMPERATURE ? (Y / N) N
NEXT PRESSURE ? (Y / N) N
NEXT COMPOUND ? (Y / N) N
```

THE LEE – KESLER EQUATION OF STATE

We have now found three equations of state, each of which provides a good description of the (P,v,T) relationships in a certain region, but there is still a range resisting our efforts: the vicinity of the critical point. Now we seek an equation of state for the description of this range. The following is expected from this equation:

1. It must describe experimental (P,v,T) relationships in a sufficiently wide range, and at the critical point satisfy

$$\left(\frac{\partial P}{\partial v}\right)_{T_c} = \left(\frac{\partial^2 P}{\partial v^2}\right)_{T_c} = 0$$

2. It must contain the second virial coefficient.
3. It must correctly describe the isothermal enthalpy correction (which will be discussed in a later chapter).

4. It must be consistent with the selected vapor-pressure equation (discussed in a later chapter).

These are very strict requirements, and a good description is to be expected only if a rather large number of parameters are included in the functional form of the equation of state. One equation of this kind is the Benedict–Webb–Rubin (BWR) equation of state. Its form, written by Lee and Kesler with reduced state variables, is

$$z = \frac{P_r \phi_r}{T_r} = 1 + \frac{B_r}{\phi_r} + \frac{C_r}{\phi_r^2} + \frac{D_r}{\phi_r^5} + \frac{c_4}{T_r^3 \phi_r^2}\left(\beta + \frac{\gamma}{\phi_r^2}\right) e^{-(\gamma/\phi_r^2)}$$

In this equation ϕ_r is the ideal reduced volume. The dimensionless form of the second virial coefficient is

$$B_r = b_1 - \frac{b_2}{T_r} - \frac{b_3}{T_r^2} - \frac{b_4}{T_r^3}$$

The third virial coefficient is

$$C_r = c_1 - \frac{c_2}{T_r} + c_3 T_r^3$$

A further coefficient is

$$D_r = d_1 + \frac{d_2}{T_r}$$

The coefficients are contained in the first column of Table 3.3. The LK equation of state is solved by iteration.

TABLE 3.3
Parameters for LK Equation of State

Coefficient	$\omega^{(0)} = 0$	$\omega^{(r)} = 0.3978$
b_1	0.1181193	0.2026579
b_2	0.265728	0.331511
b_3	0.154790	0.027655
b_4	0.030323	0.203488
c_1	0.0236744	0.0313385
c_2	0.0186984	0.0503618
c_3	0.0	0.016901
c_4	0.042724	0.041577
d_1	0.155488×10^{-4}	0.48736×10^{-4}
d_2	0.623689×10^{-4}	0.0740336×10^{-4}
β	0.65392	1.226
γ	0.060167	0.03754

Algorithm and Program: Lee – Kesler Equation of State for Simple Fluids

Write a program for the calculation of the compressibility factor, molar volume, and density of gases using the simple form of the LK equation of state. Solve the equation by Newton's method.

Algorithm.

1. Read the critical temperature and pressure.
2. Calculate B (see program VIRBNORMAL), C and D.
3. Make initial guess $\phi = 0.2T_r/P_r$
4. Calculate a Newton iteration step

$$\phi_{\text{new}} = \phi_{\text{old}} - \frac{f[\phi_{\text{old}}]}{f'[\phi_{\text{old}}]}$$

where

$$f[\phi] = 1 + \frac{B^*}{\phi} + \frac{C}{\phi^2} + \frac{D}{\phi^5} + \frac{c_n}{T_r^3\phi^2}\left(\beta + \frac{\gamma}{\phi^2}\right)e^{-\gamma/\phi^2} - \frac{P_r\phi}{T_r}$$

5. Stop the iteration if the change in ϕ between two consecutive steps is less than 10^{-5}, or $f(\phi) < 10^{-4}$, or the number of iterations exceeds 50; otherwise return to step 4.
6. Calculate the compressibility factor

$$z = \frac{P_r}{T_r}\phi$$

7. Calculate the molar volume and density.
8. Printing the results.

Program ESBWRNORMAL. A possible realization:

```
]PR#0
]LOAD ESBWRNORMAL,D2
]LIST

 10  REM :CALCULATION OF COMPRESSI
      BILITY FACTOR BY
 20  REM :LK EQUATION OF STATE
 30  REM :FOR SINGLE COMPOUND
 40  DIM B(12)
 50  FOR I = 1 TO 12
```

```
 60  READ B(I)
 70  READ DB
 80  NEXT I
 90  DATA 0.1181193,0.2026579,0.2
     65728,0.331511
100  DATA 0.15790,0.027655,0.030
     323,0.203488,0.0236744,0.031
     3385
110  DATA 0.0186984,0.0503618,0,
     0.016901,0.042724,0.04177,0.
     155488E - 4,0.48736E - 4
120  DATA 0.623689E - 4,0.0740336E
     -4,0.65392,1.226,0.060167,0.
     03754
130  INPUT "ENTER CRITICAL TEMPER
     ATURE (K) ";TC
140  INPUT "ENTER CRITICAL PRESSU
     RE (PA) ";PC
150  MR = 0.3978
160  PRINT
170  INPUT "ENTER PRESSURE (PA) "
     ;P
180  INPUT "ENTER TEMPERATURE (K)
     ";T
190  TR = T / TC
200  B = B(1) - B(2) / TR - B(3) /
     TR / TR - B(4) / TR / TR / T
     R
210  C = B(5) - B(6) / TR + B(7) /
     TR / TR / TR
220  D = B(9) + B(10) / TR
230  DEF FN F(FI) = 1 + B / FI +
     C / FI ∧ 2 + D / FI ∧ 5 + (B
     (11) + B(12) / FI ∧ 2) * EXP
     ( - B(12) / FI ∧ 2) / TR ∧ 3
     / FI ∧ 2 * B(8) - P * FI /
     PC / TR
240  DEF FN FD(FI) = - B / FI ∧
     2 - 2 * C / FI ∧ 3 - 5 * D /
     FI ∧ 6 - EXP ( - B(12) / FI
     ∧ 2) * (2 * B(8) / TR ∧ 3 /
     FI ∧ 3 * (B(11) + B(12) / FI
     ∧ 2) + 2 * B(12) * B(8) / T
     R ∧ 3 / FI ∧ 5 + 2 * B(12) *
     B(8) / TR ∧ 3 / FI ∧ 5 * (B(
     11) + B(12) / FI ∧ 2)) - P /
     PC / TR
250  FI = 0.2 * TR / P * PC
260  I = 0
```

```
270  A1 = FN F(FI)
280  A2 = FN FD(FI)
290  I = I + 1
300  IF I > 50 THEN PRINT : PRINT
      "ITERATION NOT CONVERGENT": GOTO
      400
310  F2 = FI - A1 / A2
320  IF ABS (FI - F2) < 1E - 5 OR
      ABS (A1) < 1E - 4 THEN GOTO
      350
330  FI = F2
340  GOTO 270
350  PRINT : PRINT
360  F2 = P / PC / TR * F2
370  PRINT "COMPRESSIBILITY FACTO
      R ";F2
380  PRINT "MOLAR VOLUME (M3 / MOL)
      ";F2 * 8.3144 * T / P
390  PRINT "DENSITY (MOL / M3) ";P
      / T / 8.3144 / F2
400  PRINT : PRINT
410  INPUT "NEXT TEMPERATURE ? (Y
      /N) ";A$
420  IF A$ = "Y" THEN GOTO 180
430  INPUT "NEXT PRESSURE ? (Y / N)
      ";A$
440  IF A$ = "Y" THEN GOTO 160
450  INPUT "NEXT COMPOUND ? (Y / N)
      ";A$
460  IF A$ = "Y" THEN GOTO 130
470  END
```

Example. Run the program for methane at pressures 0.5×10^5, 10^5, 10×10^5, 50×10^5, 100×10^5, and 1000×10^5 Pa and at temperatures 200, 400, and 1000 K.

```
]RUN
ENTER CRITICAL TEMPERATURE (K) 190.56
ENTER CRITICAL PRESSURE (PA) 4.6002E6

ENTER PRESSURE (PA) 0.5E5
ENTER TEMPERATURE (K) 200

COMPRESSIBILITY FACTOR .99683835
MOLAR VOLUME (M3 / MOL) .0331524511
DENSITY (MOL / M3) 30.1636822

NEXT TEMPERATURE ? (Y / N) N
NEXT PRESSURE ? (Y / N) Y
```

```
ENTER PRESSURE (PA) 1E5
ENTER TEMPERATURE (K) 200

COMPRESSIBILITY FACTOR .993663036
MOLAR VOLUME (M3 / MOL) .0165234239
DENSITY (MOL / M3) 60.5201444

NEXT TEMPERATURE ? (Y / N) N
NEXT PRESSURE ? (Y / N) N

ENTER PRESSURE (PA) 10E5
ENTER TEMPERATURE (K) 200

COMPRESSIBILITY FACTOR .933916159
MOLAR VOLUME (M3 / MOL) 1.5529905E - 03
DENSITY (MOL / M3) 643.918941

NEXT TEMPERATURE ? (Y / N) Y
ENTER TEMPERATURE (K) 400

COMPRESSIBILITY FACTOR .995240727
MOLAR VOLUME (M3 / MOL) 3.3099318E - 03
DENSITY (MOL / M3) 302.121029

NEXT TEMPERATURE ? (Y / N) N
NEXT PRESSURE ? (Y / N) Y

ENTER PRESSURE (PA) 50E5
ENTER TEMPERATURE (K) 200

COMPRESSIBILITY FACTOR .525338517
MOLAR VOLUME (M3 / MOL) 1.74714983E - 04
DENSITY (MOL / M3) 5723.60758

NEXT TEMPERATURE ? (Y / N) Y
ENTER TEMPERATURE (K) 400

COMPRESSIBILITY FACTOR .979827767
MOLAR VOLUME (M3 / MOL) 6.51734399E - 04
DENSITY (MOL / M3) 1534.36738

NEXT TEMPERATURE ? (Y / N) N
NEXT PRESSURE ? (Y / N) Y

ENTER PRESSURE (PA) 100E5
ENTER TEMPERATURE (K) 400

COMPRESSIBILITY FACTOR .969635513
MOLAR VOLUME (M3 / MOL) 3.224775E - 04
DENSITY (MOL / M3) 3100.99154

NEXT TEMPERATURE ? (Y / N) N
NEXT PRESSURE ? (Y / N) Y

ENTER PRESSURE (PA) 1000E5
ENTER TEMPERATURE (K) 200
```

```
COMPRESSIBILITY FACTOR 2.42167614
MOLAR VOLUME (M3 / MOL) 4.02695682E - 05
DENSITY (MOL / M3) 24832.6477

NEXT TEMPERATURE ? (Y / N) Y
ENTER TEMPERATURE (K) 500

COMPRESSIBILITY FACTOR 1.58925864
MOLAR VOLUME (M3 / MOL) 6.60686601E - 05
DENSITY (MOL / M3) 15135.7693

NEXT TEMPERATURE ? (Y / N) Y
ENTER TEMPERATURE (K) 1000

COMPRESSIBILITY FACTOR 1.37612315
MOLAR VOLUME (M3 / MOL) 1.14416383E - 04
DENSITY (MOL / M3) 8740.00711

NEXT TEMPERATURE ? (Y / N) N
NEXT PRESSURE ? (Y / N) N
NEXT COMPOUND ? (Y / N) N
```

THE EQUATION OF STATE OF LIQUIDS

Of the equations of state discussed above, only the LK equation of state is at all suitable for the description of the (P,v,T) relationship in the liquid phase. In fact, for the description of the behavior of liquids a special equation of state is used:

$$z = \left(0.46407 - 0.73221T_r + 0.45256T_r^2\right)P_r$$

$$-\left(0.00871 - 0.02939T_r + 0.02775T_r^2\right)P_r^2$$

The compressibility coefficient is an explicit two-variable polynomial of second degree. Its coefficients are determined by fitting to measured data. This equation of state has no physical model as background.

Algorithm and Program: Equation of State of Liquids Composed of Simple Molecules

Write a program for the calculation of the compressibility factor and molar volume of simple-molecule liquids.

Algorithm.

1. Read the critical temperature and pressure of the compound.
2. Read the actual temperature and pressure.

3. Calculate the reduced temperature and reduced pressure.
4. Calculate z.
5. Print the results.

Program ESLIQUNORMAL. A possible realization:

```
]PR#0
]LORD ESLIQUNORMAL,D2
]LIST
 10  REM :CALCULATION OF COMPRESS
      IBILITY
 20  REMM :FACTOR AND MOLAR VOLUME
      OF LIQUIDS
 30  REM :FOR SIMPLE COMPOUNDS
 40  REM :BY "POLINOM" EQUATION OF
      STATE
 50  DEF FN Z0(PR) = (.46407 - 0.
      73221 * TR + 0.45256 * TR ∧
      2) * PR - (0.00871 - 0.02939
      * TR + 0.02775 * TR ∧ 2) *
      PR ∧ 2
 60  PRINT : PRINT
 70  INPUT "ENTER CRITICAL TEMPERA
      TURE (K) ";TC
 80  INPUT "ENTER CRITICAL PRESSUR
      E (PA) ";PC
 90  PRINT : PRINT
100  INPUT "ENTER PRESSURE (PA) "
      ;P
110  PRINT
120  INPUT "ENTER TEMPERATURE (K)
      ";T
130  TR = T / TC
140  PR = P / PC
150  Z = FN Z0(PR)
160  PRINT
170  PRINT "COMPRESSIBILITY FACTO
      R ";Z
180  PRINT "MOLAR VOLUME (M3 / MOL)
      ";Z * 8.3144 * T / P
190  PRINT
200  INPUT "NEXT TEMPERATURE ? (Y
      /N) ";A$
210  IF A$ = "Y" THEN GOTO 120
220  INPUT "NEXT PRESSURE ? (Y / N)
      ";A$
230  IF A$ = "Y" THEN GOTO 100
```

```
240  INPUT "NEXT COMPOUND ? (Y / N)
      ";A$
250  IF A$ = "Y" THEN GOTO 70
260  END
```

Example. Run the program for methane at pressures 0.5×10^5, 10×10^5, 100×10^5, 500×10^5, 1000×10^5 Pa and at temperatures 100, 200, 250, 1000.

```
]RUN

ENTER CRITICAL TEMPERATURE (K) 190.56
ENTER CRITICAL PRESSURE (PA) 4.6002E6
ENTER PRESSURE (PA) 0.5E5
ENTER TEMPERATURE (K) 100

COMPRESSIBILITY FACTOR 2.22214115E - 03
MOLAR VOLUME (M3 / MOL) 3.69515407E - 05

NEXT TEMPERATURE ? (Y / N) N
NEXT PRESSURE ? (Y / N) Y
ENTER PRESSURE (PA) 10E5
ENTER TEMPERATURE (K) 100

COMPRESSIBILITY FACTOR .0444011225
MOLAR VOLUME (M3 / MOL) 3.69168692E - 05

NEXT TEMPERATURE ? (Y / N) N
NEXT PRESSURE ? (Y / N) Y
ENTER PRESSURE (PA) 100E5
ENTER TEMPERATURE (K) 100

COMPRESSIBILITY FACTOR .440060655
MOLAR VOLUME (M3 / MOL) 3.65884031E - 05

NEXT TEMPERATURE ? (Y / N) N
NEXT PRESSURE ? (Y / N) Y
ENTER PRESSURE (PA) 500E5
ENTER TEMPERATURE (K) 100

COMPRESSIBILITY FACTOR 2.11251284
MOLAR VOLUME (M3 / MOL) 3.51285535E - 05

NEXT TEMPERATURE ? (Y / N) Y
ENTER TEMPERATURE (K) 200

COMPRESSIBILITY FACTOR 1.11357156
MOLAR VOLUME (M3 / MOL) 3.70347175E - 05

NEXT TEMPERATURE ? (Y / N) Y
ENTER TEMPERATURE (K) 250

COMPRESSIBILITY FACTOR .952947507
MOLAR VOLUME (M3 / MOL) 3.96159337E - 05
```

```
NEXT TEMPERATURE ? (Y / N) N
NEXT PRESSURE ? (Y / N) Y
ENTER PRESSURE (PA) 1000E5
ENTER TEMPERATURE (K) 200

COMPRESSIBILITY FACTOR .234986021
MOLAR VOLUME (M3 / MOL) 3.90753554E - 06

NEXT TEMPERATURE ? (Y / N) Y
ENTER TEMPERATURE (K) 500

COMPRESSIBILITY FACTOR - 21.900541
MOLAR VOLUME (M3 / MOL) - 9.10449292E - 04

NEXT TEMPERATURE ? (Y / N) Y
ENTER TEMPERATURE (K) 1000

COMPRESSIBILITY FACTOR - 94.8725406
MOLAR VOLUME (M3 / MOL) - 7.88808251E - 03

NEXT TEMPERATURE ? (Y / N) N
NEXT PRESSURE ? (Y / N) N
NEXT COMPOUND ? (Y / N) N
```

NONPOLAR MOLECULES

Effect of Size and Shape of the Molecules

Our discussion concerned so far monatomic or other rotationally symmetric molecules. These can be considered in terms of forces acting between point masses. Thus, the meaning of the distance r between them—the independent variable of the pair potential—is completely evident. This is shown for methane and argon in Figure 1.6. Moreover, the figure shows two propane molecules. One CH_3 group of one of the molecules is at different distances from the two CH_3 groups and the CH_2 group of the other molecule, so that altogether nine distances are to be considered. The mutual orientation of the molecules is random, but different distances result from each orientation. An immediate consequence of this situation is that the argument leading to the law of corresponding states loses its validity. Thus, the situation is rather complicated, and this complexity is reflected macroscopically. It is not to be expected that from equations of state for simple molecules one can describe the (P,v,T) relationships with the same accuracy in the case of asymmetric molecules. The equation of state must also include factors characteristic of the size and shape of the molecule.

Several suggestions have been made for this purpose.

The Critical Compressibility Factor

The concept of critical compressibility factor has been seen already in the discussion of the van der Waals equation

$$z_c = \frac{P_c v_c}{\mathbf{R} T_c}$$

This expression is supposed by some authors to be a property of the material, characteristic of the shape and size of the molecule. In that case

$$z_c \neq \tfrac{3}{8}$$

which contradicts the van der Waals model.

Let us express the equation of state of a real gas, with the compressibility factor, in the form

$$Pv = z\mathbf{R}T$$

and rewrite this equation with reduced state parameters:

$$P_r P_c v_r v_c = z\mathbf{R} T_r T_c$$

The compressibility factor can be expressed in terms of the reduced state variables. The expression contains the critical compressibility factor z_c:

$$z = \left(\frac{P_r v_r}{T_r}\right)\left(\frac{P_c v_c}{T_c}\right) = \frac{P_r v_r}{T_r} z_c$$

Since of P_r, v_r, and T_r only two are independent, we have

$$z = \mathrm{f}[T_r, P_r; z_c]$$

The compressibility factor is a function of the reduced pressure and the reduced temperature with z_c as a parameter. This implies an extension of the theorem of corresponding states, with the introduction of a third parameter besides the critical temperature and critical pressure.

The Omega Factor

The idea of the omega factor, characteristic of the shape of the molecule, originated with Pitzer. It can be seen from Figure 3.12 that the functional form the reduced vapor pressure is sensitive to the shape of the molecule, and the more this shape deviates from rotational symmetry, the more it differs from the reduced vapor pressure of argon. Pitzer thought that the reduced pressure corresponding to the reduced temperature $T_r = 0.700$ (i.e., a temperature near

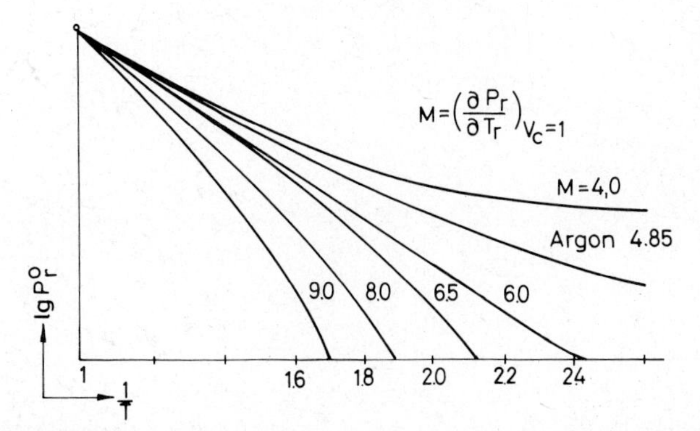

FIGURE 3.12. Reduced vapor-pressure curves for different substances.

the normal boiling point) is suitable for characterizing the shape of the molecule; he defined the dimensionless quantity called the *acentric factor* in the following way:

$$\omega = -\log P_r^0|_{T_r=0.7} - 1$$

The symbol ω is usual in the chemical literature, so this quantity will be called the *omega factor* in what follows.

According to experiment the following relationship exists between the omega factor and the critical compressibility factor:

$$z_c = 0.2905 - 0.085\omega$$

Thus, it can be said that the compressibility factor is a function of the reduced pressure and the reduced temperature with parameter ω:

$$z = f[T_r, P_r; \omega]$$

Law of Three-parameter Corresponding States

Figure 3.13 shows the change of the compressibility factor of several substances with the parameter T_r at a pressure $P_r = 1.6$, choosing ω as the independent parameter. The omega value of Ar, Xe, and CH_4 is zero, that of ethane and hydrogen sulfide 0.1, and that of heptane about 0.35.

The compressibility factor can be described with the following power series of ω:

$$z = z^{(0)} + z^{(1)}\omega + z^{(2)}\omega^2 + \cdots$$

However, in the chemically important range of ω even the quadratic term can

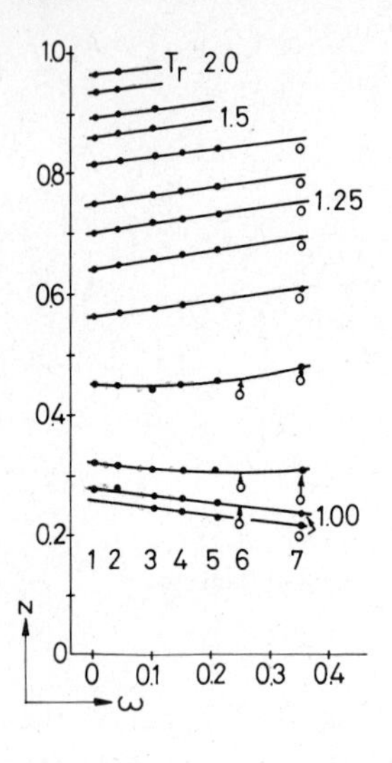

FIGURE 3.13. Compressibility factor and omega factor. 1 Ar, Xe and CH_4; 3 C_2H_6 and H_2S; 5 Benzene; 6 Ammonia; 7 ● *n*-Heptane, ○ Water.

practically be neglected. Thus, $z^{(0)}$ is the compressibility factor relevant to the family of rotationally symmetric molecules, while $z^{(1)}$ is the slope of the term linear in omega. For simple molecules the compressibility factor is a function of the reduced temperature and reduced pressure:

$$z^{(0)} = z^{(0)}[T_r, P_r]$$

and so is the factor $z^{(1)}$:

$$z^{(1)} = z^{(1)}[T_r, P_r]$$

Hence $z^{(1)}$ is a universal function, independent of the omega factor. We can write

$$z = z^{(0)} + \omega z^{(1)}$$

and this indeed is an extension of the theorem of corresponding states to the class of nonpolar molecules. Since ω is now included as a parameter in addition to the critical temperature and pressure, it is usual to speak of the law of three-parameter corresponding states.

It will be noted from the figure that ammonia and water cannot be described with the proposed linear relationship. As is discussed below, that is because they are polar substances.

Plots of two sets of curves, for $z^{(0)}$ and $z^{(1)}$, are given in Figures 3.14 and 3.15.

The shape of the $z^{(1)}$ functions is not smooth, and thus difficult to describe. This is inconvenient for computer calculations. Therefore, the experimental finding that the compressibility factor of a substances in corresponding states

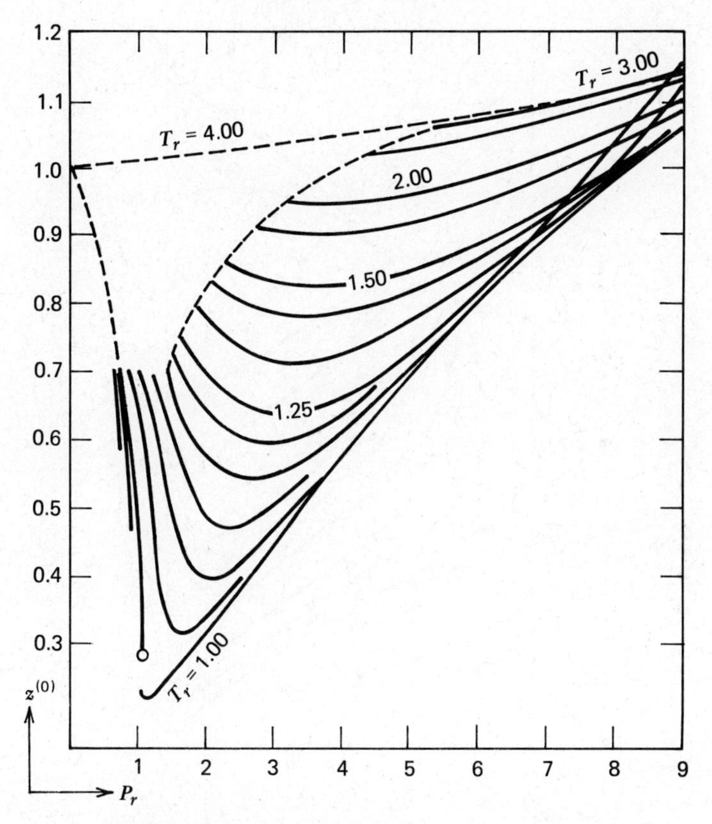

FIGURE 3.14. Compressibility factor of simple fluid.

is a linear function of the omega factor has been reformulated as follows. Let the compressibility factor of substance composed of simple molecules at given reduced variables be $z^{(0)}$, and the compressibility factor of a reference substance in the corresponding state (such as *n*-Octane) be $z^{(r)}$ (Figure 3.16). Owing to the linear relationship, we have for the compressibility factor z of a nonpolar substance with omega factor ω

$$\frac{z - z^{(0)}}{\omega} = \frac{z^{(r)} - z^{(0)}}{\omega^{(r)}}$$

whence

$$z = z^{(0)} + \left(\frac{z^{(r)} - z^{(0)}}{\omega^{(r)}} \right) \omega$$

$$= \left(1 - \frac{\omega}{\omega^{(r)}} \right) z^{(0)} + \frac{\omega}{\omega^{(r)}} z^{(r)}$$

where $\omega^{(r)}$ is the omega factor of the reference substance.

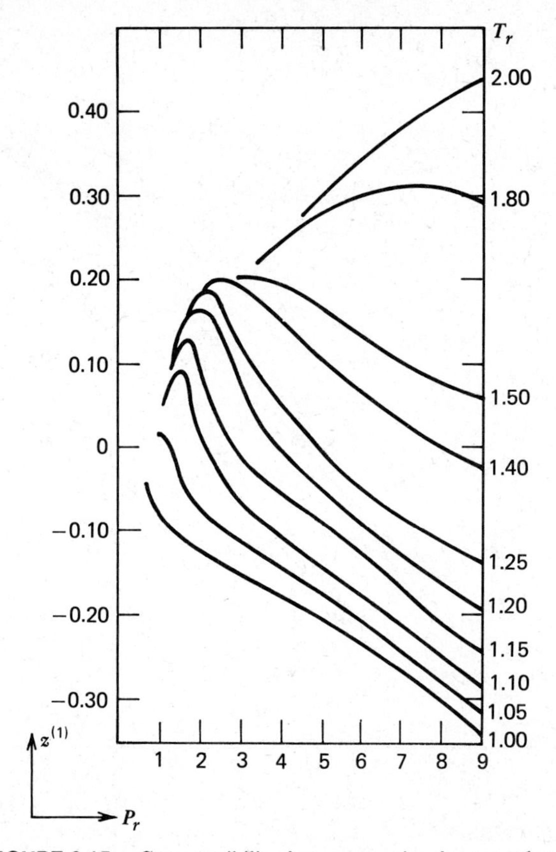

FIGURE 3.15. Compressibility-factor correction for nonpolar fluid.

Thus, the compressibility factor of a nonpolar substance characterized by ω is a linear combination of the compressibility factors of the simple substance and the reference substance relevant to the corresponding state. The function to be used for the calculation of $z^{(0)}$ and $z^{(r)}$ is the same, with a different set of parameter values.

Equations of state valid for the class of nonpolar molecules can be used in the approach expounded above.

The Virial Equation

The reduced second virial coefficient is a linear combination,

$$B_r = \left(1 - \frac{\omega}{\omega^{(r)}}\right) B_r^{(0)} + \frac{\omega}{\omega^{(r)}} B_r^{(r)}$$

where $B_r^{(0)}$ and $B_r^{(r)}$ are the reduced virial coefficients, of the simple substance

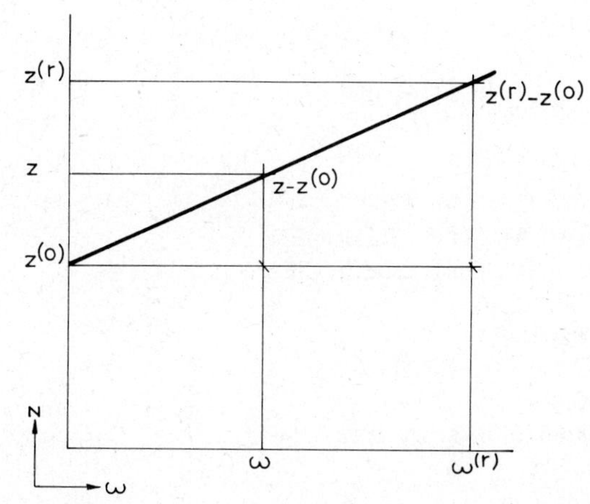

FIGURE 3.16. Linear combination of the compressibility factors of the simple fluid and the reference fluid.

and the reference substance, respectively:

$$B_r^{(0)} = b_1^{(0)} - \frac{b_2^{(0)}}{T_r} - \frac{b_3^{(0)}}{T_r^2} - \frac{b_4^{(0)}}{T_r^3}$$

$$B_r^{(r)} = b_1^{(r)} - \frac{b_2^{(r)}}{T_r} - \frac{b_3^{(r)}}{T_r^2} - \frac{b_4^{(r)}}{T_r^3}$$

Coefficients of the polynomials are contained in Table 3.3.

Algorithm and Program: Second Virial Coefficient

Write a program for the calculation of the reduced second virial coefficient. The critical temperature and omega factor are known.

Algorithm.

1. Read the critical temperature and omega factor.
2. Calculate the linear combinations of the corresponding parameters of the simple and the reference fluid using the omega factor:

$$b_i = b_i^{\text{simple}} + \frac{\omega}{\omega_v}\left(b_i^{\text{ref}} - b_i^{\text{simple}}\right)$$

   where b_i is the ith term of the polynomial describing the reduced second virial coefficient.
3. Read the temperature.
4. Calculate the reduced second virial coefficient.

5. Print the result.

Program VIRB. A possible realization:

```
]LOAD VIRB
]LIST
 10  REM :CALCULATION OF REDUCED
 20  REM : SECOND VIRIAL COEFFICIENT
 30  READ N1,N2,N3,N4
 40  READ R1,R2,R3,R4
 50  DATA 0.1181193,0.265728,0.15
      479,0.0030323
 60  DATA 0.2026579,0.331511,0.02
      7655,0.203488
 70  PRINT : PRINT
 80  INPUT "ENTER CRITICAL TEMPERA
      TURE (K) ";TC
 90  INPUT "ENTER OMEGA ";OM
100  PRINT : PRINT
110  MR = 0.3978
120  B1 = N1 + OM * (R1 - N1) / MR
130  B2 = N2 + OM * (R2 - N2) / MR
140  B3 = N3 + OM * (R3 - N3) / MR
150  B4 = N4 + OM * (R4 - N4) / MR
160  INPUT "ENTER TEMPERATURE (K)
      ";T
170  PRINT
180  TR = T / TC
190  B = B1 - B2 / TR - B3 / TR /
      TR - B4 / TR / TR / TR
200  PRINT : PRINT
210  PRINT "REDUCED TEMPERATURE "
      ;TR
220  PRINT
230  PRINT "REDUCED SECOND VIRIAL
      COEFF. ";B
240  PRINT : PRINT
250  PRINT
260  INPUT "NEXT TEMPERATURE ? (Y / N) "
      ;A$
270  IF A$ = "Y" THEN GOTO 160
280  PRINT
290  INPUT "NEXT COMPOUND ? (Y / N)
      ";A$
     IF A$ = "Y" THEN GOTO 70
310  END
```

Example. Run the program for methane and *n*-heptane at temperature intervals of 100 K from 200 to 700 K and from 300 to 1100 K.

```
]RUN

ENTER CRITICAL TEMPERATURE (K) 190.56
ENTER OMEGA 0.00406
ENTER TEMPERATURE (K) 200

REDUCED TEMPERATURE 1.0495382
REDUCED SECOND VIRIAL COEFF. -.278580454

NEXT TEMPERATURE ? (Y / N) Y
ENTER TEMPERATURE (K) 300

REDUCED TEMPERATURE 1.57430731
REDUCED SECOND VIRIAL COEFF. -.11346725

NEXT TEMPERATURE ? (Y / N) Y
ENTER TEMPERATURE (K) 400

REDUCED TEMPERATURE 2.09907641
REDUCED SECOND VIRIAL COEFF. -.0433157998

NEXT TEMPERATURE ? (Y / N) Y
ENTER TEMPERATURE (K) 500

REDUCED TEMPERATURE 2.62384551
REDUCED SECOND VIRIAL COEFF. -5.1242983E - 03

NEXT TEMPERATURE ? (Y / N) Y
ENTER TEMPERATURE (K) 600

REDUCED TEMPERATURE 3.14861461
REDUCED SECOND VIRIAL COEFF. .0187282343

NEXT TEMPERATURE ? (Y / N) Y
ENTER TEMPERATURE (K) 700

REDUCED TEMPERATURE 3.67338371
REDUCED SECOND VIRIAL COEFF. .0349830589

NEXT TEMPERATURE ? (Y / N) N
NEXT COMPOUND ? (Y / N) Y
ENTER CRITICAL TEMPERATURE (K) 540.2
ENTER OMEGA 0.3486
ENTER TEMPERATURE (K) 300

REDUCED TEMPERATURE .55534987
REDUCED SECOND VIRIAL COEFF. -1.57405177

NEXT TEMPERATURE ? (Y / N) Y
ENTER TEMPERATURE (K) 500

REDUCED TEMPERATURE .925583117
REDUCED SECOND VIRIAL COEFF. -.433162186

NEXT TEMPERATURE ? (Y / N) Y
ENTER TEMPERATURE (K) 700

REDUCED TEMPERATURE 1.29581636
REDUCED SECOND VIRIAL COEFF. -.165311553
```

```
NEXT TEMPERATURE ? (Y / N) Y
ENTER TEMPERATURE (K) 900

REDUCED TEMPERATURE 1.66604961
REDUCED SECOND VIRIAL COEFF. -.056163873

NEXT TEMPERATURE ? (Y / N) Y
ENTER TEMPERATURE (K) 1100

REDUCED TEMPERATURE 2.03628286
REDUCED SECOND VIRIAL COEFF. 1.76981801E - 03

NEXT TEMPERATURE ? (Y / N) N
NEXT COMPOUND ? (Y / N) N
```

Remarks. The change in sign of the second virial coefficient at the Boyle temperature can be seen.

Algorithm and Program: Graphical Representation of the Second Virial Coefficient

Write a program for the graphical representation of the change with temperature of the reduced virial coefficient in the reduced-temperature range 0.6–4.

Algorithm.

1. Read the omega factor.
2. Calculate and plot the reduced second virial coefficient.

Program GRBVIR. A possible realization:

```
]PR#0
]LOAD GRBVIR
]LIST
 10  REM :GRAPHICAL REPRESENTATION
      OF
 20  REM :SECOND VIRIAL COEFFICIENT
 30  READ N1,N2,N3,N4
 40  READ R1,R2,R3,R4
 50  DATA 0.1181193,0.265728,0.15
      479,0.0030323
 60  DATA 0.2026579,0.331511,0.02
      7655,0.203488
 70  HPLOT 0,0
 80  HGR
 90  HPLOT 10,50, TO 279,50
100  HPLOT 10,0 TO 10,191
110  PRINT
120  INPUT "ENTER OMEGA ";OM
```

```
130  MR  =  0.3978
140  B1  =  N1 + OM * (R1 - N1) / MR
150  B2  =  N2 + OM * (R2 - N2) / MR
160  B3  =  N3 + OM * (R3 - N3) / MR
170  B4  =  N4 + OM * (R4 - N4) / MR
180  FOR TR  =  0.6 TO 4 STEP 0.01
190  B = B1 - B2 / TR - B3 / TR /
      TR - B4 / TR / TR / TR
200  X = 70 * (TR - 0.6) + 10
210  Y = 50 - B * 70
220  IF TR  =  0.6 THEN HPLOT X,Y
230  HPLOT TO X,Y
240  NEXT TR
250  HPLOT 0,0
260  PRINT
270  INPUT "NEXT COMPOUND ? (Y / N)
      ";A$
280  IF A$ = "Y" THEN GOTO 120
290  TEXT
300  END
```

Example. Run the program with different omega factors from 0 to 0.5. Watch the monitor and explain the results.

Algorithm and Program: Virial Equation of State

Write a program for the calculation of compressibility factor, molar volume, and density of gases using the reduced form of the virial equation of state. The critical temperature, critical pressure, and omega factor of the compound are known.

Algorithm.

1. Read the critical temperature and pressure and the omega factor of the compound.
2. Read the actual temperature and pressure.
3. Calculate the reduced temperature and pressure.
4. Calculate the reduced virial coefficient (see program VIRB).
5. Solve the reduced form of the virial equation of state,

$$\left(v_r^2 - \frac{T_r}{P_r} v_r - \frac{T_r}{P_r} B_r = 0 \right)$$

for v_r.

6. Calculate the compressibility factor

$$z = \frac{P_r v_r}{T_r}$$

7. Calculate the molar volume

$$v = \frac{z\mathbf{R}T}{P}$$

and density

$$\delta = \frac{1}{v}$$

8. Print the results.

Program ESVIR. A possible realization:

```
]PR#0
]LOAD ESVIR
]LIST
 10  REM :CALCULATION OF COMPRESSI
      BILITY
 20  REM :FACTOR BY VIRIAL EQUATIO
      N OF
 30  REM :STATE OF REDUCED FORM
 40  READ N1,N2,N3,N4
 50  READ R1,R2,R3,R4
 60  DATA 0.1181193,0.265728,0.15
      479,0.0030323
 70  DATA 0.2026579,0.331511,0.02
      7655,0.203488
 80  PRINT : PRINT
 90  INPUT "ENTER CRITICAL TEMPERA
      TURE (K) ";TC
100  INPUT "ENTER CRITICAL PRESSU
      RE (PA) ";PC
110  INPUT "ENTER OMEGA ";OM
120  PRINT : PRINT
130  MR = 0.3978
140  B1 = N1 + OM * (R1 - N1) / MR
150  B2 = N2 + OM * (R2 - N2) / MR
160  B3 = N3 + OM * (R3 - N3) / MR
170  B4 = N4 + OM * (R4 - N4) / MR
180  INPUT "ENTER PRESSURE (PA) "
     ;P
190  INPUT "ENTER TEMPERATURE (K)
      ";T
```

```
200  PRINT
210  TR = T / TC
220  PR = P / PC
230  B = B1 - B2 / TR - B3 / TR /
      TR - B4 / TR / TR / TR
240  BA = TR / PR
250  IF BA ∧ 2 + 4 * BA * B < 0 THEN
      PRINT "NEGATIVE SQR ARGUMENT "
      : GOTO 330
260  VR = (BA + SQR (BA ∧ 2 + 4 *
      BA * B)) / 2
270  Z = PR * VR / TR
280  PRINT : PRINT
290  PRINT "COMPRESSIBILITY FACTOR
      : ";Z
300  V = Z * 8.3144 * T / P
310  PRINT "MOLAR VOLUME (M3 / MOL)
      ";V
320  PRINT "DENSITY (MOL / M3) ";1
      / V
330  PRINT : PRINT : PRINT
340  PRINT
350  INPUT "NEXT TEMPERATURE ? (Y / N)
      ";A$
360  IF A$ = "Y" THEN GOTO 190
370  INPUT "NEXT PRESSURE ? (Y / N)
      ";A$
380  IF A$ = "Y" THEN GOTO 180
390  PRINT
400  INPUT "NEXT COMPOUND ? (Y / N)
      ";A$
410  IF A$ = "Y" THEN GOTO 80
420  END
```

Example. Run the program for propane at pressures from 0.5×10^5 to 500×10^5 Pa and temperatures from 250 to 500 K.

```
]RUN

ENTER CRITICAL TEMPERATURE (K) 369.86
ENTER CRITICAL PRESSURE (PA) 4.2455E6
ENTER OMEGA 0.28

ENTER PRESSURE (PA) 0.5E5
ENTER TEMPERATURE (K) 300

COMPRESSIBILITY FACTOR : .991559487
MOLAR VOLUME (M3 / MOL) .0494653332
DENSITY (MOL / M3) 20.2161784
```

```
NEXT TEMPERATYRE ? (U / N) Y
ENTER TEMPERATURE (K) 400

COMPRESSIBILITY FACTOR : .99693418
MOLAR VOLUME (M3 / MOL) .0663112763
DENSITY (MOL / M3) 15.0803914

NEXT TEMPERATURE ? (Y / N) Y
ENTER TEMPERATURE (K) 500

COMPRESSIBILITY FACTOR : .998715428
MOLAR VOLUME (M3 / MOL) .0830371955
DENSITY (MOL / M3) 12.0427959

NEXT TEMPERATURE ? (Y / N) N
NEXT PRESSURE ? (Y / N) Y
ENTER PRESSURE (PA) 2E5
ENTER TEMPERATURE (K) 250

COMPRESSIBILITY FACTOR : .933264595
MOLAR VOLUME (M3 / MOL) 9.69941893E - 03
DENSITY (MOL / M3) 103.09896

NEXT TEMPERATURE ? (Y / N) Y
ENTER TEMPERATURE (K) 300

COMPRESSIBILITY FACTOR : .965320226
MOLAR VOLUME (M3 / MOL) .0120390877
DENSITY (MOL / M3) 83.0627721

NEXT TEMPERATURE ? (Y / N) Y
ENTER TEMPERATURE (K) 400

COMPRESSIBILITY FACTOR : .987621079
MOLAR VOLUME (M3 / MOL) .0164229534
DENSITY (MOL / M3) 60.8903877

NEXT TEMPERATURE ? (Y / N) N
NEXT PRESSURE ? (Y / N) Y
ENTER PRESSURE (PA) 5E5
ENTER TEMPERATURE (K) 270

COMPRESSIBILITY FACTOR : .860774769
MOLAR VOLUME (M3 / MOL) 3.8646859E - 03
DENSITY (MOL / M3) 258.753241

NEXT TEMPERATURE ? (Y / N) Y
ENTER TEMPERATURE (K) 300

COMPRESSIBILITY FACTOR : .907807895
MOLAR VOLUME (M3 / MOL) 4.52872678E - 03
DENSITY (MOL / M3) 220.812614

NEXT TEMPERATURE ? (Y / N) Y
ENTER TEMPERATURE (K) 400
```

```
COMPRESSIBILITY FACTOR : .968439742
MOLAR VOLUME (M3 / MOL) 6.44159631E - 03
DENSITY (MOL / M3) 155.241023

NEXT TEMPERATURE ? (Y / N) N
NEXT PRESSURE ? (Y / N) Y
ENTER PRESSURE (PA) 10E5
ENTER TEMPERATURE (K) 270

COMPRESSIBILITY FACTOR : .601571983
MOLAR VOLUME (M3 / MOL) 1.35046173E - 03
DENSITY (MOL / M3) 740.48748

NEXT TEMPERATURE ? (Y / N) Y
ENTER TEMPERATURE (K) 300

COMPRESSIBILITY FACTOR : .787427484
MOLAR VOLUME (M3 / MOL) 1.96409612E - 03
DENSITY (MOL / M3) 509.140052

NEXT TEMPERATURE ? (Y / N) Y
ENTER TEMPERATURE (K) 400

COMPRESSIBILITY FACTOR : .934593584
MOLAR VOLUME (M3 / MOL) 3.10823396E - 03
DENSITY (MOL / M3) 321.726103

NEXT TEMPERATURE ? (Y / N) N
NEXT PRESSURE ? (Y / N) Y
ENTER PRESSURE (PA) 30E5
ENTER TEMPERATURE (K) 250

NEGATIVE SQR ARGUMENT

NEXT TEMPERATURE ? (Y / N) Y
ENTER TEMPERATURE (K) 340

NEGATIVE SQR ARGUMENT

NEXT TEMPERATURE ? (Y / N) Y
ENTER TEMPERATURE (K) 400

COMPRESSIBILITY FACTOR : .758098335
MOLAR VOLUME (M2 / MOL) 8.40417706E-04
DENSITY (MOL / M3) 1189.8845

NEXT TEMPERATURE ? (Y / N) N
NEXT PRESSURE ? (Y / N) Y
ENTER PRESSURE (PA) 100E5
ENTER TEMPERATURE (K) 400

NEGATIVE SQR ARGUMENT

NEXT TEMPERATURE ? (Y / N) Y
ENTER TEMPERATURE (K) 500
```

```
NEGATIVE SQR ARGUMENT

NEXT TEMPERATURE ? (Y / N) N
NEXT PRESSURE ? (Y / N) Y
ENTER PRESSURE (PA) 500E5
ENTER TEMPERATURE (K) 500
NEGATIVE SQR ARGUMENT

NEXT TEMPERATURE ? (Y / N) N
NEXT PRESSURE ? (Y / N) N
NEXT COMPOUND ? (Y / N) N
```

Remark. The results show that there is no solution at high pressures and low temperatures because of the negative argument of the square root. The virial equation of state is suitable at low pressures and high temperatures only.

The Redlich – Kwong Equation of State

The RK equation in its original form is not suitable for the realization of the above development, since the parameters of any substance are determined by the critical variables. For nonpolar substances, a and b calculated from experimental data differ from those calculated with the formula, and even prove to be temperature dependent. It is remarkable that even in the case of methane, considered a simple substance, the original assumption is not true. The parameters a and b in the RK equation of state in the reduced temperature can be described however with a polynomial of second degree:

$$a = \alpha_0 + \alpha_1 T_r^{-1} + \alpha_2 T_r^{-2}$$

$$b = \beta_0 + \beta_1 T_r^{-1} + \beta_2 T_r^{-2}$$

The coefficients of the polynominal are determined by parameter estimation. Those for methane (a simple substance) and for n-octane (the reference substance) are given in Table 3.2.

Now the law of three-parameter corresponding states can be used as follows:

1. Calculate for the desired temperature, the parameters a and b for both methane and n-octane.
2. Using these values, solve the RK equation separately for each substance.
3. Form the following linear combination of the compressibility factors of methane and n-octane:

$$z = \left(1 - \frac{\omega}{0.3978}\right) z^{(0)} + \left(\frac{\omega}{0.3978}\right) z^{(r)}$$

where ω is the omega factor of the target compound. This step is justified only if methane and n-octane in corresponding states are in the same state of aggregation.

The Peng – Robinson Equation of State

The application of the RK equation of state to nonpolar molecules is feasible by taking the linear combination of the compressibility factors of methane and n-octane obtained for the corresponding state. This naturally means that a cubic equation has to be solved twice.

The PR equation of state is solved once, by suitably selecting the value of parameter a. Namely,

$$a = 0.45724 \frac{\mathbf{R}^2 T_c^2}{P_c} \alpha^2$$

$$\alpha = 1 + m\left(1 - T_r^{0.5}\right)$$

$$m = 0.37464 + 1.54226\omega - 0.26992\omega^2$$

Algorithm and Program: Peng – Robinson Equation of State

Write a program for the calculation of the compressibility factor, molar volume, and density of gases with the polynomial form of the Peng–Robinson equation of state. Solve the equation by Newton's method.

Algorithm.

1. Read the critical temperature, critical pressure, and omega factor of the compound.
2. Calculate a, b, A, B.
3. Make the initial guess $z = 6$.
4. Perform a Newtonian iteration step

$$z_{\text{new}} = z_{\text{old}} - \frac{\mathrm{f}(z_{\text{old}})}{\mathrm{f}'(z_{\text{old}})}$$

   where

$$\mathrm{f}(z) = z^3 - (1 - B)z^2 + (A - 3B^2 - 2B)z - (AB - B^2 - B^3)$$

5. Stop the iteration and go to step 7 if the number of steps exceeds 50 or the change in z between two consecutive steps is less than 10^{-5}.

6. Go to step 4.
7. Calculate the molar volume and density, or print the iteration error.
8. Print the result.

Program ESPR. A possible realization:

```
]PR#0
]LOAD ESPR
]LIST

 10  REM :CALCULATION OF COMPRESSI
      BILITY
 20  REM :FACTOR BY PENG - ROBINSON
 30  REM :EQUATION OF STATE
 40  DEF FN F(Z) = Z ∧ 3 - (1 - B
      ) * Z ∧ 2 + (A - 3 * B ∧ 2 -
      2 * B) * Z - (A * B - B ∧ 2 -
      B ∧ 3)
 50  DEF FN FD(Z) = 3 * Z ∧ 2 - 2
      * (1 - B) * Z + (A - 3 * B ∧
      2 - 2 * B)
 60  R = 8.3144
 70  PRINT : PRINT
 80  INPUT "ENTER CRITICAL TEMPERA
      TURE (K) ";TC
 90  PRINT
100  INPUT "ENTER CRITICAL PRESSU
      RE (PA) ";PC
110  INPUT "ENTER OMEGA ";OM
120  PRINT : PRINT
130  INPUT "ENTER PRESSURE (PA) "
      ;P
140  INPUT "ENTER TEMPERATURE (K)
      ";T
150  TR = T / TC
160  M = 0.37464 + 1.54226 * OM -
      0.26992 * OM ∧ 2
170  AL = 0.45724 * R * R * TC * T
      C / PC * (1 + M * (1 - SQR
      (TR))) ∧ 2
180  BL = 0.0778 * R * TC / PC
190  A = AL * P / R / R / T / T
200  B = BL * P / R / T
210  Z1 = 6
220  I = 0
230  H = FN F(Z1) / FN FD(Z1)
240  I = I + 1
250  IF I > 50 THEN PRINT : PRINT
```

```
        "ITERATION NOT CONVERGENT": GOTO
        350
260  Z2 = Z1 - H
270  IF ABS (Z2 - Z1) < 1E - 5 THEN
        GOTO 300
280  Z1 - Z2
290  GOTO 230
300  PRINT : PRINT
310  PRINT "COMPRESSIBILITY FACTOR
        ";Z2
320  PRINT
330  PRINT "MOLAR VOLUME (M3 / MOL)
        ";Z2 * R * T / P
340  PRINT "DENSITY (MOL / MS) ";P
        / Z2 / R / T
350  PRINT
360  INPUT "NEXT TEMPERATURE ? (Y / N)
        ";A$
370  IF A$ = "Y" THEN GOTO 140
380  INPUT "NEXT PRESSURE ? (Y / N)
        ";A$
390  IF A$ = "Y" THEN GOTO 130
400  INPUT "NEXT COMPOUND ? (Y / N)
        ";A$
410  IF A$ = "Y" THEN GOTO 70
420  END
```

Example. Run the program for propane in the pressure range 0.5×10^5–500×10^5 Pa and the temperature range 250–500 K.

```
]RUN

ENTER CRITICAL TEMPERATURE (K) 369.86
ENTER CRITICAL PRESSURE (PA) 4.2455E6
ENTER OMEGA 0.28

ENTER PRESSURE (PA) 0.5E5
ENTER TEMPERATURE (K) 300

COMPRESSIBILITY FACTOR .991569047
MOLAR VOLUME (M3 / MOL) .0494658101
DENSITY (MOL / M3) 20.2159835

NEXT TEMPERATURE ? (Y / N) Y
ENTER TEMPERATURE (K) 400

COMPRESSIBILITY FACTOR .996523473
MOLAR VOLUME (M3 / MOL) .0662839581
DENSITY (MOL / M3) 15.0866066

NEXT TEMPERATURE ? (Y / N) Y
ENTER TEMPERATURE (K) 500
```

```
COMPRESSIBILITY FACTOR .998437097
MOLAR VOLUME (M3 / MOL) .083014054
DENSITY (MOL / M3) 12.0461531

NEXT TEMPERATURE ? (Y / N) N
NEXT PRESSURE ? (Y / N) Y
ENTER PRESSURE (PA) 2E5
ENTER TEMPERATURE (K) 250

COMPRESSIBILITY FACTOR .941491164
MOLAR VOLUME (M3 / MOL) 9.78491767E - 03
DENSITY (MOL / M3) 102.198101

NEXT TEMPERATURE ? (Y / N) Y
ENTER TEMPERATURE (K) 300

COMPRESSIBILITY FACTOR .965663095
MOLAR VOLUME (M3 / MOL) .0120433639
DENSITY (MOL / M3) 83.0332798

NEXT TEMPERATURE ? (Y / N) Y
ENTER TEMPERATURE (K) 400

COMPRESSIBILITY FACTOR .986044732
MOLAR VOLUME (M3 / MOL) .0163967406
DENSITY (MOL / M3) 60.9877306

NEXT TEMPERATURE ? (Y / N) N
NEXT PRESSURE ? (Y / N)Y
ENTER PRESSURE (PA) 5E5
ENTER TEMPERATURE (K) 270

COMPRESSIBILITY FACTOR .875141123
MOLAR VOLUME (M2 / MOL) 3.92918761E - 03
DENSITY (MOL / M3) 254.505536

NEXT TEMPERATURE ? (Y / N) Y
ENTER TEMPERATURE (K) 300

COMPRESSIBILITY FACTOR .910613148
MOLAR VOLUME (M3 / MOL) 4.54272117E - 03
DENSITY (MOL / M3) 220.132375

NEXT TEMPERATURE ? (Y / N) Y
ENTER TEMPERATURE (K) 400

COMPRESSIBILITY FACTOR .964857449
MOLAR VOLUME (M3 / MOL) 6.41776862E - 03
DENSITY (MOL / M3) 155.817397

NEXT TEMPERATURE ? (Y / N) N
NEXT PRESSURE ? (Y / N) Y
ENTER PRESSURE (PA) 10E5
ENTER TEMPERATURE (K) 270
```

```
COMPRESSIBILITY FACTOR .703737263
MOLAR VOLUME (M3 / MOL) 1.57981134E - 03
DENSITY (MOL / M3) 632.986976

NEXT TEMPERATURE ? (Y / N) Y
ENTER TEMPERATURE (K) 300

COMPRESSIBILITY FACTOR .804917588
MOLAR VOLUME (M3 / MOL) 2.00772204E - 03
DENSITY (MOL / M3) 498.076916

NEXT TEMPERATURE ? (Y / N) Y
ENTER TEMPERATURE (K) 400

COMPRESSIBILITY FACTOR .928811761
MOLAR VOLUME (M3 / MOL) 3.089005E - 03
DENSITY (MOL / M3) 323.728838

NEXT TEMPERATURE ? (Y / N) N
NEXT PRESSURE ? (Y / N) Y
ENTER PRESSURE (PA) 30E5
ENTER TEMPERATURE (K) 250

COMPRESSIBILITY FACTOR .103688273
MOLAR VOLUME (M3 / MOL) 7.18421478E - 05
DENSITY (MOL / M3) 13919.4057

NEXT TEMPERATURE ? (Y / N) Y
ENTER TEMPERATURE (K) 340

COMPRESSIBILITY FACTOR .111299351
MOLAR VOLUME (M3 / MOL) 1.0487723E - 04
DENSITY (MOL / M3) 9534.95813

NEXT TEMPERATURE ? (Y / N) Y
ENTER TEMPERATURE (K) 400

COMPRESSIBILITY FACTOR .772909228
MOLAR VOLUME (M3 / MOL) 8.56836865E - 04
DENSITY (MOL / M3) 1167.08331

NEXT TEMPERATURE ? (Y / N) N
NEXT PRESSURE ? (Y / N) Y
ENTER PRESSURE (PA) 100E5
ENTER TEMPERATURE (K) 400

COMPRESSIBILITY FACTOR .42327398
MOLAR VOLUME (M3 / MOL) 1.40770767E - 04
DENSITY (MOL / M3) 7103.74761

NEXT TEMPERATURE ? (Y / N) N
NEXT PRESSURE ? (Y / N) N
NEXT COMPOUND ? (Y / N) N
```

Remark. The results show the performance of the Peng–Robinson equation of state.

The Lee – Kesler Equation of State

The LK equation of state can be applied in two ways to the class of nonpolar substances just discussed:

1. We can use the linear combination

$$z = \left(1 - \frac{\omega}{0.3978}\right) z^{(0)} + \left(\frac{\omega}{0.3978}\right) z^{(r)}$$

and this means that the LK equation of state has to be solved twice: with the parameter set of the simple substance and with the parameter set of the reference substance. These two sets of parameters are contained in Table 3.3.

2. Fairly good results are also obtained by forming the linear combinations of the parameters one after the other, and substituting the new parameters obtained in this way into the LK equation. In this case the LK equation has to be solved only once.

Algorithm and Program: Lee – Kesler Equation of State

Write a program for the calculation of the compressibility factor, molar volume, and density of gasses, using the LK equation of state. Solve the equation by Newton's method.

Algorithm.

1. Read the critical temperature, critical pressure, and omega factor of the compound.
2. Calculate B (see program VIRB), C, and D.
3. Make the initial guess $\phi = 0.2T_r/P_r$.
4. Calculate a Newtonian iteration step:

$$\phi_{new} = \phi_{old} - \frac{f[\phi_{old}]}{f'[\phi_{old}]}$$

where

$$f[\phi] = 1 + \frac{B}{\phi} + \frac{C}{\phi^2} + \frac{D}{\phi^5} + \frac{c_4}{T_r^3\phi^2}\left(\beta + \frac{\gamma}{\phi^2}\right) e^{-\gamma/\phi^2} - \frac{P_r\phi}{T_r}$$

5. Stop the iteration if the change in ϕ between two consecutive steps is less than 10^{-5}, or f[ϕ] is less then 10^{-4}, or the number of iterations exceeds 50; otherwise return to step 4.
6. Calculate the compressibility factor:

$$z = \frac{P_r}{T_r}\phi$$

7. Calculate the molar volume and density.
8. Print the results.

Program ESBWR. A possible realization:

```
]PR#0
]LOAD ESBWR
]LIST

 10  REM :CALCULATION OF COMPRESSI
      BILITY FACTOR BY
 20  REM :LK EQUATION OF STATE
 30  DIM N(12),R(12)
 40  DIM B(12)
 50  FOR I = 1 TO 12
 60  READ N(I)
 70  READ R(I)
 80  NEXT I
 90  DATA 0.1181193,0.2026579, 0.2
      65728,0.331511
100  DATA 0.15790,0.027655,0.030
      323,0.203488,0.0236744,0.031
      3385
110  DATA 0.0186984,0.0503618,0.
      0.016901,0.042724,0.04177,0.
      155488E - 4,0.48736E - 4
120  DATA 0.623689E - 4,0.0740336E
      - 4,0.65392,1.226,0.060167,0.
      03754
130  INPUT "ENTER CRITICAL TEMPER
      ATURE (K) ";TC
140  INPUT "ENTER CRITICAL PRESSU
      RE (PA) ";PC
150  INPUT "ENTER OMEGA ";OM
160  MR = 0.3978
170  FOR I = 1 TO 12
180  B(I) = N(I) + OM * (R(I) - N
      (I)) / MR
190  NEXT I
```

```
200  PRINT
210  INPUT "ENTER PRESSURE (PA) "
      ;P
220  INPUT "ENTER TEMPERATURE (K)
      ";T
230  TR = T / TC
240  B = B(1) - B(2) / TR - B(3) /
      TR / TR - B(4) / TR / TR / T
      R
250  C = B(5) - B(6) / TR + B(7) /
      TR / TR / TR
260  D = B(9) + B(10) / TR
270  DEF FN F(FI) = 1 + B / FI +
      C / FI ∧ 2 + D / FI ∧ 5 + (B
      (11) + B(12) / FI ∧ 2) * EXP
      ( - B(12) / FI ∧ 2) / TR ∧ 3
      / FI ∧ 2 * B(8) - P * FI /
      PC / TR
280  DEF FN FD(FI) = - B / FI ∧
      2 - 2 * C / FI ∧ 3 - 5 * D /
      FI ∧ 6 - EXP ( - B(12) / FI
      ∧ 2) * (2 * B(8) / TR ∧ 3 /
      FI ∧ 3 * (B(11) + B(12) / FI
      ∧ 2) + 2 * B(12) * B(8) / T
      R ∧ 3 / FI ∧ 5 + 2 * B(12) *
      B(8) / TR ∧ 3 / FI ∧ 5 * (B(
      11) + B(12) / FI ∧ 2)) - P /
      PC / TR
290  FI = 0.2 * TR / P * PC
300  I = 0
310  A1 = FN F(FI)
320  A2 = FN FD(FI)
330  I = I + 1
340  IF I > 50 THEN PRINT : PRINT
      "ITERATION NOT CONVERGENT": GOTO
      440
350  F2 = F1 - A1 / A2
360  IF ABS (FI - F2) < 1E - 5 OR
      ABS (A1) < 1E - 4 THEN GOTO
      390
370  FI = F2
380  GOTO 310
390  PRINT : PRINT
400  F2 = P / PC / TR * F2
410  PRINT "COMPRESSIBILITY FACTOR
      ";F2
420  PRINT "MOLAR VOLUME (M3 / MOL)
      ";F2 * 8.3144 * T / P
430  PRINT "DENSITY (MOL / M3) ";P
      / T / 8.3144 / F2
```

```
440  PRINT : PRINT
450  INPUT "NEXT TEMPERATURE ? (Y / N)
      ";A$
460  IF A$ = "Y" THEN GOTO 220
470  INPUT "NEXT PRESSURE ? (Y / N)
      ";A$
480  IF A$ = "Y" THEN GOTO 200
490  INPUT "NEXT COMPOUND ? (Y / N)
      ";A$
500  IF A$ = "Y" THEN GOTO 130
510  END
```

Example. Run the program for propane in the pressure range 0.5×10^5–500×10^5 Pa and the temperature range 250–500 K.

```
]RUN
ENTER CRITICAL TEMPERATURE (K) 369.86
ENTER CRITICAL PRESSURE (PA) 4.2455E6
ENTER OMEGA 0.28
ENTER PRESSURE (PA) 0.5E5
ENTER TEMPERATURE (K) 300

COMPRESSIBILITY FACTOR .991333287
MOLAR VOLUME (M3 / MOL) .0494540489
DENSITY (MOL / M3) 20.2207913

NEXT TEMPERATURE ? (Y / N) Y
ENTER TEMPERATURE (K) 400

COMPRESSIBILITY FACTOR .996859817
MOLAR VOLUME (M3 / MOL) .0663063301
DENSITY (MOL / M3) 15.0815163

NEXT TEMPERATURE ? (Y / N) Y
ENTER TEMPERATURE (K) 500

COMPRESSIBILITY FACTOR .998684096
MOLAR VOLUME (M3 / MOL) .0830345904
DENSITY (MOL / M3) 12.0431738

NEXT TEMPERATURE ? (Y / N) N
NEXT PRESSURE ? (Y / N) Y

ENTER PRESSURE (PA) 2E5
ENTER TEMPERATURE (K) 250

COMPRESSIBILITY FACTOR .931902824
MOLAR VOLUME (M3 / MOL) 9.68526605E - 03
DENSITY (MOL / M3) 103.249616

NEXT TEMPERATURE ? (Y / N) Y
ENTER TEMPERATURE (K) 300
```

```
COMPRESSIBILITY FACTOR .964603195
MOLAR VOLUME (M3 / MOL) .0120301452
DENSITY (MOL / M3) 83.1245162

NEXT TEMPERATURE ? (Y / N) Y
ENTER TEMPERATURE (K) 400

COMPRESSIBILITY FACTOR .987370836
MOLAR VOLUME (M3 / MOL) .0164187922
DENSITY (MOL / M3) 60.90582

NEXT TEMPERATURE ? (Y / N) N
NEXT PRESSURE ? (Y / N) Y

ENTER PRESSURE (PA) 5E5
ENTER TEMPERATURE (K) 270

COMPRESSIBILITY FACTOR -.0150515903
MOLAR VOLUME (M3 / MOL) -6.7578269E - 05
DENSITY (MOL / M3) -14797.6563

NEXT TEMPERATURE ? (Y / N) Y
ENTER TEMPERATURE (K) 300

COMPRESSIBILITY FACTOR .90725808
MOLAR VOLUME (M3 / MOL) 4.52598395E - 03
DENSITY (MOL / M3) 220.946431

NEXT TEMPERATURE ? (Y / N) Y
ENTER TEMPERATURE (K) 400

COMPRESSIBILITY FACTOR .968072264
MOLAR VOLUME (M3 / MOL) 6.43915203E - 03
DENSITY (MOL / M3) 155.299952

NEXT TEMPERATURE ? (Y / N) N
NEXT PRESSURE ? (Y / N) Y

ENTER PRESSURE (PA) 10E5
ENTER TEMPERATURE (K) 270

COMPRESSIBILITY FACTOR .28611355
MOLAR VOLUME (M3 / MOL) 6.42292874E - 04
DENSITY (MOL / M3) 1556.92215

NEXT TEMPERATURE ? (Y / N) Y
ENTER TEMPERATURE (K) 300

COMPRESSIBILITY FACTOR .0341587857
MOLAR VOLUME (M3 / MOL) 8.52029423E - 05
DENSITY (MOL / M3) 11736.6839

NEXT TEMPERATURE ? (Y / N) Y
ENTER TEMPERATURE (K) 400

COMPRESSIBILITY FACTOR .934879213
```

```
MOLAR VOLUME (M3 / MOL) 3.10918389E − 03
DENSITY (MOL / M3) 321.627808

NEXT TEMPERATURE ? (Y / N) N
NEXT PRESSURE ? (Y / N) Y

ENTER PRESSURE (PA) 30E5
ENTER TEMPERATURE (K) 250

COMPRESSIBILITY FACTOR .107759099
MOLAR VOLUME (M3 / MOL) 7.46626879E − 05
DENSITY (MOL / M3) 13393.5708

NEXT TEMPERATURE ? (Y / N) Y

ENTER TEMPERATURE (K) 340

COMPRESSIBILITY FACTOR .104195412
MOLAR VOLUME (M3 / MOL) 9.81831978E − 05
DENSITY (MOL / M3) 10185.0421

NEXT TEMPERATURE ? (Y / N) Y
ENTER TEMPERATURE (K) 400

COMPRESSIBILITY FACTOR .785163446
MOLAR VOLUME (M3 / MOL) 8.70421727E − 04
DENSITY (MOL / M3) 1148.86838

NEXT TEMPERATURE ? (Y / N) N
NEXT PRESSURE ? (Y / N) Y

ENTER PRESSURE (PA) 100E5
ENTER TEMPERATURE (K) 4:0
? EXTRA IGNORED

COMPRESSIBILITY FACTOR 17.4082563
MOLAR VOLUME (M3 / MOL) 5.78956823E − 05
DENSITY (MOL / M3) 17272.4452

NEXT TEMPERATURE ? (Y / N) Y
ENTER TEMPERATURE (K) 400

COMPRESSIBILITY FACTOR .382832488
MOLAR VOLUME (M3 / MOL) 1.27320897E − 04
DENSITY (MOL / M3) 7854.17022

NEXT TEMPERATURE ? (Y / N) N
NEXT PRESSURE ? (Y / N) N
NEXT COMPOUND ? (Y / N) N
```

Remark. The difficulty in solving the LK equation is the iteration itself. There is no absolutely sure method. It takes long experience in nonlinear-equation solving to avoid infeasible roots or iterational errors. The more sophisticated the iteration method, the more time consuming is the calculation. If you want to be sure of finding roots of the equation relevant to the liquid state, use the *regula falsi* method with a very small increment and an upper limit: $0.290 - 0.085\omega$.

Equation of State of Liquids

The equation of state of liquids is written with the omega factor:

$$z = z^{(0)} + z^{(1)}\omega$$

where $z^{(0)}$ is the polynomial pertinent to the class of simple molecules described already, while

$$z^{(1)} = -\left(0.02646 + 0.28376T_r - 0.28340T_r^2\right)P_r$$
$$-\left(0.10209 - 0.32504T_r + 0.25376T_r^2\right)P_r^2$$
$$+\left(0.00919 - 0.03016T_r + 0.02485T_r^2\right)P_r^3$$

This correction function is a two-variable polynomial, quadratic with respect to T_r and cubic with respect to P_r.

Algorithm and Program: Equation of State of Liquids

Write a program for the calculation of the compressibility factor and molar volume of liquids.

Algorithm.

1. Read the critical temperature, critical pressure, and omega factor.
2. Read the actual temperature and pressure.
3. Calculate reduced temperature and reduced pressure.
4. Calculate z using z^0 and z^1.
5. Print the results.

Program ESLIQU. A possible realization:

```
]LOAD ESLIQU
]LIST

 10  REM :CALCULATION THE COMPRESS
      IBILITY
 20  REM :FACTOR AND MOLAR VOLUME
      FOR LIQUIDS
 30  REM :BY "POLINOM" EQUATION OF
      STATE
```

```
 40  DEF FN Z0(PR) = (.46407 - 0.
      73221 * TR + 0.45256 * TR ∧
      2) * PR - (0.00871 - 0.02939
      * TR + 0.02775 * TR ∧ 2) *
      PR ∧ 2
 50  DEF FN Z1(PR) = - (0.02676 +
      0.28376 * TR - 0.2834 * TR ∧
      2) * PR - (0.10209 - 0.32504
      * TR + 0.25376 * TR ∧ 2) *
      PR ∧ 2 + (0.00919 - 0.03016 *
      TR - 0.02485 * TR ∧ 2) * PR ∧
      3
 60  PRINT : PRINT
 70  INPUT "ENTER CRITICAL TEMPERA
      TURE (K) ";TC
 80  INPUT "ENTER CRITICAL PRESSURE
      (PA) ";PC
 90  INPUT "ENTER OMEGA
      ";OM
100  PRINT : PRINT
110  INPUT "ENTER PRESSURE (PA) "
      ;P
120  PRINT
130  INPUT "ENTER TEMPERATURE (K)
      ";T
140  TR = T / TC
150  PR = P / PC
160  Z = FN Z0(PR) + OM * FN Z1
      (PR)
170  PRINT
180  PRINT "COMPRESSIBILITY FACTOR
      ";Z
190  PRINT "MOLAR VOLUME (M3 / MOL)
      ";Z * 8.3144 * T / P
200  PRINT
210  INPUT "NEXT TEMPERATURE ? (Y
      /N) "A$
220  IF A$ = "Y" THEN GOTO 130
230  INPUT "NEXT PRESSURE ? (Y / N)
      ";A$
240  IF A$ = "Y" THEN GOTO 110
250  INPUT "NEXT COMPOUND ? (Y / N)
      ";A$
260  IF A$ = "Y" THEN GOTO 70
270  END
```

Example. Run the program for butane at pressures 3.7×10^5 and 3.7×10^6 Pa and at temperatures 250 and 300 K.

```
]RUN

ENTER CRITICAL TEMPERATURE (K) 425.2
ENTER CRITICAL PRESSURE (PA) 3.7897E6
ENTER OMEGA 0.2740
ENTER PRESSURE (PA) 3.7E5
ENTER TEMPERATURE (K) 250

COMPRESSIBILITY FACTOR .0159821796
MOLAR VOLUME (M3 / MOL) 8.97852935E - 05

NEXT TEMPERATURE ? (Y / N) Y
ENTER TEMPERATURE (K) 300

COMPRESSIBILITY FACTOR .0145469571
MOLAR VOLUME (M2 / MOL) 9.80669354E - 05

NEXT TEMPERATURE ? (Y / N) N
NEXT PRESSURE ? (Y / N) Y
ENTER PRESSURE (PA) 3.7E6
ENTER TEMPERATURE (K) 250

COMPRESSIBILITY FACTOR .15492363
MOLAR VOLUME (M3 / MOL) 8.70335831E - 05

NEXT TEMPERATURE ? (Y / N) 3" " "
NEXT PRESSURE ? (Y / N) Y
ENTER PRESSURE (PA) 3.7E6
ENTER TEMPERATURE (K) 300

COMPRESSIBILITY FACTOR .137977151
MOLAR VOLUME (M3 / MOL) 9.30159914E - 05

NEXT TEMPERATURE ? (Y / N) N
NEXT PRESSURE ? (Y / N) N
NEXT COMPOUND ? (Y / N) N
```

Spline Equation of State

We have made several attempts to approximate the $z[T, P]$ state surface for the vapor phase by joining one or more two-variable polynomials. In these attempts the shape of the surface to be approximated was so very uneven, that the least-squares method created large spurious "waves" on it, and this phenomenon—especially in the calculation of excess properties using derivatives—caused errors of several orders of magnitude.

For fast and at the same time correct calculation of the compressibility factor the double cubic interpolating spline (Appendix A) seems suitable because:

1. It closely follows the surface to be described.
2. It is the smoothest function available, so it eliminates the problem of "waves."

The following form was used for the double cubic spline in the interval i, j:

$$\begin{aligned} S_\pi[T_r, P_r] = {} & a_{ij,1} + a_{ij,2}\Delta T_r + a_{ij,3}\Delta P_r \\ & + a_{ij,4}\Delta T_r \Delta P_r + a_{ij,5}\Delta T_r^2 \\ & + a_{ij,6}\Delta P_r^2 + a_{ij,7}\Delta T_r^2 \Delta P_r \\ & + a_{ij,8}\Delta T_r \Delta P_r^2 + a_{ij,9}\Delta T_r^3 \\ & + a_{ij,10}\Delta P_r^3 + a_{ij,11}\Delta T_r \Delta P_r^3 \\ & + a_{ij,12}\Delta T_r^3 \Delta P_r + a_{ij,13}\Delta T_r^2 \Delta P_r^2 \\ & + a_{ij,14}\Delta T_r^2 \Delta P_r^3 + a_{ij,15}\Delta T_r^3 \Delta P_r^2 \\ & + a_{ij,16}\Delta T_r^3 \Delta P_r^3 \end{aligned}$$

where

$$T_{r,i} \le T_r \le T_{r,i+1} \qquad P_{r,j} \le P_r \le P_{r,j+1}$$

$$\Delta T_r = T_r - T_{r,i} \qquad i = 1, \ldots, N$$

$$\Delta P_r = P_r - P_{r,j} \qquad j = 1, \ldots, M$$

These spline coefficients are stored in a direct-access backing store, so that z^0 and z^r coefficients belonging to a partitioning rectangle were put into one record, because these are always used together. To compute a compressibility factor z for a given T_r, P_r it is necessary to find the partitioning rectangle belonging to the T_r, P_r values, to bring into the core store all 32 spline coefficients pertaining to this rectangle, and to calculate the values of two double cubic polynomials.

The compressibility factor is obtained by linear combination of z^0 and z^r.

Algorithm and Program: Spline Equation of State

Write a program for the calculation of the compressibility factor of gases in the reduced-temperature range $0.95 \le T_r \le 0.97$ and the reduced-pressure range $0.4 \le P_r \le 0.6$. The 32 constants of this rectangle have been determined by a scientific library program on a large computer and are listed in statements 80–110 of program ESSPLINE (see program listing). The other constants are available from the authors.

Algorithm.

1. Read the omega factor.
2. Read the reduced pressure and temperature.
3. Calculate the differences $T_r - 0.95$, $P_r - 0.4$.
4. Calculate z^0 and z^r.
5. Calculate z.
6. Print the result.

Program ESSPLINE. A possible realization:

```
]PR#0
]LOAD ESSPLINE
]LIST

 10  REM :CALCULATION OF COMPRESSI
      BILITY
 20  REM :FACTOR BY SPLINE EQUATIO
      N OF STATE
 30  REM :IN 0.95 < TR < 0.97, 0.40 < PR
      < 0.60 REGION
 40  DIM AN(16),AR(16)
 50  FOR I = 1 TO 16
 60  READ AN(I),AR(I)
 70  NEXT I
 80  DATA 0.820599, 0.79716957, -0.5
      83688, -0.63452661, -0.740457,
      -0.5789693, 2.806982, 0.679840
      147
 90  DATA 0.685419, 0.98629081, 4.55
      1016, 7.2377491, 21.108917, 36.
      3460846, -98.355255, -160.7273
      102
100  DATA -0.602311, -1.755361557,
      -59.92575, -130.80543518, -601
      .284668, -1223.2873535, 1885.3
      35693, 4396.171875, -33.432648,
      -36.7835083
110  DATA -722.919433, -910.5063476,
      -2307.019775, -3367.1835937,
      78470.3125, 129498.5
120  PRINT : PRINT
130  INPUT "ENTER OMEGA ";OM
140  INPUT "ENTER PR ";PR
150  IF PR < 0.4 OR PR > 0.6 THEN
      PRINT "WRONG": GOTO 140
160  INPUT "ENTER TR ";TR
170  IF TR < 0.95 OR TR > 0.97 THEN
      PRINT "WRONG":GOTO 160
```

```
180  GOSUB 280
190  Z - Z0 + OM * (ZR - Z0) / 0.3
      978
200  PRINT
210  PRINT "COMPRESSIBILITY FACTOR
      ";Z
220  PRINT : PRINT
230  INPUT "NEXT TR ? (Y / N) ";A$
240  IF A$ = "Y" THEN GOTO 160
250  INPUT "NEXT PR ? (Y / N) ";A$
260  IF A$ = "Y" THEN GOTO 140
270  END
280  DT = TR - 0.95
290  DP = PR - 0.4
300  Z0 = AN(1) + AN(2) * DT + AN(
      3) * DP + AN(4) * DT * DP +
      AN(5) * DT ∧ 2 + AN(6) * DP ∧
      2 + AN(7) * DT ∧ 2 * DP
310  Z0 = Z0 + AN(8) * DT * DP ∧ 2
      + AN(9) * DT ∧ 3 + AN(10) *
      DP ∧ 3 + AN(11) * DT * DP ∧
      3 + AN(12) * DT ∧ 3 * DP + A
     N(13) * DT ∧ 2 * DP ∧ 2
320  Z0 = Z0 + AN(14) * DT ∧ 2 * D
      P ∧ 3 + AN(15) * DT ∧ 3 * DP
      ∧ 2 + AN(16) * DT ∧ 3 * DP ∧
      3
330  ZR = AR(1) + AR(2) * DT + AR(
      3) * DP + AR(4) * DT * DP +
      AR(5) * DT ∧ 2 + AR(6) * DP ∧
      2 + AR(7) * DT ∧ 2 * DP
340  ZR = ZR + AR(8) * DT * DP ∧ 2
      + AR(9) * DT ∧ 3 + AR(10) *
      DP ∧ 3 + AR(11) * DT * DP ∧
      3 + AR(12) * DT ∧ 3 * DP + A
      R(13) * DT ∧ 2 * DP ∧ 2
350  ZR = ZR + AR(14) * DT ∧ 2 * D
      P ∧ 3 + AR(15) * DT ∧ 3 * DP
      ∧ 2 + AR(16) * DT ∧ 3 * DP ∧
      3
360  RETURN
```

Example. Run the program for propane ($\omega = 0.14872$) at reduced temperatures 0.96 and 0.955 K and at reduced pressures 0.42 and 0.5 Pa.

```
]RUN

ENTER OMEGA 0.14872
ENTER PR 0.42
ENTER TR 0.96
```

```
COMPRESSIBILITY FACTOR .793777406

NEXT TR ? (Y / N) Y
ENTER TR 0.955

COMPRESSIBILITY FACTOR .796719617

NEXT TR ? (Y / N) N
NEXT PR ? (Y / N) Y
ENTER PR 0.5
ENTER TR 0.96

COMPRESSIBILITY FACTOR .689026442

NEXT TR ? (Y / N) Y
ENTER TR 0.955

COMPRESSIBILITY FACTOR .70080355

NEXT TR ? (Y / N) N
NEXT PR ? (Y / N) N
```

Remark. Storing the constants of the double cubic functions of further rectangles needs a large-capacity backing store, but the method is very fast and gives accurate results.

Molecular and Macro Parameters of the Law of Corresponding States for Nonpolar Molecules

We have seen that for simple molecules ε_0/k depends linearly on T_c and σ^3, with Nv_c a macro parameter characteristic of the material. The relationship suggested for symmetric molecules larger than methane is valid also for asymmetric molecules:

$$\frac{\varepsilon_0}{\mathbf{k}} = 82.789 + 0.282T_c$$

For cyclic molecules another relationship is valid:

$$\frac{\varepsilon_0}{\mathbf{k}} = 286.06 - 0.175T_c$$

The collision diameter is not equal to the largest dimension of the molecule, but is additive for characteristic groups. From the collision diameter calculated for the molecule the covolumen can be calculated with the relationship:

$$b = \tfrac{2}{3}\pi \mathbf{N}\sigma^3$$

POLAR MOLECULES

The dimensionless molecular parameter characteristic of polar substances is

$$\frac{\mu}{\varepsilon_0 \sigma^3}$$

where μ = the dipole moment,
ε_0 = depth of the potential well,
σ = collision diameter.

The Contribution of Polarity

A macroscopic property is sought to replace the dipole moment in the above expression.

It was shown in Figure 3.13, used to introduce Pitzer's omega factor, that the omega factors of water and *n*-heptane are identical, but the compressibility factors of the two substances are nevertheless different. The cause of difference is the polarity of water:

$$z = z^{(0)} + \omega z^{(1)} - \Delta z^{(p)}$$

where $\Delta z^{(p)}$ is the deviation caused by the polarity of the molecule.

A negative sign was used because the compressibility factor is always decreased by the polar nature of the compound. This deviation must be manifested also in a difference in the second virial coefficient between a polar and a homomorphous nonpolar substance. This is shown in Figure 3.17. The reduced second virial coefficient is

$$B_r = B_r^{(0)} + \omega B_r^{(1)} + B_r^{(p)}$$

The question is: What can be said about the dimensionless polarity contribution $B_r^{(p)}$?

According to all the observations made so far, irrespective of the size and shape of the molecule, $B_r^{(p)}$ depends only on reduced temperature. Let us express the value of $B_r^{(p)}$ at the reduced temperature $T_r = 0.6$ in the following way:

$$\phi \equiv \left(-B_r^{(p)}\right)_{T_r=0.6} = -\left(B_r - B_r^{(0)} - \omega B_r^{(1)}\right)_{T_r=0.6}$$

Let us consider this expression as a quantity characteristic of the polarity of

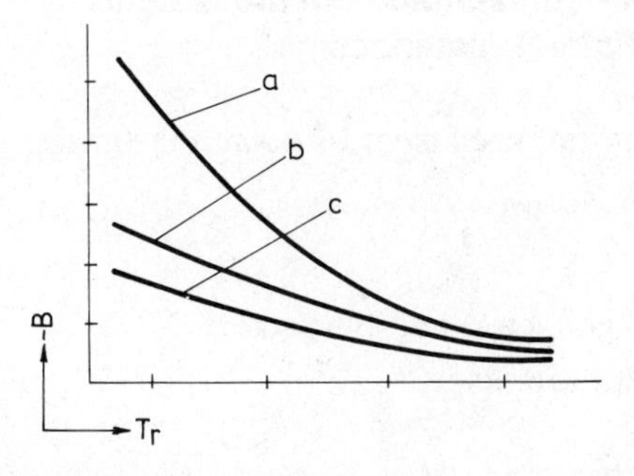

FIGURE 3.17. Second virial coefficient: (*a*) polar fluid, (*b*) nonpolar fluid, (*c*) simple fluid.

the substance, and call it the *phi factor*, on the analogy of the omega factor. Let us remember here that the omega factor was calculated from the equilibrium vapor pressure corresponding to the reduced temperature $T_r = 0.7$. It will be noted from Figure 3.17 that to take proper account of polarity a greater difference at lower temperatures than this will have to be provided for. For this purpose, the dimensionless polarity correction is described by the following power series:

$$B_r^{(p)} = \phi \sum_i \frac{b_i}{T_r^i}$$

However, statistical analysis shows that a single term of the polynomial gives satisfactory results:

$$B_r^{(p)} = \phi \frac{b}{T_r^7}$$

Let us compare this with the definition of the phi factor:

$$\phi = \left(-B_r^{(p)}\right)_{T_r = 0.6} = \phi \frac{b}{(0.6)^7}$$

The polarity correction in the reduced second virial coefficient at reduced temperature $T_r = 0.6$ is just ϕ, which is possible only if

$$b = -(0.6)^7 = -0.028$$

Accordingly, the polarity correction in the reduced second virial coefficient is

$$B_r^{(p)} = -\phi \frac{0.028}{T_r^7}$$

With this relationship it is now possible to calculate the second virial coefficient of polar components for both the virial equation and the LK equation.

Algorithm and Program: Graphical Representation of the Second Virial Coefficient of Polar Substances

Write a program for plotting the second virial coefficient of polar substances.

Algorithm.

1. Read the critical temperature, omega factor, and phi factor.
2. Read the parameters of the temperature range.

3. Calculate the second virial coefficient with the equation containing the phi factor.
4. Plot the results.

Program GRBVIRIPOLAR. A possible realization:

```
]PR#0
]LOAD GRBVIRIPOLAR,D2
]LIST

 10  REM :CALCULATION OF REDUCED
 20  REM :SECOND VIRIAL COEFFICIENT
 30  REM :AND TO DRAW IT
 40  READ N1,N2,N3,N4
 50  READ R1,R2,R3,R4
 60  DATA 0.1181193, 0.265728, 0.15
      479, 0.0030323
 70  DATA 0.2026579, 0.331511, 0.02
      7655, 0.203488
 80  PRINT : PRINT
 90  HPLOT 0,0
100  HGR
110  HPLOT 10,80 TO 279,80
120  HPLOT 10,0 TO 10,191
130  HPLOT 10,80
140  INPUT "ENTER CRITICAL TEMPER
      ATURE (K) ";TC
150  INPUT "ENTER OMEGA ";OM
150  PRINT
170  INPUT "ENTER PHI ";PHI
180  PRINT : PRINT
190  MR = 0.3978
200  B1 = N1 + OM * (R1 - N1) / MR
210  B2 = N2 + OM * (R2 - N2) / MR
220  B3 = N3 + OM * (R3 - N3) / MR
230  B4 = N4 + OM * (R4 - N4) / MR
240  INPUT "ENTER BEGINNING TEMPE
      RATURE (K) ";TB
250  INPUT "ENTER TEMPERATURE INC
      REMENT ";DT
260  INPUT "ENTER THE NUMBER OF C
      ALCULATIONS ";N
270  FOR I = 1 TO N
280  TR = (TB + (I - 1) * DT) / TC
290  B = B1 - B2 / TR - B3 / TR /
      TR - B4 / TR / TR / TR - 0.0
      28 / TR ^ 7 * PHI
300  X = (I - 1) / N * 270 + 10
```

```
310  Y = 80 - 30 * B
320  HPLOT TO X,Y
330  NEXT I
340  PRINT
350  INPUT "NEXT COMPOUND ? (Y / N)
      ";A$
360  IF A$ = "Y" THEN GOTO 130
370  TEXT
380  END
```

Example. Run the program for a substance with critical temperature 508 K, omega factor 0.3, and phi factor 0.3.

```
]RUN

ENTER CRITICAL TEMPERATURE (K) 508
ENTER OMEGA 0.3
ENTER PHI 0
ENTER BEGINNING TEMPERATURE (K) 300
ENTER TEMPERATURE INCREMENT 20
ENTER THE NUMBER OF CALCULATIONS 20

NEXT COMPOUND ? (Y / N) Y
ENTER CRITICAL TEMPERATURE (K) 508
ENTER OMEGA 0.3
ENTER PHI 0.3
ENTER BEGINNING TEMPERATURE (K) 300
ENTER TEMPERATURE INCREMENT 20
ENTER THE NUMBER OF CALCULATIONS 20

NEXT COMPOUND ? (Y / N) N
```

Molecular and Macro Parameters of Corresponding State Theory for Polar Molecules

The same relationships apply between $\varepsilon_0/\mathbf{k}$ and T_c for polar molecules as for nonpolar molecules. However, the collision diameter σ of a polar molecule differs. The deviation $\Delta\sigma$ must be taken into consideration as a correction in the calculation of the collision diameter of polar molecules:

$$\sigma_{\text{polar}} = \left(\sigma_{\text{nonpolar}}^3 + \Delta\sigma\right)^{1/3}$$

$$\Delta\sigma = 1.591 \times 10^3 \frac{\mu P_c}{T_c}$$

where μ is the dipole moment of the polar molecule.

APPENDIX A: DOUBLE CUBIC SPLINE

The definition of a double cubic spline is the following: Suppose that we are given a rectangular region

$$R: a \leq t \leq b;\ c \leq s \leq d$$

of the plane. Then if we are given two one-dimensional meshes

$$\Delta_t: a = t_0 < t_1 < \cdots < t_N = b$$

and

$$\Delta_s: c = s_0 < s_1 < \cdots < s_M = d$$

we define the resulting two-dimensional mesh

$$\Pi = \{P_{ij}\} \qquad (i = 0, 1, \ldots, N, \quad j = 0, 1, \ldots, M)$$

where $P_{ij} = (t_i, s_j)$ partitions R into a family of subrectangles

$$\{R_{ij}: t_{i-1} \leq t \leq t_i;\ s_{j-1} \leq s \leq s_j\} \qquad (i = 1, 2, \ldots, N \quad j = 1, 2, \ldots, M)$$

A simple doubly cubic spline $S_\Pi\ (t, s)$ on R with respect to Π consists of:

1. A double cubic in each rectangle R_{ij}.
2. An element of $C_2^4(R)$, that is, a function of R whose fourth-order partial derivatives, involving no more than second-order differentiation with respect to a single variable, exist and are continuous.

If the double cubic spline is such that it gives back the $\{f_{ij}\}$ values belonging to the partition $\Pi = \{P_{ij}\}$, that is,

$$S_\Pi(t_i, s_j) = f_{ij} \qquad (i = 1, 2, \ldots, N \quad j = 1, 2, \ldots, M)$$

then $S_\Pi(t, s)$ is called an interpolating spline.

The spline described above has the property that among functions which interpolate values f_{ij} and belong to the partition $\Pi = \{P_{ij}\}$, the smoothest one is the double cubic interpolating spline.

4

CONSERVATION PRINCIPLES

The possibility of *trial* is recognized by everybody. Children keep trying, and so does humanity, living in its infancy. Obviously prehistoric human often found a bough suitable for a cudgel long before they thought out a plan for its use. Trial isn't so difficult; indeed, it is inevitable. But if trial in itself results only in some consequence, whereas *experiment* attempts to decide something, the possibility of which has been premeditated, without knowing whether it is correct or not.

The multitude of phenomena occurring in the material world and mutual relationships and effects between them create complicated conditions, very difficult to grasp. Therefore it is easy to understand the endeavor in each science to observe, describe, and interpret single phenomena by themselves. To achieve this, abstractions are almost always needed, such as the decisive separation of essential from incidental, the setting apart of secondary effects, and the creation of ideal conditions. This mode of proceeding is very widely used in science; often, in fact, so-called thought-experiments are usefully discussed, which cannot actually be carried out.

THE ISOLATED BODY

The first such abstraction with which we must aquaint ourselves is that of an *isolated body*. The isolated body is constrained by a rigid wall of the first kind, described in Chapter 2; thus, whatever transformation occurs in the isolated body does not produce changes in its surroundings outside the body, and no change in the surroundings causes transformations within the isolated body.

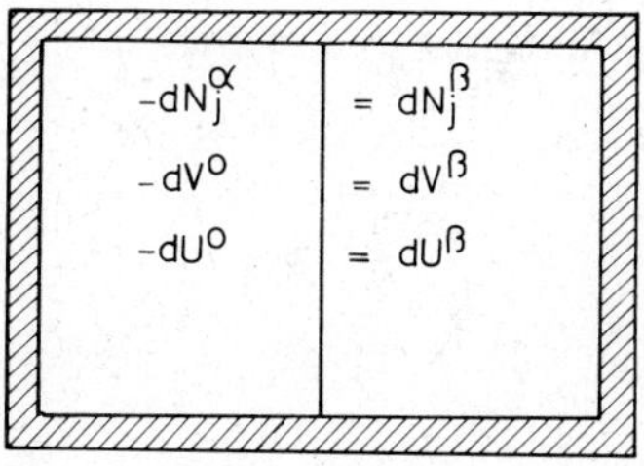

FIGURE 4.1. Isolated body divided into two subsystems.

The second abstraction enables us to say something about transformations. For this purpose the isolated body is divided by an internal wall into two subsystems (Figure 4.1). The wall is characterized by whether or not it permits, from one subsystem into the other:

1. The transfer of mass of a given component.
2. The transfer of internal energy.
3. The transfer (i.e., the change) of volume.

The transfer of volume means that the wall separating the two subsystems moves. According to three aspects the properties of the wall are classified as in Table 4.1.

Thus, transformation in the isolated body is characterized by redistribution between the subsystems of:

1. Amount of substance.
2. Energy.
3. Volume.

TABLE 4.1
Classification of Walls

Property Transfered	Transfer Prevented	Transfer Permitted
Material	$dN_j^\iota = 0$: impermeable	$dN_j^\iota \neq 0$: permeable
Internal energy	$dU^\iota = 0$: adiabatic	$dU^\iota \neq 0$: diathermal
Volume	$dV^\iota = 0$: rigid and fixed	$dV^\iota \neq 0$: movable

This means, respectively, that the ιth subsystem will undergo a change in:

1. N_j^ι, the amount of the jth component.
2. U^ι, its internal energy.
3. V^ι, its volume.

These are known chemical, physical, or geometrical quantities, satisfying the Carnap criteria. Therefore they can be assigned to the subsystems, and we can say that U^ι, V^ι, and N_j^ι are properties of the ιth subsystem. A transformation occurring in the isolated body is accompanied by a change in magnitude of the properties of all the subsystems.

It is essential that transformation in the isolated body is spontaneous, free of external interaction. However, this transformation does not last forever, but ceases after a certain time, and there is no further macro change in any of the subsystems of the isolated body. Neither is the ceasing of transformation the result of external interaction. A transformation ceases when its *cause* ceases. Then we say that the two subsystems (phases) in contact are in *equilibrium* with one another.

The fundamental problem of thermodynamics is the determination of the conditions of equilibrium. The name thermodynamics is misleading in this respect, because as a matter of fact it does not investigate the progress of changes with time, but just by the conditions of static equilibrium, which are by definition independent of time. It follows from this that conditions of equilibrium must be formulated as descriptions of the state of the body. Here the question immediately arises *what* properties and *how many* properties are needed for the characterization of the equilibrium state of a body. Before discussing this question let us agree on some terminology.

CLASSIFICATION AND DESCRIPTION OF THE THERMODYNAMICAL SYSTEMS

If a body has at least one property that is a discontinuous function of space coordinates, then this body is called *heterogeneous*. Discontinuity occurs at the boundary between subsystems. Within the subsystems there is no property that is a discontinuous function of space coordinates. Subsystems of this kind are called *phases*. To say that the body is heterogeneous amounts to saying that it has more than one phase.

The phase is *inhomogeneous* if it has at least one property that is a monotonic function of space coordinates, and is *homogeneous* if all its properties are invariant with respect to space coordinates. The homogeneity of the phase is synonymous with the statement that the contents of a macroscopic volume element can be exchanged within the phase for the contents of any other volume element of the same size, without producing a change in the macro state of the phase. For this reason it is easy to see that less data are needed for the description of a homogeneous phase, than for that of an inhomogeneous phase. Indeed, this is the third important abstraction in thermodynamics.

As concerns the answer to our earlier question, an advance has been made, because now it is evident that equilibrium between the phases presumes also equilibrium within the individual phases, that is, *equilibrium between the phases presumes that each of the equilibrium phases is homogeneous*. Thus, the previous question will be worded more conveniently in the following way: What properties and how many properties are needed for the complete description of a homogeneous phase? This question can be answered by delimiting the field of validity of thermodynamics.

Sphere of Validity of Thermodynamics

The sphere of validity of thermodynamics is delimited by the following statements concerning the transformations occurring in an isolated body:

$$dN_j^\alpha + dN_j^\beta = 0$$

$$dU^\alpha + dU^\beta = 0 \qquad j = 1, 2, \ldots, k$$

$$dV^\alpha + dV^\beta = 0$$

(Here and in the following, a Greek superscript denotes the phase.) During the transformations the loss of one phase in a given property is equal to the gain of another phase, whatever the property in question.

After integration we can write

$$N_j^\alpha + N_j^\beta = N_j^{\mathrm{t}} \text{ (const)}$$

$$U^\alpha + U^\beta = U^{\mathrm{t}} \text{ (const)}$$

$$V^\alpha + V^\beta = V^{\mathrm{t}} \text{ (const)}$$

The common conspicuous characteristic of the three properties discussed so far is additivity, in the sense that in multiphase isolated systems the amounts of the components, the energy, and the volume of the system are equal, respectively, to the sums of the amounts of components, the energies, and the volumes of the individual phases. All properties of the system for which additivity exists in the above sense are called *extensive* properties. Extensive properties are described by first-order homogeneous functions.

THE FIRST LAW OF THERMODYNAMICS

If we wish to build up the thermodynamics within the sphere of validity of the three conservation statements made above, an answer can be given to the question: what properties and how many independent properties are needed for the complete description of the state of a homogeneous phase? The answer is the first law of thermodynamics.

For the complete determination of the state of a homogeneous phase is fixing of the values of $k + 2$ extensive properties is necessary and sufficient, namely:

1. $N_1, N_2, \ldots, N_k$, the amounts of components.
2. U, the internal energy.
3. V, the volume.

The interpretation of the first law is that any other property X of the homogeneous phase is a single-valued function of the aforesaid $k + 2$ state-determining properties, that is,

$$X = X\left[U, V, N_j\right] \qquad j = 1, 2, \ldots, k$$

If this phase goes from some initial state through various changes back to its original state—that is, its original state is reproduced—then *all* its properties take on their original values. Hence all functions that describe some thermodynamical property are multivariable functions,

$$X = X[Y_1, Y_2, \ldots]$$

for which the integral along any closed curve is zero:

$$\int dX = 0$$

This is possible only if dX is a total differential:

$$dX = \left(\frac{\partial X}{\partial Y_1}\right)_{Y_2, Y_3, \ldots} dY_1 + \left(\frac{\partial X}{\partial Y_2}\right)_{Y_1, Y_3, \ldots} dY_2 + \cdots$$

It follows from the first lay that, attributing a distinguished role to energy, we have

$$U = U[X, V, N_1, N_2, \ldots, N_k]$$

A distinguished role is attributed to energy because in a conservative field of force it constituted a *potential*, while V and N_j are independent variables, *generalized coordinates*. These are physical quantities which can be changed

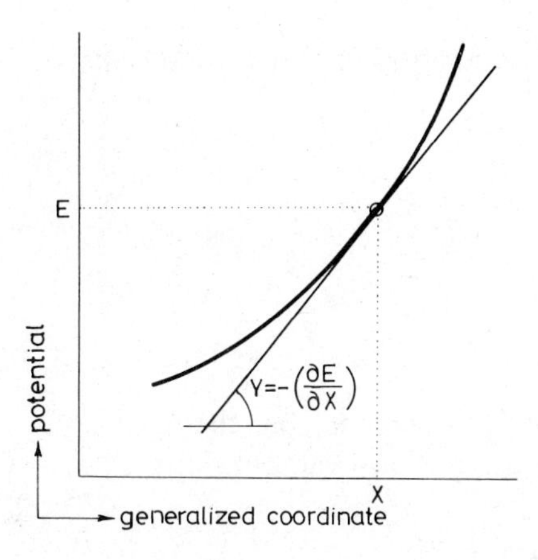

FIGURE 4.2. Potential, generalized coordinate, and generalized force.

independently of one another and are first-order homogeneous functions. In place of X, too, a physical quantity meeting this requirement must be chosen. The condition for the independent variability of the quantity X is a diathermic wall.

Figure 4.2 shows in a two-dimensional plot that the potential U is a function of a generalized coordinate. Moreover, in the state determined by the potential U and by the corresponding value of the general coordinate, a *generalized force* can also be assigned to this state. This is the partial derivative of the potential with respect to the generalized coordinate, and is represented in the figure by the slope of the tangent drawn to the curve at the given state point:

$$Y = -\left(\frac{\partial U}{\partial X}\right)_{V, N_j}$$

The generalized force is a homogeneous function of order zero.

The first law only delimits the sphere of validity of thermodynamics; it does not yet specify the causes and the directions of spontaneous processes, and does not express conditions of equilibrium. Thus it does not solve the basic problem of thermodynamics.

APPENDIX A: MATHEMATICAL RELATIONSHIPS BETWEEN THE VARIABLES OF THE MULTIVARIABLE EQUATION OF STATE

According to the first law,

$$X = X\left[U, V, N_j\right] \qquad j = 1, 2, \ldots k$$

Let us investigate the general relationships of the properties of a phase with fixed composition. We don't take into consideration the fixed parameters N_j. That is, we let

$$X = X[U, V]$$

It follows from the first law that

$$V = V[U, Z]$$

where Z is a property not yet taken into consideration. But then, evidently, according to first law the following relationship is also valid:

$$X = X[U, Z]$$

Let us form now the differential of all the three functions:

$$dX = \left(\frac{\partial X}{\partial U}\right)_V dU + \left(\frac{\partial X}{\partial V}\right)_U dV$$

$$dV = \left(\frac{\partial V}{\partial U}\right)_Z dU + \left(\frac{\partial V}{\partial Z}\right)_U dZ$$

$$dX = \left(\frac{\partial X}{\partial U}\right)_Z dU + \left(\frac{\partial X}{\partial Z}\right)_U dZ$$

The expression for dV is substituted into the first equation:

$$dX = \left(\left(\frac{\partial X}{\partial U}\right)_V + \left(\frac{\partial X}{\partial V}\right)_U \left(\frac{\partial V}{\partial U}\right)_Z\right) dU + \left(\left(\frac{\partial X}{\partial V}\right)_U \left(\frac{\partial V}{\partial Z}\right)_U\right) dZ$$

The right sides of the last two equations are bound to be equal, but then the coefficients of dU and dZ must also be equal respectively:

$$\left(\frac{\partial X}{\partial U}\right)_Z = \left(\frac{\partial X}{\partial U}\right)_V + \left(\frac{\partial X}{\partial V}\right)_U \left(\frac{\partial V}{\partial U}\right)_Z$$

$$\left(\frac{\partial X}{\partial Z}\right)_U = \left(\frac{\partial X}{\partial V}\right)_U \left(\frac{\partial V}{\partial Z}\right)_U$$

These two equations constitute the so-called chain rule.

Rewriting the first equation for condition $X = Z$, we have

$$0 = \left(\frac{\partial Z}{\partial U}\right)_V + \left(\frac{\partial Z}{\partial V}\right)_U \left(\frac{\partial V}{\partial U}\right)_Z$$

This can be written also in the following form:

$$\left(\frac{\partial V}{\partial U}\right)_Z \left(\frac{\partial Z}{\partial V}\right)_U \left(\frac{\partial U}{\partial Z}\right)_V = -1$$

The equations just deduced will be used on many occasions. It should be noted that these relationships are of quite general validity, without the restriction that U means the internal energy and V the volume.

APPENDIX B: ON A MATHEMATICAL PROPERTY OF HOMOGENEOUS LINEAR FUNCTIONS

If the function $f[x_1, x_2, \ldots x_k]$ is homogeneous of order n, this means that

$$f[\lambda x_1, \lambda x_2, \ldots, \lambda x_k] = \lambda^n f[x_1, x_2, \ldots, x_k]$$

If $n = 0$, then we have a zero-order homogeneous function. For such a function

$$f[\lambda x_1, \lambda x_2, \ldots, \lambda x_k] = f[x_1, x_2, \ldots, x_k]$$

If $n = 1$, we have a homogeneous linear function:

$$f[\lambda x_1, \lambda x_2, \ldots, \lambda x_k] = \lambda f[x_1, x_2, \ldots, x_k]$$

If $n = 2$, we have a second-order homogeneous function:

$$f[2x_1, 2x_2, \ldots, 2x_k] = \lambda^2 f[x_1, x_2, \ldots, x_k]$$

For example, the function

$$y = x_1 x_2$$

is homogeneous of order 2 because

$$(\lambda x_1)(\lambda x_2) = \lambda^2 (x_1 x_2) = \lambda^2 y$$

Let us differentiate both sides of the defining equation with respect to x_i, with the application of the chain rule on the left side:

$$\frac{\partial}{\partial(\lambda x_i)} f[\lambda x_1, \lambda x_2, \ldots, \lambda x_k] \frac{\partial(\lambda x_i)}{\partial x_i} = \lambda^n \frac{\partial}{\partial x_i} f[x_1, x_2, \ldots, x_k]$$

Carrying out the operation on the left side, that is, taking the partial derivative of the function with respect to x_i, we have

$$\lambda \dot{f}[\lambda x_1, \lambda x_2, \ldots, \lambda x_k] = \lambda^n f[x_1, x_2, \ldots, x_k]$$

whence

$$\dot{f}[\lambda x_1, \lambda x_2, \ldots, \lambda x_k] = \lambda^{n-1} f[x_1, x_2, \ldots, x_k]$$

If $n = 1$, that is, the function is homogeneous of order 1, then

$$\dot{f}[\lambda x_1, \lambda x_2, \ldots, \lambda x_k] = f[x_1, x_2, \ldots, x_k]$$

which means that the first derivative of a first-order homogeneous function is a homogeneous function of order zero. Multiplication of the independent variables by a scalar λ of a zero-order homogeneous function leaves the function invariant.

5

EQUILIBRIUM

The basic problem of thermodynamics is the problem of equilibrium. The conditions for thermodynamic equilibrium will be formulated so as to remain consistent with the concept of mechanical equilibrium.

THE SECOND LAW OF THERMODYNAMICS

The second law gives an answer to the basic problem of thermodynamics. It consists of the following series of statements:

1. All the *homogeneous* phases have an extensive property called entropy:

$$S = S[U, V, N_j] \qquad j = 1, 2, \ldots, k$$

   The entropy is a homogeneous linear function of the internal energy U, the volume V, and the amount N_j of component j of the phase.
2. The entropy S is a continuous, differentiable and monotonically increasing function of internal energy U at any fixed value of V and N_j (Figure 5.1).
3. The entropy S of an *isolated system* has a maximum in the equilibrium state.

Since the entropy is a continuous, monotonic differentiable function of internal energy, volume and amounts of components, it follows that the internal energy

$$U = U[S, V, N_j]$$

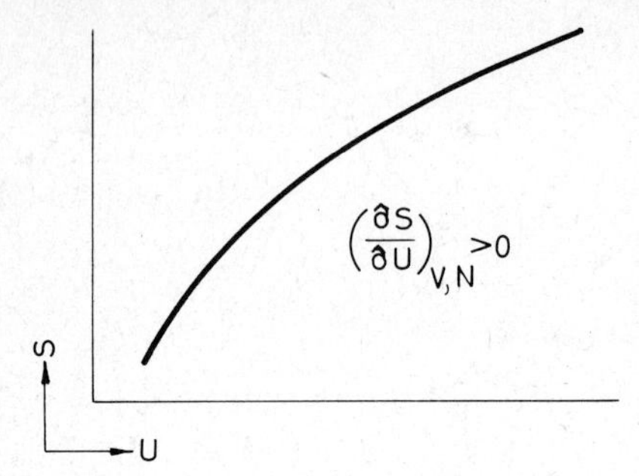

FIGURE 5.1. Entropy is a continuous, differentiable, and increasing function of internal energy.

is a single-valued, continuous, differentiable function of entropy, volume, and amounts of components. Thus, entropy is that *generalized coordinate* which was sought in the preceding chapter.

Entropy is an extensive property. The entropy of a multiphase system is therefore the sum of the entropies of the phases:

$$S = \sum_{\iota=1}^{\varphi} S^{\iota} \qquad \iota = 1, 2, \ldots, \varphi$$

where S^{ι} is the entropy of the ιth phase

$$S^{\iota} = S\left[U^{\iota}, V^{\iota}, N_j^{\iota}\right] \qquad j = 1, 2, \ldots, k$$

Thus, if all the extensive properties are multiplied by the same constant λ, the entropy is also multiplied by the same constant λ:

$$S\left[\lambda U, \lambda V, \lambda N_j\right] = \lambda S\left[U, V, N_j\right]$$

This is true for an arbitrary value of λ.

Let λ be $1/N$, where N is the total mole number:

$$N = \sum_{j=1}^{k} N_j$$

Substituting for λ, we have

$$S = S\left[\frac{U}{N}, \frac{V}{N}, \frac{N_1}{N}, \ldots, \frac{N_k}{N}\right] = \frac{1}{N} S\left[U, V, N_1, \ldots, N_k\right]$$

Molar properties are denoted by the corresponding lower case:

$$u = \frac{U}{N}$$

$$v = \frac{V}{N}$$

$$x_j = \frac{X}{N}$$

$$s = \frac{S}{N}$$

The entropy S of a body containing N moles of substance is

$$S = S\left[U, V, N_j\right] = Ns\left[u, v, x_j\right]$$

THE FUNDAMENTAL EQUATION OF THERMODYNAMICS AND THE INTENSIVE VARIABLES

Internal energy is a differentiable function of entropy, volume, and amount of component N_j:

$$U = U[S, V, N_1, N_2, \ldots, N_k]$$

The total differential of internal energy is

$$dU = \left(\frac{\partial U}{\partial S}\right)_{V, N_j} dS + \left(\frac{\partial U}{\partial V}\right)_{S, N_j} dV + \sum_{j=1}^{k} \left(\frac{\partial U}{\partial N_j}\right)_{S, V, N_i} dN_j$$

$$j = 1, 2, \ldots, k \quad i = 1, 2, \ldots, k \quad i \neq j$$

This is the fundamental equation of thermodynamics. The partial derivatives in

TABLE 5.1
Generalized Thermodynamic Forces

$\left(\frac{\partial U}{\partial S}\right)_{V, N_j} \equiv T$	Thermodynamic temperature
$-\left(\frac{\partial U}{\partial V}\right)_{S, N_j} \equiv P$	Pressure
$\left(\frac{\partial U}{\partial N_j}\right)_{S, V, N_i} \equiv \mu_j$	Chemical potential of the jth component

it have a well defined physical meaning: they are generalized forces, whose conventional names and symbols are given in Table 5.1. It will be shown that the definitions of temperature and pressure given there conform to the concept of absolute temperature and the definition of pressure in mechanics, respectively.

With the notation given in the table the fundamental equation can be written in the following form:

$$dU = T\,dS - P\,dV + \sum_{j=1}^{k} \mu_j\,dN_j$$

Dimension of Entropy

A study of the dimensions of thermodynamical variables yields the following instructive results:

$$[N_j] = \mathrm{N}$$

$$[V] = \mathrm{L}^3$$

$$[U] = \mathrm{ML}^2\Theta^2$$

$$[P] = \mathrm{ML}^{-1}\Theta^{-1}$$

$$[\mu_j] = \mathrm{ML}^2\Theta^2\mathrm{N}^{-1}$$

In the case of these properties the three basic quantities used in mechanics, mass M, length L and time Θ, and the amount of substance are sufficient. However, the factors in the product $T\,dS$, which has the dimensions of energy, cannot be expressed in terms of the four basic quantities mentioned already. Of temperature and entropy one must be considered therefore as an additional basic quantity. Historically it so happened that temperature T was chosen as basic quantity. Accordingly the dimensions of entropy are

$$[S] = \mathrm{ML}^2\Theta^{-2}\mathrm{T}^{-1}$$

The unit of entropy in System International is $\mathrm{J\,K^{-1}}$.

Thermodynamic Equations of State

Temperature, pressure, and chemical potential are all partial derivatives of a function of $S, V, N_1, \ldots, N_k$, and consequently are also functions of

$S, V, N_1, \ldots, N_k$. Thus the following functions must exist:

$$\begin{aligned} T &= T[S, V, N_j] \\ P &= P[S, V, N_j] \qquad j = 1, 2, \ldots, k \\ \mu_j &= \mu_j[S, V, N_j] \end{aligned}$$

Such functions are called *thermodynamic equations of state*, to distinguish them from the empirical equations of state, discussed in Chapter 3. Since the thermodynamic equations of state were obtained by the differentiation of a homogeneous linear function, the equations of state are homogeneous zero-order functions. This means that the multiplication of each independent variable by a scalar λ leaves the function unchanged:

$$T[\lambda S, \lambda V, \lambda N_j] = T[S, V, N_j]$$

Thus, the temperature, pressure, and chemical potential are not additive: they are not extensive properties. To distinguish them from the latter, the properties T, P, μ_j are called *intensive properties*.

ENTROPY REPRESENTATION OF THE FUNDAMENTAL EQUATION OF THERMODYNAMICS

In the foregoing the defining equation of the intensive parameters and the equation of state were obtained by the differentiation of the function $U = U[S, V, N_j]$. It was mentioned, however, that the equation

$$S = S[U, V, N_j]$$

is completely equivalent and also of fundamental importance. We can write for the total differential of entropy

$$dS = \left(\frac{\partial S}{\partial U}\right)_{V, N_j} dU + \left(\frac{\partial S}{\partial V}\right)_{U, N_j} dV + \sum_{j=1}^{k} \left(\frac{\partial S}{\partial N_j}\right)_{U, V, N_i} dN_j$$

Let us find the relations between the partial derivatives obtained in this way and the intensive properties.

From the energy representation of the fundamental equation we have, after rearrangement for the total differential of entropy,

$$dS = \frac{1}{T} dU + \frac{P}{T} dV + \sum_{j=1}^{k} \frac{\mu_j}{T} dN_j$$

TABLE 5.2
Partial Derivatives of Entropy

$\left(\dfrac{\partial S}{\partial U}\right)_{V,N_j}$	$= \dfrac{1}{T}$
$\left(\dfrac{\partial S}{\partial V}\right)_{U,N_j}$	$= \dfrac{P}{T}$
$-\left(\dfrac{\partial S}{\partial N_j}\right)_{U,V,N_i}$	$= \dfrac{M_j}{T}$

Comparing this with our preceding equation, it can be seen that the relations between the partial derivatives and the corresponding intensive properties are as given in Table 5.2.

CONDITIONS OF EQUILIBRIUM EXPRESSED IN TERMS OF INTENSIVE PROPERTIES

Let us now return to the discussion of the basic problem of thermodynamics. In statement 3 of the second law this question was answered by stating that the entropy of an isolated system in equilibrium is maximal. However, the conditions of equilibrium can be formulated also in terms of intensive properties. Let us consider the two-phase isolated system discussed already in the introduction of the first law, where the wall between the phases has special properties. The first and second laws will be applied to this system.

Thermal Equilibrium

Let the wall be impermeable to all components, rigid, and diathermal (Figure 5.2). The entropy of the system is the sum of the entropies of the phases, that is,

$$S = S^{\alpha} + S^{\beta}$$

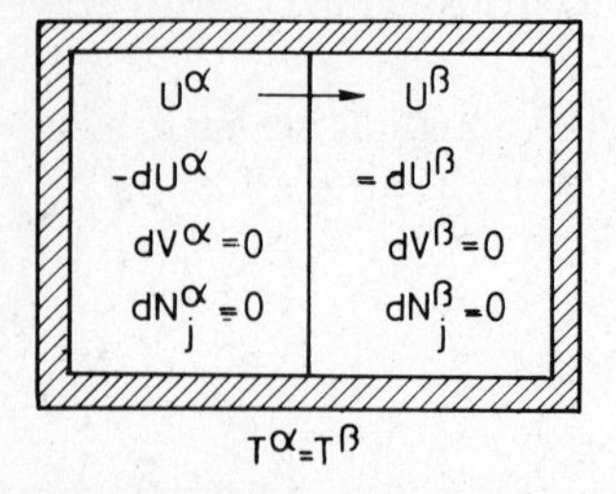

FIGURE 5.2. Isolated system, divided into two subsystems by a diathermal, rigid, impermeable wall.

whence

$$dS = dS^\alpha + dS^\beta$$

Since the volume and the mass of the phases are constant applying the entropy representation of the fundamental equation, the preceding equation can be written in the form

$$dS = \frac{1}{T^\alpha}\,dU^\alpha + \frac{1}{T^\beta}\,dU^\beta$$

The system is isolated, so that its internal energy is constant:

$$dU = dU^\alpha + dU^\beta = 0$$

Hence the energy of one of the phases can increase only at the expense of the other phase:

$$dU^\beta = -dU^\alpha$$

Taking into consideration at the same time that according to the second law the entropy of an isolated system has an extremal value at equilibrium, we have

$$dS = 0$$

$$dS = \left(\frac{1}{T^\alpha} - \frac{1}{T^\beta}\right) dU^\alpha = 0$$

which is possible only if

$$T^\alpha = T^\beta$$

that is, the temperature of the two phases is the same. If the wall is diathermal, *the condition of equilibrium is that the temperature of the two phases shall be equal*. An equilibrium of this kind is called thermal equilibrium. We have thus the condition of thermal equilibrium expressed in terms of an intensive property.

It will be remembered that in the discussion of the zeroth law the condition of thermal equilibrium was formulated as the equality of empirical temperature in the phases in contact with one another. If the temperatures are not equal and temperature is higher in the α phase than in the β phase ($T^\alpha > T^\beta$), then a change begins, according to the second law, during which the entropy of the system increases:

$$dS = \left(\frac{1}{T^\alpha} - \frac{1}{T^\beta}\right) dU^\alpha > 0$$

This is possible only if

$$dU^\alpha < 0$$

that is, if the internal energy of the phase of higher temperature decreases and energy flows through the diathermal wall from the phase of higher temperature to the phase of lower temperature, in accordance with experience. It is evident then that this energy flux continues until the temperature is equalized between the two phases.

IDENTITY OF THERMODYNAMIC TEMPERATURE AND ABSOLUTE TEMPERATURE

It will be proved that the thermodynamic temperature T used above in the discussion of thermal equilibrium is identical with the absolute temperature Θ discussed earlier.

The fundamental equation can be written in the case of $dN_j = 0$ in the following form:

$$dS = \frac{1}{T}\,dU + \frac{P}{V}\,dV$$

Presuming that our statement on the identity of thermodynamic and absolute temperatures is true, we can write that thermodynamic temperature is a function of absolute temperature:

$$T = T[\Theta]$$

The internal energy is a function of state, so that

$$dU = \left(\frac{\partial U}{\partial \Theta}\right)_V d\Theta + \left(\frac{\partial U}{\partial V}\right)_\Theta dV$$

Let us substitute this into the fundamental equation:

$$dS = \frac{1}{T}\left(\left(\frac{\partial U}{\partial \Theta}\right)_V d\Theta + \left(\frac{\partial U}{\partial V}\right)_\Theta dV\right) + \frac{P}{T}\,dV$$

$$= \frac{1}{T}\left(\frac{\partial U}{\partial \Theta}\right)_V d\Theta + \frac{1}{T}\left(\left(\frac{\partial U}{\partial V}\right) + P\right) dv$$

Since dS is a total differential, the second-order mixed derivatives are equal according to Young's theorem. For the perfect gas, used in the definition of the absolute temperature scale,

$$\left(\frac{\partial U}{\partial V}\right)_\Theta = 0$$

and thus

$$\frac{\partial^2 U}{\partial \Theta\, \partial V} = 0$$

Let us form now the second-order mixed derivatives and make them equal:

$$\frac{1}{T}\frac{\partial^2 U}{\partial V\, \partial \Theta} = -\frac{1}{T^2}\frac{dT}{d\Theta}P + \frac{1}{T}\left(\frac{\partial P}{\partial \Theta}\right)_V = 0$$

whence

$$\frac{1}{T}\frac{dT}{d\Theta} = \frac{1}{P}\left(\frac{\partial P}{\partial \Theta}\right)_V$$

Since according to the definition of absolute temperature

$$\left(\frac{\partial P}{\partial \Theta}\right)_V = \frac{P}{\Theta}$$

we can write instead of our previous equation

$$\frac{1}{T}\frac{dT}{d\Theta} = \frac{1}{\Theta}$$

After the separation of the variables we have

$$d \ln T = d \ln \Theta$$

Integration gives

$$T = c\Theta$$

where c is the constant of integration, the multiplier linearly distorting the temperature scale. Herewith it has been proved that thermodynamic temperature and absolute temperature are the same physical property, and no distinction is needed.

Simultaneous Mechanical and Thermal Equilibrium

Let us continue the investigation of equilibrium in the case of a diathermal movable wall, impermeable to all the components (Figure 5.3).

The entropy of the two-phase system is the sum of the entropies of the phases. It can be written for the entropy S of the system that

$$dS = dS^\alpha + dS^\beta$$

Since the wall is impermeable to all the components, terms representing the change in mole number can be omitted in the fundamental equation of both

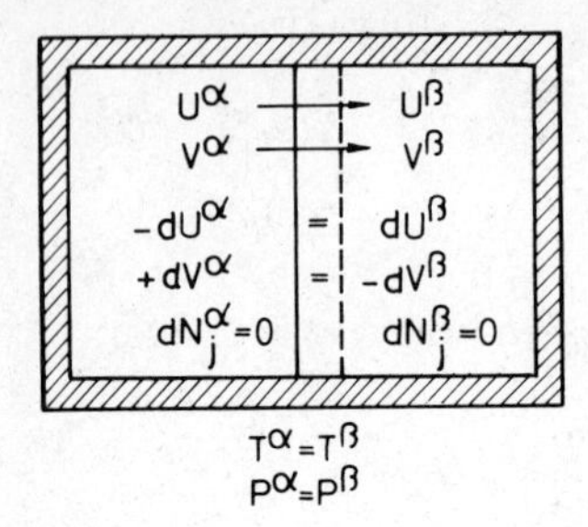

FIGURE 5.3. Isolated system divided into two subsystems by a diathermal, movable, impermeable wall.

phases. After substitution we have

$$dS = \left(\frac{1}{T^\alpha}\,dU^\alpha + \frac{P^\alpha}{T^\alpha}\,dV^\alpha\right) + \left(\frac{1}{T^\beta}\,dU^\beta + \frac{P^\beta}{T^\beta}\,dV^\beta\right)$$

The internal energy and volume of the system are constant:

$$dU^\alpha = -dU^\beta$$

$$dV^\alpha = -dV^\beta$$

According to the second law, the entropy has in the equilibrium state a constrained extremal value, a maximum:

$$dS = \left(\frac{1}{T^\alpha} - \frac{1}{T^\beta}\right) dU^\alpha + \left(\frac{P^\alpha}{T^\alpha} - \frac{P^\beta}{T^\beta}\right) dV^\alpha = 0$$

The first term on the left side is zero because of thermal equilibrium:

$$T^\alpha = T^\beta$$

The second term can be zero only if

$$P^\alpha = P^\beta$$

that is, the pressures of the phases on the two sides of the movable wall are equal. This statement is in conformity with the criterion for mechanical equilibrium. Hereby the justification for the notation introduced in Table 5.1 has been given, namely that $\partial U/\partial V = -P$.

If the temperature of the two phases is identical:

$$T^\alpha = T^\beta = T$$

but the pressure of phase α is higher than that of phase β:

$$P^\alpha > P^\beta$$

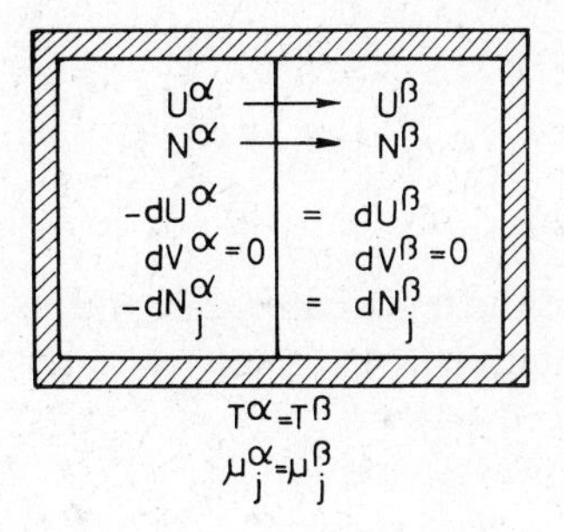

FIGURE 5.4. Isolated system, divided into two subsystems by a diathermal, rigid, permeable wall.

then only those changes can proceed by which the entropy of the system increases:

$$dS = \left(\frac{P^\alpha - P^\beta}{T} \right) dV^\alpha > 0$$

This is possible only if the volume of phase α increases:

$$dV^\alpha > 0$$

In agreement with experience, the phase of higher pressure expands.

Simultaneous Thermal and Component Equilibrium

In the cases discussed so far an impermeable wall has been assumed between the two phases. Let us assume now that the wall is not movable, is diathermal, and is permeable to a single component (Figure 5.4).

The energy of the isolated body is constant, the stock of component j is also constant, and the volume of the phases is invariant:

$$dU^\alpha = -dU^\beta$$

$$dN_j^\alpha = -dN_j^\beta$$

$$dV^\alpha = dV^\beta = 0$$

In equilibrium the entropy of the isolated body takes on an extremal value:

$$dS = \left(\frac{1}{T^\alpha} - \frac{1}{T^\beta} \right) dU^\alpha - \left(\frac{\mu_j^\alpha}{T^\alpha} - \frac{\mu_j^\beta}{T^\beta} \right) dN_j^\alpha = 0$$

The first term in the middle is zero because of thermal equilibrium, as the temperatures of the two phases are equal:

$$T^\alpha = T^\beta = T$$

Accordingly, the second term can be zero only if

$$\mu_j^\alpha = \mu_j^\beta$$

Thus, the condition of equilibrium is the equality of the temperature and of the chemical potential in the two phases. This latter criterion is separately valid for each component that can pass from one phase into the other.

If thermal equilibrium is already established, however, and if

$$\mu_j^\alpha > \mu_j^\beta$$

(i.e., the chemical potential of the component is greater in the α phase than in the β phase), then the entropy of the system increases in conjunction with component transfer:

$$dS = -\left(\frac{\mu_j^\alpha - \mu_j^\beta}{T}\right) dN_j^\alpha > 0$$

It is evident that this is possible only if

$$dN_j^\alpha < 0$$

that is, the jth component flows from the phase of greater potential to the phase of smaller potential through the wall permeable with respect to the component.

THE CONDITION OF EQUILIBRIUM: THE GENERAL CASE

Up to now the conditions of thermodynamic equilibrium have been investigated in two-phase systems, first for thermal equilibrium, then for simultaneous thermal–mechanical equilibrium, and finally for thermal–component equilibrium. Let us now investigate the problem in general for an isolated system of k components and φ phases. The wall between the phases is movable, diathermal, and permeable to each component. The entropy of the isolated body is a function of $(k + 2)\varphi$ variables:

$$S = S\left[U^\iota, V^\iota, N_j^\iota\right] \qquad \iota = 1, 2, \ldots, \varphi \quad j = 1, 2, \ldots, k$$

The condition of equilibrium is that entropy shall be maximal for each possible virtual change, that is,

$$\sum_{\iota} \sum_{j} dS_j^{\iota} = 0$$

Owing to the isolation of the system, there exist equations expressing conservation of energy, volume, and amount of each component.

$$\sum_{\iota=1}^{\varphi} U^{\iota} - U = 0 \qquad U = \text{const}$$

$$\sum_{\iota=1}^{\varphi} V^{\iota} - V = 0 \qquad V = \text{const}$$

$$\sum_{\iota=1}^{\varphi} N_j^{\iota} - N_j = 0 \qquad N_j = \text{const}$$

We seek the constrained extremal value of the function S by Lagrange's method of multipliers, and therefore, the constraint equations multiplied by the yet unknown multipliers are added to the function S. We seek the maximum of this new function:

$$S = \lambda\left(\sum U^{\iota} - U\right) + \omega\left(\sum V^{\iota} - V\right) + \sum_{j=1}^{k} \tau_j\left(\sum N_j^{\iota} - N_j\right)$$

Let us differentiate in turn with respect to each variable:

$$\frac{\partial S^{\alpha}}{\partial U^{\alpha}} + \lambda = 0 \qquad \frac{\partial S^{\alpha}}{\partial V^{\alpha}} + \omega = 0 \qquad \frac{\partial S^{\alpha}}{\partial M_j^{\alpha}} + \tau_j = 0$$

$$\frac{\partial S^{\beta}}{\partial U^{\beta}} + \lambda = 0 \qquad \frac{\partial S^{\beta}}{\partial V^{\beta}} + \omega = 0 \qquad \frac{\partial S^{\beta}}{\partial N_j^{\beta}} + \tau_j = 0 \qquad j = 1, 2, \ldots, k$$

$$\vdots \qquad\qquad \vdots \qquad\qquad \vdots$$

$$\frac{\partial S^{\varphi}}{\partial U^{\varphi}} + \lambda = 0 \qquad \frac{\partial S^{\varphi}}{\partial V^{\varphi}} + \omega = 0 \qquad \frac{\partial S^{\varphi}}{\partial N_j^{\varphi}} + \tau_j = 0$$

Since by definition

$$\frac{\partial S}{\partial U} = \frac{1}{T} \qquad \frac{\partial S}{\partial V} = \frac{P}{T} \qquad \frac{\partial S}{\partial N_j} = \frac{\mu_j}{T}$$

we can write

$$\frac{1}{T^\alpha} + \lambda = 0 \qquad \frac{P^\alpha}{T^\alpha} + \omega = 0 \qquad \frac{\mu_j^\alpha}{T^\alpha} + \tau_j = 0$$

$$\frac{1}{T^\beta} + \lambda = 0 \qquad \frac{P^\beta}{T^\beta} + \omega = 0 \qquad \frac{\mu_j^\beta}{T^\beta} + \tau_j = 0$$

$$\vdots \qquad\qquad \vdots \qquad\qquad \vdots$$

$$\frac{1}{T^\varphi} + \lambda = 0 \qquad \frac{P^\varphi}{T^\varphi} + \omega = 0 \qquad \frac{\mu_j^\varphi}{T^\varphi} + \tau_j = 0$$

Evaluating the multipliers from the equations in the first row and substituting them into the remaining equations, we have

$$\frac{1}{T^\alpha} = \frac{1}{T^\beta} = \cdots = \frac{1}{T^\varphi} = \frac{1}{T}$$

$$\frac{P^\alpha}{T^\alpha} = \frac{P^\beta}{T} = \cdots = \frac{P^\varphi}{T} = \frac{P}{T}$$

$$\frac{\mu_j^\alpha}{T} = \frac{\mu_j^\beta}{T} = \cdots = \frac{\mu_j^\varphi}{T} = \frac{\mu_j}{T}$$

The deduction shown does not say anything really new with respect to its physical content. However, what it does say is general, and has the merit of not concealing the application of Lagrange's method of multipliers (as do the deductions for the preceding special cases), but showing the method clearly and distinctly. Moreover, the physical content of the multipliers is also revealed.

THE EULER RELATION

It follows from the properties of homogeneous linear functions that for any λ

$$U\left[\lambda S, \lambda V, \lambda N_j\right] = \lambda U\left[S, V, N_j\right] \qquad j = 1, 2, \ldots, k$$

Differentiating both sides of the function with respect to λ, with the application of the chain rule, we have

$$\frac{\partial}{\partial[\lambda S]} U[\lambda S, \lambda V, \lambda N_j] \frac{\partial[\lambda S]}{\partial \lambda} + \frac{\partial}{\partial[\lambda V]} U[\lambda S, \lambda V, \lambda N_j] \frac{\partial[\lambda V]}{\partial \lambda}$$

$$+ \sum_{j=1}^{k} \frac{\partial}{\partial[\lambda N_j]} U[\lambda S, \lambda V, \lambda N_j] \frac{\partial[\lambda N_j]}{\partial \lambda} = U[S, V, N_j]$$

whence

$$\frac{\partial}{\partial[\lambda S]} U[\lambda S, \lambda V, \lambda N_j] S + \frac{\partial}{\partial[\lambda V]} U[\lambda S, \lambda V, \lambda N_j] V$$

$$+ \sum_{j=1}^{k} \frac{\partial}{\partial[\lambda N_j]} U[\lambda S, \lambda V, \lambda N_j] N_j = U[S, V, N_j]$$

Since the relation is true for any λ, it is true for the case $\lambda = 1$, so that we can write

$$\left(\frac{\partial U}{\partial S}\right) S + \left(\frac{\partial U}{\partial V}\right) V + \sum_{j=1}^{k} \left(\frac{\partial U}{\partial N_j}\right) N_j = U$$

which gives, on consideration of the definition of the intensive properties,

$$U = TS - PV + \sum_{j=1}^{k} \mu_j N_j$$

This is Euler's relation.

THE GIBBS – DUHEM RELATION: THE DEGREES OF FREEDOM OF HOMOGENEOUS PHASES

It was shown in the preceding that the conditions for equilibrium can be well formulated in terms of the intensive variables (T, P, μ_j). Now the question arises what exists between the intensive variables. The differential form of this relation can be obtained on the basis of the following considerations.

Let us differentiate the Euler relation:

$$dU = T\,dS + S\,dT - P\,dV - V\,dP + \sum_{j=1}^{k} \mu_j\,dN_j + \sum_{j=1}^{k} N_j\,d\mu_j$$

The energy form of the fundamental equation is well known:

$$dU = T\,dS - P\,dV + \sum_{j=1}^{k} \mu_j\,dN_j$$

These two latter equations can be simultaneously true only if

$$S\,dT - V\,dP + \sum_{j=1}^{k} N_j\,d\mu_j = 0$$

This is the very relation sought, and is called the Gibbs–Duhem relation.

For a homogeneous single-component phase the Gibbs–Duhem equation reduces to

$$S\,dT - V\,dP + N\,d\mu = 0$$

and for one mole of the phase to

$$d\mu = -s\,dT + v\,dP$$

which says how the chemical potential depends on the temperature and pressure. Of these three intensive properties two can change independently, and their change determines the change in the third intensive property.

Let us call the number of intensive variables that can change independently the number of degrees of freedom of the system. Accordingly, the number of degrees of freedom of a homogeneous single-component system is two.

For a homogeneous two-component phase the Gibbs–Duhem relation is

$$S\,dT - V\,dP + N_1\,d\mu_1 + N_2\,d\mu_2 = 0$$

Let us divide the equation by $N_1 + N_2$, that is to say, apply the Gibbs–Duhem relation to one mole of a two-component mixture:

$$s\,dT - v\,dP + x_1\,d\mu_1 + x_2\,d\mu_2 = 0$$

where x means mole fraction. In this equation the number of intensive variables is four, of which three can be freely chosen. The number of degrees of freedom of the two-component homogeneous phase is three, in accordance with the general principle that the number of degrees of freedom of a homogeneous phase of k components is $k + 1$.

DEGREES OF FREEDOM OF HETEROGENEOUS SYSTEMS

Let us extend the foregoing considerations to multiphase equilibrium systems, and denote the number of phases in a system of k components by φ. The Gibbs–Duhem relation is separately valid for each phase:

$$S^{\iota}dT - V^{\iota}dP + \sum_{j=1}^{k} N^{\iota}d\mu_j = 0 \qquad \iota = 1, 2, \ldots, \varphi \quad j = 1, 2, \ldots, k$$

The $k + 2$ intensive variables T, P, and μ_j have not been provided with an index corresponding to the phases in our system of equations, as indeed, according to the known criteria of phase equilibrium, these quantities must be the same for all phases. This means that the number of the equations in the system consisting of the Gibbs–Duhem equations is φ, whereas the number of variables is $k + 2$; thus the system of equations becomes determinate if $k + 2 - \varphi$ of the variables are fixed. This mathematical requirement has the physical content that, consistently with phase equilibrium, the values of

$$F_{\mathrm{G}} = k + 2 - \varphi$$

intensive properties can be freely chosen. This number F_{G} is called the Gibbs degrees of freedom for phase equilibrium.

6

THERMODYNAMIC POTENTIAL FUNCTIONS

In the preceding discussion we got acquainted with the following thermodynamic properties:

$$U, V, S, P, T, N_j, \mu_j$$

Of these, four—V, S, P, T—will be called fundamental thermodynamic properties. Of the fundamental thermodynamic properties V and S are generalized coordinates, while P and T are generalized forces. The amount of substance, N_j, is a generalized coordinate, while the chemical potential μ_j is a generalized force. The internal energy U is, physically speaking, a potential.

It was shown in the preceding chapter that the conditions for thermal, mechanical, and component equilibria can be formulated in terms of their respective generalized forces, that is, in terms of intensive properties. Here the question arises, how to formulate the conditions for equilibrium in special cases, namely, after the fixing of the temperature and/or the pressure. Potential functions corresponding to these special cases are sought, and we expect on the basis of an analogy with mechanics that these will display a minimum at special values of the forces. The special cases at issue are called:

1. Isothermal equilibrium.
2. Isobaric equilibrium.
3. Isothermal–isobaric equilibrium.

It was stated in the preceding chapter that the internal energy, volume, and amounts of chemical components remain constant in an isolated body under any transformation. The body does however have one or more unconstrained properties, which can and do change from their values in the initial state, and eventually take on equilibrium values that maximize entropy.

Consider now the body discussed, and suppose further that it is closed and has a rigid but diathermal wall. This makes it possible to change the state of the body by choosing the external body with which it is put in contact. Utilizing this possibility, let us adjust the entropy of the body to that value which it would take on in equilibrium. Then the unconstrained properties of the body will also change, to take on sooner or later their equilibrium values. In the equilibrium state the internal energy of the body will be minimized at the prescribed entropy value. This remarkable statement follows from the second law.

THE ENERGY MINIMUM PRINCIPLE

Let X be any unconstrained property of the body. According to the first law,

$$f[S, U, X] = 0$$

In accordance with the mathematical relation given in Appendix A of chapter 4,

$$\left(\frac{\partial U}{\partial X}\right)_S \left(\frac{\partial S}{\partial U}\right)_X \left(\frac{\partial X}{\partial S}\right)_U = -1$$

and after rearrangement,

$$\left(\frac{\partial U}{\partial X}\right)_S = -\frac{\left(\frac{\partial S}{\partial X}\right)_U}{\left(\frac{\partial S}{\partial U}\right)_X}$$

In equilibrium, according to the second law, the numerator of the right side is given by

$$\left(\frac{\partial S}{\partial X}\right)_U = 0$$

Hence the left side must also be zero:

$$\left(\frac{\partial U}{\partial X}\right)_S = 0$$

That is, at a given entropy value the internal energy of the system has an extremal value in the equilibrium state of the system. As can easily be seen from Figure 6.1, this extremal value is a minimum.

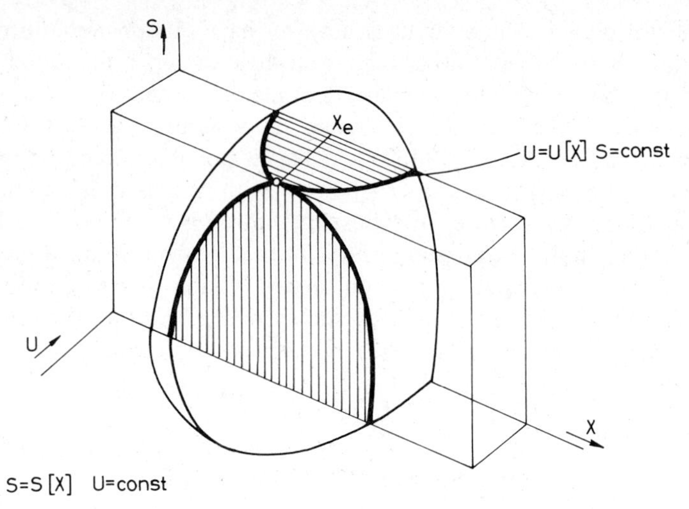

FIGURE 6.1. $[U, S, X]$ surface.

Thus, the foregoing principle can be formulated in two alternative ways:

1. An unconstrained property X of the body takes on in equilibrium a value such as to maximize the entropy at a given internal energy.
2. An unconstrained property X of the body takes on in equilibrium a value such as to minimize the internal energy of the body at a given entropy.

This latter formulation is called the *energy minimum principle*.

Entropy and temperature are two alternative physical properties for use in the description of thermal equilibrium. Entropy is a generalized coordinate, temperature a generalized force. This means that one can use, on one hand, the function

$$U = U_S[S, V, N_j]$$

or on the other hand, for the description of the same state, the function

$$U = U_T[T, V, N_j]$$

This is a coordinate transformation from the (U, S) to the (U, T) coordinate system, without the participation of the generalized coordinates V and N_j.

Transformations of this kind are systematically carried out by the Legendre transformation.

THE LEGENDRE TRANSFORMATION

Let us consider now the relation

$$y = y[x]$$

which is shown in the (x, y) plane in Figure 6.2. Let us draw a tangent to any arbitrary point of the curve. The slope of the tangent is the derivative at point P:

$$m = \frac{dy}{dx}$$

The intercept of the tangent is ψ. If point P passes along the curve $y = y[x]$, then there is a tangent for each point, and a ψ intercept for each tangent, that is,

$$\psi = \psi[m]$$

This function too describes the course of the curve, and is therefore equivalent to the function

$$y = y[x]$$

but in the second case m is the argument of the function ψ. The relation is transformed from the (y, x) plane to the (ψ, m) plane. This kind of transformation is called a Legendre transformation, that is, $\psi[m]$ is the Legendre

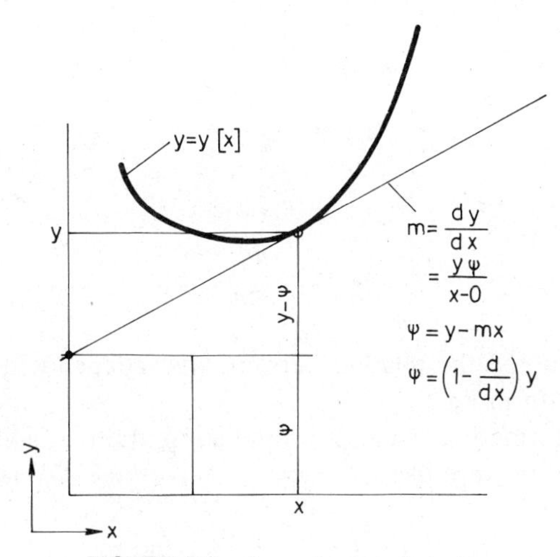

FIGURE 6.2. Legendre transformation.

transform of $y[x]$. It can be clearly seen from the figure that the derivative is

$$m = \frac{dy}{dx} = \frac{y - \psi}{x - 0}$$

whence

$$\psi = y - mx$$

If the derivative is substituted for m, and y is factored out, the following relation is obtained:

$$\psi = \left(1 - x\frac{d}{dx}\right)y$$

The expression in parentheses is the operator of the Legendre transformation. Alternatively this expression can be written in the following form:

$$\psi[m] = L\begin{bmatrix} y \\ x \end{bmatrix}$$

This means that if a Legendre transformation is made on the function $y = y[x]$, function $\psi[m]$ is obtained. Let us differentiate the relation $\psi = y - mx$:

$$d\psi = dy - m\,dx + x\,dm$$

However,

$$dy = m\,dx$$

Substituting this into the preceding expression, we have

$$d\psi = -x\,dm$$

whence

$$x = -\frac{d\psi}{dm}$$

Thus we have obtained the relation between the independent variable x and the Legendre transform ψ.

Legendre transformation can also be applied to multivariable functions, and with respect to several variables:

$$\psi = L\begin{bmatrix} U \\ X_1, X_2, \ldots, X_r \end{bmatrix}$$

or, with the Legendre transformation operator written out,

$$\psi = \left(1 - \sum^{r \le \nu} X_i \frac{\partial}{\partial X_i}\right) U$$

The generalized potential ψ generated with Legendre transformation has in the case of r independent variables the following form:

$$\psi = \psi[m_0, m_1, \ldots, m_r, X_{r+1}, X_{r+2}, \ldots, X_\nu]$$

This involves r independent generalized coordinates:

$$m_i = \left(\frac{\partial U}{\partial X_i}\right)_{X_j} \qquad i \neq j$$

whereas the independent variables $X_{r+1}, X_{r+2}, \ldots, X_p$ do not enter into the operator. After the substitution of the generalized coordinates we have

$$\psi = U - \sum_{i=0}^{r<\nu} m_i X_i$$

Let us differentiate the two expressions for ψ:

$$d\psi = \sum_{i=0}^{r} \left(\frac{\partial \psi}{\partial m_i}\right) dm_i + \sum_{i=r+1}^{\nu} \left(\frac{\partial \psi}{[\partial X_i]}\right) dX_i$$

where the brackets $[\cdot]$ indicate the original non-transformed variables, and

$$d\psi = dU - \sum_{i=0}^{r} X_i \, dm_i - \sum_{i=0}^{r} m_i \, dX_i \tag{A}$$

However, here

$$dU = \sum_{i=0}^{r} \left(\frac{\partial U}{\partial X_i}\right) dX_i + \sum_{i=r+1}^{\nu} \left(\frac{\partial U}{[\partial X_i]}\right) dX_i$$

and substituting this into the latter equation we have

$$d\psi = \sum_{i=0}^{r} \left(\frac{\partial U}{\partial X_i}\right) dX_i + \sum_{i=r+1}^{\nu} \left(\frac{\partial U}{[\partial X_i]}\right) dX_i - \sum_{i=0}^{r} X_i \, dm_i - \sum_{i=r+1}^{\nu} m_i \, dX_i \tag{B}$$

The difference of equations (A) and (B) is

$$\sum_{i=0}^{r}\left(m_i - \frac{\partial U}{\partial X_i}\right) dX_i + \sum_{i=r+1}^{\nu}\left(\frac{\partial \psi}{[\partial X_i]} - \frac{\partial U}{[\partial X_i]}\right) dX_i$$

$$-\sum\left(\frac{\partial \psi}{\partial m_i} + X_i\right) dm_i = 0$$

Since m_i and X_i are independent variables in this equation, the quantities in parentheses must vanish. The following three cases occur.

First,

$$m_i = \left(\frac{\partial U}{\partial X_i}\right)_{X_j} \qquad 0 \le i \le r$$

This does not say anything new; it corresponds to the definition of m_i. Anyway, m_i is a homogeneous zero-order function, corresponding to some intensive property. In its physical nature it is a generalized force, the partial derivative of the generalized potential in the generalized coordinate system.

Secondly,

$$X_i = -\left(\frac{\partial \psi}{\partial m_i}\right)_{m_j} \qquad 0 \le i \le r$$

This is a more general expression for the relation between the independent variable X_i and the Legendre transform ψ.

Thirdly,

$$\left(\frac{\partial \psi}{[\partial X_i]}\right)_{X_j} = \left(\frac{\partial U}{[\partial X_i]}\right)_{Xj} \qquad i > r$$

Here we have a relation between the partial derivatives on the one hand of the internal energy U, and on the other hand of the Legendre transform ψ, with respect to the same independent variable not participating in the transformation.

Let us carry out the Legendre transformation on internal energy U with respect to the following independent variables:

1. S.
2. V.
3. S and V.

The parameters N_j do not participate in the transformation in any of the three cases. After transformation let us also write out the three relations given above. The results are shown in Table 6.1.

TABLE 6.1
Generating New Thermodynamic Potentials

General Relationship	Free Energy	Enthalpy	Free Enthalpy
$\psi = L\left[\begin{matrix} U \\ X_0, X_1 \end{matrix}\right]$	$\psi = L\left[\begin{matrix} U \\ S \end{matrix}\right]$	$\psi = L\left[\begin{matrix} U \\ V \end{matrix}\right]$	$\psi = L\left[\begin{matrix} U \\ S, V \end{matrix}\right]$
$= \left(1 - \sum X_I \frac{\partial}{\partial X_i}\right) U$	$= \left(1 - S\frac{\partial}{\partial S}\right)_V U$	$= \left(1 - V\frac{\partial}{\partial V}\right)_S U$	$= \left(1 - S\frac{\partial}{\partial S} - V\frac{\partial}{\partial V}\right) U$
$m_I = \left(\frac{\partial U}{\partial X_i}\right)_{X_j}$	$T = \left(\frac{\partial U}{\partial S}\right)_V$	$P = -\left(\frac{\partial U}{\partial V}\right)_S$	
$\psi = U - \sum m_I X_i$	$F \equiv U - TS$	$H \equiv U + PV$	$G \equiv U - TS + PV$
$X_i = -\left(\frac{\partial \psi}{\partial m_i}\right)_{m_j}$	$\left(\frac{\partial F}{\partial T}\right)_V = -S$	$\left(\frac{\partial H}{\partial P}\right)_S = V$	$\left(\frac{\partial G}{\partial T}\right)_P = -S$
			$\left(\frac{\partial G}{\partial P}\right)_S = V$
$\left(\frac{\partial \psi}{[\partial X_i]}\right)_{X_j} = \left(\frac{\partial U}{[\partial X_i]}\right)_{X_j}$	$\left(\frac{\partial F}{\partial V}\right)_T = \left(\frac{\partial U}{\partial V}\right)_S = -P$	$\left(\frac{\partial H}{\partial S}\right)_P = \left(\frac{\partial U}{\partial S}\right)_V = T$	$\left(\frac{\partial G}{\partial V}\right)_T = \left(\frac{\partial U}{\partial V}\right)_S = -P$
			$\left(\frac{\partial G}{\partial S}\right)_P = \left(\frac{\partial U}{\partial S}\right)_V = T$

With the Legendre transformation three thermodynamic potentials have been generated. These are the free energy, enthalpy, and free enthalpy. The question now is, what is the physical content of these three potential functions? Do they comply with the requirements for expressing equilibrium in special cases, that is, at constrained values of the forces, by minimization?

FREE ENERGY

Let us investigate the problem of equilibrium in the special case in which the body has rigid walls and is in thermal contact with a heat bath. The *heat bath* is a body of constant volume and constant composition, and of such large mass that its temperature is not appreciably changed by the energy transfers involved in equilibration. The combination of the two-phase body and the heat bath will be called the complex system. The complex system is isolated from its surroundings. Whatever its history, the condition for the establishment of equilibrium as expressed by the energy minimum principle is that the total integral energy of the body and the heat bath shall be minimized:

$$d(U + U^{\sigma}) = 0$$

$$d^2(U + U^{\sigma}) > 0$$

The total entropy is constant, which follows from the fact that the complex system is thermally isolated from the surroundings and is in equilibrium:

$$d(S + S^{\sigma}) = 0$$

where the superscript σ refers to the heat bath. The same temperature prevails in the phases of the body and in the heat bath, according to the already known condition for thermal equilibrium:

$$T = T^{\sigma}$$

On this basis, we now ask, what is the condition for any other kind of equilibrium in the body, once thermal equilibrium is established?

To determine this, let us return to our initial equation, which can be written in the following form:

$$d(U + U^{\sigma}) = dU + T^{\sigma} dS^{\sigma} = 0$$

Since the entropy of the equilibrated complex system is constant,

$$dS^{\sigma} = -dS$$

and we can write

$$dU - T^o dS = 0$$

However, the temperature T in the complex system is constant, so that factoring out the differential operator, we have

$$d(U - T^o S) = d(U - TS)$$

Though the expression in parentheses contains the properties of the *body*, it should not be forgotten that it yields the internal energy of the *complex system*. We know, however, that in equilibrium this is bound to be a minimum, that is,

$$d^2(U - TS) > 0$$

This says that the function $U - TS \equiv F$—that is, the free energy found with the Legendre transformation—is minimized in the equilibrium state. The equilibrium value of any unconstrained property of the body minimizes the free energy of the body at fixed temperature.

Let us now investigate the special case of an isothermal body of constant volume in which the phases are separated. The condition for equilibrium is the minimization of the free energy of the body:

$$dF = \sum_{\iota=1}^{\varphi} dF^{\iota} = 0$$

The free energy of one open phase is

$$dF^{\iota} = S^{\iota} dT^{\iota} - P^{\iota} dV^{\iota} + \sum_{j=1}^{k} \mu_j^{\iota} dN_j^{\iota} \qquad \iota = 1, 2, \ldots, \varphi \quad j = 1, 2, \ldots, k$$

It was stipulated that all the changes in the body are isothermal; hence $dT^{\iota} = 0$, and the total volume of the body is constant:

$$dV^{\iota} = \sum_{\iota=1}^{\varphi} dV^{\iota} = 0$$

Moreover, the amount of all the components in the body is constant:

$$dN_j = \sum_{\iota=1}^{\varphi} dN_j^{\iota} = 0$$

Under these constraints, the free energy of the body is to be minimized:

$$F = \sum_{\iota=1}^{\varphi} \sum_{j=1}^{k} \mu_j^{\iota} N_j^{\iota} = \min$$

where

$$\mu_j^{\iota} = f\left[T, P, N_j^{\iota}\right]$$

At the constrained minimum of free energy, it follows trivially that

$$\mu_j^{(\iota)} = \mu_j^{(\iota+1)}$$

It follows that there are two approaches to the calculation of the equilibrium composition:

1. On the basis of the equality $\mu_j^{\iota} = \mu_j^{\iota+1}$.
2. On the basis of minimum free energy.

Though the routes to the two solutions are different, they necessarily yield identical equilibrium compositions. In both approaches a suitable explicit expression for the chemical potential must be used. For the present we can say nothing about such expressions, but we will return to them in Chapter 11 on mixtures. Problems of vapor–liquid equilibrium will be discussed in Chapter 12.

THE ENTHALPY

Let us bring the body investigated into contact, through an adiabatic, impermeable, movable wall, with a manostat. The manostat is a system of such large mass that its pressure is not changed appreciably by the resulting changes of its volume. The combination of the body investigated and the manostat will be called the complex system. The complex system is isolated from its surroundings.

Applying the energy minimum principle, we can write for the complex system

$$d(U + U^{\sigma}) = 0$$

$$d^2(U + U^{\sigma}) > 0$$

the superscript referring in this case to the manostat.

The evident condition of isobaric equilibrium is the identity of pressure in the complex system:

$$P = P^{\sigma}$$

On account of the movable wall,

$$d(V + V^{\sigma}) = 0$$

On the other hand, because of the adiabatic wall,

$$dS = 0$$

$$dS^{\sigma} = 0$$

For the internal energy of the manostat we have that

$$dU^{\sigma} = -P\,dV^{\sigma}$$

$$= P\,dV$$

The total internal energy of the complex system is invariant:

$$d(U + U^{\sigma}) = dU + P\,dV = 0$$

$$= d(U + PV) = 0$$

$$= dH = 0$$

and so

$$d^2H > 0$$

where

$$H \equiv U + PV$$

is the enthalpy found with the Legendre transformation.

It can be stated that the equilibrium value of any unconstrained property of the body minimizes the enthalpy of the body at constant pressure.

We do not give here an example of the application of the enthalpy function. This will be done in great detail in Chapter 8.

THE FREE ENTHALPY

Let us investigate now a body which is simultaneously in contact with a heat bath and a manostat. The wall adjoining the manostat is movable, and the wall adjoining the heat bath is diathermal.

The energy minimum principle written for the complex system is

$$d(U + U^{\sigma}) = 0$$

$$d^2(U + U^{\sigma}) > 0$$

The two evident conditions of isothermal–isobaric equilibrium are that temperatures and pressures shall be respectively identical in the complex system:

$$T = T^{\sigma}$$

$$P = P^{\sigma}$$

On the other hand,

$$dS + dS^{\sigma} = 0$$

$$dV + dV^{\sigma} = 0$$

The internal energy of the heat-bath is

$$\begin{aligned} dU^{\sigma} &= T\,dS^{\sigma} - P\,dV^{\sigma} \\ &= -T\,dS + P\,dV \end{aligned}$$

The total internal energy of the complex system is

$$\begin{aligned} d(U + U^{\sigma}) &= dU - T\,dS + P\,dV = 0 \\ &= d(U - TS + PV) = 0 \\ &= d(H - TS) = 0 \\ &= dG = 0 \end{aligned}$$

and thus

$$d^2G > 0$$

The free enthalpy obtained by Legendre transformation is thus seen to be

$$G \equiv H - TS$$

The equilibrium value of some unconstrained internal variable X of the body minimizes the free enthalpy of the body at given constant pressure and constant temperature.

Now let us investigate the special case in which a chemical reaction proceeds in the closed body under isothermal, isobaric conditions. What will be the condition of chemical equilibrium? We can say that the minimum of free enthalpy

$$dG = 0$$

We will return to the special case of chemical equilibrium in Chapter 13.

Isobaric and Isochoric Molar Heat Capacities

From the thermodynamic functions introduced in the preceding discussion further functions can be defined. Of these the following two are of particular importance:

1. Heat capacity at constant volume,

$$C_V = \left(\frac{\partial U}{\partial T}\right)_V$$

2. Heat capacity at constant pressure,

$$C_p = \left(\frac{\partial H}{\partial T}\right)_P$$

Molar heat capacities are properties that in practice are easy to handle, calorimetrically measurable, and calculable from spectroscopic data, so they will be often used in the following.

RELATIONS BETWEEN THERMODYNAMIC FUNCTIONS

In the preceding discussion we got acquainted with four thermodynamic potentials

$$U, H, F, G$$

with four fundamental variables

$$V, P, S, T$$

TABLE 6.2
Thermodynamic Potentials

Integrated Form	Differential Form
$U = U[S, V, N_j]$	$dU = T\,dS - P\,dV + \sum \mu_j\,dN_j$
$H \equiv U + PV = H[S, P, N_j]$	$dH = T\,dS + V\,dP + \sum \mu_j\,dN_j$
$F \equiv U - TS = F[T, V, N_j]$	$dF = -S\,dT - P\,dV + \sum \mu_j\,dN_j$
$G \equiv H - TS = G[T, P, N_j]$	$dG = -S\,dT + V\,dP + \sum \mu_j\,dN_j$

TABLE 6.3
Partial Derivatives of the Potentials

U	H	F	G
$T = \left(\frac{\partial U}{\partial S}\right)_{V,N_j}$	$T = \left(\frac{\partial H}{\partial S}\right)_{P,N_j}$	$-S = \left(\frac{\partial F}{\partial T}\right)_{V,N_j}$	$-S = \left(\frac{\partial G}{\partial T}\right)_{P,N_j}$
$-P = \left(\frac{\partial U}{\partial V}\right)_{S,N_j}$	$V = \left(\frac{\partial H}{\partial P}\right)_{S,N_j}$	$-P = \left(\frac{\partial F}{\partial V}\right)_{T,N_j}$	$V = \left(\frac{\partial G}{\partial P}\right)_{T,N_j}$
$\mu_j = \left(\frac{\partial U}{\partial N_j}\right)_{V,S,N_i}$	$\mu_j = \left(\frac{\partial H}{\partial N_j}\right)_{P,S,N_i}$	$\mu_j = \left(\frac{\partial F}{\partial N_j}\right)_{V,T,N_i}$	$\mu_j = \left(\frac{\partial G}{\partial N_j}\right)_{P,T,N_i}$

and with the amounts of components and the chemical potentials,

$$N_j, \mu_j \qquad j = 1, 2, \ldots, k$$

Several relationships were found between these characteristics. Let us summarize these in tables.

The Conjugate Variables

It is shown clearly by Table 6.2 that

$$P \text{ and } V$$
$$T \text{ and } S$$
$$\mu_j \text{ and } N_j$$

occur together. Of these pairs, (p, V) is in close relationship with mechanical equilibrium and energy transfer in the form of mechanical work, (T, S) with thermal equilibrium, and (μ_j, N_j) with phase equilibrium and component transfer.

Table 6.3 contains the partial derivatives of the potentials. Simple comparison with Table 6.2 yields four further important relations (Table 6.4)

TABLE 6.4
Relations Between Potentials

$\left(\frac{\partial U}{\partial S}\right)_{V,N_j} = \left(\frac{\partial H}{\partial S}\right)_{P,N_j}$	$\left(\frac{\partial F}{\partial T}\right)_{V,N_j} = \left(\frac{\partial G}{\partial T}\right)_{P,N_j}$
$\left(\frac{\partial U}{\partial V}\right)_{S,N_j} = \left(\frac{\partial F}{\partial V}\right)_{T,N_j}$	$\left(\frac{\partial H}{\partial P}\right)_{S,N_j} = \left(\frac{\partial G}{\partial P}\right)_{T,N_j}$

TABLE 6.5
Maxwell Relations

$\left(\frac{\partial T}{\partial V}\right)_{S,N_j} = -\left(\frac{\partial P}{\partial S}\right)_{V,N_j}$	$\left(\frac{\partial S}{\partial V}\right)_{T,N_j} = \left(\frac{\partial P}{\partial T}\right)_{V,N_j}$
$\left(\frac{\partial T}{\partial P}\right)_{S,N_j} = \left(\frac{\partial V}{\partial S}\right)_{P,N_j}$	$\left(\frac{\partial S}{\partial P}\right)_{T,N_j} = -\left(\frac{\partial V}{\partial T}\right)_{P,N_j}$

The Maxwell Relations

Further relations can be obtained between the four fundamental variables if we consider that the order of differentiation can be transposed in the taking of second derivatives. Thus, for example, we have from the first column of the second table

$$\frac{\partial}{\partial V}T = \frac{\partial}{\partial V}\left(\frac{\partial U}{\partial S}\right) = \frac{\partial^2 U}{\partial V\,\partial S}$$

and

$$-\frac{\partial}{\partial S}P = \frac{\partial}{\partial S}\left(\frac{\partial U}{\partial V}\right) = \frac{\partial^2 U}{\partial S\,\partial V}$$

whence

$$\left(\frac{\partial T}{\partial V}\right)_S = -\left(\frac{\partial P}{\partial S}\right)_V$$

From each of the other three columns another expression can be obtained by a similar procedure. Thus finally we have the four *Maxwell relations* between the derivatives containing only the four fundamental variables (Table 6.5). The symmetry of the expressions is again evident.

Owing to their subordinate role, Maxwell relations containing also chemical potentials are not discussed here.

Practical Rule for Remembering Thermodynamic Relations

With the aid of Figure 6.3 thermodynamic relations can be kept in mind or reproduced without lengthy deduction.

The sides of the square are labeled with the energy functions, the corners of the square with the fundamental variables. Conjugate variables are always located at opposite corners. Thus each energy function is flanked by two fundamental variables: the two of which it is a function. The two further conjugate variables contained in the energy equation are automatically obtained, and for these the sign given in the figure is valid. For example, in the case of the free energy, T and V flank F, so the independent variables are T

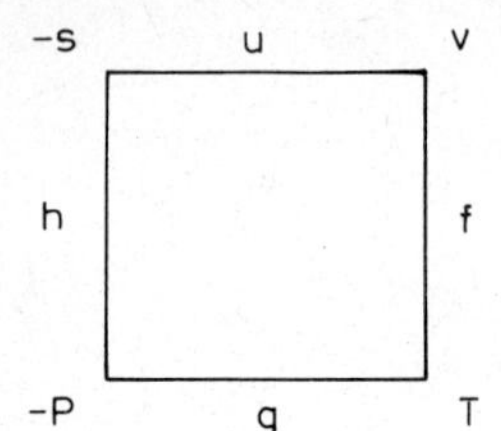

FIGURE 6.3. The magic square.

and V, while the conjugate variables with the indicated sign are $-S$ and $-p$; thus the relation sought is

$$dF = -S\,dT - P\,dV$$

in agreement with the Table 6.2.

The relations in the Table 6.3 are obtained with the following rule: Each energy function has two derivatives with respect to the fundamental variables. These variables flank the energy function. According to whether we differentiate with respect to one or the other variable, the differential quotient is always equal to the conjugate variable at the opposite corner, the sign of which is decisive (Figure 6.4).

The relations in the Table 6.4 are obtained with the same rule (Figure 6.5).

If we are looking for the Maxwell relations, only the four fundamental variables at the corners of the square are taken into consideration. The variables contained in the partial-derivative on one side of the Maxwell relation are written in the order in which they are encountered when passing around the square (Figure 6.6). The variables contained in the partial-derivative on the other side of the Maxwell relation are written in the order in which they are encountered when starting from the fourth corner of the square (the one not yet used) and passing in the opposite direction around the square, taking the sign into consideration.

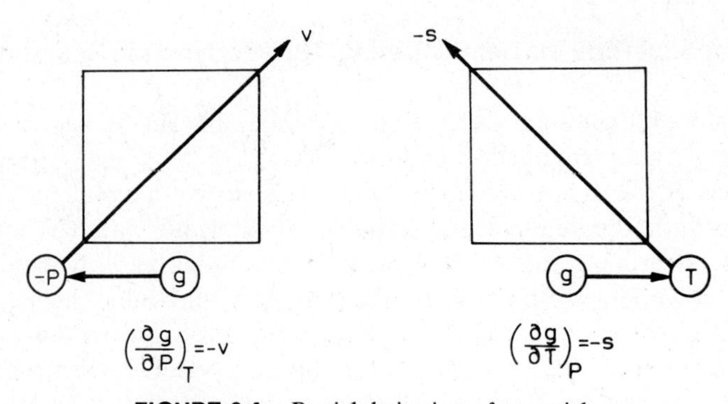

FIGURE 6.4. Partial derivatives of potentials.

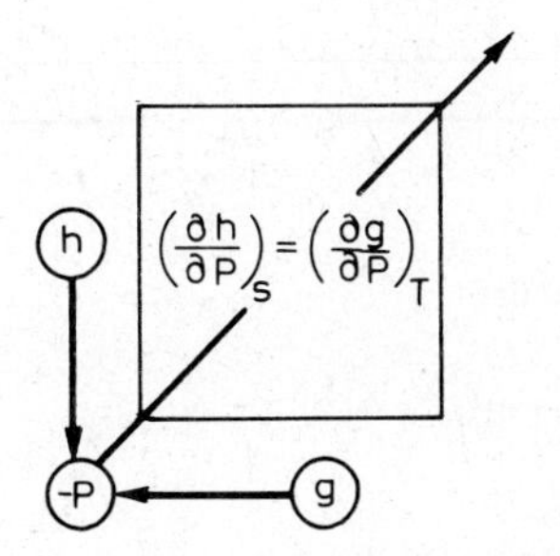

FIGURE 6.5. Relations between potentials.

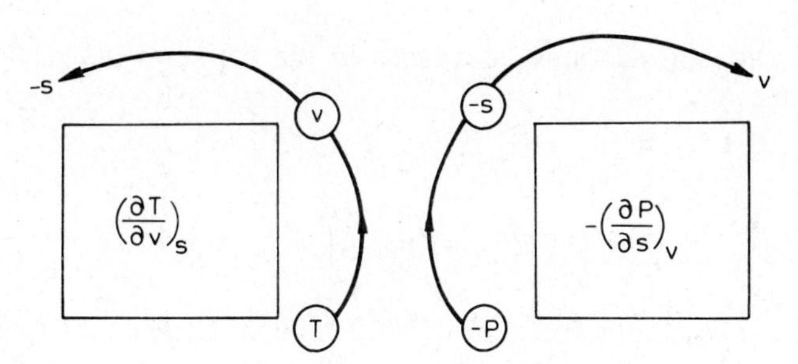

FIGURE 6.6. Maxwell relations.

Generation of Thermodynamic Relations

The few rules of thumb discussed in the foregoing do not give a sufficiently general answer to the question how to generate a desired thermodynamic relation. Remember that for the eight thermodynamic variables U, F, H, G, T, S, P, V, there are $8 \times 7 \times 6 = 336$ kinds of partial derivatives of the type

$$\left(\frac{\partial X}{\partial Y}\right)_Z$$

and any of these can be expressed in several ways, depending on the variables in terms of which the relation is to be generated. For example, the partial derivative

$$\left(\frac{\partial P}{\partial V}\right)_H$$

expressed in terms of experimentally measurable quantities, has the following form:

$$\left(\frac{\partial P}{\partial V}\right)_H = \frac{C_p}{T\left(\frac{\partial V}{\partial T}\right)_P^2 + C_P\left(\frac{\partial V}{\partial P}\right)_T - V\left(\frac{\partial V}{\partial T}\right)_T}$$

The generation of relations of this kind needs great inventiveness. From a mathematical point of view the question is one of transformation of coordinates.

The Transformation of Coordinates

For any such transformation the following theorem is valid:

$$\int\int \mathrm{f}[x, y]\, dx\, dy = \int\int \mathrm{f}[\mathrm{A}[u, v], \mathrm{B}[u, v]]\, J\, du\, dv$$

where the following relations exist between the variables in the (X, Y) and (U, V) planes:

$$x = \mathrm{A}[u, v]$$

$$y = \mathrm{B}[u, v]$$

and J in the integrand is the value of the Jacobian determinant, defined as

$$J \equiv \begin{vmatrix} \dfrac{\partial \mathrm{A}}{\partial u} & \dfrac{\partial \mathrm{B}}{\partial u} \\ \dfrac{\partial \mathrm{A}}{\partial v} & \dfrac{\partial \mathrm{B}}{\partial v} \end{vmatrix}$$

On the left side of the equation for the transformation we have the differential element of area $dx\, dy$; on its right side, the differential element of area $du\, dv$. It will be shown that the Jacobian determinant actually gives the relation between these differential elements of area. With this in mind, let us inspect Figure 6.7*a*, showing the differential element $dx\, dy$ in the (X, Y) plane. Naturally, in this representation it is a rectangle.

Because of the functional relation

$$x = \mathrm{A}[u, v]$$

$$y = \mathrm{B}[u, v]$$

the differentials of x and y are

$$dx = \left(\frac{\partial \mathrm{A}}{\partial u}\right)_v du + \left(\frac{\partial \mathrm{A}}{\partial v}\right)_u dv$$

$$dy = \left(\frac{\partial \mathrm{B}}{\partial u}\right)_u du + \left(\frac{\partial \mathrm{B}}{\partial v}\right)_u dv$$

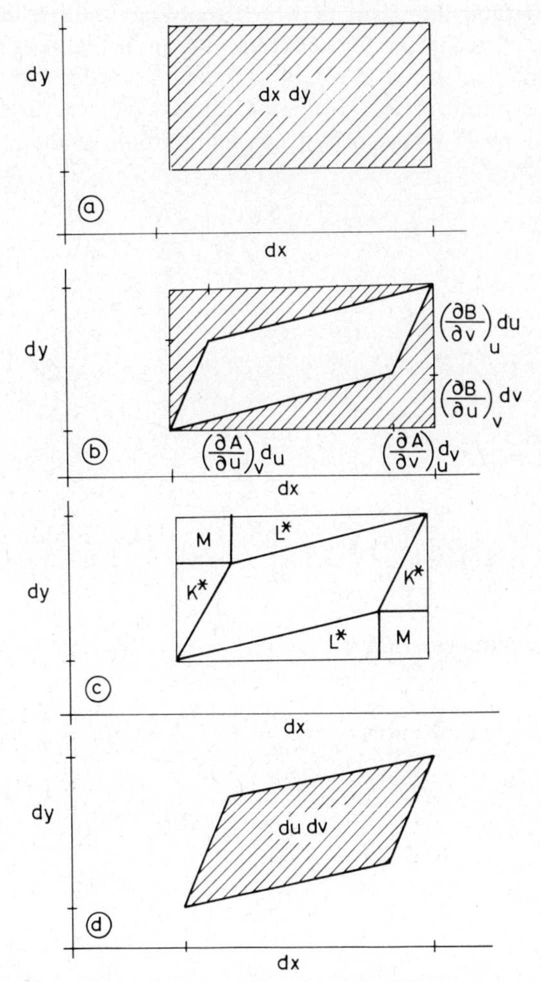

FIGURE 6.7. Transformation from the (U, V) plane to the (X, Y) plane.

and hence the differential element of area (omitting now the index for the invariant quantity) is

$$dx\,dy = \left(\left(\frac{\partial \mathrm{A}}{\partial u}\right) du + \left(\frac{\partial \mathrm{A}}{\partial v}\right) dv\right)\left(\left(\frac{\partial \mathrm{B}}{\partial u}\right) du + \left(\frac{\partial \mathrm{B}}{\partial v}\right) dv\right)$$

$$= \left(\frac{\partial \mathrm{A}}{\partial u}\right) du \left(\frac{\partial \mathrm{B}}{\partial u}\right) du + \left(\frac{\partial \mathrm{A}}{\partial u}\right) du \left(\frac{\partial \mathrm{B}}{\partial v}\right) dv$$

$$+ \left(\frac{\partial \mathrm{A}}{\partial v}\right) dv \left(\frac{\partial \mathrm{B}}{\partial u}\right) du + \left(\frac{\partial \mathrm{A}}{\partial v}\right) dv \left(\frac{\partial \mathrm{B}}{\partial v}\right) dv$$

The differential element of area $dv\,du$ is not a rectangle in the (X, Y) plane, but a parallelogram, the edges of which represent the partial increments at constant v and u, according to the interpretation in Figure 6.7*b*.

The differential element $dv\,du$ is obtained according to Figure 6.7*c* by subtracting the outlined area from the already known area $dx\,dy$. This is obtained according to Figure 6.7*d*, using its notation, in the following way:

$$\text{outlined area} = 2K^* + 2L^* + 2M$$

where

$$K = 2K^* = \left(\frac{\partial \mathrm{A}}{\partial v}\right) dv \left(\frac{\partial \mathrm{B}}{\partial v}\right) dv$$

$$L = 2L^* = \left(\frac{\partial \mathrm{A}}{\partial u}\right) du \left(\frac{\partial \mathrm{B}}{\partial u}\right) du$$

$$2M = \left(\frac{\partial \mathrm{A}}{\partial v}\right) dv \left(\frac{\partial \mathrm{B}}{\partial u}\right) du + \left(\frac{\partial \mathrm{A}}{\partial v}\right) dv \left(\frac{\partial \mathrm{B}}{\partial u}\right) du$$

The area of the parallelogram is

$$dx\,dy - \text{outlined area} = \left(\left(\frac{\partial \mathrm{A}}{\partial u}\right)\left(\frac{\partial \mathrm{B}}{\partial v}\right) - \left(\frac{\partial \mathrm{A}}{\partial v}\right)\left(\frac{\partial \mathrm{B}}{\partial u}\right)\right) du\,dv$$

$$= \begin{vmatrix} \left(\frac{\partial \mathrm{A}}{\partial u}\right) & \left(\frac{\partial \mathrm{B}}{\partial u}\right) \\ \left(\frac{\partial \mathrm{A}}{\partial v}\right) & \left(\frac{\partial \mathrm{B}}{\partial v}\right) \end{vmatrix} du\,dv = J\,du\,dv$$

The differential element of area $du\,dv$ in the (U, V) plane is transformed with the Jacobian determinant to the differential element of area $dx\,dy$ in the (X, Y) plane.

For the writing of the Jacobian determinant the following convention is introduced:

$$\begin{vmatrix} \left(\frac{\partial \mathrm{A}}{\partial u}\right)_v & \left(\frac{\partial \mathrm{B}}{\partial u}\right)_v \\ \left(\frac{\partial \mathrm{A}}{\partial v}\right)_u & \left(\frac{\partial \mathrm{B}}{\partial v}\right)_u \end{vmatrix} \equiv J\left[\frac{\mathrm{A}, \mathrm{B}}{u, v}\right] \equiv \frac{[\mathrm{A}, \mathrm{B}]}{[u, v]}$$

Using this convention, it is easy to carry out operations with Jacobian determinants. For example,

$$\frac{[A, B]}{[u, v]} \cdot \frac{[u, v]}{[x, y]} = \frac{[A, B]}{[x, y]}$$

$$\frac{[A, B]}{[A, B]} = 1$$

$$\frac{[A, A]}{[x, y]} = 0$$

$$\frac{[A, \lambda]}{[x, y]} = 0$$

where λ is constant;

$$\frac{[A, B]}{[x, y]} = -\frac{[B, A]}{[x, y]} = -\frac{[A, B]}{[y, x]}$$

With this convention a partial derivative can be written in the following form:

$$\left(\frac{\partial A}{\partial B}\right)_C = \frac{[A, C]}{[B, C]}$$

If a multivariable function f exists such that

$$f[A, B, C] = 0$$

then the total differential of A is

$$dA = \left(\frac{\partial A}{\partial B}\right)_C dB + \left(\frac{\partial A}{\partial C}\right)_B dC$$

and this can be written with the new notation as follows:

$$dA = \frac{[A, C]}{[B, C]} dB + \frac{[A, B]}{[C, B]} dC$$

After rearrangement we have

$$[B, C]\, dA = [A, C]\, dB + \frac{[B, C][A, B]}{[C, B]} dC$$

TABLE 6.6
Thermodynamic Potentials

Differential Form	Jacobian Form
$dU = T\,dS - P\,dV$	$[U, X] = T[S, X] - P[V, X]$
$dH = T\,dS + V\,dP$	$[H, X] = T[S, X] + V[P, X]$
$dF = -S\,dT - P\,dV$	$[F, X] = -S[T, X] - P[V, X]$
$dG = -S\,dT + V\,dP$	$[G, X] = -S[T, X] + V[P, X]$

and finally

$$[\mathrm{B},\mathrm{C}]\,d\mathrm{A} + [\mathrm{C},\mathrm{A}]\,d\mathrm{B} + [\mathrm{A},\mathrm{B}]\,d\mathrm{C} = 0$$

Jacobian algebra can be used for writing the total differential in operator form:

$$d\mathrm{A} = b\,d\mathrm{B} + c\,d\mathrm{C}$$

and we can operate on this equation with an arbitrary thermodynamic variable X. We can write

$$[\mathrm{A}, X] = b[\mathrm{B}, X] + c[\mathrm{C}, X]$$

Let us use this expression for the rewriting of the thermodynamic potentials in the Jacobian form. The results are shown in Table 6.6.

As a first application let us rewrite the known Maxwell relation in the Jacobian form:

$$\left(\frac{\partial T}{\partial P}\right)_S = \left(\frac{\partial V}{\partial S}\right)_P$$

$$\frac{[T,S]}{[P,S]} = \frac{[V,P]}{[S,P]} = \frac{[P,V]}{[P,S]}$$

Hence

$$[T,S] = [P,V]$$

and similarly

$$[S,P] = -[V,P]$$

Knowing the Jacobian forms of the thermodynamic potentials and of the Maxwell relations, the generation of the desired thermodynamic relation will consist of a series of simple Jacobian algebraic manipulations. Two applications are given below.

Application

Let us seek for the partial derivative $(\partial H/\partial G)_T$ a relationship containing measurable quantities on the right side. We first rewrite the partial derivative in the Jacobian form:

$$\left(\frac{\partial H}{\partial G}\right)_T = \frac{[H,T]}{[G,T]}$$

Let us substitute the Jacobian forms of the two potential functions:

$$\left(\frac{\partial H}{\partial G}\right)_T = \frac{T[S,T] + V[P,T]}{-S[T,T] + V[P,T]}$$

Since

$$[T,T] = 0$$

the sum of the following two fractions is obtained:

$$\left(\frac{\partial H}{\partial G}\right)_T = \frac{T[S,T]}{V[P,T]} + \frac{V[P,T]}{V[P,T]}$$

However,

$$\frac{V[P,T]}{V[P,T]} = 1$$

Thus, the partial is

$$\left(\frac{\partial H}{\partial G}\right)_T = \frac{T[S,T]}{V[P,T]} + 1$$

To eliminate the entropy let us use the Jacobian form of the Maxwell relation,

$$[S,T] = [V,P]$$

The partial is

$$\left(\frac{\partial H}{\partial G}\right)_T = \frac{T[V,P]}{V[P,T]} + 1$$

Here we already have measurable variables. An obvious transformation is

$$[P,T] = -[T,P]$$

because then we have for the partial

$$\left(\frac{\partial H}{\partial G}\right)_T = \frac{T[V,P]}{-V[T,P]} + 1$$

where

$$\frac{[V,P]}{[T,P]} = \left(\frac{\partial V}{\partial T}\right)_P$$

After substitution the relationship sought is obtained:

$$\left(\frac{\partial H}{\partial G}\right)_T = -\frac{T}{V}\left(\frac{\partial V}{\partial T}\right)_P + 1$$

Application

How does the enthalpy of a homogeneous phase change with pressure in an isothermal process?

Let us rewrite the partial $(\partial H/\partial P)_T$ in the Jacobian form:

$$\left(\frac{\partial H}{\partial P}\right)_T = \frac{[H,T]}{[P,T]} = \frac{T[S,T] + V[P,T]}{[P,T]}$$

$$= \frac{T[S,T]}{[P,T]} + \frac{V[P,T]}{[P,T]}$$

$$= \frac{T[S,T]}{[P,T]} + V$$

The Maxwell relation is used for the elimination of $[S,T]$:

$$[S,T] = [V,P]$$

We can write for the partial

$$\left(\frac{\partial H}{\partial P}\right)_T = \frac{T[V,P]}{[P,T]} + V$$

$$= \frac{T[V,P]}{-[T,P]} + V$$

The first term on the right-hand side contains the Jacobian form of a partial. Rewriting yields the relationship sought:

$$\left(\frac{\partial H}{\partial P}\right)_T = -T\left(\frac{\partial V}{\partial T}\right)_P + V$$

7

ELEMENTS OF STATISTICAL THERMODYNAMICS

In the first part of this book the apparatus of classical thermodynamics was built up. Several thermodynamic properties were introduced, and the relations between thermodynamic variables were elaborated. Though these relations are quite general, they say nothing about the value that some prescribed chemical property X of a given chemical substance takes on, say at prescribed temperature T and in a prescribed volume v. This is due to the fact that in classical thermodynamics no restrictions were made concerning the structure of the substance. Thereby we ignored any information relating to the particles forming the phase. The reason for this renunciation is that taking into consideration in the equations of motion all the molecules forming the phase—of a number corresponding to their classical degrees of freedom—would be a hopeless task. Another possibility must be found for the preservation of information. This is a new venture.

The basis of the venture is quantum mechanics. This makes it possible to form an idea of the energy of the molecules. Quantum mechanics says that a given molecule can be in such and such quantum state, having a definite energy eigenvalue.

Statistical–physical considerations are also involved in the venture. Indeed, when we renounce describing the particles with equations of motion, an uncertainty is created with respect to processes on the micro level. However, the role of the abandoned coordinates is taken over on the macro level by the coordinate known from classical thermodynamics, to which the name *entropy* has been given.

The role of entropy is to express the uncertainty that arose on the micro level from the abandonment of the molecular coordinates. We refrain from taking account of individual particles, so as to be able to take account of the ensemble of particles.

Quantum mechanics gives adequate information on the structure and energetic conditions of the molecules. Knowledge arising from this source has to be "integrated" in a way leading to the thermodynamic properties of a continuous homogeneous phase. As there are a great number of molecules in the continuous phase, the employment of the apparatus of probability theory and mathematical statistics in this "integration" was envisaged from the outset. Thus we arrived at a style of investigation different from the methods of classical thermodynamics, which is nevertheless consistent with classical thermodynamics. That must be so, as the aim of the investigations is the same in classical thermodynamics and in statistical thermodynamics. This requirement must be interpreted as follows. In classical thermodynamics the state of a homogeneous one-component phase is given by its temperature T, volume V, and amount of substance N. Let X be any property:

$$X = X[T, V, N]$$

According to the statistical concept the expected value of the probability variable representing the same property is given by the following relation:

$$\langle X \rangle = \sum_i P_i X_i$$

$$X_i = X[\varepsilon_i, T, n]$$

where $P_i = P[\varepsilon_i, T, n]$ = probability of quantum state i,
X_i = value of property X in quantum state i,
ε_i = energy eigenvalue in quantum state i,
n = number of molecules forming the amount N.

Classical thermodynamics and statistical thermodynamics are made to correspond on the following basis. If n, the number of molecules forming the body, is sufficiently large, then the state developing on the macro level is the manifestation of statistical phenomena on the micro level. For this reason

$$X = \langle X \rangle$$

This correspondence is of theoretical use if the probability P_i of quantum state i can be given on the basis of the statistical phenomena on the micro level. This can be actually done. Let us perform for that purpose a thought experiment.

CANONICAL ENSEMBLE

Let us take N bodies, each enclosed by a diathermal rigid wall of completely constant volume, containing a constant amount of substance. Let us place these bodies in a heat bath. Thermal equilibrium is established. Bodies in thermal equilibrium constitute a *canonical ensemble*. In the canonical ensemble the temperature of all the bodies is identical, but their energies are not identical. Let the numbers of bodies in the canonical ensemble possessing energies $E_1, E_2, E_3, \ldots$ be respectively $N_1, N_2, N_3, \ldots$.

In the second part of the experiment, suppose we change the diathermal walls between the equilibrized bodies into adiabatic walls. With this change the energies $E_1, E_2, E_3, \ldots$ of the bodies at the same temperature are fixed.

Actually the equilibrium state established in the bath is fixed. In fact we have to do with two equilibria:

1. The bodies are in equilibrium with one another.
2. Molecules within a body are in equilibrium.

First we will deal with the canonical ensemble.

Energy and Probability. Let us count the bodies in the canonical ensemble possessing an energy E_j. Their number is N_j. The probability that the energy of a body in the canonical ensemble is E_j can be written in the following form:

$$P_i = \lim_{N \to \infty} \frac{N_i}{N}$$

In the canonical ensemble the expected energy of one body is the following average

$$\langle E \rangle = \sum_i P_i E_i$$

the expected entropy is

$$\langle S \rangle = \sum_i P_i S_i$$

According to our initial assumption a body of energy E_i occurs with a probability P_i in the canonical ensemble, and this probability depends only on E_i:

$$P_i = P[E_i]$$

The probability that the energy of a body is E_k in the canonical ensemble is

$$P_k = P[E_k]$$

These two bodies are simultaneously, but independently of one another, in states i and k, respectively. The probability of the simultaneous occurrence of these two states is a function on one hand of the product of the two independent probabilities, and on the other hand of the sum of the two energies:

$$P_i P_k = P[E_i]P[E_k] = P[E_i + E_k]$$

This latter demand is satisfied by the exponential function.

Mathematical Remark. Let us seek a function that satisfies the following relationship:

$$\mathrm{f}[x] \cdot \mathrm{f}[y] = \mathrm{f}[x + y]$$

Differentiate with respect to y:

$$\mathrm{f}[x]\frac{d\mathrm{f}[y]}{dy} = \frac{\partial \mathrm{f}[x + y]}{\partial y} = \frac{\partial \mathrm{f}[x + y]}{\partial (x + y)} \cdot \frac{\partial (x + y)}{\partial y} = \frac{d\mathrm{f}[x + y]}{d(x + y)} \cdot 1$$

and then with respect to x:

$$\mathrm{f}[y]\frac{d\mathrm{f}[x]}{dx} = \frac{\partial \mathrm{f}[x + y]}{\partial x} = \frac{\partial \mathrm{f}[x + y]}{\partial (x + y)} \cdot \frac{\partial (x + y)}{\partial y} = \frac{d\mathrm{f}[x + y]}{d(x + y)} \cdot 1$$

The right-hand sides are identical; thus the left-hand sides are also identical:

$$\mathrm{f}[x]\frac{d\mathrm{f}[y]}{dy} = \mathrm{f}[y]\frac{d\mathrm{f}[x]}{dx}$$

After the separation of the variables the left-hand side does not depend on x, and the right-hand side does not depend on y, so each separately is equal to the same constant k:

$$\frac{1}{\mathrm{f}[y]} \cdot \frac{d\mathrm{f}[y]}{dy} = \frac{1}{\mathrm{f}[x]} \cdot \frac{d\mathrm{f}[x]}{dx} = -k$$

The negative sign of the constant expresses the requirement that

$$\frac{dP_i}{dE_i} < 0$$

with β defined to be positive.

The probability that a body of the canonical ensemble has the energy E_i is

$$P_i = (\text{const})\, e^{-\beta E_i}$$

where the negative sign in the exponent indicates that the probability P_i decreases with increasing energy E_i.

Entropy and Probability. Entropy is a homogeneous linear function of the energy therefore the relationship between entropy and probability must be

$$S_i = -k \ln P_i$$

and the expected value of the entropy of a body of the canonical ensemble is

$$S = \langle S \rangle = -k \sum_j P_j \ln P_j$$

We will prove now that the constant in this relationship is the Boltzmann's constant.

Let us see a system consisting of two parts with equal volume separated in the initial state. The system contents on mole of ideal gas in the leftside volume at the initial state. Removing the wall between the two parts the gas expands. Entropy change of expansion calculated by the relationship of the classical thermodynamics is

$$\Delta S = S_2 - S_1 = k \ln \frac{V_2}{V_1} = k \ln 2$$

where k is the Boltzmann's constant.

Let us now calculate the entropy change by the probability model. The probability of the presence of a given molecule of the leftside part at the initial respectively final state is

$$P_1 = 1$$

$$P_2 = \tfrac{1}{2}$$

The entropy in this two states

$$S_1 = -k \cdot 1 \cdot \ln 1 = 0$$

$$S_2 = -k\left(\tfrac{1}{2} \ln \tfrac{1}{2} + \tfrac{1}{2} \ln \tfrac{1}{2}\right) = k \ln 2$$

the entropy change of expansion

$$\Delta S = S_2 - S_1 = k \ln 2$$

The two results are equal only if the constant of the probability model is the Boltzmann's constant.

Energy Distribution. The probability that a body in the ensemble has energy E_i is

$$P_i = \text{const}\, e^{-\beta E_i}$$

$$= \frac{e^{-\beta E_i}}{Q}$$

where the negative sign in the exponent indicates that probability P_i decreases with increasing energy E_i. Of the two parameters of the exponential function, the physical content of the constant Q will be shown immediately. Summation of both sides gives

$$\sum P_i = 1 = \frac{\sum e^{-\beta E_i}}{Q}$$

and hence

$$Q = \sum e^{-\beta E_i}$$

Thus, the probability that the energy of a body in the canonical ensemble is E_i can be written in the following form:

$$P_i = \frac{e^{-\beta E_i}}{\sum e^{-\beta E_i}}$$

The expression $e^{-\beta E_i}$ is called the *Boltzmann factor*. It is called a factor because the fraction of bodies with energy E_i is proportional to it. The constant in the denominator is the sum of the Boltzmann factors, and is called the *partition function*. With this notation Boltzmann's energy distribution in the canonical ensemble can be written in the following form:

$$P_i = \frac{e^{-\beta E_i}}{Q}$$

Taking the logarithm of this relation, we have

$$\ln P_i = -\beta E_i - \ln Q$$

By further rearrangement two useful forms of this relation can be obtained:

$$E_i = -\frac{1}{\beta}(\ln P_i + \ln Q)$$

$$\ln Q = -\ln P_i - \beta E_i$$

We do not know anything yet about the physical content of the parameter β.

Boltzmann Distribution Law. The internal energy and entropy of a homogeneous phase of constant volume and composition are

$$U = \sum_i P_i E_i$$

$$S = -\mathbf{k}\sum_i P_i \ln P_i$$

Let us substitute the statistical expressions for internal energy and entropy into the following phenomenological relation, pertinent to a body of constant volume and composition:

$$dU = T\,dS$$

We have

$$d\left(\sum_i P_i E_i\right) = T\,d\left(-\mathbf{k}\sum_i P_i \ln P_i\right)$$

$$d\left(\sum_i P_i\left(-\frac{1}{\beta}(\ln P_i + \ln Q)\right)\right) = -\mathbf{k}T\,d\left(\sum_i P_i \ln P_i\right)$$

$$-\frac{1}{\beta}\,d\left(\sum_i P_i \ln P_i + \ln Q \sum P_i\right) = -\mathbf{k}T\,d\left(\sum_i P_i \ln P_i\right)$$

However, since $\Sigma_i\, dP_i = 0$, on the left side $\ln Q \Sigma_i\, dP_i = 0$, and thus

$$-\frac{1}{\beta}\,d\left(\sum_i P_i \ln P_i\right) = -\mathbf{k}T\,d\left(\sum_i P_i \ln P_i\right)$$

It follows from this that

$$\frac{1}{\beta} = \mathbf{k}T$$

and the final form of the Boltzmann distribution law is

$$P_i = \frac{e^{-E_i/\mathbf{k}T}}{\sum_i e^{-E_i/\mathbf{k}T}}.$$

Free Energy. The expected value of the free energy of a body in the canonical ensemble, since we have to do with a macro body, is determined by its temperature, its volume, and its amount of substance:

$$\langle F \rangle = F[T, V, N] = F$$

It was shown in Chapter 6 that the free energy of a body at given temperature and volume, containing a given amount of substance, is minimal in the equilibrium state:

$$dF = 0$$

$$d^2F > 0$$

The definition of free energy is

$$F = U - TS$$

Substitute into it the statistical expression for the internal energy and the entropy:

$$F = \sum_i P_i E_i - T\left(-\mathbf{k}\sum_i P_i \ln P_i\right)$$

After factoring we have

$$F = \sum_i P_i(E_i + \mathbf{k}T \ln P_i)$$

Let us introduce the following notation:

$$\mathbf{k}T = \frac{1}{\beta}$$

$$F = \sum_i P_i \frac{1}{\beta}(\beta E_i + \ln P_i)$$

Differentiation with respect to P_i gives

$$dF = \frac{1}{\beta}\sum(\beta E_i + \ln P_i + 1)\, dP_i$$

Let us select from the summation the bodies in states i and k:

$$dF = \frac{1}{\beta}((\beta E_i + \ln P_i + 1)\, dP_i + (\beta E_k + \ln P_k + 1)\, dP_k)$$

Since the sum of probabilities is one

$$dP_i + dP_k = 0$$

$$\frac{dF}{dP_i} = \frac{1}{\beta}(\beta E_i + \ln P_i - \beta E_k - \ln P_k)$$

We have shown the following relations:

$$\ln P_i = -\beta E_i - \ln Q$$

$$\ln P_k = -\beta E_k - \ln Q$$

Substituting these into our equation, we have

$$\frac{dF}{dP_i} = (\beta E_i - \beta E_i - \ln Q - \beta E_k - \beta E_k + \ln Q) = 0$$

The free energy therefore has an extremal value. This extremum is a minimum, since the second derivative of the free energy is positive:

$$\frac{d^2F}{dP_i^2} = \frac{d}{dP_i}\left(\frac{1}{\beta}(\beta E_i + \ln P_i - \beta E_k - \ln P_k)\right)$$

$$= \frac{1}{\beta}\left(\frac{1}{P_i} + \frac{1}{P_k}\right) > 0$$

QUANTUMSTATES AND MICROSTATES

Let us now remove a body from the canonical ensemble. This can be done without a change in state of the body, as the body has been surrounded after equilibration with an adiabatic wall. The energy of the body is

$$E_j = n \cdot \langle \varepsilon \rangle$$

where n is the number of molecules and $\langle \varepsilon \rangle$ is the expected value of the total energy of one molecule. This total energy is

$$\langle \varepsilon \rangle = \sum p_i \varepsilon_i$$

where ε_i = energy eigenvalue of the ith quantum state of the molecule,
p_i = probability that a molecule is in the ith quantum state.

The probability that the energy of a molecule is ε_i can be written in the following form:

$$p_i = \lim_{n \to \infty} \frac{n_i}{n}$$

Quantum States. The expression *quantum state* normally refers to one molecule. The energy ε of the molecule is the sum of the energies of its different kinds of motion. If the number of the atoms in the molecule is a, then the molecule has $3a$ degrees of freedom according to the laws of mechanics, discussed in Chapter 1. Of these 3 appertain to the translational motion in space of the molecule as a whole. Of the remaining $3a - 3$ degrees of freedom r appertain to rotation: the molecule rotates as a whole, and this accounts for 3 degrees of freedom, but besides this certain atomic groups may rotate also within the molecule about some bond. The degrees of freedom of internal rotation depend in the given case on the structure of the molecule. Thus, if a total of r degrees of freedom are allotted to rotation, there remain $3a - 3 - r$ degrees of freedom for vibration. All the degrees of freedom are quantized. The quantum state of a molecule can be defined by a total of $3a$ quantum numbers, classified according to the kinds of motions as discussed above. When we speak of the ith quantum state, this means the quantum state defined by a number of $3a$ integers contained in the ith combination. The energy eigenvalue belonging to this quantum state is ε_i.

Several quantum states may belong to the same energy state. We say in this case that the energy state is *degenerate*. The number of quantum states belonging to the same energy state is called its *statistical weight*. We recall in this connection Chapter 1.

Microstates. The expression *microstate* refers to n molecules of the investigated body in the macrostate E_j, and gives the quantum state of these molecules. Here two principles of conservation must be taken into account: Only those microstates are considered to which:

1. n molecules belong.
2. A total energy E_j belongs.

It will be shown now, for equidistant energy states, how the microstates of n molecules possessing E_j total energy can be counted. For example, let $E_j = 14$ and $n = 9$. Denoting the molecules by m and the energy quanta by ϵ, each microstate can be described by a series of letters m and ϵ such as

$$m\,\epsilon\,m\,\epsilon\,\epsilon\,\epsilon\,m\,m\,\epsilon\,\epsilon\,m\,\epsilon\,m\,m\,\epsilon\,\epsilon\,\epsilon\,\epsilon$$

This series of symbols means that the molecule standing first possesses one, the second three, the third zero, the fourth two, the fifth one, the sixth zero, the seventh four, the eighth two, and the ninth one energy quantum. Each symbol series beginning with m, containing nine m's and fourteen ϵ's corresponds to one microstate.

What we are interested in is how many microstates can be distinguished in the case of given E_j and n. This question will be answered by a combinatorial analysis.

If the molecules can be distinguished, then any of the n molecules can be put first. This means that we can choose it in n ways. Once we have chosen it, we can choose only from the remaining $n - 1$ molecules for the second place. Thus for the first two places we can choose in a total of $n(n - 1)$ ways. For the first three places $n(n - 1)(n - 2)$ ways of choosing are possible, and so on. For the arrangement of all the molecules the number of possibilities is

$$n(n-1)(n-2)\cdots(3)(2)(1) = n!$$

So far we have discussed only the possibilities for the ordered arrangement of distinguishable molecules; and we have not yet considered in how many ways the energy quantum E_j can be divided between them. An alternative formulation of this question is, in how many ways the symbol series consisting of symbols m and ϵ can be arranged.

Actually $n + E - 1$ symbols have to be permuted. One is subtracted from the number of symbols, because in every case the symbol m is put first. The number of the permutations of freely permutable symbols is then $(n + E - 1)!$

Let us stop here for a moment. Permutation is relevant to things distinguishable from one another, but consider whether it makes sense, from the point of view of a microstate, to distinguish the three energy quanta of a molecule possessing three energy quanta. This has evidently no physical sense, and therefore results in an overestimation of the number of possible microstates by a factor of $E!$, which is the number of the permutations of the E distinguishable energy quanta. Similar arguments can be made for the arrangements of the n symbols m: it is evident that the exchange of two symbols m does not bring about a new microstate, as it is completely indifferent which of the molecules stands at a given place of the symbol series. From this point of view the number of microstates has been again overestimated by a factor, which is just $(n - 1)!$, since m freely permutable symbols have $(n - 1)!$ ways of arrangement.

We have seen that $(n + E - 1)!$ is an overestimation of the number of microstates, the factor of overestimation being on one part $E!$ and on the other $(n - 1)!$. Therefore, divided by these factors, we find for the number of possible microstates

$$W = \frac{(E + n - 1)!}{E!(n - 1)!}$$

To summarize: W tells in how many ways n indistinguishable molecules can be arranged in n distinguishable places, and a total of E indistinguishable energy quanta can be arranged on these molecules. A microstate is characterized by a series of n integers, specifying how many energy quanta each successive molecule possesses.

The number series characteristic of a microstate is written in square brackets.

For example, let $n = 3$ and $E_j = 3$. What is the number of possible microstates?

Three energy quanta have to be distributed between three distinguishable molecules. Possible microstates are as follows:

1. All three energy quanta are taken up by a single molecule: a, b, or c. The possible microstates are

$$[3,0,0]$$

$$[0,3,0]$$

$$[0,0,3]$$

2. Two energy quanta are taken up by one molecule, one energy quantum by another molecule, and so no energy is left for the third molecule. The possible microstates are

$$[2,1,0]$$

$$[2,0,1]$$

$$[1,2,0]$$

$$[0,2,1]$$

$$[1,0,2]$$

$$[0,1,2]$$

3. The three energy quanta are divided between the three molecules:

$$[1,1,1]$$

After counting, we have that the number W of possible microstates is 10:

$$W = \frac{(3+3-1)!}{3!\,2!} = \frac{1 \times 2 \times 3 \times 4 \times 5}{(1 \times 2 \times 3)(1 \times 2)} = \frac{4 \times 5}{1 \times 2} = 10$$

Distribution. We wish to form a notion of the molecular motion in a body consisting of molecules. We have seen now that a body, without changing its macrostate (E_j = const, n = const, V = const), nevertheless changes, and we say of this change that the body goes from one microstate to another microstate. The setting up of a correspondence between an equilibrium state on the macro level and continual change on the micro level is realized through the microstates. We cannot say anything about which microstate is realized and when, but we can declare that each of the W microstates relevant to a given body is realized with the same probability. This is the postulate of statistical thermodynamics. The probability of realization of any of the microstates is

$$p_W = \frac{1}{W}$$

Putting this expression of probability in the equation of entropy

$$S - \mathbf{k}\sum_{1}^{W} \frac{1}{W}\ln\frac{1}{W} = -\mathbf{k}\frac{W}{W}\ln\frac{1}{W}$$

we get the Boltzmann's relationship of entropy

$$S = \mathbf{k}\ln W$$

When speaking here of the probability of microstates, it must be remembered that order also plays a role in the distinguishing of microstates. However, the question can be raised, how many microstates exist, in each of which an identical number n_i of molecules are in the quantum state ε_i without regard to order. In that case a microstate can be characterized by a number series whose first member is the number of molecules in the energy level ε_0, whose second is the number of molecules in the energy level ε_1, and so on, ending with the number $\bar{n}$ of molecules in the highest possible energy level. If we wish to express, using these variables, the conservation of energy and the conservation of the number of molecules we can write

$$E_j = \sum_i n_i \varepsilon_i$$

$$n = \sum n_i$$

where $0 \leq \varepsilon_i \leq E_j$. Of those microstates that are characterized by a given number series $\{n_d\}$, we can say that they represent the same energy distribution. The series of numbers characteristic of a distribution is put in braces.

Now let the number of microstates representing distribution d be W_d. The probability of distribution d is then, in the case where there are W possible microstates,

$$p_d = \frac{W_d}{W}$$

If we know the distribution $\{n_0, n_1, n_2, \ldots, \bar{n}\}$, then we can give the number of microstates in which this distribution is realized. The number of permutations of n molecules is $n!$. This is an overestimation, since n_0 molecules can be arranged in the energy level ε_0 in $n_0!$ ways, but the molecules cannot be distinguished. Similarly, n_1 molecules are indistinguishable at the ε_1 energy level, so that $n_1!$ microstates are in the same distribution. Finally, the number of microstates realizing distribution d is

$$W_d = \frac{n!}{n_0!n_1! \cdots \bar{n}!}$$

where $\bar{n}$ is the number of molecules in the highest possible energy level.

As an example, among the microstates of the system discussed in the preceding example ($n = 3$, $E_j = 3$), we find those of identical distribution. How many distributions are possible, what is the probability of the single distributions? We have the following enumeration:

Microstates	Distributions $\{n_0, n_1, n_2, n_3\}$	Number of Microstates	Probability of Distribution
[3, 0, 0] [0, 3, 0] [0, 0, 3]	{2, 0, 0, 1}	$\frac{1 \times 2 \times 3}{(1 \times 2) \times 1 \times 1 \times 1} = 3$	$\frac{3}{10}$
[2, 1, 0] [2, 0, 1] [1, 2, 0] [1, 0, 2] [0, 1, 2] [0, 2, 1]	{1, 1, 1, 0}	$\frac{1 \times 2 \times 3}{1 \times 1 \times 1 \times 1} = 6$	$\frac{6}{10}$
[1, 1, 1]	{0, 3, 0, 0}	$\frac{1 \times 2 \times 3}{1 \times (1 \times 2 \times 3) \times 1 \times 1} = 1$	$\frac{1}{10}$

Algorithm and Program: Number of Microstates

Write a program to calculate the number of possible microstates and arrangements of particles if the number of particles and the total energy content of the system are given. The energy levels are supposed to be equally spaced. The maximum number of energy levels is 6.

Algorithm.

1. Read the number of particles and the energy content of the system relative to the first level.

2. Calculate each possible distribution of the particles on the six different levels. Recommended method: Suppose some particle to be on the sixth level, and calculate the maximum possible number of particles having the remaining energy (the energy which was not allocated to the sixth level). Continue this process till the first level is reached. To find whether the distributions is acceptable, calculate its energy content and compare with the total energy of the system.
3. Calculate the number of microstates. To a given distribution more than one arrangement of particles can belong. Each arrangement is called a microstate. The number of microstates in a distribution is

$$W = \frac{n!}{n_0!n_1! \cdots n_j!}$$

where n = the total number of particles,
n_i = number of particles on the ith energy level,
j = number of energy levels, in this case 6.

4. Print the result.

Program NODIST. A possible realization:

```
]LOAD NODIST
]LIST

 10  REM :CALCULATION OF THE POSSIBLE
      DISTRIBUTIONS
 20  REM :THE TOTAL NUMBER OF PARTICLES
      AND
 30  REM :THE ENERGY CONTENT RELATIVE
      TO THE
 40  REM :FIRST ENERGY LEVEL IS
      KNOWN. THE MAXIMUM NUMBER
      OF LEVELS IS 6; ENERGY
      LEVELS ARE EQUALLY SPACED.
      THE ENERGY OF A PARTICLE ON
      THE FIRST LEVEL IS 0, ON THE
 50  REM :SECOND LEVEL 1,... ON THE
      6TH LEVEL 5.
 60  PRINT : PRINT
 70  INPUT "ENTER THE TOTAL NUMBER
      OF PARTICLES ";NT
 80  INPUT "ENTER TOTAL ENERGY ";E
      T
 90  S = 0
100  SE  =  0:SP  =  0:SD  =  0
110  PRINT : PRINT
```

```
120  PRINT "L0  L1  L2  L3  L4
      L5  NMS"
130  PRINT " = = = = = = = = = = = = = = =
      = = = = = = = = = = =
      = = = = = = = = = = = = = = "
140  BI = INT (ET / 5)
150  FOR I = 0 TO BI
160  BJ = INT ((ET - 5 * I) / 4)
170  FOR J = 0 TO BJ
180  BK = INT ((ET - I * 5 - J *
      4) / 3)
190  FOR K = 0 TO BK
200  BL = INT ((ET - I * 5 - J *
      4 - K * 3) / 2)
210  FOR L = 0 TO BL
220  BM = INT (ET - I * 5 - J * 4
      - K * 3 - L * 2)
230  FOR M = 0 TO BM
240  N = NT - I - J - K - L - M
250  IF N < 0 THEN GOTO 500
260  E = 5 * I + 4 * J + 3 * K + 2
      * L + M
270  IF E < > ET THEN GOTO 500
280  FF = NT
290   GOSUB 740
300  NF = FR
310  FF = I
320   GOSUB 740
330  FI = FR:FF = J
340   GOSUB 740
350  FJ = FR:FF = K
360   GOSUB 740
370  FK = FR:FF = L
380   GOSUB 740
390  FL = FR:FF = M
400   GOSUB 740
410  FM = FR: FF = N
420   GOSUB 740
430  NM = NF / FI / FJ / FK / FL /
      FM / FR
440  S = S + 1
450  SD = SD + NM
460  SE = SE + E * NM
470  SP = SP + NT * NM
480  PRINT TAB( 2);N; TAB( 7);M;
      TAB( 12);L; TAB( 17);K; TAB
      (22);J; TAB( 27);I; TAB( 32);
      NM
490  PRINT
```

```
500  NEXT M
510  NEXT L
520  NEXT K
530  NEXT J
540  NEXT I
550  PRINT : PRINT
560  PRINT : PRINT
570  PRINT "THE NUMBER OF DISTRIBUTIONS
      ";S
580  PRINT
590  PRINT "THE SUM OF PARTICLES"
600  PRINT "(NO. OF PARTICLES * NO.
      OF MICROSTATES)"
610  PRINT "   "SP
620  PRINT
630  PRINT "THE SUM OF ENERGY "
640  PRINT "(NO. OF MICROSTATES * ENERGY
      OF THE SYSTEM)"
650  PRINT "   "SE
660  PRINT
670  PRINT "THE NUMBER OF MICROSTATES
      ";SD
680  PRINT : PRINT
690  INPUT "NEXT ENERGY SUM ? (Y /
      N) ";A$
700  IF A$ = "Y" THEN GOTO 80
710  INPUT "NEXT NO. OF PARTICLES
      ? (Y / N) ";A$
720  IF A$ = "Y" THEN GOTO 70
730  END
740  FR = 1
750  IF FF = 0 THEN GOTO 790
760  FOR II = 1 TO FF
770  FR = FR * II
780  NEXT II
790 RETURN
```

Example. Run the program for 3 and 4 particles, with the total energy of the system 1, 2, 3, 4, and 5.

```
]RUN

ENTER THE TOTAL NUMBER OF PARTICLES 3
ENTER TOTAL ENERGY 1
```

L0	L1	L2	L3	L4	L5	NMS
2	1	0	0	0	0	3

THE NUMBER OF DISTRIBUTIONS 1

THE SUM OF PARTICLES
(NO. OF PARTICLES * NO. OF MICROSTATES)
9

THE SUM OF ENERGY
(NO. OF MICROSTATES * ENERGY OF THE SYSTEM)
3

THE NUMBER OF MICROSTATES 3

NEXT ENERGY SUM ? (Y / N) Y
ENTER TOTAL ENERGY 2

L0	L1	L2	L3	L4	L5	NMS
1	2	0	0	0	0	3
2	0	1	0	0	0	3

THE NUMBER OF DISTRIBUTIONS 2

THE SUM OF PARTICLES
(NO. OF PARTICLES * NO. OF MICROSTATES)
18

THE SUM OF ENERGY
(NO. OF MICROSTATES * ENERGY OF THE SYSTEM)
12

THE NUMBER OF MICROSTATES 6

NEXT ENERGY SUM ? (Y / N) Y
ENTER TOTAL ENERGY 3

L0	L1	L2	L3	L4	L5	NMS
0	3	0	0	0	0	1
1	1	1	0	0	0	6
2	0	0	1	0	0	3

THE NUMBER OF DISTRIBUTIONS 3

THE SUM OF PARTICLES
(NO. OF PARTICLES * NO. OF MICROSTATES)
30

THE SUM OF ENERGY
(NO. OF MICROSTATES * ENERGY OF THE SYSTEM)
30

THE NUMBER OF MICROSTATES 10

NEXT ENERGY SUM ? (Y / N) Y

ENTER TOTAL ENERGY 4

L0	L1	L2	L3	L4	L5	NMS
0	2	1	0	0	0	3
1	0	2	0	0	0	3
1	1	0	1	0	0	6
2	0	0	0	1	0	3

THE NUMBER OF DISTRIBUTIONS 4

THE SUM OF PARTICLES
(NO. OF PARTICLES * NO. OF MICROSTATES)
45

THE SUM OF ENERGY
(NO. OF MICROSTATES * ENERGY OF THE SYSTEM)
60

THE NUMBER OF MICROSTATES 15

NEXT ENERGY SUM ? (Y / N) N
NEXT NO. OF PARTICLES ? (Y / N) Y
ENTER THE TOTAL NUMBER OF PARTICLES 4
ENTER TOTAL ENERGY 1

L0	L1	L2	L3	L4	L5	NMS
3	1	0	0	0	0	4

THE NUMBER OF DISTRIBUTIONS 1

THE SUM OF PARTICLES
(NO. OF PARTICLES * NO. OF MICROSTATES)
16

THE SUM OF ENERGY
(NO. OF MICROSTATES * ENERGY OF THE SYSTEM)
4

THE NUMBER OF MICROSTATES 4

NEXT ENERGY SUM ? (Y / N) Y
ENTER TOTAL ENERGY 2

L0	L1	L2	L3	L4	L5	NMS
2	2	0	0	0	0	6
3	0	1	0	0	0	4

THE NUMBER OF DISTRIBUTIONS 2

THE SUM OF PARTICLES
(NO. OF PARTICLES * NO. OF MICROSTATES)
40

THE SUM OF ENERGY
(NO. OF MICROSTATES * ENERGY OF THE SYSTEM)
20

THE NUMBER OF MICROSTATES 10

NEXT ENERGY SUM ? (Y / N) Y
ENTER TOTAL ENERGY 3

L0	L1	L2	L3	L4	L5	NMS
1	3	0	0	0	0	4
2	1	1	0	0	0	12
3	0	0	1	0	0	4

THE NUMBER OF DISTRIBUTIONS 3

THE SUM OF PARTICLES
(NO. OF PARTICLES * NO. OF MICROSTATES)
80

THE SUM OF ENERGY
(NO. OF MICROSTATES * ENERGY OF THE SYSTEM)
60

THE NUMBER OF MICROSTATES 20

NEXT ENERGY SUM ? (Y / N) Y
ENTER TOTAL ENERGY 4

L0	L1	L2	L3	L4	L5	NMS
0	4	0	0	0	0	1
1	2	1	0	0	0	12
2	0	2	0	0	0	6
2	1	0	1	0	0	12
3	0	0	0	0	1	04

THE NUMBER OF DISTRIBUTIONS 5

THE SUM OF PARTICLES
(NO. OF PARTICLES * NO. OF MICROSTATES)
140

THE SUM OF ENERGY
(NO. OF MICROSTATES * ENERGY OF THE SYSTEM)
140

THE NUMBER OF MICROSTATES 35

NEXT ENERGY SUM ? (Y / N) N
NEXT NO. OF PARTICLES ? (Y / N) N

Remarks. The probability of a distribution is proportional to the number of microstates in it.

Algorithm and Program: Most Probable Distributions

Write a program to determine the number of possible distributions of a given number of particles if the total energy of the system is given, and choose the most probable ten distributions.

Algorithm.

1. Read the number of particles and total energy of the system.
2. Calculate the possible distributions (see program NODIST)
3. Calculate the natural logarithm of the number of microstates (W) (see program NODIST)
4. Select the most probable ten distributions.
5. Print the results.

Program POSDIST. A possible realization:

```
]PR#0
]LOAD POSDIST
]LIST

 10  REM :CALCULATION OF THE POSSIBLE
      DISTRIBUTIONS
 20  REM :AND PRINTING OF THE MOST
      PROBABLE TEN DISTRIBUTIONS
 30  DIM RE(10,7)
 40   REM :THE TOTAL NUMBER OF PARTICLES
      AND
 50  REM :THE ENERGY CONTENT RELATIVE
      TO THE
 60  REM :FIRST ENERGY LEVEL IS KNOWN.
      THE NUMBER OF LEVEL S IS 6,
      ENERGY LEVELS ARE EQUALLY
      SPACED. THE ENERGY OF A PARTICLE
      ON THE FIRST LEVEL IS 0, ON THE
 70  REM :SECOND LEVEL 1 ...ON THE
      6-TH LEVEL 5.
 80  PRINT : PRINT
 90  INPUT "ENTER THE TOTAL NUMBER
      OF PARTICLES";NT
100  INPUT "ENTER TOTAL ENERGY ";
      ET
110  S = 0
120  FF = NT
130  GOSUB 770
140  FT = FR
150  BI = INT (ET / 5)
```

```
160  FOR I = 0 TO BI
170  BJ - INT ((ET - 5 * I) / 4)
180  FOR J = 0 TO BJ
190  BK = INT ((ET - I * 5 - J *
      4) / 3)
200  FOR K = 0 TO BK
210  BL = INT ((ET - I * 5 - J *
      4 - K * 3) / 2)
220  FOR L = 0 TO BL
230  BM = INT (ET - I * 5 - J * 4
      - K * 3 - L * 2)
240  FOR M = 0 TO BM
250  N = NT - I - J - K - L - M
260  IF N < 0 THEN GOTO 570
270  E = 5 * I + 4 * J + 3 * K + 2
      * L + M
280  IF E < > ET THEN GOTO 570
290  FF = I: GOSUB 770
300  FI = FR
310  FF = J: GOSUB 770
320  FJ = FR
330  FF = K: GOSUB 770
340  FK = FR
350  FF = L: GOSUB 770
360  FL = FR
370  FF = M: GOSUB 770
380  FM = FR
390  FF = N: GOSUB 770
400  NF = FR
410  W = FT - FI - FJ - FK - FL -
      FM - NF
420  S = S + 1
430  IF S > 10 THEN GOTO 480
440  RE(S,1) = N: RE(S,2) = M: RE(S,
      3) = L
450  RE(S,4) = K: RE(S,5) = J: RE(S,
      6) = I
460  RE (S,7) = W
470  GOTO 570
480  U = 1000
490  FOR II = 1 TO 10
500  IF RE (II,7) < U THEN U = RE(
      II,7): P = II
510  NEXT II
520  IF W < U THEN GOTO 570
530  RE(P, 1) = N: RE(P,2) = M: RE(P,
      3) = L
```

```
540  RE(P,4) = K: RE(P,5) = J: RE(P,
      6) = I
550  RE(P,7) = W
560  GOTO 570
570  NEXT M
580  NEXT L
590  NEXT K
600  NEXT J
610  NEXT I
620  PRINT : PRINT
630  PRINT "E0  E1  E2  E3  E4
      E5 W"
640  PRINT "===================
      ====================="
650  FOR I = 1 TO 10
660  PRINT TAB(2); RE(I,1); TAB(
      7); RE(I,2); TAB( 12); RE(I,3);
      TAB(17); RE(I,4); TAB(22);
      RE(I,5); TAB(27); RE(I,6); TAB(
      30); RE(I,7)
670  PRINT
680  NEXT I
690  PRINT : PRINT
700  PRINT "THE NUMBER OF DISTRIB
      UTIONS ";S
710  PRINT : PRINT
720  INPUT "NEXT ENERGY SUM ? (Y /
      N) ";A$
730  IF A$ = "Y" THEN GOTO 100
740  INPUT "NEXT NO. OF PARTICLES
      ? (Y / N) ";A$
750  IF A$ = "Y" THEN GOTO 90
760  END
770  FR = 0
780  IF FF = 0 THEN GOTO 830
790  L1 = 1 / LOG (10)
800  FOR JJ = 1 TO FF
810  FR = FR + L1 * LOG(JJ)
820  NEXT JJ
830  RETURN
```

Example. Run the program for particle numbers 10, 20 and total energies 10, 14, 16.

```
]RUN

ENTER THE TOTAL NUMBER OF PARTICLES 10
ENTER TOTAL ENERGY 10
```

E0	E1	E2	E3	E4	E5	W
6	1	1	1	1	0	3.70243054
5	2	1	2	0	0	3.8785218
5	2	2	0	1	0	3.8785218
3	4	3	0	0	0	3.62324929
5	3	1	0	0	1	3.70243054
5	3	0	1	1	0	3.70243054
4	4	1	0	1	0	3.79934055
3	5	1	1	0	0	3.70243054
4	3	2	1	0	0	4.10037055
5	1	3	1	0	0	3.70243054

THE NUMBER OF DISTRIBUTIONS 30

NEXT ENERGY SUM ? (Y / N) Y
ENTER TOTAL ENERGY 14

E0	E1	E2	E3	E4	E5	W
4	2	1	2	1	0	4.5774918
4	2	2	0	2	0	4.27646181
4	3	1	0	1	1	4.40140054
3	2	3	2	0	0	4.40140054
3	4	1	1	0	1	4.40140054
4	1	3	1	1	0	4.40140054
4	2	2	1	0	1	4.5774918
3	3	2	1	1	0	4.70243054
2	4	2	2	0	0	4.27646181
3	3	3	0	0	1	4.22530929

THE NUMBER OF DISTRIBUTIONS 63

NEXT ENERGY SUM ? (Y / N) Y
ENTER TOTAL ENERGY 16

E0	E1	E2	E3	E4	E5	W
4	2	1	1	1	1	4.8785218
3	3	2	0	1	1	4.70243054
3	2	3	1	0	1	4.70243054
3	2	2	2	1	0	4.8785218
2	3	3	1	1	0	4.70243054
4	1	2	1	2	0	4.5774918
3	3	1	1	2	0	4.70243054
3	3	1	2	0	1	4.70243054
2	4	2	1	0	1	4.5774918
2	4	1	2	1	0	4.5774918

THE NUMBER OF DISTRIBUTIONS 83

NEXT ENERGY SUM ? (Y / N) N
NEXT NO. OF PARTICLES ? (Y / N) Y

ENTER THE TOTAL NUMBER OF PARTICLES 20
ENTER TOTAL ENERGY 10

E0	E1	E2	E3	E4	E5	W
14	4	1	0	1	0	6.06550502
11	8	1	0	0	0	6.17944837
12	6	2	0	0	0	6.54742515
13	4	3	0	0	0	6.4334818
14	2	4	0	0	0	5.76447502
14	4	0	2	0	0	5.76447502
12	7	0	1	0	0	6.00335712
13	5	1	1	0	0	6.51266305
14	3	2	1	0	0	6.36653501
13	6	0	0	1	0	5.7345118

THE NUMBER OF DISTRIBUTIONS 30

NEXT ENERGY SUM ? (Y / N) Y
ENTER TOTAL ENERGY 14

E0	E1	E2	E3	E4	E5	W
10	7	2	1	0	0	7.82290105
12	3	4	1	0	0	7.54742515
12	4	2	2	0	0	7.72351642
13	3	2	1	1	0	7.51266305
10	6	4	0	0	0	7.58881784
12	5	1	1	1	0	7.6266064
11	6	1	2	0	0	7.6266064
11	5	3	1	0	0	7.9276364
11	6	2	0	1	0	7.6266064
12	4	3	0	1	0	7.54742515

THE NUMBER OF DISTRIBUTIONS 70

NEXT ENERGY SUM ? (Y / N) Y
ENTER TOTAL ENERGY 16

E0	E1	E2	E3	E4	E5	W
10	6	2	2	0	0	8.36696909
9	7	3	1	0	0	8.34577979
12	3	3	1	1	0	8.14948515
11	4	3	2	0	0	8.32557641
12	4	2	1	0	1	8.02454641
10	6	3	0	1	0	8.19087783
12	4	1	2	1	0	8.02454641
10	7	1	1	1	0	8.12393104
11	5	2	1	1	0	8.40475765
10	5	4	1	0	0	8.36696909

```
THE NUMBER OF DISTRIBUTIONS 101

NEXT ENERGY SUM ? (Y / N) N
NEXT NO. OF PARTICLES ? (Y / N) N
```

Remark. There is a big difference in W between distributions. The probability of a distribution is proportional to W. Don't forget W in the program is the logarithm of W.

Algorithm and Program: Oscillator

Write a program to calculate the most probable quantum state of a harmonic oscillator.

Algorithm.

1. Read $h\nu$ for the oscillator.
2. Read the number of energy levels.
3. Calculate the energy of the oscillator,

$$\varepsilon_v = \left(v + \tfrac{1}{2}\right)h\nu$$

   where v is the quantum number.
4. Calculate Boltzmann factor.

$$B = e^{-\varepsilon_v}$$

5. Calculate the probability of quantum state v:

$$p_v = \frac{B_v}{Q} \qquad Q = \sum_v B_v$$

6. Print the result.

Program OSCI. A possible realization:

```
]LOAD OSCI
]LIST

 10  REM :CALCULATION OF PROBABILITY
      OF
 20  REM : POSITION OF OSCILLATOR
 30  PRINT : PRINT
 40  REM : THE MAXIMUM NUMBER OF ENERGY
      LEVELS IS 20
 50  DIM E(20), B(20), P(20)
 60  INPUT "ENTER H * NU ";HN
 70  INPUT "ENTER THE NUMBER OF ENERGY
      LEVELS ";N
 80  Q = 0
 90   FOR I = 0 TO N
```

```
100  E(I) = (I + 0.5) * HN
110  B(I) = EXP ( - E(I))
120  Q = Q + B(I)
130  NEXT I
140  FOR I = 0 TO N
150  P(I) = B(I) / Q
160  NEXT I
170  PRINT : PRINT
180  PRINT " V E BF
     PV"
190  PRINT "===================
     ======================="
200  FOR I = 0 TO N
210  PRINT TAB(2); I; TAB(5); E(
     I); TAB(10);B (I); TAB(26);
     P(I)
220  PRINT
230  NEXT I
240  PRINT "===================
     ======================="
250  PRINT
260  PRINT TAB( 10);"Q = "; Q
270  PRINT : PRINT
280  INPUT "NEXT NO. OF ENERGY LEVELS
     ? (Y / N) "; A$
290  IF A$ = "Y" THEN GOTO 70
300  INPUT "NEXT H * NU ? (Y / N) "; A
     $
310  IF A$ = "Y" THEN GOTO 60
320  END
```

Example. Run the program for $h\nu = 1$, 0.1, 0.01, and for numbers of energy levels (quantum states) 6 and 10.

```
]RUN

ENTER H * NU 1
ENTER THE NUMBER OF ENERGY LEVELS 6
```

V	E	BF	PV
0	.5	.60653066	.632697504
1	1.5	.22313016	.232756404
2	2.5	.0820849986	.0856262959
3	3.5	.0301973834	.0315001539
4	4.5	.0111089965	.011588259
5	5.5	4.08677144E − 03	4.26308225E − 03
6	6.5	1.50343919E − 03	1.56830032E − 03

Q = .958642409

NEXT NO. OF ENERGY LEVELS ? (Y / N) Y
ENTER THE NUMBER OF ENERGY LEVELS 10

V	E	BF	PV
0	.5	.60653066	.632131116
1	1.5	.22313016	.232548042
2	2.5	.0820849986	.0855496437
3	3.5	.0301973834	.0314719551
4	4.5	.0111089965	.0115778853
5	5.5	4.08677144E − 03	4.25926596E − 03
6	6.5	1.50343919E − 03	1.56689638E − 03
7	7.5	5.5308437E − 04	5.76428965E − 04
8	8.5	2.03468369E − 04	2.12056366E − 04
9	9.5	7.48518299E − 05	7.80111773E − 05
10	10.5	2.75364494E − 05	2.86987083E − 05

Q = .95950135

NEXT NO. OF ENERGY LEVELS ? (Y / N) N
NEXT H * NU ? (Y / N) Y
ENTER H * NU 0.1
ENTER THE NUMBER OF ENERGY LEVELS 6

V	E	BF	PV
0	.05	.951229425	.189034175
1	.15	.860707976	.171045195
2	.25	.778800783	.154768093
3	.35	.70468809	.140039961
4	.45	.637628152	.126713397
5	.55	.57694981	.114655023
6	.65	.522045777	.103744155

Q = 5.03205002

NEXT NO. OF ENERGY LEVELS ? (Y / N) Y
ENTER THE NUMBER OF ENERGY LEVELS 10

V	E	BF	PV
0	.05	.951229425	.142644967
1	.15	.860707976	.129070503
2	.25	.778800783	.116787821
3	.35	.70468809	.10567399
4	.45	.637628152	.0956177805
5	.55	.57694981	.0865185457
6	.65	.522045777	.0782852175
7	.75	.472366553	.0708353941
8	.85	.427414932	.064094515
9	.95	.386741024	.0579951155
10	1.05	.349937749	.0524761506

Q = 6.66851027

NEXT NO. OF ENERGY LEVELS ? (Y / N) N

```
NEXT H * NU ? (Y / N) Y
ENTER H * NU 0.01
ENTER THE NUMBER OF ENERGY LEVELS 6
```

V	E	BF	PV
0	5E − 03	.995012479	.147178353
1	.015	.98511194	.145713904
2	.025	.975309912	.144264026
3	.035	.965605416	.142828575
4	.045	.955997482	.141407407
5	.055	.946485148	.14000038
6	.065	.937067464	.138607353

Q = 6.76058985

```
NEXT NO. OF ENERGY LEVELS ? (Y / N) Y
ENTER THE NUMBER OF ENERGY LEVELS 10
```

V	E	BF	PV
0	5E − 03	.995012479	.0955223314
1	.015	.98511194	.0945718683
2	.025	.975309912	.0936308625
3	.035	.965605416	.0926992199
4	.045	.955997482	.0917768472
5	.055	.946485148	.0908636523
6	.065	.937067464	.0899595439
7	.075	.927743486	.0890644315
8	.085	.918512284	.0881782256
9	.095	.909372934	.0873008375
10	.105	.900324523	.0864321797

Q = 10.4165431

```
NEXT NO. OF ENERGY LEVELS ? (Y / N) N
NEXT H * NU ? (Y / N) N
```

Remark. From the result it can be seen that the higher the energy level, the less the probability of its occurrence.

Algorithm and Program: Rotator

Write a program to determine the most probable quantum state of the rotator.

Algorithm.

1. Read the rotation constant B and the number of energy levels, J.
2. Calculate the rotator energy

$$\varepsilon_{rj} = J(J + 1)B$$

3. Calculate the Boltzmann factor

$$B_J = (2J + 1)e^{-\varepsilon_{rj}}$$

4. Calculate the probability of quantum state J.
5. Print the results.

Program ROTA. A possible realization:

```
]PR#0
]LOAD ROTA
]LIST

 10  REM :CALCULATION OF PROBABILITY
      OF
 20  REM :POSITION OF ROTATOR
 30  PRINT : PRINT
 40  REM :THE MAXIMUM NUMBER OF ENERGY
      LEVELS IS 20
 50  DIM E(20),B(20),P(20)
 60  INPUT "ENTER BE ";BE
 70  INPUT "ENTER THE NUMBER OF ENERGY
      LEVELS ";N
 80  Q = 0
 90   FOR I = 0 TO N
100  E(I) = BE * I * (I + 1)
110  B(I) = EXP ( - E(I)) * (2 *
      I + 1)
120  Q = Q + B(I)
130  NEXT I
140  FOR I = 0 TO N
150  P(I) = B(I) / Q
160  NEXT I
170  PRINT : PRINT
180  PRINT " J J * (J + 1) 2 * J + 1"
190  PRINT "==================
      ======================="
200  FOR I = 0 TO N
210  PRINT TAB(2); I; TAB(8); I *
      (I + 1); TAB(15); 2 * I + 1
220  PRINT
230  NEXT I
240  PRINT "==================
      ======================="
250  PRINT : PRINT : PRINT
260  PRINT " J BF
      PJ"
270  PRINT "==================
      ======================="
```

```
280  FOR I = 0 TO N
290  PRINT TAB(2); I; TAB(7); B(
     I); TAB(23); P(I)
300  PRINT
310  NEXT I
320  PRINT "===================
     ====================="
330  PRINT TAB( 6);"Q = ";Q
340  PRINT : PRINT
350  INPUT "NEXT NO. OF ENERGY LEVELS
     ? (Y / N) "; A$
360  IF A$ = "Y" THEN GOTO 70
370  INPUT "NEXT BE ? (Y / N) "; A
     $
380  IF A$ = "Y" THEN GOTO 60
390  END
```

Example. Run the program for $B = 0.1$, $J_{max} = 6$.

```
]RUN

ENTER BE 0.1
ENTER THE NUMBER OF ENERGY LEVELS 6
```

J	J * (J + 1)	2 * J + 1
0	0	1
1	2	3
2	6	5
3	12	7
4	20	9
5	30	11
6	42	13

J	BF	PJ
0	1	.0973783061
1	2.45619226	.239179842
2	2.74405818	.267211737
3	2.10835949	.205308475
4	1.21801755	.118608486
5	.547657752	.0533299842
6	.194942498	.0189831703

Q = 10.2692277

```
NEXT NO. OF ENERGY LEVELS ? (Y / N) N
NEXT BE ? (Y / N) N
```

PARTITION FUNCTION AND THERMODYNAMIC PROPERTIES

The partition function is a dimensionless quantity. Its importance lies in the fact that it expresses the relation between the macro and molecular properties of the system. Suppose:

1. The partition function can be calculated from molecular data.
2. The relation between the partition function and macro properties is known.

Then we can get by calculation, with the aid of the partition function, from molecular knowledge to macro properties.

First the possibility of this step will be proved.

Internal Energy

The statistical expression for the molar internal energy is

$$u = \sum_i P_i E_i$$

Let us substitute here the Boltzmann distribution for P_i:

$$u = \frac{\sum_i \exp(-\beta E_i)}{Q} E_i$$

$$= \frac{\sum_i E_i \exp(-\beta E_i)}{\sum_i \exp(-\beta E_i)}$$

Observe that

$$\frac{\partial}{\partial \beta} \exp(-\beta E_i) = -E_i \exp(-\beta E_i)$$

On the other hand, the numerator in the expression for u, expressed in terms of the partition function, is

$$\sum_i \frac{\partial}{\partial \beta} \exp(-\beta E_i) = \left(\frac{\partial Q}{\partial \beta}\right)_{v,N} = -\sum_i E_i \exp(-\beta E_i)$$

Substituting this, we have for the molar internal energy

$$u = -\frac{1}{Q}\left(\frac{\partial Q}{\partial \beta}\right)_{v,N} = -\left(\frac{\partial \ln Q}{\partial \beta}\right)_{v,N}$$

$$= \mathbf{k}T^2\left(\frac{\partial \ln Q}{\partial T}\right)_{v,N}$$

Entropy

The statistical expression for the molar entropy is

$$s = -\mathbf{k}\sum_i P_i \ln P_i$$

Let us use the expression for $\ln P_i$ just elaborated:

$$s = -\mathbf{k}\sum_i P_i\left(-\frac{E_i}{\mathbf{k}T} - \ln Q\right)$$

$$= \frac{1}{T}\sum_i P_i E_i + \mathbf{k}\ln Q \sum P_i$$

However

$$u = \sum_i P_i E_i = \mathbf{k}T^2\left(\frac{\partial \ln Q}{\partial T}\right)_{v,N}$$

and

$$\sum_i P_i = 1$$

Substituting these, we obtain finally for the molar entropy

$$s = \mathbf{k}\ln Q + \mathbf{k}T\left(\frac{\partial \ln Q}{\partial T}\right)_{v,N}$$

$$= \mathbf{k}\ln Q + \frac{u}{T}$$

Free Energy

The definition of free energy is

$$f = u - Ts$$

Into this, the expressions for the internal energy and entropy given in terms of the partition function are substituted:

$$f = \mathbf{k}T^2\left(\frac{\partial \ln Q}{\partial T}\right)_{v,N} - T\left(\mathbf{k}\ln Q + \mathbf{k}T\left(\frac{\partial \ln Q}{\partial T}\right)_{v,N}\right)$$

whence the molar free energy is

$$f = -\mathbf{k}T \ln Q$$

Pressure

The thermodynamic relationship between pressure and free energy is

$$P = -\left(\frac{\partial f}{\partial v}\right)_T = -\frac{\partial}{\partial v}(-\mathbf{k}T \ln Q)$$

The pressure can be expressed in terms of the partition function:

$$P = \mathbf{k}T\left(\frac{\partial \ln Q}{\partial v}\right)_T$$

Enthalpy

It follows from the defining equation of enthalpy that

$$h = u + Pv$$

$$= kT^2\left(\frac{\partial \ln Q}{\partial T}\right)_{v,N} + kT\left(\frac{\partial \ln Q}{\partial v}\right)_T v$$

Free Enthalpy

There is the following relationship between free enthalpy and free energy:

$$g = f + Pv$$

whence the molar free enthalpy is

$$g = -\mathbf{k}T \ln Q + Pv$$

The molar free enthalpy of the perfect gas is

$$g = -kT \ln Q + RT$$

Algorithm and Program: Partition Function

Write a program to calculate the internal energy, entropy, and free energy from the partition function. The form of the partition function is

$$q = 1 + \exp\left(-\frac{\varepsilon}{\mathbf{k}T}\right) + \exp\left(-\frac{2\varepsilon}{\mathbf{k}T}\right) + \exp\left(-\frac{3\varepsilon}{\mathbf{k}T}\right)$$

The thermodynamic properties are

$$u = \mathbf{R}T^2\left(\frac{\partial \ln q}{\partial T}\right)_v$$

$$s = \mathbf{R}\ln q + \mathbf{R}T\left(\frac{\partial \ln q}{\partial T}\right)_v$$

$$f = -\mathbf{R}T\ln q$$

$$\frac{\partial \ln q}{\partial T} = \frac{1}{q}\frac{\partial q}{\partial T}$$

Algorithm.

1. Reading the energy difference ε between quantum levels.
2. Reading the temperature.
3. Calculate internal energy, entropy, and free energy.
4. Print the results.

Program PARTFUNC. A possible realization:

```
]LOAD PARTFUNC
]LIST

 10  REM :CALCULATION OF INTERNAL
      ENERGY,
 20  REM :ENTROPY AND FREE ENERGY
      FROM PARTITION FUNCTION
 30  K = 1.381E - 23
 40  PRINT : PRINT
 50  INPUT "ENTER THE ENERGY OF
      MOLECULE (J) ";E
 60  INPUT "ENTER THE TEMPERATURE
      (K) ";T
 70  DEF FN A(E) = E / K / T
 80  DEF FN B(E) = EXP ( - FN A
      (E))
 90   DEF FN C(E) = EXP ( - 2 * FN
      A(E))
100  DEF FN D(E) = EXP ( - 3 *
      FN A(E))
110  DEF FN Q(E) = 1 + FN B(E) +
      FN C(E) + FN D(E)
120  DEF FN F(E) = LOG ( FN Q(E
      ))
130  DEF FN G(E) = FN B(E) + FN
```

```
    C(E) * 2 + FN D(E) * 3
140 DEF FN H(E) = E / K / T / T
150 DEF FN I(E) = 1 / FN Q(E) *
    FN H(E) * FN G(E)
160 R = 8.3144
170 U = R * T * T * FN I(E)
180 S = R * FN F(E) + R * T * FN
    I(E)
190 F = - R * T * FN F(E)
200 PRINT : PRINT
210 PRINT "MOLAR INTERNAL ENERGY
    ";U;" J / MOL"
220 PRINT
230 PRINT "MOLAR ENTROPY ";S;" J
    /MOL / K"
240 PRINT
250 PRINT "MOLAR FREE ENERGY ";F
    ;" J / MOL"
260 PRINT : PRINT
270 INPUT "NEXT TEMPERATURE ? (Y
    /N) ";A$
280 IF A$ = "Y" THEN GOTO 60
290 INPUT "NEXT ENERGY ? (Y / N) "
    ;A$
300 IF A$ = "Y" THEN GOTO 50
310 END
```

Example. Run the program for molecules with energy 5×10^{-21} and 10^{-21} J at temperature 300 and 1000 K.

```
]RUN

ENTER THE ENERGY OF MOLECULE (J) 5E - 21
ENTER THE TEMPERATURE (K) 300

MOLAR INTERNAL ENERGY 1187.62992 J / MOL
MOLAR ENTROPY 6.84721307 J / MOL K
MOLAR FREE ENERGY -866.533998 J / MOL

NEXT TEMPERATURE ? (Y / N) Y
ENTER THE TEMPERATURE (K) 1000

MOLAR INTERNAL ENERGY 3201.24776 J / MOL
MOLAR ENTROPY 10.8809696 J / MOL K
MOLAR FREE ENERGY -7679.72181 J / MOL

NEXT TEMPERATURE ? (Y / N) N
NEXT ENERGY ? (Y / N) 1E

]RUN
```

```
ENTER THE ENERGY OF MOLECULE (J) 1E - 21
ENTER THE TEMPERATURE (K) 300

MOLAR INTERNAL ENERGY 724.369213 J / MOL
MOLAR ENTROPY 11.2307731 J / MOL K
MOLAR FREE ENERGY -2644.86271 J / MOL

NEXT TEMPERATURE ? (Y / N) Y
ENTER THE TEMPERATURE (K) 1000

MOLAR INTERNAL ENERGY 848.670906 J / MOL
MOLAR ENTROPY 11.4990191 J / MOL K
MOLAR FREE ENERGY -10650.3482 J / MOL

NEXT TEMPERATURE ? (Y / N) N
NEXT ENERGY ? (Y / N) N
```

FACTORIZATION OF THE PARTITION FUNCTION

We have shown in the foregoing a series of relations between the molar partition function and thermodynamic properties. The key issue—the precondition for the calculation of thermodynamic properties—is the knowledge of the canonical partition function. In the calculation of the partition function, factorization is the most essential manipulation.

Factorization means the following: If in a given thermodynamical state of the body the energy E_i contributes to its energy, is the sum over two subsystems:

$$E_i = E_{ai} + E_{bi}$$

and if this relationship is independent of further subsystems present in the system, then

$$\begin{aligned} Q_i &= e^{-E_i/\mathbf{k}T} \\ &= e^{-(E_{ai}+E_{bi})/\mathbf{k}T} \\ &= e^{-E_{ai}/\mathbf{k}T} e^{-E_{bi}/\mathbf{k}T} \\ &= Q_{ai} Q_{bi} \end{aligned}$$

The total partition function is the product of the partition functions of the subsystems.

The energy of a molecule is the sum of the energy contributions of various forms of motion:

$$E = E_{\mathrm{t}} + E_{\mathrm{v}} + E_{\mathrm{r}} + E_{\mathrm{e}} + E_0$$

According to the rules of factoring, the total partition function is the following product:

$$Q = Q_t Q_v Q_r Q_e Q_0$$

where Q_t = partition function of translational motion,

Q_v = partition function of vibrations within the molecule,

Q_r = partition function of the rotations of the molecule,

Q_e = partition function of electron excitation,

Q_0 = partition function of the ground state separated from the forms of motion listed above.

If the body consists of N subsystems (molecules), these can be divided into two groups, according to the physical nature of the phenomenon: localized and nonlocalized subsystems. A localized subsystem is fixed in the molecule as a rotator or oscillator. However, the translational motion of the molecule is nonlocalized if it moves freely in the gas space. The factorization of the

TABLE 7.1
Contributions to the Thermodynamic Properties of Perfect Gases

	Translation	Rotation (Diatomic)	Rotation (Nonlinear)	Vibration
ε	$\frac{h^2}{8m}\left(\frac{n_x^2}{a^2}+\frac{n_y^2}{b^2}+\frac{n_z^2}{c^2}\right)$	$\frac{J(J+1)h^2}{8\pi^2 I}$		$\upsilon h\nu$
q	$\left(\frac{2\pi m\mathbf{k}T}{h^2}\right)^{3/2}\frac{\mathbf{k}T}{P}$	$\frac{8\pi^2 I\mathbf{k}T}{h^2}=\frac{T}{\Theta_r}$	$\frac{\pi^{1/2}}{\sigma}\left(\frac{T^3}{\Theta_A\Theta_B\Theta_C}\right)^{1/2}$	$\frac{1}{1-e^{-h\nu/\mathbf{k}T}}=\frac{1}{1-e^{-\Theta_v/T}}$
Q	$\left(\frac{2\pi m\mathbf{k}T}{h^2}\right)^{3/2}\frac{\upsilon^N e^N}{N^N}$	q_r^N	q_r^N	Q_v^N
u^*	$\frac{3}{2}\mathbf{R}T$	$\mathbf{R}T$	$\frac{3}{2}\mathbf{R}T$	$\mathbf{R}T\frac{\Theta_v/T}{(e^{\Theta_v/T}-1)}$
c_V^*	$\frac{3}{2}\mathbf{R}$	$\mathbf{R}$	$\frac{3}{2}\mathbf{R}$	$\mathbf{R}\left(\frac{\Theta_v}{T}\right)^2\frac{e^{\Theta_v/T}}{(e^{\Theta_v/T}-1)^2}$
h^*	$\frac{5}{2}\mathbf{R}T$	$\mathbf{R}T$	$\frac{3}{2}\mathbf{R}T$	$\mathbf{R}T\frac{\Theta_v/T}{e^{\Theta_v/T}-1}$
c_P^*	$\frac{5}{2}\mathbf{R}$	$\mathbf{R}$	$\frac{3}{2}\mathbf{R}$	
s^*	$\mathbf{R}(\frac{5}{2}+\ln q_t)$	$\mathbf{R}\ln\left(\frac{eT}{\sigma\Theta_r}\right)$	$\mathbf{R}\ln q_r$	$\mathbf{R}\frac{\Theta_v/T}{e^{\Theta_v/T}-1}-\ln(1-e^{\Theta_v/T}))$
f^*	$-\mathbf{R}T(1+\ln q_t)$			
g^*	$-\mathbf{R}T\ln q_t$	$-\mathbf{R}T\ln\left(\frac{T}{\sigma\Theta_r}\right)$	$-\mathbf{R}Tq_r$	$\mathbf{R}T\ln(1-e^{-\Theta_v/T})$

partition function in the case of localized subsystems is

$$Q = q^N$$

where q is the molecular partition function. We will show on the other hand that the partition function of translational motion is

$$Q_t = \frac{q_t^N}{N!}$$

where q_t is the molecular partition function of translational motion.

In Table 7.1 we give in advance the partition functions of translation, vibration, and rotation, as well as their contributions to various thermodynamic properties.

PARTITION FUNCTION OF TRANSLATIONAL MOTION

It was shown in Chapter 1 that the energy eigenvalues of one-dimensional translational motion are

$$\varepsilon_t = \frac{\mathbf{h}^2 n^2}{8ma^2}$$

The molecular partition function is

$$q_t \sum_{n=1}^{\infty} \exp\left(-\frac{\mathbf{h}^2 n^2}{8ma^2} \cdot \frac{1}{\mathbf{k}T}\right) = \sum_{n=1}^{\infty} e^{-ln^2}$$

If the coefficient of n^2 in the exponent is sufficiently small, then it is permitted to integrate rather than sum

$$q_t = \int_0^{\infty} e^{-ln^2}\, dn$$

$$= \frac{1}{2}\left(\frac{\pi}{l}\right)^{1/2}$$

$$= \left(\frac{2\pi m \mathbf{k}T}{\mathbf{h}^2}\right)^{1/2} a$$

The molecular partition function for three-dimensional translational motion is

$$q_t = \left(\frac{2\pi m \mathbf{k}T}{\mathbf{h}^2}\right)^{3/2} v = (qv)$$

where the following denotation has been introduced:

$$q = \left(\frac{2\pi m\mathbf{k}T}{\mathbf{h}^2}\right)^{3/2}$$

The partition function is proportional to the volume v available to the molecule. Alternatively, we can say that the partition function of a molecule localized in volume v is qv. It is not important, from this point of view, whether or not there is another molecule in the available volume. Let us assume that in each of two neighboring cells of volume v there is one molecule, and each of the molecules can move only in its own cell.

The partition coefficient of the system consisting of two cells is

$$(qv)(qv) = q^2v^2$$

However, if the wall between the two cells is removed and both molecules can move in the total volume $2v$, the partition coefficient of the system consisting of the two mobile molecules will be

$$q_{\mathrm{m}}^2 = \frac{(q2v)^2}{2!}$$

We have divided by two (factorial two), because the two localized molecules can be arranged in two ways, but on the other hand the two mobile molecules cannot be distinguished. Thus $(q2v)^2$ is just a factor of 2 larger than the correct q_{m}^2.

The partition function of three molecules is after mobilization

$$q_{\mathrm{m}}^3 = \frac{(q3v)^3}{3!}$$

We have $3! = 6$ in the denominator, because three localized molecules can be arranged in six different ways, but the three mobile molecules cannot be distinguished. Thus $(q3v)^3$ is just six times larger than the partition function of the system consisting of localized molecules.

Following the same argument, in the case of N molecules,

$$q_{\mathrm{m}}^N = \frac{(qNv)^N}{N!}$$

$$= \frac{q^N N^N v^N}{N!}$$

According to Stirling's formula,

$$\frac{N^N}{N!} \approx e^N$$

After the elimination of the factorial the partition function of the system consisting of N molecules is

$$q_{\mathrm{m}}^N = q^N e^N v^N$$

Taking the Nth root on both sides,

$$q_{\mathrm{m}} = (qv)e$$

We have that the partition function of the mobile molecule is higher by a factor e than that of the localized molecule.

The total partition function of translational motion, since we have N mobile molecules, is

$$Q_{\mathrm{t}} = q^N \frac{v^N}{N!}$$

Substituting for q, we have

$$Q_{\mathrm{t}} = \left(\frac{2\pi m \mathbf{k} T}{\mathbf{h}^2}\right)^{3N/2} \frac{v^N}{N!}$$

Using Stirling's formula,

$$Q_{\mathrm{t}} = \left(\frac{2\pi m \mathbf{k} T}{\mathbf{h}^2}\right)^{3N/2} \frac{v^N e^N}{N^N}$$

This expression gives the total partition function for translational motion.

Algorithm and Program: Translational Partition Function

Write a program for the calculation of the value of translational molecular partition function of a perfect gas. The translational molecular partition function is substituting the EOS of the ideal gas

$$q_{\mathrm{t}} = \left(\frac{2\pi m \mathbf{k} T}{h^2}\right)^{3/2} \frac{\mathbf{k} T}{P}$$

where the mass of a molecule is

$$m = \frac{M}{\mathbf{N}}$$

Algorithm.

1. Read the molecular mass.
2. Read the temperature and pressure.
3. Calculate the natural logarithm of q_t.
4. Calculate q_t.
5. Print the results.

Program PARTFUNCTRANS. A possible realization:

```
]LOAD PARTFUNCTRANS
]LIST

 10  REM :CALCULATION OF TRANSLATI
      ONAL PARTITION FUNCTION FOR
      A PERFECT GAS
 20  PRINT : PRINT
 30  INPUT "ENTER THE MOLECULAR MA
      SS (KG / MOL) ";M
 40  PRINT
 50  INPUT "ENTER PRESSURE (PA) ";
      P
 60  INPUT "ENTER THE TEMPERATURE
      (K) ";T
 70  H = 6.626176E - 34
 80  K = 1.380662E - 23
 90  NA = 6.022045E - 23
100  QT = 3 / 2 * ( LOG (2) + LOG
     (3.1415) + LOG (M) - LOG (
     NA) + LOG (K) + LOG (T) -
     2 * LOG (H)) + LOG (K) + LOG
     (T) - LOG (P)
110  QT = EXP (QT)
120  PRINT : PRINT
130  INVERSE
140  PRINT "VALUE OF MOLECULAR
      PARTITION FUNCTION :"
150  PRINT TAB( 15);QT
160  NORMAL
170  PRINT : PRINT
180  INPUT "NEXT TEMPERATURE ? (Y
      /N) ";A$
190  IF A$ = "Y" THEN GOTO 60
200  INPUT "NEXT PRESSURE ? (Y / N)
      ";A$
210  IF A$ = "Y" THEN GOTO 50
220  INPUT "NEXT COMPOUND ? (Y / N)
      ";A$
```

```
230  IF A$ = "Y" THEN GOTO 20
240  END
```

Example. Run the program for hydrogen, chlorine, and hydrochloric acid at temperatures 298, 500, 1000, 1500, K and pressure 100000 Pa.

```
]RUN

ENTER THE MOLECULAR MASS (KG / MOL) 2.016E - 3
ENTER PRESSURE (PA) 100000
ENTER THE TEMPERATURE (K) 298

VALUE OF MOLECULAR PARTITION FUNCTION :
 113851.98

NEXT TEMPERATURE ? (Y / N) Y
ENTER THE TEMPERATURE (K) 500

VALUE OF MOLECULAR PARTITION FUNCTION :
 415169.014

NEXT TEMPERATURE ? (Y / N) Y
ENTER THE TEMPERATURE (K) 1000

VALUE OF MOLECULAR PARTITION FUNCTION :
 2348550.61

NEXT TEMPERATURE ? (Y / N) Y
ENTER THE TEMPERATURE (K) 1500

VALUE OF MOLECULAR PARTITION FUNCTION :
 6471844.46

NEXT TEMPERATURE ? (Y / N) N
NEXT PRESSURE ? (Y / N) N
NEXT COMPOUND ? (Y / N) Y

ENTER THE MOLECULAR MASS (KG / MOL) 70.906E - 3
ENTER PRESSURE (PA) 100000
ENTER THE TEMPERATURE (K) 298

VALUE OF MOLECULAR PARTITION FUNCTION :
 23748121.6

NEXT TEMPERATURE ? (Y / N) Y
ENTER THE TEMPERATURE (K) 500

VALUE OF MOLECULAR PARTITION FUNCTION :
 86599148.7

NEXT TEMPERATURE ? (Y / N) Y
ENTER THE TEMPERATURE (K) 1000

VALUE OF MOLECULAR PARTITION FUNCTION :
 489878764

NEXT TEMPERATURE ? (Y / N) Y
ENTER THE TEMPERATURE (K) 1500
```

```
VALUE OF MOLECULAR PARTITION FUNCTION :
 1.34994708E + 09

NEXT TEMPERATURE ? (Y / N) N
NEXT PRESSURE ? (Y / N) N
NEXT COMPOUND ? (Y / N) Y

ENTER THE MOLECULAR MASS (KG / MOL) 36.46E - 3
ENTER PRESSURE (PA) 100000
ENTER THE TEMPERATURE (K) 298

VALUE OF MOLECULAR PARTITION FUNCTION :
 8756484.61

NEXT TEMPERATURE ? (Y / N) Y
ENTER THE TEMPERATURE (K) 500

VALUE OF MOLECULAR PARTITION FUNCTION :
 31931118.6

NEXT TEMPERATURE ? (Y / N) Y
ENTER THE TEMPERATURE (K) 1000

VALUE OF MOLECULAR PARTITION FUNCTION :
 180629685

NEXT TEMPERATURE ? (Y / N) Y
ENTER THE TEMPERATURE (K) 1500

VALUE OF MOLECULAR PARTITION FUNCTION :
 497756880

NEXT TEMPERATURE ? (Y / N) N
NEXT PRESSURE ? (Y / N) Y
ENTER PRESSURE (PA) 1000000
ENTER THE TEMPERATURE (K) 1500

VALUE OF MOLECULAR PARTITION FUNCTION :
 49775688.3

NEXT TEMPERATURE ? (Y / N) N
NEXT PRESSURE ? (Y / N) N
NEXT COMPOUND ? (Y / N) N
```

Remark. The calculated values here are the translational contribution to the partition function of the molecule.

Algorithm and Program: Translational Contribution to Thermodynamic Properties

Write a program to calculate the translational contribution to the thermodynamic properties C_p^*, h^*, s^*, g^* of ideal gases. The corresponding formulas are listed in first column of Table 7.1.

Algorithm.

1. Read the molecular mass.
2. Read the temperature and pressure.
3. Calculate the natural logarithm of the translational molecular partition function (see program PARTFUNCTRANS).
4. Calculate the translational c_p^*, h^*, s^*, g^*.
5. Print the results.

Program TRANSCONTR. A possible realization:

```
]PR#0
]LOAD TRANSCONTR
]LIST

 10  REM :CALCULATION OF TRANSLATI
      ONAL CONTRIBUTIONS TO THE
      THERMODYNAMIC PROPERTIES OF
      PERFECT GAS
 20  PRINT : PRINT
 30  INPUT "ENTER THE MOLECULAR MA
      SS (KG / MOL) ";M
 40  PRINT
 50  INPUT "ENTER PRESSURE (PA) ";
      P
 60  INPUT "ENTER THE TEMPERATURE
      (K) ";T
 70  H = 6.626176E - 34
 80  K = 1.380662E - 23
 90  NA = 6.022045E23
100  R = 8.3144
110  QT = 3 / 2 * ( LOG (2) + LOG
      (3.1415) + LOG (M) - LOG (
      NA) + LOG (K) + LOG (T) -
      2 * LOG (H)) + LOG (K) + LOG
      (T) - LOG (P)
120  PRINT : PRINT
130  PRINT "CP0 (J / MOL K)"; TAB(
      25); 5 / 2 * R
140  PRINT "ENTHALPY (J / MOL)"; TAB(
      25); 5 / 2 * R * T
150  PRINT "ENTROPY (J / MOL K)"; TAB(
      25); R * (5 / 2 + QT)
160  PRINT "FREE ENERGY (J / MOL)";
      TAB( 25); - R * T * QT
```

```
170  PRINT : PRINT
180  INPUT "NEXT TEMPERATURE ? (Y
      /N) ";A$
190  IF A$ = "Y" THEN GOTO 60
200  INPUT "NEXT PRESSURE ? (Y / N)
      ";A$
210  IF A$ = "Y" THEN GOTO 50
220  INPUT "NEXT COMPOUND ? (Y / N)
      ";A$
230  IF A$ = "Y" THEN GOTO 20
240  END
```

Example. Run the program for hydrogen molecules, chlorine molecules, chlorine atoms, and hydrochloric acid molecules at pressure 100,000 Pa and temperatures 298, 500, 1000 K.

```
]RUN

ENTER THE MOLECULAR MASS (KG / MOL) 2.016E - 3
ENTER PRESSURE (PA) 100000
ENTER THE TEMPERATURE (K) 298

                    CPO (J / MOL K)               20.786
                    ENTHALPY (J / MOL)            6194.228
                    ENTROPY (J / MOL K)           117.587686
                    FREE ENERGY (J / MOL)         -28846.9025

NEXT TEMPERATURE ? (Y / N) Y
ENTER THE TEMPERATURE (K) 500

                    CPO (J / MOL K)               20.786
                    ENTHALPY (J / MOL)            10393
                    ENTROPY (J / MOL K)           128.344745
                    FREE ENERGY (J / MOL)         -53779.3724

NEXT TEMPERATURE ? (Y / N) Y
ENTER THE TEMPERATURE (K) 1000

                    CPO (J / MOL K)               20.786
                    ENTHALPY (J / MOL)            20786
                    ENTROPY (J / MOL K)           142.752502
                    FREE ENERGY (J / MOL)         -121966.502

NEXT TEMPERATURE ? (Y / N) N
NEXT PRESSURE ? (Y / N) Y
ENTER PRESSURE (PA)
?REENTER
```

```
ENTER PRESSURE (PA) 1000000
ENTER THE TEMPERATURE (K) 1000

          CPO (J / MOL K)            20.786
          ENTHALPY (J / MOL)         20786
          ENTROPY (J / MOL K)        123.607889
          FREE ENERGY (J / MOL)      -102821.889

NEXT TEMPERATURE ? (Y / N) N
NEXT PRESSURE ? (Y / N) N
NEXT COMPOUND ? (Y / N) Y

ENTER THE MOLECULAR MASS (KG / MOL) 70.906E - 3
ENTER PRESSURE (PA) 100000
ENTER THE TEMPERATURE (K) 298

          CPO (J / MOL K)            20.786
          ENTHALPY (J / MOL)         6194.228
          ENTROPY (J / MOL K)        161.989572
          FREE ENERGY (J / MOL)      -42078.6643

NEXT TEMPERATURE ? (Y / N) Y
ENTER THE TEMPERATURE (K) 500

          CPO (J / MOL K)            20.786
          ENTHALPY (J / MOL)         10393
          ENTROPY (J / MOL K)        172.74663
          FREE ENERGY (J / MOL)      -75980.3152

NEXT TEMPERATURE ? (Y / N) Y
ENTER THE TEMPERATURE (K) 1000

          CPO (J / MOL K)            20.786
          ENTHALPY (J / MOL)         20786
          ENTROPY (J / MOL K)        187.154388
          FREE ENERGY (J / MOL)      -166368.388

NEXT TEMPERATURE ? (Y / N) N
NEXT PRESSURE ? (Y / N) N
NEXT COMPOUND ? (Y / N) Y

ENTER THE MOLECULAR MASS (KG / MOL) 35.453E - 3
ENTER PRESSURE (PA) 100000
ENTER THE TEMPERATURE (K) 298

          CPO (J / MOL K)            20.786
          ENTHALPY (J / MOL)         6194.228
          ENTROPY (J / MOL K)        153.344917
          FREE ENERGY (J / MOL)      -39502.5573
```

```
NEXT TEMPERATURE ? (Y / N) N
NEXT PRESSURE ? (Y / N) N
NEXT COMPOUND ? (Y / N) Y

ENTER THE MOLECULAR MASS (KG / MOL) 36.469E - 3
ENTER PRESSURE (PA) 100000
ENTER THE TEMPERATURE (K) 298

                    CPO (J / MOL K)               20.786
                    ENTHALPY (J / MOL)            6194.228
                    ENTROPY (J / MOL K)           153.697299
                    FREE ENERGY (J / MOL)         -39607.567

NEXT TEMPERATURE ? (Y / N) Y
ENTER THE TEMPERATURE (K) 500

                    CPO (J / MOL K)               20.786
                    ENTHALPY (J / MOL)            10393
                    ENTROPY (J / MOL K)           164.454357
                    FREE ENERGY (J / MOL)         -71834.1787

NEXT TEMPERATURE ? (Y / N) Y
ENTER THE TEMPERATURE (K) 1000

                    CPO (J / MOL K)               20.786
                    ENTHALPY (J / MOL)            20786
                    ENTROPY (J / MOL K)           178.862115
                    FREE ENERGY (J / MOL)         -158076.115

NEXT TEMPERATURE ? (Y / N) N
NEXT PRESSURE ? (Y / N) N
NEXT COMPOUND ? (Y / N) N
```

PARTITION FUNCTION OF THE HARMONIC OSCILLATOR

The energy eigenvalues of the harmonic oscillator are

$$\varepsilon_v = \left(v + \tfrac{1}{2}\right)\mathbf{h}\nu \qquad v = 0, 1, 2, \ldots$$

where ν is the normal frequency of the oscillator. Accordingly the molecular partition function is

$$q_{\text{v}} = e^{-\mathbf{h}\nu/\mathbf{k}T}\left(1 + e^{-\mathbf{h}\nu/\mathbf{k}T} + e^{-2\mathbf{h}\nu/\mathbf{k}T} + \cdots\right) = \sum_{v=0}^{\infty} e^{-(v+\frac{1}{2})\mathbf{h}\nu/\mathbf{k}T}$$

If we omit the term belonging to the zero-point energy of the molecule, then the partition function is

$$q_v = \sum_{v=0}^{\infty} e^{-v\mathbf{h}\nu/\mathbf{k}T}$$

Let us introduce the following abbreviation:

$$x = e^{-\mathbf{h}\nu/\mathbf{k}T}$$

With this notation the partition function can be written in the following form (if $x < 1$):

$$\begin{aligned} q_v &= (1 + x + x^2 + \cdots) \\ &= \frac{1}{1 - x} \\ &= \frac{1}{1 - e^{-\mathbf{h}\nu/\mathbf{k}T}} \end{aligned}$$

Let us define the characteristic temperature in the following way:

$$\frac{\mathbf{h}\nu}{\mathbf{k}} = \Theta_v$$

At the characteristic temperature the energy of the vibrational quantum is $k\Theta_v$. Hence, the vibrational partition function is

$$q_v = \frac{1}{1 - e^{-\Theta_v/T}}$$

The above can be generalized to polyatomic molecules. In them, vibrations can occur in several modes, and each normal mode with its own frequency is taken into account as a linear harmonic oscillator.

Vibrational contributions are summarized in Table 7.1

Algorithm and Program: Vibrational Partition Function

Write a program to calculate the vibrational contribution to the partition function of a perfect gas. The form of the function can be seen in Table 7.1.

Algorithm.

1. Read the characteristic temperature of vibration.
2. Read the actual temperature.
3. Calculate the value of the function.
4. Print the results.

Program PARTFUNCVIBR. A possible realization:

```
]PR#0
]LOAD PARTFUNCVIBR
]LIST

 10  REM :CALCULATION OF VIBRATIONAL
      PARTITION FUNCTION
 20  PRINT : PRINT
 30  PRINT "ENTER CHARACTERISTIC
      TEMPERATURE"
 40  INPUT "OF VIBRATION (K) ";TV
 50  PRINT
 60  INPUT "ENTER TEMPERATURE (K)
      ";T
 70  QV = 1 / (1 - EXP ( - TV / T)
      )
 80  PRINT : PRINT
 90  PRINT "VALUE OF VIBRATIONAL"
100  PRINT "PARTITION FUNCTION ";
      QV
110  PRINT : PRINT
120  INPUT "NEXT TEMPERATURE ? (Y
      /N) ";A$
130  IF A$ = "Y" THEN GOTO 50
140  INPUT "NEXT COMPOUND ? (Y / N)
      ";A$
150  IF A$ = "Y" THEN GOTO 20
160  END
```

Example. Run the program for hydrogen, chlorine, and hydrochloric acid at temperatures 1000, 2000, and 5000 K.

```
]RUN

ENTER CHARACTERISTIC TEMPERATURE
OF VIBRATION (K) 6338.3
ENTER TEMPERATURE (K) 1000

VALUE OF VIBRATIONAL
PARTITION FUNCTION 1.00177043

NEXT TEMPERATURE ? (Y / N) Y

ENTER TEMPERATURE (K) 2000

VALUE OF VIBRATIONAL
PARTITION FUNCTION 1.04388418

NEXT TEMPERATURE ? (Y / N) Y

ENTER TEMPERATURE (K) 5000

VALUE OF VIBRATIONAL
PARTITION FUNCTION 1.39176818
```

```
NEXT TEMPERATURE ? (Y / N) N
NEXT COMPOUND ? (Y / N) Y

ENTER CHARACTERISTIC TEMPERATURE
OF VIBRATION (K) 807.3
ENTER TEMPERATURE (K) 1000

VALUE OF VIBRATIONAL
PARTITION FUNCTION 1.80525229

NEXT TEMPERATURE ? (Y / N) Y

ENTER TEMPERATURE (K) 2000

VALUE OF VIBRATIONAL
PARTITION FUNCTION 3.01094029

NEXT TEMPERATURE ? (Y / N) Y

ENTER TEMPERATURE (K) 5000

VALUE OF VIBRATIONAL
PARTITION FUNCTION 6.70693362

NEXT TEMPERATURE ? (Y / N) N
NEXT COMPOUND ? (Y / N) Y

ENTER CHARACTERISTIC TEMPERATURE
OF VIBRATION (K) 4301.38
ENTER TEMPERATURE (K) 1000

VALUE OF VIBRATIONAL
PARTITION FUNCTION 1.01373597

NEXT TEMPERATURE ? (Y / N) Y

ENTER TEMPERATURE (K) 2000

VALUE OF VIBRATIONAL
PARTITION FUNCTION 1.1317387

NEXT TEMPERATURE ? (Y / N) Y

ENTER TEMPERATURE (K) 5000

VALUE OF VIBRATIONAL
PARTITION FUNCTION 1.73323835

NEXT TEMPERATURE ? (Y / N) N
NEXT COMPOUND ? (Y / N) N
```

Algorithm and Program: Vibrational Contribution to Thermodynamic Properties

Write a program to calculate the vibrational contribution to the thermodynamic properties of the perfect gas. The corresponding formulas are listed in Table 7.1.

Algorithm.

1. Read the characteristic temperature of vibration.
2. Read the temperature.
3. Calculate C^*_{pv}, h^*_v, s^*_v, g^*_v.
4. Print the results.

Program VIBRCONTR. A possible realization:

```
]PR#0
]LOAD VIBRCONTR
]LIST

 10  REM : CALCULATION OF VIBRATIONAL
      CONTRIBUTIONS TO THE THERMO
      DYNAMIC PROPERTIES
 20  PRINT : PRINT
 30  PRINT "ENTER CHARACTERISTIC
      TEMPERATURE"
 40  INPUT "OF VIBRATION (K) ";TV
 50  PRINT
 60  INPUT "ENTER TEMPERATURE (K)
      ";T
 70  R = 8.3144
 80  CV = R * (TV / T) ∧ 2 * ( EXP
     (TV / T) / ( EXP (TV / T) -
     1) ∧ 2)
 90  HV = R * T * TV / T / ( EXP (T
     V / T) - 1)
100  SV = R * (TV / T / ( EXP (TV /
     T) - 1) - LOG(1 - EXP ( -
     TV / T)))
110  GV = R * T * LOG (1 - EXP (
     - TV / T))
120  PRINT : PRINT
130  PRINT "CPO (J / MOL K)"; TAB(
     25);CV
140  PRINT "ENTHALPY (J / MOL)"; TAB(
     25);HV
150  PRINT "ENTROPY (J / MOL K)"; TAB(
     25);SV
160  PRINT "FREE ENERGY (J / MOL)";
     TAB( 25);GV
170  PRINT : PRINT
180  INPUT "NEXT TEMPERATURE ? (Y
     /N) "'; A$
190  IF A$ = "Y" THEN GOTO 50
200  INPUT "NEXT COMPOUND ? (Y / N)
```

```
    "; A$
210  IF A$ = "Y" THEN GOTO 20
220  END
```

Example. Run the program for molecules of hydrogen, chlorine, and hydrochloric acid at temperatures 298, 1000, and 500 K.

```
]RUN

ENTER CHARACTERISTIC TEMPERATURE
OF VIBRATION (K) 6338.3
ENTER TEMPERATURE (K) 298

        CPO (J / MOL K)             2.17837939E - 06
        ENTHALPY (J / MOL)          3.05206132E - 05
        ENTROPY (J / MOL K)         1.05489691E - 07
        FREE ENERGY (J / MOL)       -9.15314683E - 07

NEXT TEMPERATURE ? (Y / N) Y

ENTER TEMPERATURE (K) 1000

          CPO (J / MOL K)             .592412482
          ENTHALPY (J / MOL)          93.3003344
          ENTROPY (J / MOL K)         .108007408
          FREE ENERGY (J / MOL)       -14.7070732

NEXT TEMPERATURE ? (Y / N) Y

ENTER TEMPERATURE (K) 5000

          CPO (J / MOL K)             7.28505025
          ENTHALPY (J / MOL)          20645.8547
          ENTROPY (J / MOL K)         6.87770381
          FREE ENERGY (J / MOL)       -13742.6644

NEXT TEMPERATURE ? (Y / N) N
NEXT COMPOUND ? (Y / N) Y

ENTER CHARACTERISTIC TEMPERATURE
OF VIBRATION (K) 807.3
ENTER TEMPERATURE (K) 298

          CPO (J / MOL K)             4.66446794
          ENTHALPY (J / MOL)          478.925306
          ENTROPY (J / MOL K)         2.18016661
          FREE ENERGY (J / MOL)       -170.764344

NEXT TEMPERATURE ? (Y / N) Y
```

```
ENTER TEMPERATURE (K) 1000

          CPO (J / MOL K)             7.87717864
          ENTHALPY (J / MOL)          5405.02661
          ENTROPY (J / MOL K)         10.3163457
          FREE ENERGY (J / MOL)       -4911.31904

NEXT TEMPERATURE ? (Y / N) Y

ENTER TEMPERATURE (K) 5000

          CPO (J / MOL K)             8.29636093
          ENTHALPY (J / MOL)          38306.1661
          ENTROPY (J / MOL K)         23.4847159
          FREE ENERGY (J / MOL)       -79117.4134

NEXT TEMPERATURE ? (Y / N) N
NEXT COMPOUND ? (Y / N) Y

ENTER CHARACTERISTIC TEMPERATURE
OF VIBRATION (K) 4301.3
ENTER TEMPERATURE (K) 298

        CPO (J / MOL K)             9.33336837E - 04
        ENTHALPY (J / MOL)          .0192695232
        ENTROPY (J / MOL K)         6.91416906E - 05
        FREE ENERGY (J / MOL)       -1.33470062E - 03

NEXT TEMPERATURE ? (Y / N) Y

ENTER TEMPERATURE (K) 001000

          CPO (J / MOL K)             2.14215156
          ENTHALPY (J / MOL)          491.275523
          ENTROPY (J / MOL K)         .604713732
          FREE ENERGY (J / MOL)       -113.438209

NEXT TEMPERATURE ? (Y / N) Y

ENTER TEMPERATURE (K) 5000

          CPO (J / MOL K)             7.82007595
          ENTHALPY (J / MOL)          26223.3313
          ENTROPY (J / MOL K)         9.81761342
          FREE ENERGY (J / MOL)       -22864.7358

NEXT TEMPERATURE ? (Y / N) N
NEXT COMPOUND ? (Y / N) N
```

PARTITION FUNCTION OF THE ROTATOR

In the elaboration of the rotational partition function of the diatomic molecule we must also start from the results of quantum-mechanical calculations. In Chapter 1 the following result was obtained for the energy eigenvalues of the rotator:

$$\varepsilon_r = J(J+1)\frac{\mathbf{h}^2}{8\pi^2 I}$$

where J = rotational quantum number,
$\mathbf{h}$ = Planck's constant,
I = moment of inertia.

However, an important circumstance was not taken into consideration. The energy eigenvalue actually depends only on the value of J, but the wave function also depends on the quantum number m, the value of which can be any integer between $-J$ and J, zero included. This means that if we do not take explicit account of the quantum number m, then the rotational energy level $\varepsilon_r(J)$ is $(2J+1)$-fold degenerate.

Hence the molecular partition function of the rotator is

$$q_r = \sum_{J=0}^{\infty} (2J+1)\exp\left(-J(J+1)\frac{\mathbf{h}^2}{8\pi^2 I\mathbf{k}T}\right)$$

Summation can be replaced by integration if the multiplier of $J(J+1)$ is sufficiently small, that is, if the moment of inertia I is sufficiently large. (This is the case with most of the molecules, with the exception of H_2, HD, and D_2.)

Let us introduce the following notation:

$$b = \frac{\mathbf{h}^2}{8\pi^2 I\mathbf{k}T}$$

We also introduce a new variable:

$$z = J + \tfrac{1}{2}$$

With this notation the summation can be rewritten as an integration:

$$q_r = \int_0^{\infty} e^{-bz^2} 2z\,dz = \frac{1}{b}$$

Thus, molecular partition function is

$$q_r = \frac{8\pi^2 I\mathbf{k}T}{\mathbf{h}^2} = \frac{T}{\Theta_r}$$

where Θ_r is the characteristic temperature of rotation.

This expression is true only if the two atoms of the diatomic molecule are different, so that exchange of the two atoms in a given orientation gives a new configuration. Molecules consisting of identical atoms, that is, symmetric molecules, possess half as many rotational states as asymmetric molecules. Thus the rotational partition function of symmetric diatomic molecules is

$$q_r = \frac{8\pi^2 I \mathbf{k} T}{2\mathbf{h}^2}$$

where the factor 2 appearing in the denominator is the actual value of the *symmetry number* σ.

For polyatomic molecules the rotational partition function is

$$q_r = \frac{\pi^2}{\sigma}\left(\frac{T^3}{\Theta_A \Theta_B \Theta_C}\right)^{1/2}$$

and this relation involves the three characteristic temperatures:

$$\Theta_A = \frac{\mathbf{h}^2}{8\pi^2 I_A \mathbf{k}}$$

$$\Theta_B = \frac{\mathbf{h}^2}{8\pi^2 I_B \mathbf{k}}$$

$$\Theta_C = \frac{\mathbf{h}^2}{8\pi^2 I_C \mathbf{k}}$$

where I_A, I_B, and I_C are the three principal moments of inertia of the polyatomic molecule. A few symmetry-number values by way of example:

Substance	σ
Ar	0
HCl	1
H_2	2
NH_3	3
CH_4	12

Algorithm and Program: Rotational Partition Function

Write a program for the calculation of the rotational partition function. The form of the function can be seen in Table 7.1.

Algorithm.

1. Read the moment of inertia and symmetry number (see the result of program NYOM in Chapter 1).
2. Read the actual temperature.
3. Calculate the value of the function.
4. Print the results.

Program PARTFUNCROT. A possible realization:

```
]PR#0
]LOAD PARTFUNCROT
]LIST

 10  REM :CALCULATION OF ROTATIONAL
      PARTITION FUNCTION
 20  PRINT : PRINT
 30  PRINT "ENTER MOMENT OF INERTIA"
 40  INPUT " * 10 ∧ + 20 (KG * M ∧ 2)
      ";I
 50  INPUT "ENTER SYMMETRY NUMBER
      (1 / 2) ";SI
 60  PRINT
 70  INPUT "ENTER TEMPERATURE (K)
      ";T
 80  PI = 3.1415
 90  H = 6.626176E - 34
100  K = 1.380662E - 23
110  TA = 2 * LOG (H) - LOG (8) -
      2 * LOG (PI) - LOG (I) - LOG
      (K) - LOG (1E - 20)
120  TA = EXP (TA)
130  QR = T / SI / TA
140  PRINT : PRINT
150  PRINT "VALUE OF ROTATIONAL"
160  PRINT "PARTITION FUNCTION ";
      QR
170  PRINT : PRINT
180  INPUT "NEXT TEMPERATURE ? (Y
      /N) "; A$
190  IF A $ = "Y" THEN GOTO 60
200  INPUT "NEXT COMPOUND ? (Y / N)
```

```
        "; A$
210  IF A$ = "Y" THEN GOTO 20
220  END
```

Example. Run the program for molecules of hydrogen, chlorine, and hydrochloric acid at temperatures 298, 1000, and 1500 K.

```
]RUN

ENTER MOMENT OF INERTIA
 * 10 ∧ + 20 (KG * M ∧ 2) 4.5995E - 28
ENTER SYMMETRY NUMBER (1 / 2) 2
ENTER TEMPERATURE (K) 298

VALUE OF ROTATIONAL
PARTITION FUNCTION 1.70146273

NEXT TEMPERATURE ? (Y / N) Y

ENTER TEMPERATURE (K) 1000

VALUE OF ROTATIONAL
PARTITION FUNCTION 5.70960649

NEXT TEMPERATURE ? (Y / N) Y

ENTER TEMPERATURE (K) 1500

VALUE OF ROTATIONAL
PARTITION FUNCTION 8.56440973

NEXT TEMPERATURE ? (Y / N) N
NEXT COMPOUND ? (Y / N) Y

ENTER MOMENT OF INERTIA
* 10 ∧ + 20 (KG * M ∧ 2) 1.5684E - 25
ENTER SYMMETRY NUMBER (1 / 2) 2
ENTER TEMPERATURE (K) 298

VALUE OF ROTATIONAL
PARTITION FUNCTION 580.187881
NEXT TEMPERATURE ? (Y / N) Y

ENTER TEMPERATURE (K) 1000

VALUE OF ROTATIONAL
PARTITION FUNCTION 1946.9392

NEXT TEMPERATURE ? (Y / N) Y

ENTER TEMPERATURE (K) 1500

VALUE OF ROTATIONAL
PARTITION FUNCTION 2920.40879

NEXT TEMPERATURE ? (Y / N) N
NEXT COMPOUND ? (Y / N) Y
```

```
ENTER MOMENT OF INERTIA
 * 10 ^ + 20 (KG * M ^ 2) 2.6419E - 27
ENTER SYMMETRY NUMBER (1 / 2) 1

ENTER TEMPERATURE (K) 298

VALUE OF ROTATIONAL
PARTITION FUNCTION 19.546013

NEXT TEMPERATURE ? (Y / N) Y

ENTER TEMPERATURE (K) 1000

VALUE OF ROTATIONAL
PARTITION FUNCTION 65.5906477

NEXT TEMPERATURE ? (Y / N) Y

ENTER TEMPERATURE (K) 1500

VALUE OF ROTATIONAL
PARTITION FUNCTION 98.3859714

NEXT TEMPERATURE ? (Y / N) N
NEXT COMPOUND ? (Y / N) N
```

Algorithm and Program: Rotational Contribution to Thermodynamic Properties

Write a program to calculate the rotational contribution to the thermodynamic properties of ideal gases. The corresponding formulas are listed in Table 7.1.

Algorithm.

1. Read the moment of inertia and symmetry number.
2. Read the actual temperature.
3. Calculate the rotational characteristic temperature

$$\Theta = \frac{\mathbf{h}^2}{8\pi^2 I \mathbf{k}}$$

4. Calculate C_{pr}^*, h_r^*, S_r^*, g_r^*.
5. Print the results.

Program ROTCONTR. A possible realization:

```
]PR#0
]LOAD ROTCONTR
]LIST

 10  REM :CALCULATION OF ROTATIONAL
       CONTRIBUTIONS TO THE THERMO
```

```
     DYNAMIC PROPERTIES
 20  PRINT : PRINT
 30  PRINT "ENTER MOMENT OF INERTIA"
 40  INPUT " * 10 ∧ + 20 (KG * M ∧ 2)
      ";I
 50  INPUT "ENTER SYMMETRY NUMBER
      (1 / 2) ";SI
 60  PRINT
 70  INPUT "ENTER TEMPERATURE (K)
      ";T
 80  PI = 3.1415
 90  H = 6.626176E - 34
100  K = 1.380662E - 23
110  R = 8.3144
120  E = 2.71828183
130  TA = 2 * LOG (H) - LOG (8) -
      2 * LOG (PI) - LOG (I) - LOG
      (K) - LOG (1E - 20)
140  TA = EXP (TA)
150  CR = R
160  HR = R * T
170  SR = R * LOG (E ~ T / SI / T
      A)
180  RG = - R * T * LOG (T / SI /
      TA)
190  PRINT : PRINT
200  PRINT "CP0 (J / MOL K) "; TAB(
      25);CR
210  PRINT "ENTHALPY (J / MOL) "; TAB(
      25);HR
220  PRINT "ENTROPY (J / MOL K)"; TAB(
      25);SR
230  PRINT "FREE ENERGY (J / MOL)";
      TAB(25);RG
240  PRINT : PRINT
250  INPUT "NEXT TEMPERATURE ? (Y
      /N) ";A$
260  IF A$ = "Y" THEN GOTO 60
270  INPUT "NEXT COMPOUND ? (Y / N)
      ";A$
280  IF A$ = "Y" THEN GOTO 20
290  END
```

Example. Run the program for molecules of hydrogen, chlorine, and hydrochloric acid at temperatures 298, 1000, and 5000 K.

```
]RUN

ENTER MOMENT OF INERTIA
 * 10 ∧ + 20 (KG * M ∧ 2) 4.59953E - 28
```

```
ENTER SYMMETRY NUMBER (1 / 2) 2
ENTER TEMPERATURE (K) 1000

          CP0 (J / MOL K)              8.3144
          ENTHALPY (J / MOL)           8314.4
          ENTROPY (J / MOL K)          22.7993871
          FREE ENERGY (J / MOL)        -14484.9871

NEXT TEMPERATURE ? (Y / N) Y
ENTER TEMPERATURE (K) 5000

          CPO (J / MOL K)              8.3144
          ENTHALPY (J / MOL)           41572
          ENTROPY (J / MOL K)          36.1808977
          FREE ENERGY (J / MOL)        -139332.488

NEXT TEMPERATURE ? (Y / N) Y
ENTER TEMPERATURE (K) 298

          CP0 (J / MOL K)              8.3144
          ENTHALPY (J / MOL)           2477.6912
          ENTROPY (J / MOL K)          12.7334607
          FREE ENERGY (J / MOL)        -1316.88009

NEXT TEMPERATURE ? (Y / N) N
NEXT COMPOUND (Y / N) Y
ENTER MOMENT OF INERTIA
 * 10 ∧ + 20 (KG * M ∧ 2) 1.5684E - 25
ENTER SYMMETRY NUMBER (1 / 2) 2
ENTER TEMPERATURE (K) 298

          CP0 (J / MOL K)              8.3144
          ENTHALPY (J / MOL)           2477.6912
          ENTROPY (J / MOL K)          61.2218537
          FREE ENERGY (J / MOL)        -15766.4212

NEXT TEMPERATURE ? (Y / N) Y
ENTER TEMPERATURE (K) 1000

          CP0 (J / MOL K)              8.3144
          ENTHALPY (J / MOL)           8314.4
          ENTROPY (J / MOL K)          71.2877802
          FREE ENERGY (J / MOL)        -62973.3801

NEXT TEMPERATURE ? (Y / N) Y
```

```
ENTER TEMPERATURE (K) 5000

          CP0 (J / MOL K)                8.3144
          ENTHALPY (J / MOL)             41572
          ENTROPY (J / MOL K)            84.6692908
          FREE ENERGY (J / MOL)          -381774.454

NEXT TEMPERATURE ? (Y / N) N
NEXT COMPOUND ? (Y / N) Y

ENTER MOMENT OF INERTIA
* 10 ∧ + 20 (KG * M ∧ 2) 2.611909E - 27
ENTER SYMMETRY NUMBER (1 / 2) 1
ENTER TEMPERATURE (K) 298

          CP0 (J / MOL K)                8.3144
          ENTHALPY (J / MOL)             2477.6912
          ENTROPY (J / MOL K)            32.9362847
          FREE ENERGY (J / MOL)          -7337.32163

NEXT TEMPERATURE ? (Y / N) Y

ENTER TEMPERATURE (K) 1000

          CP0 (J / MOL K)                8.3144
          ENTHALPY (J / MOL)             8314.4
          ENTROPY (J / MOL K)            43.0022111
          FREE ENERGY (J / MOL)          -3487.811

NEXT TEMPERATURE ? (Y / N) Y

TEMPERATURE (K) 5000

          CP0 (J / MOL K)                8.3144
          ENTHALPY (J / MOL)             41572
          ENTROPY (J / MOL K)            56.3837216
          FREE ENERGY (J / MOL)          -240346.608

NEXT TEMPERATURE ? (Y / N) Y
NEXT COMPOUND ? (Y / N) N
```

Algorithm and Program: Rotational Contribution for Polyatomic Molecules

Write a program to calculate the rotational contribution to the thermodynamic properties of an ideal gas of polyatomic molecules. The rotational characteristic temperature is

$$\Theta = \Theta_A \Theta_B \Theta_C$$

The rotational partition function is

$$q_r = \frac{\pi^{1/2}}{\sigma}\left(\frac{T^3}{\Theta}\right)^{1/2}$$

where σ is the symmetry number.

Algorithm.

1. Read the rotational characteristic temperature of the polyatomic molecule.
2. Read the symmetry number.
3. Read the temperature.
4. Calculate g_r.
5. Calculate c_{pr}^*, h_r^*, s_r^*, g_r^*.
6. Print the results.

Program ROTCONTRPOLY. A possible realization:

```
]PR#0
]LOAD ROTCONTRPOLY
]LIST

 10  REM :CALCULATION OF ROTATIONAL
      CONTRIBUTIONS TO THE THERMO
      DYNAMIC PROPERTIES FOR POLY
      ATOMIC MOLECULES
 20  PRINT : PRINT
 30  PRINT "ENTER ROTATIONAL CHARA
      CTERISTIC"
 40  PRINT "TEMPERATURE OF POLYATO
      MIC"
 50  INPUT "MOLECULE (K) ";TA
 60  PRINT
 70  INPUT "ENTER SYMMETRY NUMBER
      ";SI
 80  PRINT
 90  INPUT "ENTER TEMPERATURE (K)
      ";T
100  PI = 3.1415
110  H = 6.626176E - 34
120  K = 1.380662E - 23
130  R = 8.3144
140  E = 2.71828183
150  CR = 3 * R / 2
160  HR = 3 * R * T / 2
```

```
170  SR = R * LOG ( SQR (PI) / SI
      * SQR (T ∧ 3 * E ∧ 3 / TA)
180  RG = - R * T * LOG ( SQR (P
      I) / SI * SQR ( T ∧ 3 / TA))
190  PRINT : PRINT
200  PRINT "CP0 (J / MOL K) "; TAB(
      25);CR
210  PRINT "ENTHALPY (J / MOL)"; TAB(
      25);HR
220  PRINT "ENTROPY (J / MOL K)"; TAB(
      25);SR
230  PRINT "FREE ENERGY (J / MOL)";
      TAB(25);RG
240  PRINT : PRINT
250  INPUT "NEXT TEMPERATURE ? (Y
      /N) ";A$
260  IF A$ = "Y" THEN GOTO 80
270  INPUT "NEXT COMPOUND ? (Y / N)
      ";A$
280  IF A$ = "Y" THEN GOTO 20
290  END
```

Example. Run the program for ammonia ($\sigma = 3$, $\Theta = 1876$) at temperatures 298, 500, 1000, and 1500 K.

```
]RUN

ENTER ROTATIONAL CHARACTERISTIC
TEMPERATURE OF POLYATOMIC
MOLECULE (K) 1876

ENTER SYMMETRY NUMBER 3

ENTER TEMPERATURE (K) 298

                CP0 (J / MOL K)            12.4716
                ENTHALPY (J / MOL)         3716.5368
                ENTROPY (J / MOL K)        47.8155289
                FREE ENERGY (J / MOL)      -10532.4908

NEXT TEMPERATURE ? (Y / N) Y

ENTER TEMPERATURE (K) 1000

                CP0 (J / MOL K)            12.4716
                ENTHALPY (J / MOL)         12471.6
                ENTROPY (J / MOL K)        62.9144185
                FREE ENERGY (J / MOL)      -50442.8185

NEXT TEMPERATURE ? (Y / N) Y
```

```
ENTER TEMPERATURE (K) 5000

            CP0 (J / MOL K)              12.4716
            ENTHALPY (J / MOL)           62358
            ENTROPY (J / MOL K)          82.9866843
            FREE ENERGY (J / MOL)        -352575.422

NEXT TEMPERATURE ? (Y / N) N
NEXT COMPOUND ? (Y / N) N
```

Algorithm and Program: Calculation of Thermodynamic Properties of Ideal Gases from Spectroscopic Data

Write a program to calculate the ideal-gas heat capacity, enthalpy, entropy, and free energy from the molecular mass, vibrational characteristic temperature, moment of inertia, and symmetry number.

Algorithm.

1. Read the molecular mass, vibrational characteristic temperature, moment of inertia, and symmetry number.
2. Read the pressure and temperature.
3. Calculate the translational contribution as in program TRANSCONTR.
4. Calculate the vibrational contribution as in program VIBRCONTR.
5. Calculate the rotational contribution as in program ROTCONTR.
6. Add the contributions.
7. Print the results.

Program THERMOSPECT. A possible realization:

```
]PR#0
]LOAD THERMOSPACT
]LIST

 10  REM :CALCULATION OF THERMODYN
      AMIC PROPERTIES OF IDEAL GAS
      FROM SPECTROSCOPIC DATA (LI
      NEAR MOLECULES ONLY)
 20  PRINT : PRINT
 30  INPUT "ENTER THE MOLECULAR
      MASS (KG / MOL) ";M
 40  PRINT
 50  PRINT "ENTER CHARACTERISTIC
      TEMPERATURE"
 60  INPUT "OF VIBRATION (K) ";TV
 70  PRINT
```

```
 80  PRINT "ENTER MOMENT OF INERTIA"
 90  INPUT " * 10 ∧ + 20 (KG * M ∧ 2)
      ";I
100  PRINT
110  INPUT "ENTER SYMMETRY NUMBER
      (1 / 2) ";SI
120  PRINT
130  INPUT "ENTER PRESSURE (PA) "
      ;P
140  INPUT "ENTER THE TEMPERATURE
      (K) ";T
150  H = 6.626176E - 34
160  K = 1.380662E - 23
170  NA = 6.022045E - 23
180  R = 8.3144
190  PI = 3.1415
200  E = 2.71828183
210  QT = 3 / 2 / * ( LOG (2) + LOG
      (3.1415) + LOG (M) - LOG (
      NA) + LOG (K) + LOG (T) -
      2 * LOG (H)) + LOG (K) + LOG
      (T) - LOG (P)
220  CT = 5 * R / 2
230  HT = 5 * R * T / 2
240  ST = R * (5 / 2 + QT)
250GT = - R * T * QT
260  CV = R * (TV / T) ∧ 2 * ( EXP
      (TV / T) / ( EXP (TV / T) -
      1) ∧ 2)
270  HV = R * T * TV / T / ( EXP (
      TV / T) - 1)
280  SV = R * (TV / T / ( EXP (TV /
      T) - 1) - LOG (1 - EXP ( -
      TV / T )))
290  GV = R * T * LOG (1 - EXP (
      - TV / T ))
300  TA = 2 * LOG (H) - LOG (8) -
      2 * LOG (PI) - LOG (I) - LOG
      (K) - LOG (1E - 20)
310  TA = EXP (TA)
320  CR = R
330  HR = R * T
340  SR = R * LOG (E * T / SI / T
      A)
350  RG = - R * T * LOG (T / SI /
      TA)
360  C = CT + CV + CR
370  H = HT + HV + HR
380  S = ST + SV + SR
```

```
390  G = GT + GV + RG
400  PRINT : PRINT
410  PRINT "**********************
      *****************"
420  PRINT "CP0 (J / MOL K)"; TAB(
      25);C
430  PRINT "ENTHALPY (J / MOL)";TAB(
      25);H
440  PRINT "ENTROPY (J / MOL K)"; TAB(
      25);S
450  PRINT "FREE ENERGY (J / MOL)";
      TAB(25);G
460  PRINT "**********************
      *****************"
470  PRINT : PRINT
480  INPUT "NEXT TEMPERATURE ? (Y
      /N) ";A$
490  IF A$ = "Y" THEN GOTO 140
500  INPUT "NEXT PRESSURE ? (Y / N)
      ";A$
510  IF A$ = "Y" THEN GOTO 130
520  INPUT "NEXT COMPOUND ? (Y / N)
      ";A$
530  IF A$ = "Y" THEN GOTO 20
540  END
```

Example. Run the program for hydrogen, chlorine, and hydrochloric acid at temperatures 298 and 100 K and at pressures 100,000 and 1,000,000 Pa.

```
]RUN

ENTER THE MOLECULAR MASS (KG / MOL) 2.016

ENTER CHARACTERISTIC TEMPERATURE
OF VIBRATION (K) 6338.3

ENTER MOMENT OF INERTIA
 * 10 ∧ + 20 (KG * M ∧ 2) 4.599535E - 28

ENTER SYMMETRY NUMBER (1 / 2) 2

ENTER PRESSURE (PA) 100000
ENTER THE TEMPERATURE (K) 298

***************************************

          CP0 (J / MOL K)            29.1004022
          ENTHALPY (J / MOL)         8671.91923
          ENTROPY (J / MOL K)        216.471917
          FREE ENERGY (J / MOL)      -55836.7119

***************************************
```

```
NEXT TEMPERATURE ? (Y / N) Y
ENTER THE TEMPERATURE (K) 1000

****************************************

          CP0 (J / MOL K)                29.6928125
          ENTHALPY (J / MOL)             29193.7003
          ENTROPY (J / MOL K)            251.810667
          FREE ENERGY (J / MOL)          -222616.966

****************************************
NEXT TEMPERATURE ? (Y / N) N
NEXT PRESSURE ? (Y / N) Y
ENTER PRESSURE (PA) 1000000
ENTER THE TEMPERATURE (K) 1000

****************************************

          CP0 (J / MOL K)                29.6928125
          ENTHALPY (J / MOL              29193.7003
          ENTROPY (J / MOL K)            232.666053
          FREE ENERGY (J / MOL)          -203472.353

****************************************

NEXT TEMPERATURE ? (Y / N) N
NEXT PRESSURE ? (Y / N) N
NEXT COMPOUND ? (Y / N) Y

ENTER THE MOLECULAR MASS (KG / MOL) 70.906E - 3

ENTER CHARACTERISTIC TEMPERATURE
OF VIBRATION (K) 807.3

ENTER MOMENT OF INERTIA
 * 10 ^ + 20 (KG * M ^ 2) 1.5684E - 25

ENTER SYMMETRY NUMBER (1 / 2) 2

ENTER PRESSURE (PA) 100000
ENTER THE TEMPERATURE (K) 298

****************************************

          CP0 (J / MOL K)                33.7648679
          ENTHALPY (J / MOL)             9150.84451
          ENTROPY (J / MOL K)            225.391592
          FREE ENERGY (J / MOL)          -58015.8499

****************************************
NEXT TEMPERATURE ? (Y / N) Y
ENTER THE TEMPERATURE (K) 1000
****************************************
```

```
              CP0 (J / MOL K)                36.9775787
              ENTHALPY (J / MOL)             34505.4266
              ENTROPY (J / MOL K)            268.758514
              FREE ENERGY (J / MOL)          -234253.087

****************************************

NEXT TEMPERATURE ? (Y / N) N
NEXT PRESSURE ? (Y / N) N
NEXT COMPOUND ? (Y / N) Y

ENTER THE MOLECULAR MASS (KG / MOL) 36.461E - 3

ENTER CHARACTERISTIC TEMPERATURE
OF VIBRATION (K) 4301.3

ENTER MOMENT OF INERTIA
 * 10 ∧ + 20 (KG * M ∧ 2) 2.6419E - 27

ENTER SYMMETRY NUMBER (1 / 2) 1

ENTER PRESSURE (PA) 100000
ENTER THE TEMPERATURE (K) 298

****************************************

              CP0 (J / MOL K)                29.1013333
              ENTHALPY (J / MOL)             8671.93847
              ENTROPY (J / MOL K)            186.725842
              FREE ENERGY (J / MOL)          -46972.3623

****************************************

NEXT TEMPERATURE ? (Y / N) Y
ENTER THE TEMPERATURE (K) 1000

****************************************

              CP0 (J / MOL K)                31.2425516
              ENTHALPY (J / OL)              29591.6755
              ENTROPY (J / MOL K)            222.561229
              FREE ENERGY (J / MOL)          -192969.553

****************************************

NEXT TEMPERATURE ? (Y / N) N
NEXT PRESSURE ? (Y / N) N
NEXT COMPOUND ? (Y / N) Y
```

Remarks. If you summarize the results in table form you can see that the ideal-gas heat capacity and enthalpy depend only on the mass of molecules and on the temperature, not on the pressure. The entropy and free energy depend on the pressure too.

THE THIRD LAW OF THERMODYNAMICS

It has been shown already earlier that the molar entropy expressed in terms of the partition function is

$$s = \mathbf{k} \ln Q + \frac{u}{T}$$

$$= \mathbf{k}\left(\ln Q + \frac{u}{\mathbf{k}T}\right)$$

where u is the molar internal energy.

Let us decompose the energy into zero-point energy and thermal energy:

$$u = u_0 + u_{\text{th}}$$

The partition function can be factored also in this way:

$$Q = Q_0 Q_{\text{th}}$$

$$= e^{-u_0/\mathbf{k}T} Q_{\text{th}}$$

Written in logarithmic form, this is

$$\ln Q = \ln Q_{\text{th}} - \frac{u_0}{\mathbf{k}T}$$

and substituting this in the expression for the entropy, we have

$$s = \mathbf{k}\left(\ln Q_{\text{th}} - \frac{u_0}{\mathbf{k}T} + \frac{u}{\mathbf{k}T}\right)$$

However, at absolute zero the thermal energy is by definition zero. Molecules (other than helium atoms) are arranged at absolute zero temperature into crystals, and their vibrational quantum number can only be $v = 0$. The total energy u is equal to the zero-point energy

$$u = u_0$$

For the same reason the partition function of thermal motion is

$$Q_{\text{th}} = 1$$

Substituting this back into the expression of entropy:

$$s = \mathbf{k}\left(\ln Q_{\text{th}} - \frac{u_0}{\mathbf{k}T} + \frac{u_0}{\mathbf{k}T}\right)$$

$$= \mathbf{k} \ln 1$$

This means that the entropy of the simple crystalline body at absolute zero temperature is equal to zero:

$$s_0 = 0$$

This statement cannot be deduced in phenomenological thermodynamics, where it is therefore called the *third law of thermodynamics*.

This law specifies an important reference state.

8

THE ENTHALPY

We return to the physical meaning of enthalpy. We have already seen one aspect of it in the discussion of thermodynamic potentials: an arbitrary thermodynamic body reaches, at constant pressure, an equilibrium state, which minimizes enthalpy. Now we will discuss a one-component homogeneous phase, and consider instead the change in enthalpy with changing state of this phase. Two problems will be discussed. First the relation between internal energy and enthalpy will be investigated. Following this, we will treat the throttling effect.

ENTHALPY AND INTERNAL ENERGY

Let us consider two bodies of identical volume and temperature in the initial state: each one mole of an ideal gas. Let the wall of one of the bodies be rigid, and the other body be in equilibrium with a manostat at pressure P.

Let us now place these two bodies in a bath at a temperature higher by dT than the first. Owing to this, the temperature of both bodies increases, which is accompanied in the first body by an increase in pressure dP at constant volume, and in the second by an increase in volume dV at constant pressure.

The increase in internal energy, du, is the same in both cases, since the internal energy of ideal gases is independent of pressure and volume. This increase in energy is provided in both cases by the bath. However, in the second case the increase in enthalpy of the body is

$$du + P\,dv = d(u + Pv) = dh$$

Since du is the same as in the first case, here there is an additional positive contribution $P\,dv$, which similarly must be provided by the bath. This energy is expended as work done in increasing the volume of the body against the pressure P. Thus, enthalpy comprises that indispensable volumetric work, which is done by the body on the manostat.

Using the defining equation of molar heat capacity, our preceding equation takes the following form:

$$\left(\frac{\partial h}{\partial T}\right)_P^* = \left(\frac{\partial u}{\partial T}\right)_v^* + P\left(\frac{\partial v}{\partial T}\right)_P^*$$

In the second term on the right-hand side, as we are dealing with an ideal gas,

$$\left(\frac{\partial v}{\partial T}\right)^* = \frac{v}{T}$$

Substituting this, we have

$$c_P^* = c_v^* + P\frac{v}{T} = c_v^* + \mathbf{R}$$

That is, in the case of an ideal gas the difference between the isobaric and isochoric molar heat capacities is just **R**, the universal gas constant. In the case of real gases the difference between the two molar heat capacities is

$$c_P - c_v = -T\left(\frac{\partial v}{\partial T}\right)_P^2 \left(\frac{\partial P}{\partial v}\right)_T = \Delta c$$

The ratio of the two is

$$\frac{c_P}{c_v} = -\left(\frac{\partial v}{\partial T}\right)_P \left(\frac{\partial T}{\partial P}\right)_v \left(\frac{\partial P}{\partial v}\right)_s = \kappa$$

With an adequate equation of state Δc and κ can be calculated for a given case.

THE THROTTLING EFFECT

Consider now the case of flowing gas. The cause of the flow is a pressure drop: gas flows from a region of higher pressure to a region of lower pressure. We will use a model of batch-type flow. This model involves a manostat of pressure P_1 and a manostat of pressure P_2, and the manostat of higher pressure pumps one mole of gas through a porous plug into the manostat of lower pressure (Figure 8.1). The whole system is surrounded by an adiabatic wall. The substance is in a homogeneous gaseous phase on both sides of the porous plug.

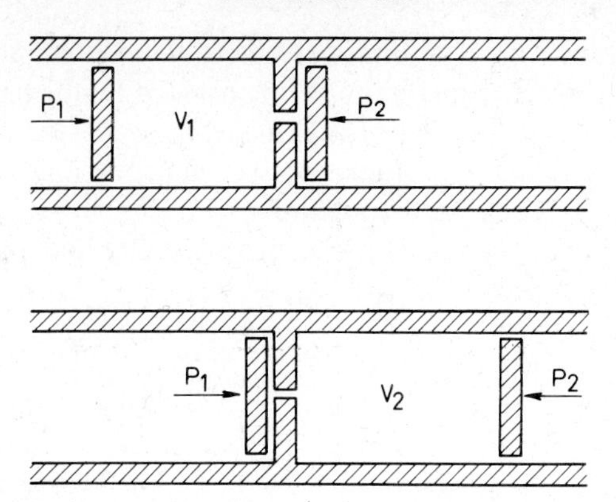

FIGURE 8.1. Transfer of one mole of gas through a porous plug.

The manostat at pressure P_1 performs work P_1v_1 on one mole of gas, while pressing it through the porous plug. The gas performs work $-P_2v_2$ on the manostat at pressure P_2. The transfer work is the difference of these two, and since the wall is adiabatic, this can be supplied only by the mole of gas at the expense of its internal energy, that is,

$$u_2 - u_1 = P_1v_1 - P_2v_2$$

After rearrangement we have

$$u_2 + P_2v_2 = u_1 + P_1v_1$$

so that the flow of gas is an isenthalpic process:

$$h_2 = h_1$$

Flow of gas through a porous plug can be realized in practice; in the case of a large pressure drop the phenomenon is called throttling. Let us investigate the magnitude of the differential throttling effect, defined by the following partial derivative:

$$\mu = \left(\frac{\partial T}{\partial P}\right)_h$$

$$= \frac{1}{c_P}\left(T\left(\frac{\partial v}{\partial T}\right)_P - v\right)$$

The value of the differential throttling effect for given values of P and T can be calculated from the molar heat capacity with a suitable equation of state. Its sign is determined by the expression in the outer parentheses on the right-hand side, because $c_P > 0$ always.

For an ideal gas

$$\left(\frac{\partial v}{\partial T}\right)_P^* = \frac{v}{T}$$

and hence

$$\left(T\left(\frac{v}{T}\right) - v\right)^* = 0$$

It follows from this that the differential throttling effect of an ideal gas is zero, that is, the temperature of an ideal gas does not change when passing through a throttle. The case is different with real gases. The adiabatic expansion of a real gas passing through a throttle may result in a decrease or increase of gas temperature, or possibly no change, depending on whether the differential throttling effect is positive, negative, or just zero in the temperature and pressure ranges concerned.

Let us transform the condition of zero throttling effect:

$$\mu = \frac{1}{c_P}\left(T\left(\frac{\partial v}{\partial T}\right)_P - v\right) = 0$$

$$= \frac{1}{c_P}\left(-T\frac{(\partial P/\partial T)_v}{(\partial P/\partial v)_T} - v\right) = 0$$

This is possible if the expression in large parentheses is zero:

$$T\left(\frac{\partial P}{\partial T}\right)_v + v\left(\frac{\partial P}{\partial v}\right)_T = 0$$

Rewritten in reduced variables, this condition has the following form:

$$T_r\left(\frac{\partial P_r}{\partial T_r}\right)_{v_r} + v_r\left(\frac{\partial P_r}{\partial v_r}\right)_{T_r} = 0$$

Calculating the partials from a suitable equation of state and substituting them in the equation above, we obtain the function $P_r[T_r]$ satisfying the condition of isothermal throttling. The function obtained by a fit to experimental data (not calculated from any equation of state) is the following polynomial:

$$P_r = -36.275 + 71.598T_r - 41.567T_r^2 + 11.826T_r^3 - 1.672T_r^4 + 0.0912T_r^5$$

A plot is shown in Figure 8.2. This is called the *inversion curve*. Within the inversion curve, the differential throttling effect is positive ($\mu > 0$) and thus gas passing through the throttle is cooled. Outside of the inversion curve the

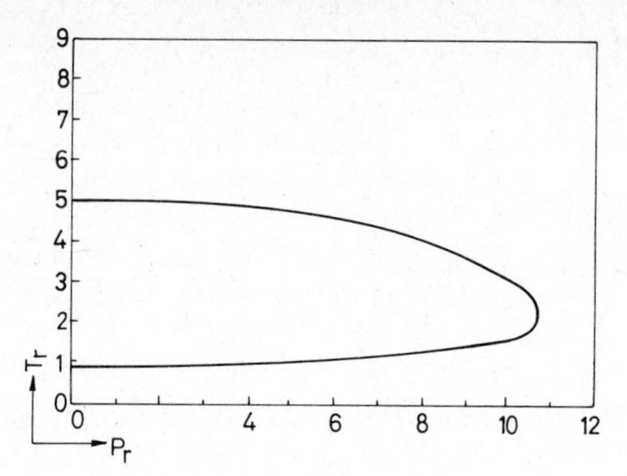

FIGURE 8.2. Inversion curve.

throttling effect is negative ($\mu < 0$), that is, gas passing through the throttle is warmed.

A further study of the figure shows that the temperature range of the positive differential throttling effect narrows with increasing pressure, and there exists a maximal pressure above which the differential throttling effect is always negative. This is called the *inversion pressure*.

A gas can be cooled (and liquefied) by expansion through a throttle only in a range in which differential throttling effect is positive, that is, in which expansion is accompanied by decrease in temperature.

CALCULATION OF ENTHALPY

The discussion of thermodynamics that considers matter as a continuum gives completely general relations between thermodynamic properties. These relations say nothing about the actual magnitudes of the properties, and indeed, cannot contain such information, because these magnitudes depend on the specific molecular structure of the body.

Thus, the specific content of a general thermodynamic relation must be given by constitutive equations. We have already met with such equations. The equations of state expressing the (P,v,T) relation are constitutive equations. The empirical equations of state are connected with intermolecular forces. In discussing them we gave the temperature and pressure ranges, and the class of components, in which they are valid.

Similarly, the work of intramolecular forces and the energy of the single molecules must be given by constitutive equations, so that the energy of the phases consisting of these molecules, and the magnitudes of other thermodynamic properties, can be calculated. This is dealt with in Chapter 7 on statistical mechanics.

The estimation of an arbitrary physical property of a chemical substance (element, compound, or mixture) in a given state is an important practical problem, because data from direct measurements on the given state are seldom available. In solving this problem a computer is conveniently used, as it is helpful for two technical tasks:

1. Storage of available data.
2. Calculation of the desired data from the available data, with the aid of programs.

In this chapter the calculation of data will be restricted to the estimation of enthalpy, because in chemical calculations this is the quantity estimated most often. What we say here, will also be methodologically usable in the next chapter for other properties.

Before we get to the heart of the matter, a remark must be made. The word "estimation" has a certain pejorative flavor; it suggests an activity undertaken for want of something better. This is not the case, as some properties of some substances can be estimated with arbitrary accuracy. However, chemists, physicists, and engineers are satisfied in practical work to know the magnitude of the property to a certain accuracy. This is no compromise, as a higher accuracy would make no sense in the given situation. To determine from available data some property of a substance at the prescribed accuracy with the lowest expenditure of work (costs): this is estimation. The estimation of enthalpy will be dealt with in this sense.

The molar enthalpy is a function only of temperature and pressure:

$$h = \frac{dH}{dN} = h[P, T]$$

The enthalpy of a selected substance can thus be given in a two-dimensional table. Obviously this table will become larger as its temperature or pressure range is extended, or as its intervals are made smaller. In this connection, note that ordinarily the enthalpy value in an arbitrary state can only be calculated with some interpolation formula from the data in the table. If we require, for example, that this shall be linear interpolation, then the table will have to be larger than if we used quadratic interpolation, within the same limits of error.

Such a table was constructed long ago for water, and in recent times two-dimensional tables have been constructed also for the most important substances in the chemical industry. The storage of these tables by a computer is technically possible, and the calculation of the enthalpy for an arbitrary state with a preprogrammed formula does not cause any difficulty. Nevertheless, this technique would be a misuse of the possibilities offered by the computer.

Both the storage of data and the application of interpolation processes cost money. The more data are stored, the more expensive is the storage, but the simpler is the interpolation process that can be used. On the other hand, the sparser is the table, the more complicated is the interpolation process that must be used, and the more costly is interpolation. Storage and computing processes compete with one another, but their joint cost gives the cost of the estimation of the desired data. In practice it is not always easy to find the "cheapest" solution.

Here we propose a possible solution. This is based on the consideration that the molar enthalpy of the desired state (T, P) is taken as an enthalpy difference with respect to a reference state (T_0, P_u). This difference can be regarded as the result of two changes. The first is an isobaric increase in temperature from the reference state to temperature T. Here the molar enthalpy is $h(T, P_u)$. The second change is isothermal, from pressure P_u to P. Here the enthalpy is $h(T, P)$. If we select the reference pressure P_u so that at the chosen temperature the gas shall behave as an ideal gas, then the isobaric change in temperature depends only on the temperature limits, and the enthalpy correction because of the isothermal change in pressure always means the deviation from the enthalpy of the ideal gas:

$$h[T, P] = h^*[T] + \Delta h_{\text{corr}}[T, P]$$

What we have to do now is to discuss separately the calculation of the two terms on the right-hand side.

Algorithm and Program: Creation, Reading, and Maintenance of a Thermodynamic Database

Do not skip the following discussion if you intend to run your own program.

The algorithms in this and next chapters need many physical properties of the individual compounds. These quantities are available in textbooks, monographs, and desk drawers, but the units and reference states used by different sources are not the same. So it can happen that calculations with a good algorithm do not produce comparable results. The collection of the needed data and their conversion to the right units and reference state are very hard work if we have to do it again for every problem. Another disadvantage of this method of data collection is that it requires a large effort in keyboarding, and this is also a source of error. To avoid problems it is strongly recommended to build up a thermodynamic database in permanent storage, in which each record describes one compound, containing the same type of data in the same units and with the same reference state. Then one needs to collect the data only once; storing and retrieving them is the business of the computer. The user need only know the location of the corresponding record in the database and not the data themselves. To demonstrate the utility of such a database we show here the creation and use of one. For our special purposes it will be enough to create a database in which the records contain the following 14 data:

1. Molecular mass (kg/mole).
2. Critical pressure (Pa).
3. Critical temperature (K).

4. Omega factor.
5. Normal boiling point (K).
6. Enthalpy of vaporization at normal boiling point (J/mole).
7. The constant term in the ideal-gas enthalpy polynomial (J/mole).
8. The coefficient of the linear term (J/mole K).
9. The next coefficient (J/mole K^2).
10. The next coefficient (J/mole K^3).
11. The next coefficient (J/mole K^4).
12. The next coefficient (J/mole K^5).
13. The constant I_s in the ideal-gas entropy equation (J/mole K).
14. Critical volume (m^3/mole).
15. Empty.

The 21 compounds chosen are:

Sequential Number	Compound
1	Methane
2	Ethane
3	Propane
4	Butane
5	*n*-Pentane
6	*n*-Hexane
7	*n*-Heptane
8	*n*-Octane
9	*n*-Nonane
10	Chloroform
11	1,1-Dichlorethane
12	1,2-Dichlorethane
13	Methanol
14	Ethanol
15	Water
16	Acetic acid
17	Hydrogen
18	Carbon monoxide
19	Carbon dioxide
20	Carbon
21	Oxygen

The source of data was the database of the authors, which contains 42 properties of 260 compounds collected over the past years. A part of this database can be seen in the Appendix. To enlarge the little database of this chapter you can use the program WRDAT. To print the contents of the database use REDAT. To modify some data in a specified record use MODDAT. The name of the database file is PHDAT. The content of the database is listed after the description of program REDAT.

Structure of Program WRDAT.

1. Open the file PHDAT.
2. Close the file PHDAT.
3. Read the 15 constants from the keyboard and store them in an array.
4. Find the end of the file PHDAT.
5. Write the 15 constants to the diskette file as a record after the last record.
6. If another compound is wanted, go to step 3.
7. Close of file PHDAT.

Listing of Program WRDAT.

```
]LOAD WRDAT
]LIST

 10  REM :EVALUATION OF PHYSICAL
      PROPERTY
 20  REM :DATA FILE
 30  D$ =  CHR$ (4)
 40  PRINT D$;"OPEN PHDAT"
 50  PRINT D$;"CLOSE PHDAT"
 60  PRINT D$;"APPEND PHDAT"
 80  DIM A(16)
 90  INPUT "ENTER MOLECULAR MASS
      (KG / MOL)"; A(1)
100  INPUT "ENTER CRITICAL PRESURE
      (PA)"; A(2)
110  INPUT "ENTER CRITICAL TEMPER
      ATURE (K)"; A(3)
120  INPUT "ENTER OMEGA "; A(4)
130  INPUT "ENTER NORMAL BOILING
      POINT (K)"; A(5)
140  INPUT "ENTER ENTHALPY OF
      VAPORIZATION ON NORMAL BP (J / MO
      L) "; A(6)
150  PRINT "ENTER THE 6 CONSTANTS
      OF IDEAL GAS ENTHALPY POLYN
      OMIAL "
160  FOR I  =  1 TO 6
170  INPUT A(6 + I)
```

```
180  NEXT I
190  INPUT "ENTER THE CONSTANT OF
      IDEAL-GAS ENTROPY EQUATION (J
      /MOL K) "; A(13)
191  INPUT "ENTER CRIT. MOLAR VOL
      UME "; A(14)
192  INPUT "ENTER SOLUBILITY PARA
      METER "; A(15)
193  PRINT D$;"APPEND PHDAT"
195  PRINT D$;"WRITE PHDAT"
200  FOR I = 1 TO 14
210  PRINT A(I);",";
220  NEXT I
230  PRINT A(15)
235  PRINT D$;"PR#0"
240  PRINT : PRINT
250  INPUT "NEXT COMPOUND ? (Y / N)"
      ; A$
260  IF A$ = "Y" THEN GOTO 90
265  PRINT D$;"CLOSE PHDAT"
270  END
```

Structure of Program REDAT.

1. Read the sequential number of the compound from the keyboard.
2. Open, position, and read the diskette file PHDAT.
3. Display the data on the screen.

Listing of Program REDAT.

```
]LOAD REDAT
]LIST

 10  DIM A(15)
 20 INPUT "ENTER THE SERIAL NUMBER
     OF COMPOUND" ;IS
 30 GOSUB 100
 40 FOR I = 1 TO 15
 50 PRINT TAB(3); I; TAB(10); A(
     I)
 60 NEXT I
 70 INPUT "NEXT READING ? (Y / N) ";
     A$
 80 IF A$ = "Y" THEN GOTO 20
 90 END
100 REM :READING OF SPECIFIED
     RECORD OF PHDAT FILE INTO ARRAY
     A(13)
```

```
110 REM :ARRAY A HAS TO BE DECLA
      RED IN THE CALLING PROGRAM
120 D$ =  CHR$ (4)
130 PRINT D$;"OPEN PHDAT"
140 PRINT D$;"POSITION PHDAT,R";
      IS - 1
150 PRINT D$;"READ PHDAT"
160 INPUT A(1), A(2), A(3), A(4), A(5),
      A(6), A(7), A(8), A(9), A(10),
      A(11), A(12), A(13), A(14),
      A(15)
170 PRINT D$; "CLOSE PHDAT"
180 PRINT D$; "IN#0"
190 RETURN
```

Listing of the Content of the File PHDAT.

```
]RUN
ENTER THE SERIAL NUMBER OF COMPOUND 1

                        1       .016043
                        2       4600200
                        3       190.56
                        4       4.0611E - 03
                        5       111.66
                        6       8185.2
                        7       -67143
                        8       37.916
                        9       -.034191
                       10       9.0781E - 05
                       11       -5.9724E - 08
                       12       1.3804E - 11
                       13       -19.613
                       14       9.85E - 05
                       15       0

NEXT READING ? (Y / N) Y
ENTER THE SERIAL NUMBER OF COMPOUND 2

                        1       .03007
                        2       4894000
                        3       305.46
                        4       .094016
                        5       184.96
                        6       14725
                        7       -69207
                        8       34.36
                        9       -9.7285E - 03
```

```
10        1.2752E - 04
11        -1.0199E - 07
12        2.6509E - 11
13        25.911
14        1.463E - 04
15        0

NEXT READING ? (Y / N) Y
ENTER THE SERIAL NUMBER OF COMPOUND 3

 1        .044096
 2        4245500
 3        369.86
 4        .14872
 5        231.15
 6        18786
 7        -81693
 8        33.162
 9        .022072
10        1.5005E - 04
11        -1.3423E - 07
12        3.6675E - 11
13        52.405
14        2.03E - 04
15        0

NEXT READING ? (Y / N) Y
ENTER THE SERIAL NUMBER OF COMPOUND 4

 1        .058124
 2        3799700
 3        425.2
 4        .19673
 5        272.66
 6        22408
 7        -95168
 8        .49614
 9        .19038
10        -6.4465E - 05
11        1.0262E - 08
12        -3.7172E - 13
13        202.24
14        2.5501E - 04
15        0

NEXT READING ? (Y / N) Y
ENTER THE SERIAL NUMBER OF COMPOUND 5
```

1	.072151
2	3333600
3	461
4	.21547
5	301
6	24459
7	−115760
8	−9.51090001
9	.25534
10	−9.6135E − 05
11	1.8132E − 08
12	−9.34430001E − 13
13	258.04
14	3.1102E − 04
15	0

NEXT READING ? (Y / N) Y
ENTER THE SERIAL NUMBER OF COMPOUND 6

1	.086178
2	3029600
3	507.9
4	.29638
5	341.9
6	28872
7	−123000
8	−8.3281
9	.29943
10	−1.138E − 04
11	2.1738E − 08
12	−1.1599E − 12
13	272.44
14	3.6801E − 04
15	0

NEXT READING ? (Y / N) Y
ENTER THE SERIAL NUMBER OF COMPOUND 7

1	.10021
2	2735800
3	540.2
4	.34869
5	371.59
6	31715
7	−137090
8	−9.7035
9	.34791

```
10        -1.3323E - 04
11        2.565E - 08
12        -1.3794E - 12
13        293.27
14        4.2601E - 04
15        0

NEXT READING ? (Y / N) Y
ENTER THE SERIAL NUMBER OF COMPOUND 8

1         .11423
2         2492600
3         569.4
4         .394
5         398.83
6         34390
7         -151650
8         -10.708
9         .39552
10        -1.5186E - 04
11        2.9201E - 08
12        -1.5733E - 12
13        312.02
14        4.9002E - 04
15        0

NEXT READING ? (Y / N) Y
ENTER THE SERIAL NUMBER OF COMPOUND 9

1         .12826
2         2279800
3         595
4         .44047
5         423.95
6         36928
7         -166170
8         -11.621
9         .44293
10        -1.7036E - 04
11        3.2755E - 08
12        -1.7644E - 12
13        330.35
14        5.4303E - 04
15        0

NEXT READING ? (Y / N) Y
ENTER THE SERIAL NUMBER OF COMPOUND 10
```

```
1       .1194
2       5471600
3       536.56
4       .21118
5       334.46
6       29726
7       -97810
8       20.073
9       .10898
10      -8.8127E - 05
11      4.0051E - 08
12      -7.65610001E - 12
13      126.81
14      2.4E - 04
15      0
```

NEXT READING ? (Y / N) Y
ENTER THE SERIAL NUMBER OF COMPOUND 11

```
1       .09896
2       5066200
3       523.16
4       .23467
5       330.45
6       28721
7       -153850
8       11.98
9       .13782
10      -7.6408E - 05
11      2.4344E - 08
12      -3.1694E - 12
13      163.47
14      2.3276E - 04
15      0
```

NEXT READING ? (Y / N) Y
ENTER THE SERIAL NUMBER OF COMPOUND 12

```
1       .09896
2       5370000
3       561
4       .27413
5       356.65
6       32029
7       -155180
8       29.671
9       .10176
```

```
                    10      -4.5156E - 05
                    11      1.1935E - 08
                    12      -1.4167E - 12
                    13      85.06
                    14      2.2001E - 04
                    15      0

NEXT READING ? (Y / N) Y
ENTER THE SERIAL NUMBER OF COMPOUND 13

                     1      .03204
                     2      8095900
                     3      512.56
                     4      .61445
                     5      337.86
                     6      35278
                     7      -190190
                     8      35.706
                     9      -.018947
                    10      1.0284E - 04
                    11      -8.3728E - 08
                    12      2.3623E - 11
                    13      36.755
                    14      1.178E - 04
                    15      0

NEXT READING ? (Y / N) Y
ENTER THE SERIAL NUMBER OF COMPOUND 14

                     1      .04607
                     2      6383500
                     3      516.26
                     4      .74909
                     5      351.46
                     6      38770
                     7      -216020
                     8      17.829
                     9      .07427
                    10      3.1084E - 05
                    11      -5.0547E - 08
                    12      1.7031E - 11
                    13      134.41
                    14      1.669E - 04
                    15      0

NEXT READING ? (Y / N) Y
ENTER THE SERIAL NUMBER OF COMPOUND 15
```

1	.01802
2	22119000
3	647.36
4	.3208
5	373.16
6	40671
7	−239130
8	34.487
9	−7.1269E − 03
10	1.578E − 05
11	−8.9159E − 09
12	1.8704E − 12
13	−5.3633
14	5.63E − 05
15	0

NEXT READING ? (Y / N) Y
ENTER THE SERIAL NUMBER OF COMPOUND 16

1	.060053
2	5785700
3	594.75
4	.454
5	391.66
6	23697
7	−419610
8	13.246
9	.096082
10	−4.2441E − 06
11	−3.1652E − 08
12	1.3573E − 11
13	151.51
14	1.71E − 04
15	0

NEXT READING ? (Y / N) Y
ENTER THE SERIAL NUMBER OF COMPOUND 17

1	2.061E − 03
2	1297000
3	32.96
4	−.22447
5	20.36
6	1092.8
7	57.767
8	26.99
9	5.9639E − 03

```
                    10          -8.0202E - 06
                    11          5.3626E - 09
                    12          -1.229E - 12
                    13          -25.782
                    14          6.414E - 05
                    15          0

NEXT READING ? (Y / N) Y
ENTER THE SERIAL NUMBER OF COMPOUND 18

                     1          .02801
                     2          3505800
                     3          133.06
                     4          .03491
                     5          8166
                     6          6045.7
                     7          -114000
                     8          30.065
                     9          -4.833E - 03
                    10          8.4603E - 06
                    11          -3.8499E - 09
                    12          5.6088E - 13
                    13          28.501
                    14          9.31E - 05
                    15          0

NEXT READING ? (Y / N) Y
ENTER THE SERIAL NUMBER OF COMPOUND 19

                     1          .04401
                     2          7386600
                     3          304.26
                     4          .3974
                     5          194.66
                     6          17166
                     7          -392980
                     8          21.073
                     9          .033522
                    10          -1.5807E - 05
                    11          3.7272E - 09
                    12          -2.5401E - 13
                    13          75.668
                    14          9.4E - 05
                    15          0

NEXT READING ? (Y / N) Y
ENTER THE SERIAL NUMBER OF COMPOUND 20
```

```
                    1     .01201
                    2     0
                    3     0
                    4     0
                    5     3927
                    6     0
                    7     114.91
                    8     -2.3996
                    9     .018431
                   10     3.2112E - 06
                   11     -9.7413E - 09
                   12     3.2732E - 12
                   13     8.2834
                   14     0
                   15     0

NEXT READING ? (Y / N) Y
ENTER THE SERIAL NUMBER OF COMPOUND 21
                    1     .032
                    2     5046000
                    3     154.66
                    4     .018044
                    5     90.16
                    6     6820.3
                    7     -73.029
                    8     30.459
                    9     -8.9909E - 03
                   10     2.0954E - 05
                   11     -1.4465E - 08
                   12     3.4786E - 12
                   13     34.534
                   14     7.337E - 05
                   15     0
NEXT READING ? (Y / N) N
```

Structure of Program MODDAT.

1. Read the sequential number of record you would like to modify.
2. Open PHDAT; locate and read the record.
3. Read the new data and their locations in the record.
4. Write the new record to the diskette file.
5. Close the file.

Listing of Program MODDAT.

```
]PR#0
]LOAD MODDAT
```

```
]LIST

 10  REM :MODIFICATION OF ONE NUMBER
      IN PHDAT FILE
 20  DIM A(15)
 30  INPUT "ENTER THE SEQUENTIAL
      NUMBER";IS
 40  D$ = CHR$ (4)
 50  PRINT D$; "OPEN PHDAT"
 60  PRINT D$;"POSITION PHDAT,R";I
      S - 1
 70  PRINT D$;"READ PHDAT"
 80  INPUT A(1), A(2), A(3), A(4), A(5),
      A(6), A(7), A(8), A(9), A(10),
      A(11), A(12), A(13), A(14), A(15)
 90  PRINT D$;"CLOSE PHDAT"
100  PRINT D$;"INO#0"
110  INPUT "ENTER THE POSITION IN
      THE RECORD ";I
120  INPUT "ENTER THE NEW VALUE";
      N
130  A(I) = N
140  INPUT "OTHER MODIF. IN THIS
      RECORD (Y / N) ";A$
150  IF A$ = "Y" THEN GOTO 110
160  PRINT D$;"OPEN PHDAT"
170  PRINT D$;"POSITION PHDAT,R";
      IS - 1
180  PRINT D$;"WRITE PHDAT"
190  FOR I = 1 TO 14
200  PRINT A(I);",";
210  NEXT I
220  PRINT A(15)
230  PRINT D$;"CLOSE PHDAT"
240  PRINT D$;"PR#0"
250  INPUT "NEXT MODIFICATION ?
      (Y / N)";A$
260  IF A$ = "Y" THEN GOTO 30
270  END
```

Remark. The opening, locating, and reading of the diskette file PHDAT can also be accomplished separately in a subroutine named SRREDAT. This subroutine will be used in programs that need to read the database.

Listing of Subroutine SRREDAT.

```
]PR#0
]LOAD SRREDAT
```

```
]LIST

2000  REM :READING OF A SPECIFIED
       RECORD OF PHDAT FILE INTO
       ARRAY A DIMENSIONED A(15)
2010  REM :THE SEQUENTIAL NUMBER
       OF THE COMPOUND IN THE CALLING
       PROGRAM HAS TO BE "IS"
2020  REM :ARRAY A HAS TO BE DECL
       ARED IN THE CALLING PROGRAM
2030  D$ = CHR$ (4)
2040  PRINT D$;"OPEN PHDAT"
2050  PRINT D$;"POSITION PHDAT, R"
       ;IS - 1
2060  PRINT D$;"READ PHDAT"
2070  INPUT A(1), A(2), A(3), A(4),
       A(5), A(6), A(7), A(8), A(9), A(10),
       A(11), A(12), A(13), A(14), A(15)
2080  PRINT D$;"CLOSE PHDAT"
2090  PRINT D$;"IN#0"
2100  RETURN
```

ENTHALPY OF THE IDEAL GAS

In the chapter on statistical mechanics we gave the method for calculating the thermodynamic properties, including the enthalpy, of ideal gases. We said also that owing to the great time demand of calculations of this kind, calculation results have been published (and are continually being published) in tabular form for the substances occurring most frequently. This led to the idea that for substances for which calculations have already been published, instead of using correct statistical-mechanical formulas, the results of the calculating can be described by simple, accurate approximating functions for the thermodynamic properties such as the enthalpy.

We insist that the general form of the function shall be the same for any substance; only the constants of the function can be component-specific. Moreover, we demand sufficient accuracy to set up an enthalpy balance of satisfactory accuracy. Experience shows that the temperature dependence of the enthalpy of any ideal gas can be described in the temperature interval of practical interest with satisfactory accuracy by a polynomial of fifth degree, so that only the coefficients of the polynomial need be stored. These coefficients are obtained from the published data by the least-squares method. We seek the temperature dependence of molar enthalpy in the following form:

$$h^*[T] = a_0 + a_1 T + a_2 T^2 + a_3 T^3 + a_4 T^4 + a_5 T^5$$

According to the definition of molar heat capacity,

$$c_P^* = \left(\frac{\partial h^*}{\partial T}\right)_P$$

$$= a_1 + 2a_2T + 3a_3T^2 + 4a_4T^3 + 5a_5T^4$$

and because of the relation for the entropy

$$ds = \frac{dh}{T}$$

we have for molar entropy

$$s^*[T, P_u] = a_1 \ln T + 2a_2T + \tfrac{3}{2}a_3T^2 + \tfrac{4}{3}a_4T^3 + \tfrac{5}{4}a_5T^4 + I_s$$

where the meaning of the coefficients is evident, and I_s is the integration constant for the entropy polynomial.

The tables based on statistical-mechanical calculations contain series of data for all three properties, so that it seems convenient to perform parameter estimation simultaneously for the three series of data. Thereby coherent coefficients can be obtained for the computation of the three properties.

The Reference State of Enthalpy

The processing of the data of an enthalpy table yields a polynomial of fifth degree, containing no constant term. In the perfect gas at absolute zero the molar enthalpy is zero. This is one possible choice for the reference state; however, other conventions are also possible, depending on the object of the calculations. Our convention is the following:

1. The standard substance is an ideal gas.
2. The reference temperature is 0 K.
3. The temperature dependence of the molar enthalpy of the ideal gas is given by a polynomial of fifth degree.
4. The molar enthalpy of any element at 0 K temperature is zero.
5. The molar enthalpy of any compound at 0 K temperature is the enthalpy of formation.

Algorithm and Program: Ideal Gas Enthalpy

Write a program to calculate the enthalpy of ideal gases with a polynomial of fifth degree. Pick up the data from your database.

Algorithm.

1. Read the data-file sequential number of the desired compound.
2. Call the database – reader routine.
3. Read the temperature.
4. Calculate the value of the polynomial by the Horner scheme. (This is a very quick calculation method, because it does not need raising to a power.)
5. Print the result.

Program IDENTH. A possible realization:

```
]LOAD IDENTH
]LIST

 10  REM :CALCULATION OF IDEAL
      GAS ENTHALPY
 20  REM :USING THE "PHDAT" DATAFILE
 30  PRINT : PRINT
 40  DIM A(15)
 50  INPUT "ENTER THE SEQUENTIAL NU
      MBER OF THE COMPOUND IN THE
      DATAFILE ";IS
 60  GOSUB 190
 70  PRINT
 80  INPUT "ENTER THE TEMPERATURE
      (K) ";T
 90  H0 = ((((A(12) * T + A(11)) *
      T + A(10)) * T + A(9)) * T +
      A(8)) * T + A(7)
100  PRINT
110  PRINT "THE MOLAR ENTHALPY OF
      IDEAL GAS IS "
120  PRINT " ";HO;"
      (J / MOL) "
130  PRINT
140  INPUT "NEXT TEMPERATURE ? (Y
      /N) "; A$
150  IF A$ = "Y" THEN GOTO 70
160  INPUT "NEXT COMPOUND ? (Y / N)
      "; A$
170  IF A$ = "Y" THEN GOTO 50
180  END
190  REM :READING OF SPECIFIED
      RECORD OF PHDAT FILE INTO
      ARRAY A(15)
200  REM :THE SEQUENTIAL NUMBER
```

```
    OF THE COMPOUND IN THE CALLING
    PROGRAM HAS TO BE "IS"
210 REM :ARRAY A HAS TO BE DECLA
    RED IN THE CALLING PROGRAM
220 D$ = CHR$ (4)
230 PRINT D$; "OPEN PHDAT"
240 PRINT D$; "POSITION PHDAT, R";
    IS - 1
250 PRINT D$; "READ PHDAT"
260 INPUT A(1), A(2), A(3), A(4), A(5),
    A(6), A(7), A(8), A(9), A(10),
    A(11), A(12), A(13), A(14), A(15)
270 PRINT D$; "CLOSE PHDAT"
280 PRINT D$; "IN#0"
290 RETURN
```

Example. Calculate the ideal-gas enthalpy of methane and propane at temperatures 400 and 500 K.

```
]RUN

ENTER THE SEQUENTIAL NUMBER OF THE COMPOUND IN THE DATAFILE 1

ENTER THE TEMPERATURE (K) 400

THE MOLAR ENTHALPY OF IDEAL GAS IS
 -53024.7574 (J / MOL)

NEXT TEMPERATURE ? (Y / N) Y

ENTER THE TEMPERATURE (K) 500

THE MOLAR ENTHALPY OF IDEAL GAS IS
 -48686.5 (J / MOL)

NEXT TEMPERATURE ? (Y / N) N
NEXT COMPOUND ? (Y / N) Y

ENTER THE SEQUENTIAL NUMBER OF THE COMPOUND IN THE DATAFILE 3

ENTER THE TEMPERATURE (K) 400

THE MOLAR ENTHALPY OF IDEAL GAS IS
 -58354.216 (J / MOL)

NEXT TEMPERATURE ? (Y / N) Y

ENTER THE TEMPERATURE (K) 500

THE MOLAR ENTHALPY OF IDEAL GAS IS
 -48081.0313 (J / MOL)

NEXT TEMPERATURE ? (Y / N) N
NEXT COMPOUND ? (Y / N) N
```

Remarks. The enthalpy at zero temperature is the enthalpy of formation extrapolated to 0 K. This is the reference point of the enthalpy in our data-system. (The advantage of this reference state will be seen in Chapter 13 on chemical-reaction equilibrium.)

Algorithm and Program: Ideal Gas Constant-Pressure Heat Capacity

Write a program the calculate the constant-pressure heat capacity of an ideal gas based on the fifth-degree polynomial equation for enthalpy.

Algorithm.

1. Read the data-file sequential numberof the desired compound.
2. Call the datebase – reader routine.
3. Read the temperature.
4. Calculate the C_p polynomial by the Horner scheme.
5. Print the results.

Program IDCP. A possible realization:

```
]PR#0
]LOAD IDCP
]LIST

 10  REM :CALCULATION OF IDEAL
      GAS CP
 20  DIM A(15)
 30  PRINT : PRINT
 40  INPUT "ENTER THE SEQUENTIAL
      NUMBER IN DATAFILE"; IS
 50  GOSUB 180
 60  PRINT
 70  INPUT "ENTER THE TEMPERATURE
      (K)"; T
 80  CP = (((5 * A(12) * T + 4 * A(
      11)) * T + 3 * A(10)) * T +
      2 * A(9)) * T + A(8)
 90  PRINT
100  PRINT : PRINT
110  PRINT "THE CONSTANT PRESSURE
     IDEAL GAS HEAT CAPACITY "
      ; CP; "(J / MOL K) "
120  PRINT
130  INPUT "NEXT TEMPERATURE ? (Y
      /N) "; A$
```

```
140  IF A$ = "Y" THEN GOTO 70
150  INPUT "NEXT COMPOUND ? (Y / N)
      "; A$
160  IF A$ = "Y" THEN GOTO 40
170  END
180  REM :READING OF SPECIFIED
      RECORD OF PHDAT FILE INTO ARRAY
      A(15)
190  REM :THE SEQUENTIAL NUMBER
      OF THE COMPOUND IN THE CALLING
      PROGRAM HAS TO BE "IS"
200  REM :ARRAY A HAS TO BE DECLA
      RED IN THE CALLING PROGRAM
210  D$ = CHR$ (4)
220  PRINT D$; "OPEN PHDAT"
230  PRINT D$; "POSITION PHDAT, R"
      IS - 1
240  PRINT D$; "READ PHDAT"
250  INPUT A(1), A(2), A(3), A(4), A(5),
      A(6), A(7), A(8), A(9), A(10),
      A(11), A(12), A(13), A(14), A(15)
260  PRINT D$; "CLOSE PHDAT"
270  PRINT D$; "IN#0"
280  RETURN
```

Example. Run the program for methane, propane, and hydrogen at temperatures 400 and 500 K.

```
]RUN
ENTER THE SEQUENTIAL NUMBER IN DATAFILE 1
ENTER THE TEMPERATURE (K) 400

THE CONSTANT PRESSURE IDEAL GAS HEAT CAPACITY 40.615648
(J / MOL K)
NEXT TEMPERATURE ? (Y / N) Y
ENTER THE TEMPERATURE (K) 500

THE CONSTANT PRESSURE IDEAL GAS HEAT CAPACITY 46.2625
(J / MOL K)

NEXT TEMPERATURE ? (Y / N) N
NEXT COMPOUND ? (Y / N) Y

ENTER THE SEQUENTIAL NUMBER IN DATAFILE 3
ENTER THE TEMPERATURE (K) 400

THE CONSTANT PRESSURE IDEAL GAS HEAT CAPACITY 93.17512
(J / MOL K)

NEXT TEMPERATURE ? (Y / N) Y
ENTER THE TEMPERATURE (K) 500
```

THE CONSTANT PRESSURE IDEAL GAS HEAT CAPACITY 112.117437 (J / MOL K)

NEXT TEMPERATURE ? (Y / N) N
NEXT COMPOUND ? (Y / N) N

PRESSURE CORRECTION OF ENTHALPY

Let us return now to the equation

$$h[T, P] = h^*[T] + \Delta h_{\text{corr}}[T, P]$$

The explicit form of the first term on the right-hand side of this equation has already been discussed. Something useful on the explicit form of the second term can be said if the integration in the following equation has been performed in some way:

$$\Delta h_{\text{corr}} = \int_0^P \left(v - T\left(\frac{\partial v}{\partial T}\right)_P \right) dP$$

We will now prove that for substances in corresponding states this correction is the same. We start from the equation of state of real gases:

$$Pv = z\mathbf{R}T$$

where the compressibility factor z is a function of temperature and pressure. The molar volume of real gases is

$$v = \frac{z\mathbf{R}T}{P}$$

The partial derivative of molar volume with respect to temperature is

$$\left(\frac{\partial v}{\partial T}\right)_P = \frac{z\mathbf{R}}{P} + \frac{\mathbf{R}T}{P}\left(\frac{\partial z}{\partial T}\right)_P$$

Let us substitute this expression into the partial derivative of enthalpy with respect to pressure:

$$\left(\frac{\partial h}{\partial P}\right)_T = v - T\left(\frac{\partial v}{\partial T}\right)_P$$

$$= \frac{z\mathbf{R}T}{P} - \frac{z\mathbf{R}T}{P} - \frac{\mathbf{R}T^2}{P}\left(\frac{\partial z}{\partial T}\right)_P$$

$$= -\frac{\mathbf{R}T^2}{P}\left(\frac{\partial z}{\partial T}\right)_P$$

This relation can be rewritten in terms of reduced parameters:

$$\frac{1}{T_c}\left(\frac{\partial h}{\partial P_r}\right) = -\frac{\mathbf{R}T_r^2}{P_r}\left(\frac{\partial z}{\partial T_r}\right)_{P_r}$$

Separation of the variables and integration yields

$$\frac{h^* - h}{T_c} = \mathbf{R}T_r^2 \int_0^{P_r}\left(\frac{\partial z}{\partial T_r}\right)_{P_r}\frac{dP_r}{P_r}$$

This means that the isothermal pressure correction for real gases in corresponding states is the same.

Let us substitute here by way of example the Peng–Robinson equation of state. The enthalpy correction is

$$\Delta h_{\text{corr}} = \mathbf{R}\left(T(z-1) - 2.08T_c(1+m)\left(1 + m\left(1 - T_r^{0.5}\right)\right)\ln\frac{z - 2.414B}{z + 0.414B}\right)$$

where

$$B = \frac{bP}{\mathbf{R}T}$$

$$m = 0.37464 + 1.54226\omega - 0.26992\omega^2$$

Practically everything depends on what we can say about the compressibility factor z or about its partial derivative. Let us consider in this respect Table 8.1.

In the range of validity of the three-parameter corresponding-states theory, the isothermal pressure correction of the enthalpy is

$$\Delta h[T_r, P_r, \omega] = \left(1 - \frac{\omega}{\omega^{(r)}}\right)\Delta h^{(0)}[T_r, P_r] + \frac{\omega}{\omega^{(r)}}\Delta h^{(r)}[T_r, P_r]$$

where $\Delta h^{(0)}$ = correction for simple, rotationally symmetric molecules,
$\Delta h^{(r)}$ = enthalpy correction of the reference substance for nonpolar molecules,
ω = omega factor,
$\omega^{(r)}$ = omega factor of the reference substance.

It has been mentioned already in Chapter 3 on equations of state that when the LK equation is used, the linear occurrence of the compressibility factor presumes that the equation has been solved twice: once for the simple substance, and then for the reference substance. However, the equation need

TABLE 8.1
Isothermal Pressure Correction of Enthalpy

Virial equation:

$$PT\left(\frac{dB}{dT} - \frac{B}{T}\right)$$

Redlich–Kwong:

$$\mathbf{R}T\left((1 - z) + \frac{3A^2}{2B}\ln\left(1 - \frac{BP}{z}\right)\right)$$

Peng–Robinson:

$$\mathbf{R}\left(T(z - 1) - 2.08T_c(1 + m)\left(1 + m\left(1 - T_r^{0.5}\right)\right)\ln\frac{z + 2.414B}{z - 0.414B}\right)$$

Lee–Kesler:

$$\mathbf{R}\left(T_rT_c(z - 1) - \frac{b_2 + 2b_3/T_r + 3b_4/T_r^2}{T_r\phi_r} - \frac{c_2 - 3c_3/T_r^2}{2T_r\phi_r^2} + \frac{d_2}{5T_r\phi_r^5} + 3E\right)$$

Note: The meaning of the parameters and their numerical values are to be found in Chapter 3, dealing with equations of state.

be solved only once if it is written with coefficients formed by linear combination of the coefficients for the simple substance and the reference substance.

Algorithm and Program: Pressure Correction of Enthalpy and Real-Gas Enthalpy

Write a program to calculate the enthalpy departure and the real-gas enthalpy from the Peng–Robinson equation of state.

Algorithm.

1. Read the data-file sequential number of the desired compound.
2. Call the database – reader routine.
3. Read the actual pressure and temperature.
4. Calculate the compressibility factor from the Peng–Robinson equation of state with Newton's method as in program ESPR in Chapter 3.
5. Calculate the pressure correction of enthalpy, Δh, derived from the Peng–Robinson equation of state. This equation uses the compressibility factor calculated in the previous step, the temperature, and the pressure.
6. Calculate the ideal-gas enthalpy (see program IDENTH in this chapter).
7. Calculate the real-gas enthalpy from the ideal-gas enthalpy and the pressure correction of enthalpy.
8. Print the results.

Program REENTH. A possible realization:

```
]PR#0
]LOAD REENTH
]LIST

 10  REM :CALCULATION OF ENTHALPY
      DEPARTURE AND REAL GAS ENTHA
      LPY BY PENG-ROBINSON EQUATION
      OF STATE
 20  DIM A(15)
 30  PRINT : PRINT
 40  INPUT "ENTER THE SEQUENTIAL
      NUMBER OF COMPOUND IN DATA
      FILE "; IS
 50  GOSUB 430
 60  INPUT "ENTER THE PRESSURE (PA
      ) "; P
 70  INPUT "ENTER THE TEMPERATURE
      (K) "; T
 80  DEF FN F(Z) = Z ∧ 3 - (1 - B)
      * Z ∧ 2 + (A - 3 * B ∧ 2 -
      2 * B) * Z - (A * - B ∧ 2 -
      B ∧ 3)
 90  DEF FN FD(Z) = 3 * Z ∧ 2 - 2
      * (1 - B) * Z + (A - 3 * B ∧
      2 - 2 * B)
100  R = 8.3144
110  TC = A(3)
120  PC = A(2)
130  TR = T / TC
140  OM = A(4)
150   M = 0.37464 + 1.54226 * OM -
      0.26992 * OM ∧ 2
160  AL = 0.45724 * R * R * TC * T
      C / PC * (1 + M * (1 - SQR
      (TR))) ∧ 2
170  BL = 0.0778 * R * TC / PC
180  A = AL * P / R / R / T / T
190  B = BL * P / R / T
200  Z1 = 6
210  I = 0
220  H = FN F(Z1) / FN FD(Z1)
230  I = I + 1
240  IF I > 50 THEN PRINT : PRINT
      "ITERATION NOT CONVERGENT": GOTO
      330
250  Z2 = Z1 - H
260  IF ABS (Z2 - Z1) < 1 E - 5 THEN
```

```
     GOTO 290
270  Z1 = Z2
280  GOTO 220
290  DH = R * ((T * (Z2 - 1) - 2.0
     8 * TC * (1 + M) * (1 + M *
     (1 - SQR (TR))) * LOG ((Z2
     + 2.414 * B) / (Z2 - 0.414 *
     B))))
300  PRINT : PRINT
310  PRINT "THE ENTHALPY DEPARTURE
     "; DH; " (J / MOL)"
320  H0 = ((((A(12) * T + A(11)) *
     T + A(10)) * T + A(9)) * T +
     A(8)) * T + A(7)
330  H = DH + H0
340  PRINT "THE REAL GAS ENTHALPY
     IS "; H; " (J / MOL)"
350  PRINT
360  INPUT "NEXT TEMPERATURE ? (Y
     /N) "; A$
370  IF A$ = "Y" THEN GOTO 70
380  INPUT "NEXT PRESSURE ? (Y / N)
     "; A$
390  IF A$ = "Y" THEN GOTO 60
400  INPUT "NEXT COMPOUND ? (Y / N)
     "; A$
410  IF A$ = "Y" THEN GOTO 40
420  END
430  D$ = CHR$ (4)
440  PRINT D$; "OPEN PHDAT"
450  PRINT D$; "POSITION PHDAT, R";
     IS - 1
460  PRINT D$; "READ PHDAT"
470  INPUT A(1), A(2), A(3), A(4), A(5),
     A(6), A(7), A(8), A(9), A(10),
     A(11), A(12), A(13), A(14), A(15)
480  PRINT D$; "CLOSE PHDAT"
490  PRINT D$; "IN#0"
500  RETURN
```

Example. Run the program for methane and propane at temperatures 400 and 500 K, and at pressures 0.5×10^5, 10^5, 10×10^5, 100×10^5, and 1000×10^5 Pa.

```
]RUN

ENTER THE SEQUENTIAL NUMBER OF COMPOUND IN DATA FILE 1
ENTER THE PRESSURE (PA) 0.5E5
ENTER THE TEMPERATURE (K) 400
```

```
THE ENTHALPY DEPARTURE  -5.53443247 (J / MOL)
THE REAL GAS ENTHALPY IS  -53030.2919 (J / MOL)

NEXT TEMPERATURE ? (Y / N) Y
ENTER THE TEMPERATURE (K) 500

THE ENTHALPY DEPARTURE  -3.57764173 (J / MOL)
THE REAL GAS ENTHALPY IS  -48690.0777 (J / MOL)

NEXT TEMPERATURE ? (Y / N) N
NEXT PRESSURE ? (Y / N) Y
ENTER THE PRESSURE (PA) 1E5
ENTER THE TEMPERATURE (K) 400

THE ENTHALPY DEPARTURE  -11.0642699 (J / MOL)
THE REAL GAS ENTHALPY IS  -53035.8217 (J / MOL)

NEXT TEMPERATURE ? (Y / N) Y
ENTER THE TEMPERATURE (K) 500

THE ENTHALPY DEPARTURE  -7.15082597 (J / MOL)
THE REAL GAS ENTHALPY IS  -48693.6509 (J / MOL)

NEXT TEMPERATURE ? (Y / N) N
NEXT PRESSURE ? (Y / N) Y
ENTER THE PRESSURE (PA) 10E5
ENTER THE TEMPERATURE (K) 400

THE ENTHALPY DEPARTURE  -109.795634 (J / MOL)
THE REAL GAS ENTHALPY IS  -53134.5531 (J / MOL)

NEXT TEMPERATURE ? (Y / N) Y
ENTER THE TEMPERATURE (K) 500

THE ENTHALPY DEPARTURE  -70.7028607 (J / MOL)
THE REAL GAS ENTHALPY IS  -48757.2029 (J / MOL)

NEXT TEMPERATURE ? (Y / N) N
NEXT PRESSURE ? (Y / N) Y
ENTER THE PRESSURE (PA) 100E5
ENTER THE TEMPERATURE (K) 400

THE ENTHALPY DEPARTURE  -994.259915 (J / MOL)
THE REAL GAS ENTHALPY IS  -54019.0174 (J / MOL)

NEXT TEMPERATURE ? (Y / N) Y
ENTER THE TEMPERATURE (K) 500

THE ENTHALPY DEPARTURE  -624.962851 (J / MOL)
THE REAL GAS ENTHALPY IS  -49311.4629 (J / MOL)

NEXT TEMPERATURE ? (Y / N) N
NEXT PRESSURE ? (Y / N) Y
ENTER THE PRESSURE (PA) 1000E5
ENTER THE TEMPERATURE (K) 400
```

```
THE ENTHALPY DEPARTURE -2232.00483 (J / MOL)
THE REAL GAS ENTHALPY IS -55256.7623 (J / MOL)

NEXT TEMPERATURE ? (Y / N) N
NEXT PRESSURE ? (Y / N) N
NEXT COMPOUND ? (Y / N) Y
ENTER THE SEQUENTIAL NUMBER OF COMPOUND IN DATA FILE 3
ENTER THE PRESSURE (PA) 0.5E5
ENTER THE TEMPERATURE (K) 400

THE ENTHALPY DEPARTURE -35.7560108 (J / MOL)
THE REAL GAS ENTHALPY IS -58389.972 (J / MOL)

NEXT TEMPERATURE ? (Y / N) Y
ENTER THE TEMPERATURE (K) 500

THE ENTHALPY DEPARTURE -24.8655532 (J / MOL)
THE REAL GAS ENTHALPY IS -48105.8968 (J / MOL)

NEXT TEMPERATURE ? (Y / N) N
NEXT PRESSURE ? (Y / N) Y
ENTER THE PRESSURE (PA) 1E5
ENTER THE TEMPERATURE (K) 400

THE ENTHALPY DEPARTURE -71.6719653 (J / MOL)
THE REAL GAS ENTHALPY IS -58425.888 (J / MOL)

NEXT TEMPERATURE ? (Y / N) Y
ENTER THE TEMPERATURE (K) 500

THE ENTHALPY DEPARTURE -49.7618808 (J / MOL)
THE REAL GAS ENTHALPY IS -48130.7931 (J / MOL)

NEXT TEMPERATURE ? (Y / N) N
NEXT PRESSURE ? (Y / N) Y
ENTER THE PRESSURE (PA) 10E5
ENTER THE TEMPERATURE (K) 400

THE ENTHALPY DEPARTURE -748.198915 (J / MOL)
THE REAL GAS ENTHALPY IS -59102.4149 (J / MOL)
NEXT TEMPERATURE ? (Y / N) Y
ENTER THE TEMPERATURE (K) 500

THE ENTHALPY DEPARTURE -503.110929 (J / MOL)
THE REAL GAS ENTHALPY IS -48584.1422 (J / MOL)

NEXT TEMPERATURE ? (Y / N) N
NEXT PRESSURE ? (Y / N) Y
ENTER THE PRESSURE (PA) 100E5
ENTER THE TEMPERATURE (K) 400

THE ENTHALPY DEPARTURE -10714.1308 (J / MOL)
THE REAL GAS ENTHALPY IS -69068.3469 (J / MOL)

NEXT TEMPERATURE ? (Y / N) Y
ENTER THE TEMPERATURE (K) 500
```

```
THE ENTHALPY DEPARTURE  -5214.87599 (J / MOL)
THE REAL GAS ENTHALPY IS  -53295.9073 (J / MOL)

NEXT TEMPERATURE ? (Y / N) N
NEXT PRESSURE ? (Y / N) Y
ENTER THE PRESSURE (PA) 1000E5
ENTER THE TEMPERATURE (K) 400

THE ENTHALPY DEPARTURE  -10192.387 (J / MOL)
THE REAL GAS ENTHALPY IS  -68546.603 (J / MOL)

NEXT TEMPERATURE ? (Y / N) Y
ENTER THE TEMPERATURE (K) 500

THE ENTHALPY DEPARTURE  -8189.35101 (J / MOL)
THE REAL GAS ENTHALPY IS  -56270.3823 (J / MOL)

NEXT TEMPERATURE ? (Y / N) N
NEXT PRESSURE ? (Y / N) N
NEXT COMPOUND ? (Y / N) N
```

Remark. You can see that the enthalpy departure is always negative, in other words, the real-gas enthalpy is always less then the corresponding ideal-gas enthalpy.

CALCULATION OF THE ENTHALPY OF MULTICOMPONENT HOMOGENEOUS PHASES: CONCEPT OF THE PSEUDO-ONE-COMPONENT PHASE

The problem in the case of multicomponent mixtures is to trace back the enthalpy calculation

$$h[T, P, x_j] = h^*[T, x_j] + \Delta h_{\text{corr}}[T, P, x_j]$$

to the following relation

$$h[T, P, x_j] = h^*_{\text{ps}}[T] + \Delta h_{\text{corr,ps}}[T, P]$$

This means that instead of the multicomponent mixture we seek to treat a pseudo-one-component phase behaving in the pressure–temperature–enthalpy space like the multicomponent mixture.

It causes no difficulty to interpret the first term on the right-hand side as describing a pseudo-one-component ideal gas:

$$h^*_{\text{ps}} = \sum_{i=0}^{5} a_{i,\text{ps}} T^i \qquad i = 0, 1, 2, \ldots, 5$$

TABLE 8.2
Mixing Rules

Omega factor of the pseudocomponent:

$$\omega_{ps} = \sum_{j=1}^{k} x_j \omega_j \qquad j = 1, 2, \ldots, k$$

where ω_j = omega factor of the jth component,
x_j = mole fraction of the jth component.

Pseudocritical compressibility factor:

$$z_{c,ps} = \sum_{j=1}^{k} x_j z_{c,j}$$

where $z_{c,j}$ = critical compressibility factor of component j.

Pseudocritical volume:

$$v_{c,ps} = \frac{1}{8} \sum_{j=1}^{k} \sum_{i=1}^{k} x_i x_j (v_{c,i}^{1/3} + v_{c,j}^{1/3})^3$$

where $v_{c,i}$ = critical volume of component i,
$v_{c,j}$ = critical volume of component j.

Pseudocritical temperature:

$$T_{c,ps} = \frac{1-k}{8 v_{c,ps}} \sum_{j=1}^{k} \sum_{i=1}^{k} x_i x_j (v_{c,i}^{1/3} + v_{c,j}^{1/3})^3 \cdot \sqrt{T_{c,i} T_{c,j}}$$

where

$$1 - k = \sum_{j=1}^{k} \sum_{i=1}^{k} x_i x_j \frac{8\sqrt{v_{c,i} v_{c,j}}}{\left(v_{c,i}^{1/3} + v_{c,j}^{1/3}\right)^3}$$

and $T_{c,i}$ = critical temperature of component i,
$T_{c,j}$ = critical temperature of component j.

Pseudocritical pressure:

$$P_{c,ps} = z_{c,ps} \frac{\mathbf{R} T_{c,ps}}{v_{c,ps}}$$

Pseudo phi factor:

$$\varphi_{ps} = \sum_{j=1}^{k} \sum_{i=1}^{k} x_i x_j \frac{\varphi_i + \varphi_j}{2}$$

but if $\varphi_i = 0$ or $\varphi_j = 0$, then $(\varphi_i + \varphi_j)/2 = 0$, that is, if there is only one polar component in the mixture, then its effect has to be considered only once. Here φ_i is the phi factor of component i.

with

$$a_{i,\mathrm{ps}} = \sum_{j=1}^{k} a_{ji} x_j \qquad j = 1, 2, \ldots, k$$

where $a_{i,\mathrm{ps}}$ = ith coefficient of the enthalpy polynomial of the pseudo-one-component phase,

a_{ji} = ith coefficient of the enthalpy polynomial of the jth component,

x_j = mole fraction of the jth component in the multicomponent mixture,

k = number of components in the multicomponent mixture.

The enthalpy correction is calculated on the basis of corresponding-states theory:

$$\Delta h_{\mathrm{corr}} = \Delta h_{\mathrm{corr}}\left[T, P, T_{cj}, P_{cj}, x_j, \omega_j\right] \qquad j = 1, 2, \ldots, k$$

In the one-component case the function is the same:

$$\Delta h_{\mathrm{corr}} = \Delta h_{\mathrm{corr}}[T, P, T_c, P_c, \omega]$$

The question is how to carry out the following parameter aggregation:

$$\Delta h_{\mathrm{corr}}\left[T, P, T_{cj}, P_{cj}, x_j, \omega_j\right] = \Delta h_{\mathrm{corr}}\left[T, P, T_{c,\mathrm{ps}}, P_{c,\mathrm{ps}}, \omega_{\mathrm{ps}}\right]$$

On the basis of experience the aggregation (mixing) rules in Table 8.2 can be recommended.

When the pseudo-critical properties of the pseudo-one-component phase are known, the course of enthalpy calculation is the same as for one-component substances.

9

CALCULATION OF THE THERMODYNAMIC PROPERTIES OF A ONE-COMPONENT HOMOGENEOUS PHASE

A one-component homogeneous (gas or liquid) phase has two degrees of freedom. Let us consider now those thermodynamic properties of which any two determine the state of the one-component body:

Generalized coordinate	s	v
Generalized force	T	P
Potential	u	h

We have taken only six properties, omitting the free energy and free enthalpy. When the six properties are known, the latter can be calculated from their defining equations. The present question is, if the values of any two of the six listed properties are fixed, how to calculate the values of the other four. We have the following fifteen possible pairs of independent variables:

TP	Tv	Th	Tu	Ts
	Pv	Ph	Pu	PS
		vh	vu	vs
			hu	hs
				vs

Before discussing these cases one by one, let us recall Chapter 8 on the calculation of enthalpy. The considerations used there are true also if properties other than P and T are chosen as the independent variables. The formulation

$$X = X[Y, Z]$$

where X, Y and Z are arbitrary intensive or molar extensive properties, is of general validity.

THE BASIC CASE: P, T ARE KNOWN

For the calculations it must first of all be decided in which phase the substance is present. When P and T are known, this can be done with the following reasoning. The equilibrium vapor pressure $P°$ is a function of temperature, since the two-phase, one-component equilibrium system has one degree of freedom (Figure 9.1)

$$P° = P°[T]$$

where $P°$ is a known function. The actual state is:

1. A gas phase if $P < P°[T]$.
2. A liquid phase if $P > P°[T]$.
3. A mixture of two phases if the pressure is equal to the vapor pressure ($P = P°[T]$).

In the last case the state of the system is not unequivocally determined, as ε can take on arbitrary values between 0 and 1, without disturbing the equilibrium (Figure 9.2).

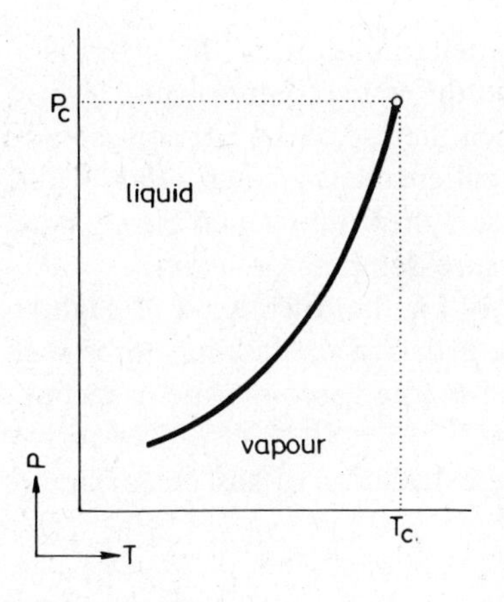

FIGURE 9.1. Equilibrium vapor pressure.

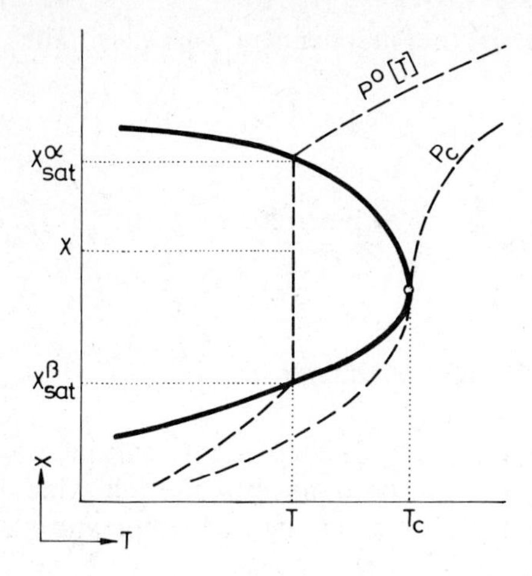

FIGURE 9.2. Vapor–liquid system. X is an extensive property; the vapor ratio is defined as $\varepsilon = (X - X^{\beta}_{\text{sat}})/(X^{\alpha}_{\text{sat}} - X^{\beta}_{\text{sat}})$.

Further calculations involve also z, following from the appropriate equation of state, as well as the isothermal pressure corrections Δh and Δs, so that a function $P°$ consistent with the equation of state is to be used here.

The vapor-pressure function consistent with the Redlich–Kwong equation of state is not sufficiently accurate, and therefore we recommend the use of the Lee-Kesler formula originating from the LK equation of state:

$$\ln P_r° = 5.92714 - \frac{6.09648}{T_r} - 1.28862 \ln T_r - 0.169347 T_r^6$$

$$+ \omega \left(15.2518 - \frac{15.6875}{T_r} - 13.4721 \ln T_r + 0.43577 T_r^6 \right)$$

It can be seen from this criterion that the determination of the phase is a critical operation. If the absolute value of the difference between pressure P and calculated $P°[T]$ does not exceed 10%, then the use of the LK equation of state is always justified in the following. If the difference is greater—that is, the body is certainly homogeneous and single-phase—then other equations of state can also be used, according to the actual pressure–temperature values.

A quadratic virial equation can be used only for the calculation of slightly imperfect gases ($v_r > 2$). The Redlich–Kwong equation of state has three real roots; of these only the use of that belonging to the gaseous state is recommended.

The LK equation of state is suitable for the calculation of gaseous or liquid phase alike.

Calculation of the Compressibility Coefficient

1. Using the quadratic virial equation, the compressibility coefficient can be calculated with the following formula:

$$z = 1 + \frac{B[T]}{\mathbf{R}T}$$

where the second virial coefficient $B[T]$ is obtained with the calculation formula contained in Table 9.1.

2. Using the Redlich–Kwong equation, from the roots of the following cubic equation one chooses the one relevant to the gaseous state:

$$z^3 - z^2 + \left(A_{\text{RK}}P - B_{\text{RK}}P - B_{\text{RK}}^2 p^2\right)z - \left(A_{\text{RK}}B_{\text{RK}}P^2\right) = 0$$

where the parameters $A_{\text{RK}}[T]$, $B_{\text{RK}}[T]$ are calculated with the formula contained in Table 9.1.

3. For the use of the Peng–Robinson equation of state see the algorithms and programs of this chapter.

4. When using the Lee–Kesler equation, the following nonlinear equation is first solved:

$$\frac{P_r\phi_r}{T_r} - \left(1 + \frac{B_r}{\phi_r} + \frac{C_r}{\phi_r^2} + \frac{D_r}{\phi_r^5} + \frac{C_4}{T_r^3\phi_r^2}\left(\beta + \frac{\gamma}{\phi_r^2}\right)\exp\left(-\frac{\gamma}{\phi_r^2}\right)\right) = 0$$

where ϕ_r is the so-called ideal reduced volume:

$$\phi_r = \frac{vP_c}{\mathbf{R}T_c}$$

Then the compressibility coefficient is obtained:

$$z = \frac{P_r\phi_r}{T_r}$$

TABLE 9.1
Thermodynamic Functions Derived from Four Equations of State

	Virial EOS	Cubic EOS a. Redlich-Kwong b. Peng-Robinson
$\pi[T]$	$B = \frac{\mathbf{R}T_c}{P_c} B_v$ $\frac{dB}{dT} = \frac{\mathbf{R}T_c}{P_c}\left(\frac{dB_v}{dT}\right)$	a. $A_{RK} = \left(\frac{a}{R^2T^{2.5}}\right)^{0.5}$ $B_{RK} = \left(\frac{b}{RT}\right)$ b. $A_{PR} = \left(\frac{aP}{R^2T^2}\right)$ $B_{PR} = \left(\frac{bP}{\mathbf{R}T}\right)$
z	$1 + \frac{BP}{RT}$ $1 + \frac{B_v}{\phi_r}$	a. $z^3 - z^2 + (A^2P - B^2P^2 - BP)z - (A^2BP^2) = 0$ b. $z^3 - (1 - B)z^2 + (A - 3B^2 - 2B)z - (AB - B^2 - B^3) = 0$
Δh	$PT\left(\frac{dB}{dT} - \frac{B}{T}\right)$	a. $\mathbf{R}T\left((1 - z) + \frac{3A^2}{2B}\ln\left(1 + \frac{BP}{z}\right)\right)$ b. $\mathbf{R}\left(T(z - 1) - 2.08T_c(1 + m)\left(1 + m\left(1 - T_r^{0.5}\right)\right)\ln\frac{z + 2.414B}{z - 0.414B}\right.$
Δs	$P\frac{dB}{dT}$	a. $\mathbf{R}\left(\frac{A}{2B}\ln\left(1 + \frac{BP}{z}\right) - \ln(z - BP)\right)$ b. $\mathbf{R}\left(\ln(z - B) - 2.08m\left(T_r^{0.5}(m + 1) - m\right)\right)\ln\frac{z + 2.414B}{z - 0.414B}$
$\ln\frac{f}{P}$	$\frac{1}{\mathbf{R}}\frac{BP}{T}$	a. $(z - 1) - \ln(z - BP) - \frac{A^2}{B}\ln\left(1 + \frac{BP}{Z}\right)$ b. $z - 1 - \ln(z - B) - \frac{A}{2\sqrt{2}B}\ln\frac{z + 2.414B}{z - 0.414B}$

The parameters of the equation are calculated with the formulas contained in Table 9.1. (See also Table 3.2.) The dependence of the compressibility coefficient on the parameter T_r can be clearly observed in Figure 9.3 in the plots of z versus P_r.

5. The spline "equation of state" described in Chapter 3 is now used only for calculating the compressibility factor. For computation of other properties (enthalpy, entropy, heat capacity, fugacity coefficient) the compressibility factors computed by the spline method are substituted into the correction function derived from the BWR equation of state, and the resulting values of Δh, Δs, and $\ln(f/p)$ are used for further calculations.

Lee-Kesler EOS

$$B = b_1 - b_2/T_r - b_3/T_r^2 - b_4/T_r^3$$

$$C = c_1 - c_2/T_r - c_3/T_r^3$$

$$D = d_1 - d_2/T_r$$

$$E = \frac{c_4}{2T_r^2\gamma}\left(\beta + 1 - \left(\beta + 1 + \frac{\gamma}{\phi_r^2}\right)\exp\left(\frac{\gamma}{\phi_r^2}\right)\right)$$

$$\left(\frac{P_r\phi_r}{T_r}\right) = 1 + \frac{B}{\phi_r} + \frac{C}{\phi_r^2} + \frac{D}{\phi_r^5} + \frac{c_4}{T_r^3\phi_r^2}\left(\beta + \frac{\gamma}{\phi_r^2}\right)\exp\left(-\frac{\gamma}{\phi_r^2}\right)$$

$$\mathbf{R}T_rT_c\left(z - 1 - \frac{b_2 + 2b_3/T_r + 3b_4/T_r^2}{T_r\phi_r} - \frac{c_2 - 3c_3/T_r^2}{2T_r\phi_r^2} + \frac{d_2}{5T_r\phi_r^5}\right) + 3E$$

$$\mathbf{R}\left(\ln Z - \frac{b_1 + b_3/T_r^2 + 2b_4/T_r^3}{\phi_r} - \frac{c_1 - 2c_3/T_r^3}{2\phi_r^2} - \frac{d_1}{5\phi_r^5} + 2E\right)$$

$$(z - 1) - \ln z + \frac{B}{\phi_r} + \frac{C}{2\phi_r^2} + \frac{D}{5\phi_r^5} + E$$

Calculation of *v*, *u*, *h*, *s*

The *molar volume* can be calculated with the following relation:

$$v = \frac{\mathbf{R}Tz[P,T]}{P}$$

where the compressibility coefficient is calculated as described above.

The *molar enthalpy* can be separated into two terms:

$$h[T,P] = h^*[T] + \Delta h[T,P]$$

where the first term on the right-hand side, the enthalpy h^* of the ideal gas,

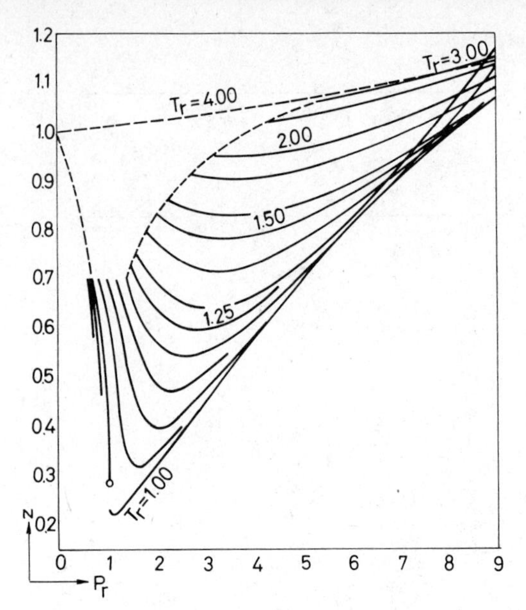

FIGURE 9.3. Compressibility factor z versus P_r.

can be calculated as a fifth-degree polynomial in the temperature:

$$h^*[T] = \sum_{i=0}^{5} a_i T^i$$

and the second term on the right-hand side, the isothermal pressure correction to the enthalpy,

$$\Delta h = h - h^*$$

can be calculated with the arithmetic expressions in Table 9.1:

$$\Delta h = \Delta h[T, P, z[T, P]]$$

or

$$\Delta h = \Delta h[T_r, \phi_r[T_r, P_r], z[T_r, \phi_r]]$$

The calculation of the compressibility coefficient and of $\phi_r[P_r, T_r]$ have been discussed already. The behavior of the enthalpy correction can be clearly observed in Figure 9.4, in the plots of $\Delta h/RT_c$ versus P_r, for various values of T_r.

The *molar internal energy* can be divided into two terms:

$$u[T, P] = u^*[T] + \Delta u[T, P]$$

where the first term on the right-hand side, the internal energy $u^*[T]$ of the ideal gas, can be calculated from the enthalpy $h^*[T]$ of the ideal gas:

$$u^*[T] = h^*[T] - \mathbf{R}T$$

and the second term on the right-hand side, $\Delta u[T, P]$, is the isothermal pressure correction of the internal energy. On the basis of the equation of definition of enthalpy this can be written in the following form:

$$\Delta u = \Delta h - P(v - v^*)$$

$$= \Delta h - Pv + \mathbf{R}T$$

where Δh = isothermal pressure correction of enthalpy,
v^* = volume of the ideal gas, and
$Pv = \mathbf{R}T\, z[T, P]$.

Hence, the molar internal energy is

$$u = u[T, P] = h^*[T] + \Delta h[T, P, z[T, P]] - \mathbf{R}Tz[T, P]$$

Thus the calculation of the molar internal energy is reduced to the calculation of the compressibility coefficient z and of the isothermal enthalpy correction Δh.

The *molar entropy* can be separated into three terms:

$$s[T, P] = s^*[T] - \mathbf{R}\ln\frac{P}{P_u} + \Delta s[T, P].$$

Here s^*, the first term on the right-hand side, is the entropy of the ideal gas at the reference pressure P_u. This is dependent only on temperature, and takes the form

$$s^*[T] = a_1 \ln T + 2a_2 T + \tfrac{3}{2}a_3 T^2 + \tfrac{4}{3}a_4 T^3 + \tfrac{5}{4}a_5 T^4 + I_s$$

where the coefficients are identical with those in the enthalpy polynomial, and I_s constant of integration. The second term on the right-hand side is the difference between the entropy of the ideal gas at pressure P and its entropy at the same temperature but at the reference pressure:

$$S^*[T, P] - S^*[T] = -\mathbf{R} \ln \frac{P}{P_u}$$

Algorithm and Program: Vapor Pressure

Write a program for the calculation of the vapor pressure from the temperature.

Algorithm.

1. Read the sequential number of the desired compound in the data file.
2. Call the database-reader routine.
3. Read the actual temperature.
4. Calculate the reduced temperature.
5. Calculate the reduced saturation pressure and the saturation pressure.
6. Print the results.

Program VAPPRES. A possible realization:

```
]LOAD VAPPRES
]LIST

 10  REM :CALCULATION OF VAPOR
      PRESSURE
 20  DIM A(15)
 30  INPUT "ENTER SEQUENTIAL NUMBER
      IN DATA FILE "; IS
 40  GOSUB 220
 50  PRINT : PRINT
 60  INPUT "ENTER THE TEMPERATURE (K) "; T
 70  PC = A(2)
 80  TC = A(3)
 90  OM = A(4)
100  TB = A(5)
110  TR = T / TC
120  PS = 5.92714 - 6.09648 / TR -
      1.28862 * LOG (TR) + 0.1693
      47 * TR ∧ 6 + OM * (15.2518 -
      15.6875 / TR - 13.4721 * LOG
      (TR) + 0.43577 * TR ∧ 6)
130  PS = EXP (PS) * PC
140  PRINT : PRINT
150  PRINT "VAPOR PRESSURE "; PS;
      " (PA)"
```

```
160  PRINT : PRINT
170  INPUT "NEXT TEMPERATURE (Y / N)
      "; A$
180  IF A$ = "Y" THEN GOTO 60
190  INPUT "NEXT COMPOUND ? (Y / N)
      "; A$
200  IF A$ = "Y" THEN GOTO 30
210  END
220  REM :READING OF SPECIFIED
      RECORD OF PHDAT FILE INTO ARRAY
      A(15)
230  REM :THE SEQUENTIAL NUMBER
      OF THE COMPOUND IN THE CALLING
      PROGRAM HAS TO BE "IS"
240  REM :ARRAY A HAS TO BE DECLA
      RED IN THE CALLING PROGRAM
250  D$ = CHR$ (4)
260  PRINT D$; "OPEN PHDAT"
270  PRINT D$; "POSITION PHDAT, R";
      IS - 1
280  PRINT D$; "READ PHDAT"
290  INPUT A(1), A(2), A(3), A(4), A(5),
      A(6), A(7), A(8), A(9), A(10),
      A(11), A(12), A(13), A(14), A(15)
300  PRINT D$; "CLOSE PHDAT"
310  PRINT D$; "IN#0"
320  RETURN
```

Example. Run the program for methane and propane at temperatures 100, 130, 150, 200, 300 K.

```
]RUN
ENTER SEQUENTIAL NUMBER IN DATA FILE 1
ENTER THE TEMPERATURE (K) 100

VAPOR PRESSURE 34946.8569 (PA)

NEXT TEMPERATURE (Y / N) Y
ENTER THE TEMPERATURE (K) 130

VAPOR PRESSURE 373914.922 (PA)

NEXT TEMPERATURE (Y / N) Y
ENTER THE TEMPERATURE (K) 150

VAPOR PRESSURE 1052830.79 (PA)

NEXT TEMPERATURE (Y / N) Y
ENTER THE TEMPERATURE (K) 200

VAPOR PRESSURE 6107448.69 (PA)

NEXT TEMPERATURE (Y / N) Y
```

```
ENTER THE TEMPERATURE (K) 300

VAPOR PRESSURE 269814064 (PA)

NEXT TEMPERATURE (Y / N) N
NEXT COMPOUND ? (Y / N) Y
ENTER SEQUENTIAL NUMBER IN DATA FILE 3
ENTER THE TEMPERATURE (K) 100

VAPOR PRESSURE .032890987 (PA)

NEXT TEMPERATURE (Y / N) Y
ENTER THE TEMPERATURE (K) 130

VAPOR PRESSURE 18.4826848 (PA)

NEXT TEMPERATURE (Y / N) Y
ENTER THE TEMPERATURE (K) 150

VAPOR PRESSURE 282.592445 (PA)

NEXT TEMPERATURE (Y / N) Y
ENTER THE TEMPERATURE (K) 200

VAPOR PRESSURE 19889.6319 (PA)

NEXT TEMPERATURE (Y / N) Y
ENTER THE TEMPERATURE (K) 300

VAPOR PRESSURE 1004653.39 (PA)

NEXT TEMPERATURE (Y / N) N
NEXT COMPOUND ? (Y / N) N
```

Algorithm and Program: Graphical Representation of Vapor Pressure

Write a program to plot the vapor pressure as a function of the temperature.

Algorithm.

1. Enter graphic mode.
2. Read the sequential number of the compound in the data file.
3. Call the database-reader routine.
4. Calculate and plot the vapor pressure in a loop. The loop index is the reduced temperature.

Program GRVAPPRES. A possible realization:

```
]LOAD GRVAPPRES
]LIST

 10  REM :GRAPHICAL REPRESENTATION
```

```
        OF VAPOR PRESSURE IN THE
        RANGE 0.5 < TR < 1
 20  HPLOT 0,0
 30  HGR
 40  HCOLOR = 7
 50  HPLOT 10,155 TO 279,155
 60  HPLOT 10,0 TO 10,191
 70  DIM A(15)
 80  INPUT "ENTER SEQUENTIAL NUMBER
        IN DATA FILE "; IS
 90  GOSUB 340
100  PRINT : PRINT
110  PC = A(2)
120  TC = A(3)
130  OM - A(4)
140  TB = A(5)
150  FOR TR = 0.5 TO 1 STEP 0.005
160  PA = 5.92714 - 6.09648 / TR -
        1.28862 * LOG (TR) + 0.1693
        47 * TR ∧ 6 + OM * (15.2518 -
        15.6875 / TR - 13.4721 * LOG
        (TR) + 0.43577 * TR ∧ 6)
170  PS = EXP (PS)
180  X = (TR - 0.5) * 520 + 10
190  Y = 150 - 140 * PS
200  IF TR = 0.5 THEN HPLOT X,Y
210  HPLOT TO X,Y
220  NEXT TR
230  PB = 101325 / PC
240  HCOLOR = 2
250  HPLOT 10,150 - 140 * PB
260  HPLOT 10,150 - 140 * PB TO 2
        70,150 - 140 * PB
270  HPLOT 114,0
280  HPLOT 114,100 TO 114,160
290  HCOLOR = 7
300  INPUT "NEXT COMPOUND ? (Y / N)
        "; A$
310  IF A$ = "Y" THEN GOTO 80
320  TEXT
330  END
340  REM :READING OF SPECIFIED
        RECORD OF PHDAT FILE INTO ARRAY
        A(15)
350  REM :THE SEQUENTIAL NUMBER
        OF THE COMPOUND IN THE CALLING
        PROGRAM HAS TO BE "IS"
360  REM :ARRAY A HRS TO BE DECLA
        RED IN THE CALLING PROGRAM
```

```
370  D$ = CHR$ (4)
380  PRINT D$; "OPEN PHDAT"
390  PRINT D$; "POSITION PHDAT, R";
      IS - 1
400  PRINT D$; "READ PHDAT"
410  INPUT A(1), A(2), A(3), A(4), A(5),
      A(6), A(7), A(8), A(9), A(10),
      A(11), A(12), A(13), A(14), A(15)
420  PRINT D$; "CLOSE PHDAT"
430  PRINT D$; "IN#0"
440  RETURN
```

Example. Run the program for methane and propane.

```
]RUN
ENTER SEQUENTIAL NUMBER IN DATA FILE 1

NEXT COMPOUND ? (Y / N) Y
ENTER SEQUENTIAL NUMBER IN DATA FILE 3

NEXT COMPOUND ? (Y / N) N
```

Remark. Watch the monitor and explain the curves.

Algorithm and Program: Determination of the Phase

Write a program to determine whether the actual phase is gas or liquid.

Algorithm.

1. Read the sequential number of the desired compound in the datafile.
2. Call the database-reader routine.
3. Read the actual temperature and pressure.
4. Calculate the saturation pressure from the temperature (see program VAPPRES).
5. Compare the actual pressure and the saturation pressure.
6. Print the result.

Program PHASE? A possible realization:

```
]LOAD PHASE?
]LIST

 10  REM :DETERMINATION OF THE PHASE
      (LIQUID / GAS / TWO PHASE)
 20  DIM A(15)
 30  INPUT "ENTER SEQUENTIAL NUMBER
```

```
        IN DATA FILE "; IS
 40  GOSUB 260
 50  PRINT : PRINT
 60  INPUT "ENTER THE PRESSURE (PA)
        "; P
 70  INPUT "ENTER THE TEMPERATURE
        (K) "; T
 80  PC  =  A(2)
 90  TC  =  A(3)
100  OM  =  A(4)
110  TB  =  A(5)
120  TR  =  T / TC
130  PS  =  5.92714 - 6.09648 / TR -
        1.28862 * LOG (TR) + 0.1693
        47 * TR ∧ 6 + OM * (15.2518 -
        15.6875 / TR - 13.4721 * LOG
        (TR) + 0.43577 * TR ∧ 6)
140  PS  =  EXP (PS) * PC
150  PRINT : PRINT
160  IF PS  >  P THEN PRINT "VAPOR
        "; PRINT
170  IF PS  <  P THEN PRINT "LIQUID
        ": PRINT
180  IF PS  =  P THEN PRINT "LIQUID-
        VAPOR PHASE": PRINT
190  INPUT "NEXT TEMPERATURE (Y / N)
        "; A$
200  IF A$ = "Y" THEN GOTO 70
210  INPUT "NEXT PRESSURE ? (Y / N)
        "; A$
220  IF A$ = "Y" THEN GOTO 60
230  INPUT "NEXT COMPOUND ? (Y / N)
        "; A$
240  IF A$ = "Y" THEN GOTO 30
250  END
260  REM :READING OF SPECIFIED
        RECORD OF PHDAT FILE INTO ARRAY
        A(15)
270  REM :THE SEQUENTIAL NUMBER OF
        THE COMPOUND IN THE CALLING
        PROGRAM HAS TO BE "IS"
280  REM :ARRAY A HAS TO BE DECLA
        RED IN THE CALLING PROGRAM
290  D$ =  CHR$ (4)
300  PRINT D$; "OPEN PHDAT"
310  PRINT D$; "POSITION PHDAT, R";
        IS - 1
320  PRINT D$; "READ PHDAT"
330  INPUT A(1), A(2), A(3), A(4), A(5),
```

```
        A(6), A(7), A(8), A(9), A(10),
        A(11), A(12), A(13), A(14), A(15)
340  PRINT D$; "CLOSE PHDAT"
350  PRINT D$; "IN#O"
360  RETURN
```

Example. Run the program for methane and propane at temperatures 100, 150, 200, and 300 K and at pressures 101,325 and 1,013,250 Pa.

```
]RUN
ENTER SEQUENTIAL NUMBER IN DATA FILE 1
ENTER THE PRESSURE (PA) 101325
ENTER THE TEMPERATURE (K) 100

LIQUID

NEXT TEMPERATURE (Y / N) Y
ENTER THE TEMPERATURE (K) 150

VAPOR

NEXT TEMPERATURE (Y / N) Y
ENTER THE TEMPERATURE (K) 200

VAPOR

NEXT TEMPERATURE (Y / N) Y
ENTER THE TEMPERATURE (K) 300

VAPOR

NEXT TEMPERATURE (Y / N) N
NEXT PRESSURE ? (Y / N) Y
ENTER THE PRESSURE (PA) 1013250
ENTER THE TEMPERATURE (K) 100

LIQUID

NEXT TEMPERATURE (Y / N) Y
ENTER THE TEMPERATURE (K) 150

VAPOR

NEXT TEMPERATURE (Y / N) Y
ENTER THE TEMPERATURE (K) 200

VAPOR

NEXT TEMPERATURE (Y / N) Y
ENTER THE TEMPERATURE (K) 300

VAPOR

NEXT TEMPERATURE (Y / N) N
NEXT PRESSURE ? (Y / N) Y
ENTER THE PRESSURE (PA) 10132500
ENTER THE TEMPERATURE (K) 300
```

```
VAPOR

NEXT TEMPERATURE (Y / N) N
NEXT PRESSURE ? (Y / N) N
NEXT COMPOUND ? (Y / N) Y
ENTER SEQUENTIAL NUMBER IN DATA FILE 3
ENTER THE PRESSURE (PA) 101325
ENTER THE TEMPERATURE (K) 100

LIQUID

NEXT TEMPERATURE (Y / N) Y
ENTER THE TEMPERATURE (K) 150

LIQUID

NEXT TEMPERATURE (Y / N) Y
ENTER THE TEMPERATURE (K) 200

LIQUID

NEXT TEMPERATURE (Y / N) Y
ENTER THE TEMPERATURE (K) 300

VAPOR

NEXT TEMPERATURE (Y / N) N
NEXT PRESSURE ? (Y / N) Y
ENTER THE PRESSURE (PA) 1013250
ENTER THE TEMPERATURE (K) 100

LIQUID

NEXT TEMPERATURE (Y / N) Y
ENTER THE TEMPERATURE (K) 150

LIQUID

NEXT TEMPERATURE (Y / N) Y
ENTER THE TEMPERATURE (K) 200

LIQUID

NEXT TEMPERATURE (Y / N) Y
ENTER THE TEMPERATURE (K) 300

LIQUID

NEXT TEMPERATURE (Y / N) N
NEXT PRESSURE ? (Y / N) N
NEXT COMPOUND ? (Y / N) N
```

Algorithm and Program: Internal Energy of Real Gases

Write a program for the calculation of the internal energy of a real gas, using the fifth-degree polynomial equation for the ideal gas enthalpy.

Algorithm.

1. Read the sequential number of the desired compound.
2. Call the datafile-reader routine.
3. Read the temperature and pressure
4. Solve the Peng–Robinson equation of state (see program ESPR in Chapter 3).
5. Calculate the real-gas enthalpy (see program REENTH in Chapter 8).
6. Calculate the real-gas internal energy $U = H - pV$.
7. Print the result.

Program REU. A possible realization:

```
]PR#0
]LOAD REU
]LIST

 10  REM :CALCULATION OF INTERNAL
      ENERGY OF REAL GAS BY PENG-
      ROBINSON EQUATION OF STATE
 20  DIM A(15)
 30  PRINT : PRINT
 40  INPUT "ENTER THE SEQUENTIAL
      NUMBER OF COMPOUND IN DATA FILE
      "; IS
 50  GOSUB 430
 60  INPUT "ENTER THE PRESSURE (PA)
      "; P
 70  INPUT "ENTER THE TEMPERATURE
      (K) "; T
 80  DEF FN F(Z) = Z ∧ 3 - (1 - B
      ) * Z ∧ 2 + (A - 3 * B ∧ 2 -
      2 * B) * Z - (A * B - B ∧ 2 -
      B ∧ 3)
 90  DEF FN FD(Z) = 3 * Z ∧ 2 - 2
      * (1 - B) * Z + (A - 3 * B ∧
      2 - 2 * B)
100  R = 8.3144
110  TC = A(3)
120  PC = A(2)
130  TR = T / TC
140  OM = A(4)
150  M = 0.37464 + 1.54226 * OM -
      0.26992 * OM ∧ 2
160  AL = 0.45724 * R * R * TC * T
      C / PC * (1 + M * (1 - SQR
      (TR))) ∧ 2
```

```
170  BL = 0.0778 * R * TC / PC
180  A = AL * P / R / R / T / T
190  B = BL * P / R / T
200  Z1 = 6
210  I = 0
220  H = FN F(Z1) / FN FD(Z1)
230  I = I + 1
240  IF I > 50 THEN PRINT : PRINT
      "ITERATION NOT CONVERGENT": GOTO
      320
250  Z2 = Z1 - H
260  IF ABS (Z2 - Z1) < 1E - 5 THEN
      GOTO 290
270  Z1 = Z2
280  GOTO 220
290  DH = R * ((T * (Z2 - 1) - 2.0
      8 * TC * (1 + M) * (1 + M *
      (1 - SQR (TR))) * LOG ((Z2
      +2.414 * B) / (Z2 - 0.414 *
      B))))
300  PRINT : PRINT
310  H0 = ((((A(12) * T + A(11)) *
      T + A(10)) * T + A(9)) * T +
      A(8)) * T + A(7)
320  H = DH + H0
330  U = H - Z2 * R * T
340  PRINT "REAL GAS INTERNAL ENERGY
      "; U; " (J / MOL)"
350  PRINT
360  INPUT "NEXT TEMPERATURE ? (Y
      / N) "; A$
370  IF A$ = "Y" THEN GOTO 70
380  INPUT "NEXT PRESSURE ? (Y / N)
      "; A$
390  IF A$ = "Y" THEN GOTO 60
400  INPUT "NEXT COMPOUND ? (Y / N)
      "; A$
410  IF A$ = "Y" THEN GOTO 40
420  END
430  D$ = CHR$ (4)
440  PRINT D$; "OPEN PHDAT"
450  PRINT D$; "POSITION PHDAT, R";
      IS - 1
460  PRINT D$; "READ PHDAT"
470  INPUT A(1), A(2), A(3), A(4), A(5),
      A(6), A(7), A(8), A(9), A(10),
      A(11), A(12), A(13), A(14), A(15)
480  PRINT D$; "CLOSE PHDAT"
490  PRINT D$; "IN#O"
500  RETURN
```

Example. Run the program for methane and propane at pressures 0.5×10^5, 10×10^5, 100×10^5, 500×10^5, and 1000×10^5 Pa and at temperature 400 and 500 K.

```
]RUN

ENTER THE SEQUENTIAL NUMBER OF COMPOUND IN DATA FILE 1
ENTER THE PRESSURE (PA) 0.5E5
ENTER THE TEMPERATURE (K) 400

REAL GAS INTERNAL ENERGY -56354.8155 (J / MOL)

NEXT TEMPERATURE ? (Y / N) N
NEXT PRESSURE ? (Y / N) Y
ENTER THE PRESSURE (PA) 10E5
ENTER THE TEMPERATURE (K) 400

REAL GAS INTERNAL ENERGY -56436.4111 (J / MOL)

NEXT TEMPERATURE ? (Y / N) Y
ENTER THE TEMPERATURE (K) 600

REAL GAS INTERNAL ENERGY -48797.3895 (J / MOL)

NEXT TEMPERATURE ? (Y / N) N
NEXT PRESSURE ? (Y / N) Y
ENTER THE PRESSURE (PA) 100E5
ENTER THE TEMPERATURE (K) 400

REAL GAS INTERNAL ENERGY -57186.1936 (J / MOL)

NEXT TEMPERATURE ? (Y / N) Y
ENTER THE TEMPERATURE (K) 500

REAL GAS INTERNAL ENERGY -53443.855 (J / MOL)

NEXT TEMPERATURE ? (Y / N) N
NEXT PRESSURE ? (Y / N) Y
ENTER THE PRESSURE (PA) 1000E5
ENTER THE TEMPERATURE (K) 400

REAL GAS INTERNAL ENERGY -60362.4939 (J / MOL)

NEXT TEMPERATURE ? (Y / N) Y
ENTER THE TEMPERATURE (K) 500

REAL GAS INTERNAL ENERGY -56078.5721 (J / MOL)

NEXT TEMPERATURE ? (Y / N) N
NEXT PRESSURE ? (Y / N) N
NEXT COMPOUND ? (Y / N) 3

]RUN

ENTER THE SEQUENTIAL NUMBER OF COMPOUND IN DATA FILE 3
ENTER THE PRESSURE (PA) 0.5E5
```

```
ENTER THE TEMPERATURE (K) 400

REAL GAS INTERNAL ENERGY -61703.946 (J / MOL)

NEXT TEMPERATURE ? (Y / N) N
NEXT PRESSURE ? (Y / N) Y
ENTER THE PRESSURE (PA) 10E5
ENTER THE TEMPERATURE (K) 400

REAL GAS INTERNAL ENERGY -62186.4177 (J / MOL)

NEXT TEMPERATURE ? (Y / N) Y
ENTER THE TEMPERATURE (K) 500

REAL GAS INTERNAL ENERGY -52599.4303 (J / MOL)

NEXT TEMPERATURE ? (Y / N) N
NEXT PRESSURE ? (Y / N) Y
ENTER THE PRESSURE (PA) 100E5
ENTER THE TEMPERATURE (K) 400

REAL GAS INTERNAL ENERGY -70435.8844 (J / MOL)

NEXT TEMPERATURE ? (Y / N) Y
ENTER THE TEMPERATURE (K) 500

REAL GAS INTERNAL ENERGY -56359.5197 (J / MOL)

NEXT TEMPERATURE ? (Y / N) N
NEXT PRESSURE ? (Y / N) Y
ENTER THE PRESSURE (PA) 1000E5
ENTER THE TEMPERATURE (K) 400

REAL GAS INTERNAL ENERGY -75914.9015 (J / MOL)

NEXT TEMPERATURE ? (Y / N) Y
ENTER THE TEMPERATURE (K) 500

REAL GAS INTERNAL ENERGY -64414.9873 (J / MOL)

NEXT TEMPERATURE ? (Y / N) N
NEXT PRESSURE ? (Y / N) N
NEXT COMPOUND ? (Y / N) N
```

Algorithm and Program: Ideal Gas Entropy

Write a program to calculate the entropy of ideal gases using the equation derived from the fifth-degree polynomial equation for the ideal gas enthalpy.

Algorithm.

1. Read the datafile sequential number of the desired compound.
2. Call the database-reader routine.
3. Read the actual temperature.

4. Read the actual pressure.
5. Calculate the value of $s°$.
6. Print the results.

Program IDENTR. A possible realization:

```
]PR#0
]LOAD IDENTR
]LIST

 10  REM :CALCULATION OF IDEAL GAS
      ENTROPY
 20  DIM A(15)
 30  PRINT
 40  INPUT "ENTER THE SEQUENTIAL
      NUMBER IN DATA FILE "; IS
 50  GOSUB 230
 60  PRINT
 70  INPUT "ENTER THE PRESSURE (PA)
      "; P
 80  PRINT
 90  INPUT "ENTER TEMPERATURE (K)
      "; T
100  PRINT : PRINT
110  PU = 101325
120  R = 8.3144
130  S0 = A(13) + A(8) * LOG (T) +
     (((5 / 4 * A(12) * T + 4 / 3
     * A(11)) * T + 3 / 2 * A(10
     )) * T + 2 * A(9)) * T - R *
     LOG (P / PU)
140  PRINT "THE IDEAL GAS ENTROPY
     "; S0; " (J / MOL / K)"
150  PRINT
160  INPUT "NEXT TEMPERATURE ? (Y
     / N) "; A$
170  IF A$ = "Y" THEN GOTO 80
180  INPUT "NEXT PRESSURE ? (Y / N)
     "; A$
190  IF A$ = "Y" THEN GOTO 60
200  INPUT "NEXT COMPOUND ? (Y / N)
     "; A$
210  IF A$ = "Y" THEN GOTO 30
220  END
230  REM :READING OF SPECIFIED
     RECORD OF PHDAT FILE INTO ARRAY
     A(5)
240  REM :THE SEQUENTIAL NUMBER
```

```
         OF THE COMPOUND IN THE CALLING
         PROGRAM HAS TO BE "IS"
250  REM :ARRAY A HAS TO BE DECLA
        RED IN THE CALLING PROGRAM
260  D$ = CHR$ (4)
270  PRINT D$; "OPEN PHDAT"
280  PRINT D$; "POSITION PHDAT, R";
        IS - 1
290  PRINT D$; "READ PHDAT"
300  INPUT A(1), A(2), A(3), A(4), A(5),
         A(6), A(7), A(8), A(9), A(10),
         A(11), A(12), A(13), A(14), A(15)
310  PRINT D$; "CLOSE PHDAT"
320  PRINT D$; "IN#0"
330  RETURN
```

Example. Run the program for methane and propane at temperatures 400 and 500 K, and pressures 50,000, 1,000,000, and 5,000,000 Pa.

```
]RUN

ENTER THE SEQUENTIAL NUMBER IN DATA FILE 1
ENTER THE PRESSURE (PA) 50000
ENTER TEMPERATURE (K) 400

THE IDEAL GAS ENTROPY 203.211835 (J / MOL K)

NEXT TEMPERATURE ? (Y / N) Y
ENTER TEMPERATURE (K) 500

THE IDEAL GAS ENTROPY 212.868939 (J / MOL K)

NEXT TEMPERATURE ? (Y / N) N
NEXT PRESSURE ? (Y / N) Y
ENTER THE PRESSURE (PA) 1000000
ENTER TEMPERATURE (K) 400

THE IDEAL GAS ENTROPY 178.304119 (J / MOL K)

NEXT TEMPERATURE ? (Y / N) Y
ENTER TEMPERATURE (K) 500

THE IDEAL GAS ENTROPY 187.961222 (J / MOL K)
NEXT TEMPERATURE ? (Y / N) N
NEXT PRESSURE ? (Y / N) Y
ENTER THE PRESSURE (PA) 5000000
ENTER TEMPERATURE (K) 400

THE IDEAL GAS ENTROPY 164.922608 (J / MOL K)
NEXT TEMPERATURE ? (Y / N) Y
ENTER TEMPERATURE (K) 500
THE IDEAL GAS ENTROPY 174.579711 (J / MOL K)
```

```
NEXT TEMPERATURE ? (Y / N) N
NEXT PRESSURE ? (Y / N) N
NEXT COMPOUND ? (Y / N) Y
ENTER THE SEQUENTIAL NUMBER IN DATA FILE 3
ENTER THE PRESSURE (PA) 50000
ENTER TEMPERATURE (K) 400

THE IDEAL GAS ENTROPY 300.355399 (J / MOL K)

NEXT TEMPERATURE ? (Y / N) Y

ENTER TEMPERATURE (K) 500

THE IDEAL GAS ENTROPY 323.200697 (J / MOL K)

NEXT TEMPERATURE ? (Y / N) N
NEXT PRESSURE ? (Y / N) Y
ENTER THE PRESSURE (PA) 1000000
ENTER TEMPERATURE (K) 400

THE IDEAL GAS ENTROPY 275.447683 (J / MOL K)

NEXT TEMPERATURE ? (Y / N) Y
ENTER TEMPERATURE (K) 500

THE IDEAL GAS ENTROPY 298.29298 (J / MOL K)

NEXT TEMPERATURE ? (Y / N) N
NEXT PRESSURE ? (Y / N) Y
ENTER THE PRESSURE (PA) 5000000
ENTER TEMPERATURE (K) 400

THE IDEAL GAS ENTROPY 262.066172 (J / MOL K)
NEXT TEMPERATURE ? (Y / N) Y
ENTER TEMPERATURE (K) 500

THE IDEAL GAS ENTROPY 284.91147 (J / MOL K)

NEXT TEMPERATURE ? (Y / N) N
NEXT PRESSURE ? (Y / N) N
NEXT COMPOUND ? (Y / N) N
```

Remark. Notice the change in the entropy with the temperature and pressure.

Algorithm and Program: Graphical Representation of the Ideal-Gas Entropy

Write a program to plot the function $s°(T)$ at constant pressure.

Algorithm.

1. Read the datafile sequential number of the compound.
2. Call the database-reader routine.

3. Read the pressure.
4. Read the beginning temperature, the temperature increment, and the number of calculations.
5. Plot the results.

Program GRIDENTR. A possible realization.

```
]PR#0
]LOAD GRIDENTR
]LIST

 10  REM :CALCULATION OF IDEAL
      GAS ENTROPY
 20  DIM A(15)
 30  PRINT
 40  HPLOT 0,0
 50  HGR
 60  HPLOT 10,155 TO 279,155
 70  HPLOT 10,0 TO 10,191
 80  INPUT "ENTER THE SEQUENTIAL
      NUMBER IN DATA FILE ", IS
 90  GOSUB 330
100  PRINT
110  INPUT "ENTER THE PRESSURE (PA)
      "; P
120  PRINT
130  INPUT "ENTER THE BEGINNING
      TEMPERATURE (K) "; TB
140  INPUT "ENTER THE TEMPERATURE
      INCREMENT "; DT
150  INPUT "ENTER THE NUMBER OF
      CALCULATIONS "; N
160  PRINT : PRINT
170  R = 8.3144
180  PU = 101325
190  FOR T = TB TO N * DT + TB STEP
      DT
200  S0 = A(13) + A(8) * LOG (T) +
      (((5 / 4 * A(12) * T + 4 / 3
      * A(11)) * T + 3 / 2 * A(10
      )) * T + 2 * A(9)) * T - R *
      LOG (P / PU)
210  X = (T - TB) / (N * DT) * 230
      + 10
220  Y = 220 - S0 / A(8) / 2 * 30
230  IF T = TB THEN HPLOT X,Y
240  HPLOT TO X,Y
```

```
250  NEXT T
260  PRINT
270  INPUT "NEXT PRESSURE ? (Y / N)
      "; A$
280  IF A$ = "Y" THEN GOTO 100
290  INPUT "NEXT COMPOUND ? (Y / N)
      "; A$
300  IF A$ = "Y" THEN GOTO 80
310  TEXT
320  END
330  REM :READING OF THE SPECIFIED
      RECORD OF PHDAT FILE INTO
      ARRAY A(15)
340  REM :THE SEQUENTIAL NUMBER OF
      THE COMPOUND IN THE CALLING
      PROGRAM HAS TO BE "IS"
350  REM :ARRAY A HAS TO BE DECLA
      RED IN THE CALLING PROGRAM
360  D$ =  CHR$ (4)
370  PRINT D$; "OPEN PHDAT"
380  PRINT D$; "POSITION PHDAT, R";
      IS - 1
390  PRINT D$; "READ PHDAT"
400  INPUT A(1), A(2), A(3), A(4), A(5),
      A(6), A(7), A(8), A(9), A(10),
      A(11), A(12), A(13), A(14), A(15)
410  PRINT D$; "CLOSE PHDAT"
420  PRINT D$; "IN/V0"
430  RETURN
```

Example. Run the program for methane and propane with beginning temperature 200 K, temperature increment 20 K, number of calculations 50, and pressures 101,325 and 1,013,250 Pa.

```
]RUN

ENTER THE SEQUENTIAL NUMBER IN DATA FILE 1
ENTER THE PRESSURE (PA) 101325
ENTER THE BEGINNING TEMPERATURE (K) 200
ENTER THE TEMPERATURE INCREMENT 20
ENTER THE NUMBER OF CALCULATIONS 50

NEXT PRESSURE ? (Y / N) Y
ENTER THE PRESSURE (PA) 1013250
ENTER THE BEGINNING TEMPERATURE (K) 200
ENTER THE TEMPERATURE INCREMENT 20
ENTER THE NUMBER OF CALCULATIONS 50

NEXT PRESSURE ? (Y / N) N
NEXT COMPOUND ? (Y / N) Y
```

```
ENTER THE SEQUENTIAL NUMBER IN DATA FILE 3
ENTER THE PRESSURE (PA) 101325
ENTER THE BEGINNING TEMPERATURE (K) 200
ENTER THE TEMPERATURE INCREMENT 20
ENTER THE NUMBER OF CALCULATIONS 50

NEXT PRESSURE ? (Y / N) Y
ENTER THE PRESSURE (PA) 1013250
ENTER THE BEGINNING TEMPERATURE (K) 200
ENTER THE TEMPERATURE INCREMENT 20
ENTER THE NUMBER OF CALCULATIONS 50

NEXT PRESSURE ? (Y / N) N
NEXT COMPOUND ? (Y / N) N
```

Remark. Watch the screen and explain the results.

Algorithm and Program: Entropy of Real Gases

Write a program for the calculation of the entropy of a real gas. The actual temperature and pressure are known. For the calculation of the entropy correction use the Peng–Robinson equation of state.

Algorithm.

1. Read the sequential number of the desired compound in the datafile.
2. Call the database reader routine.
3. Read the actual temperature and pressure.
4. Calculate the compressibility factor with the Peng–Robinson equation of state (see program ESPR in Chapter 3).
5. Calculate the entropy correction.
6. Calculate the ideal gas entropy using the equation derived from the fifth-degree polynomial equation for the ideal gas enthalpy (see program IDENTR in this chapter).
7. Calculate the real-gas entropy from the ideal gas entropy and the entropy correction.
8. Print the results.

Program REENTR. A possible realization:

```
]PR#0
]LOAD REENTR
]LIST

 10  REM :CALCULATION OF ENTROPY
```

```
     DEPARTURE AND THE REAL GAS
     ENTROPY
 20  DIM A(15)
 30  PRINT : PRINT
 40  INPUT "ENTER THE SEQUENTIAL
     NUMBER OF COMPOUND IN DATAFILE
     "; IS
 50  GOSUB 440
 60  INPUT "ENTER THE PRESSURE (PA)
     "; P
 70  INPUT "ENTER THE TEMPERATURE
     (K) "; T
 80  DEF FN F(Z) = Z ∧ 3 - (1 - B
     ) * Z ∧ 2 + (A - 3 * B ∧ 2 -
     2 * B) * Z - (A * B - B ∧ 2 -
     B ∧ 3)
 90  DEF FN FD(Z) = 3 * Z ∧ 2 - 2
     * (1 - B) * Z + (A - 3 * B ∧
     2 - 2 * B)
100  R = 8.3144
110  TC = A(3)
120  PC - A(2)
130  TR = T / TC
140  OM = A(4)
150  M - 0.37464 + 1.54226 * OM -
     0.26992 * OM ∧ 2
160  AL = 0.45724 * R * R * TC * T
     C / PC * (1 + M * (1 - SQR
     (TR))) ∧ 2
170  BL = 0.0778 * R * TC / PC
180  A = AL * P / R / R / T / T
190  B = BL * P / R / T
200  Z1 = 6
210  I = 0
220  H = FN F(Z1) / FN FD(Z1)
230  I = I + 1
240  IF I > 50 THEN PRINT : PRINT
     "ITERATION NOT CONVERGENT": GOTO
     340
250  Z2 = Z1 - H
260  IF ABS (Z2 - Z1) < 1E - 5 THEN
     GOTO 290
270  Z1 = Z2
280  GOTO 220
290  DS = R * ( LOG (Z2 - B) - 2.0
     8 * M * (1 / SQR (TR) * (M +
     1) - M) * LOG ((Z2 + 2.414 *
     B) / (Z2 - 0.414 * B)))
300  PRINT : PRINT
```

```
310  PRINT "THE ENTROPY DEPARTURE
      "; DS; " (J / MOL K)"
320  PU = 101325
330  SO = A(13) + A(8) * LOG (T) +
      (((5 / 4 * A(12) * T + 4 / 3
      * A(11)) * T + 3 / 2 * A(10
      )) * T + 2 * A(9)) * T - R *
      LOG (P / PU)
340  S - DS + SO
350  PRINT "THE REAL GAS ENTROPY
      IS "; S; " (J / MOL K)"
360  PRINT
370  INPUT "NEXT TEMPERATURE ? (Y
      /N) "; A$
380  IF A$ = "Y" THEN GOTO 70
390  INPUT "NEXT PRESSURE ? (Y / N)
      "; A$
400  IF A$ = "Y" THEN GOTO 60
410  INPUT "NEXT COMPOUND ? (Y / N)
      "; A$
420  IF A$ = "Y" THEN GOTO 40
430  END
440  D$ = CHR$ (4)
450  PRINT D$; "OPEN PHDAT"
460  PRINT D$; "POSITION PHDAT, R";
      IS - 1
470  PRINT D$; "READ PHDAT"
480  INPUT A(1), A(2), A(3), A(4), A(5),
      A(6), A(7), A(8), A(9), A(10),
      A(11), A(12), A(13), A(14), A(15)
490  PRINT D$; "CLOSE PHDAT"
500  PRINT D$; "IN#0"
510  RETURN
```

Example. Run the program for methane and propane at pressures 0.5×10^5, 10×10^5, 100×10^5, and 1000×10^5 Pa, and at temperatures 400 and 500 K.

```
]RUN

ENTER THE SEQUENTIAL NUMBER OF COMPOUND IN DATA FILE 1
ENTER THE PRESSURE (PA) 0.5E5
ENTER THE TEMPERATURE (K) 400

THE ENTROPY DEPARTURE -.0107369469 (J / MOL K)
THE REAL GAS ENTROPY IS 203.201098 (J / MOL K)

NEXT TEMPERATURE ? (Y / N) Y
ENTER THE TEMPERATURE (K) 500

THE ENTROPY DEPARTURE -6.33085654E - 03 (J / MOL K)
```

```
THE REAL GAS ENTROPY IS 212.862608 (J / MOL K)

NEXT TEMPERATURE ? (Y / N) N
NEXT PRESSURE ? (Y / N) Y
ENTER THE PRESSURE (PA) 10E5
ENTER THE TEMPERATURE (K) 400

THE ENTROPY DEPARTURE -.213535143 (J / MOL K)
THE REAL GAS ENTROPY IS 178.090584 (J / MOL K)

NEXT TEMPERATURE ? (Y / N) Y
ENTER THE TEMPERATURE (K) 500

THE ENTROPY DEPARTURE -.125493332 (J / MOL K)
THE REAL GAS ENTROPY IS 187.835729 (J / MOL K)

NEXT TEMPERATURE ? (Y / N) N
NEXT PRESSURE ? (Y / N) Y
ENTER THE PRESSURE (PA) 100E5
ENTER THE TEMPERATURE (K) 400

THE ENTROPY DEPARTURE -1.97611869 (J / MOL K)
THE REAL GAS ENTROPY IS 157.183387 (J / MOL K)

NEXT TEMPERATURE ? (Y / N) Y
ENTER THE TEMPERATURE (K) 500

THE ENTROPY DEPARTURE -1.14347314 (J / MOL K)
THE REAL GAS ENTROPY IS 167.673135 (J / MOL K)

NEXT TEMPERATURE ? (Y / N) N
NEXT PRESSURE ? (Y / N) Y
ENTER THE PRESSURE (PA) 1000E5
ENTER THE TEMPERATURE (K) 400

THE ENTROPY DEPARTURE -6.63056997 (J / MOL K)
THE REAL GAS ENTROPY IS 133.384322 (J / MOL K)

NEXT TEMPERATURE ? (Y / N) Y
ENTER THE TEMPERATURE (K) 500

THE ENTROPY DEPARTURE -4.65153881 (J / MOL K)
THE REAL GAS ENTROPY IS 145.020456 (J / MOL K)

NEXT TEMPERATURE ? (Y / N) N
NEXT PRESSURE ? (Y / N)
NEXT COMPOUND ? (Y / N) N
```

Algorithm and Program: Constant-Pressure Real-Gas Heat Capacity

Write a program for the calculation of the constant-pressure heat capacity of real gases. For the calculation of the heat-capacity correction use the Peng–Robinson equation of state.

Algorithm.

1. Read the sequential number of the desired compound in the datafile.
2. Call the database-reader routine.
3. Read the actual temperature and pressure.
4. Calculate the compressibility factor with the Peng–Robinson equation of state (see program **ESPR** in Chapter 3).
5. Calculate the constant-pressure heat-capacity correction.
6. Calculate the ideal-gas constant-pressure heat capacity.
7. Calculate the real-gas constant-pressure heat capacity.
8. Print the results.

Program RECP. A possible realization:

```
]PR#0
]LOAD RECP
]LIST

 10  REM :CALCULATION OF ISOBAR
      HEAT CAPACITY DEPARTURE AND CP
      BY PENG-ROBINSON EQUATION OF
      STATE
 20  DIM A (15)
 30  PRINT : PRINT
 40  INPUT "ENTER THE SEQUENTIAL
      NUMBER OF COMPOUND IN DATAFILE
      "; IS
 50  GOSUB 470
 60  INPUT "ENTER THE PRESSURE (PA
     ) "; P
 70  INPUT "ENTER THE TEMPERATURE
     (K) "; T
 80  DEF FN F (Z) = Z ∧ 3 - (1 - B
     ) * Z ∧ 2 + (A - 3 * B ∧ 2 -
     2 * B) * Z - (A * B - B ∧ 2 -
     B ∧ 3)
 90  DEF FN FD(Z) = 3 * Z ∧ 2 - 2
     * (1 - B) * Z + (A - 3 * B ∧
     2 - 2 * B)
100  R = 8.3144
110  TC = A(3)
120  PC = A(2)
130  TR = T / TC
140  OM = A(4)
150  M = 0.37464 + 1.54226 * OM -
     0.26992 * OM ∧ 2
160  AL = 0.45724 * R * R * TC * T
```

```
     C / PC * (1 + M * (1 - SQR
     (TR))) ∧ 2
170  BL = 0.0778 * R * TC / PC
180  A = AL * P / R / R / T / T
190  B = BL * P / R / T
200  L = 1 + M * (1 - SQR (TR))
210  Z1 = 6
220  I = 0
230  H = FN F(Z1) / FN FD(Z1)
240  I = I + 1
250  IF I > 50 THEN PRINT : PRINT
     "ITERATION NOT CONVERGENT": GOTO
     370
260  Z2 = Z1 - H
270  IF ABS (Z2 - Z1) < 1E - 5 THEN
     GOTO 300
280  Z1 = Z2
290  GOTO 230
300  DZ = B * Z2 ∧ 2 - (6 * B ∧ 2 +
     2 * B - 2 * A - A * M * SQR
     (TR) / L) * Z2 + 3 * B ∧ 3 +
     2 * B ∧ 2 - 3 * A * B - A *
     B * M * SQR (TR) / L
310  DZ = DZ / T / (3 * Z2 ∧ 2 + 2
     * (B - 1) * Z2 - 3 * B ∧ 2 -
     2 * B + A)
320  DC = R * (Z2 - 1 + T * DZ + 2
     .08 * (1 + M) / TR * ((1 + M
     * (1 - SQR (TR))) * ((T *
     DZ + 0.414 * B) / (Z2 - 0.41
     4 * B) - (T * DZ - 2.414 * B
     ) / (Z2 + 2.414 * B)) + M /
     2 * SQR (TR) * LOG ((Z2 +
     2.414 * B) / (Z2 - .414 * B))))
330  PRINT : PRINT
340  PRINT "THE CP DEPARTURE IS "
     ; DC; " (J / MOL K)"
350  PU = 101325
360  CP = (((5 * A(12) * T + 4 * A
     (11)) * T + 3 * A(10)) * T +
     2 * A(9)) * T + A(8)
370  CP = DC + CP
380  PRINT "THE REAL GAS CP IS ";
     CP; " (J / MOL K)"
390  PRINT
400  INPUT "NEXT TEMPERATURE ? (Y
     /N) "; A$
410  IF A$ = "Y" THEN GOTO 70
420  INPUT "NEXT PRESSURE ? (Y / N)
```

```
        ";A$
430  IF A$ = "Y" THEN GOTO 60
440  INPUT "NEXT COMPOUND ? (Y / N)
        "; A$
450  IF A$ = "Y" THEN GOTO 40
460  END
470  D$ = CHR$ (4)
480  PRINT D$; "OPEN PHDAT"
490  PRINT D$; "POSITION PHDAT, R";
      IS - 1
500  PRINT D$; "READ PHDAT"
510  INPUT A(1), A(2), A(3), A(4), A(5),
      A(6), A(7), A(8), A(9), A(10),
      A(11), A(12), A(13), A(14), A(15)
520  PRINT D$; "CLOSE PHDAT"
530  PRINT D$; "IN#0"
540  RETURN
```

Example. Run the program for hydrogen and propane at pressures 101325, 1,013,250, 10,132,500, and 101,325,000 Pa at temperatures 400 and 500 K.

```
]RUN

ENTER THE SEQUENTIAL NUMBER OF COMPOUND IN DATAFILE 1
ENTER THE PRESSURE (PA) 101325
ENTER THE TEMPERATURE (K) 400

THE CP DEPARTURE IS .0508091639 (J / MOL K)
THE REAL GAS CP IS 40.6664572 (J / MOL K)

NEXT TEMPERATURE ? (Y / N) Y
ENTER THE TEMPERATURE (K) 500

THE CP DEPARTURE IS .030839626 (J / MOL K)
THE REAL GAS CP IS 46.2933396 (J / MOL K)

NEXT TEMPERATURE ? (Y / N) N
NEXT PRESSURE ? (Y / N) Y
ENTER THE PRESSURE (PA) 1013250
ENTER THE TEMPERATURE (K) 400

THE CP DEPARTURE IS .510352384 (J / MOL K)
THE REAL GAS CP IS 41.1260004 (J / MOL K)

NEXT TEMPERATURE ? (Y / N) Y
ENTER THE TEMPERATURE (K) 500

THE CP DEPARTURE IS .306647919 (J / MOL K)
THE REAL GAS CP IS 46.5691479 (J / MOL K)

NEXT TEMPERATURE ? (Y / N) N
NEXT PRESSURE ? (Y / N) Y
```

```
ENTER THE PRESSURE (PA) 10132500
ENTER THE TEMPERATURE (K) 400

THE CP DEPARTURE IS 4.98593963 (J / MOL K)
THE REAL GAS CP IS 45.6015877 (J / MOL K)

NEXT TEMPERATURE ? (Y / N) Y
ENTER THE TEMPERATURE (K) 500

THE CP DEPARTURE IS 2.81305556 (J / MOL K)
THE REAL GAS CP IS 49.0755555 (J / MOL K)

NEXT TEMPERATURE ? (Y / N) N
NEXT PRESSURE ? (Y / N) Y
ENTER THE PRESSURE (PA) 101325000
ENTER THE TEMPERATURE (K) 400

THE CP DEPARTURE IS 9.85952145 (J / MOL K)
THE REAL GAS CP IS 50.4751694 (J / MOL K)

NEXT TEMPERATURE ? (Y / N) Y
ENTER THE TEMPERATURE (K) 500

THE CP DEPARTURE IS 7.82554317 (J / MOL K)
THE REAL GAS CP IS 54.0880431 (J / MOL K)

NEXT TEMPERATURE ? (Y / N) N
NEXT PRESSURE ? (Y / N) N
NEXT COMPOUND ? (Y / N) Y
ENTER THE SEQUENTIAL NUMBER OF COMPOUND IN DATA FILE 3
ENTER THE PRESSURE (PA) 101325
ENTER THE TEMPERATURE (K) 400

THE CP DEPARTURE IS .285542134 (J / MOL K)
THE REAL GAS CP IS 93.4606621 (J / MOL K)

NEXT TEMPERATURE ? (Y / N) Y
ENTER THE TEMPERATURE (K) 500

THE CP DEPARTURE IS .172171453 (J / MOL K)
THE REAL GAS CP IS 112.289609 (J / MOL K)

NEXT TEMPERATURE ? (Y / N) N
NEXT PRESSURE ? (Y / N) Y
ENTER THE PRESSURE (PA) 1013250
ENTER THE TEMPERATURE (K) 400

THE CP DEPARTURE IS 3.37743611 (J / MOL K)
THE REAL GAS CP IS 96.5525561 (J / MOL K)

NEXT TEMPERATURE ? (Y / N) Y
ENTER THE TEMPERATURE (K) 500

THE CP DEPARTURE IS 1.84078172 (J / MOL K)
THE REAL GAS CP IS 113.958219 (J / MOL K)

NEXT TEMPERATURE ? (Y / N) N
NEXT PRESSURE ? (Y / N) Y
```

```
ENTER THE PRESSURE (PA) 10132500
ENTER THE TEMPERATURE (K) 400

THE CP DEPARTURE IS 73.9781334 (J / MOL K)
THE REAL GAS CP IS 167.153253 (J / MOL K)

NEXT TEMPERATURE ? (Y / N) Y
ENTER THE TEMPERATURE (K) 500

THE CP DEPARTURE IS 28.8368295 (J / MOL K)
THE REAL GAS CP IS 140.954267 (J / MOL K)

NEXT TEMPERATURE ? (Y / N) N
NEXT PRESSURE ? (Y / N) Y
ENTER THE PRESSURE (PA) 101325000
ENTER THE TEMPERATURE (K) 400

THE CP DEPARTURE IS 20.7590611 (J / MOL K)
THE REAL GAS CP IS 113.934181 (J / MOL K)

NEXT TEMPERATURE ? (Y / N) Y
ENTER THE TEMPERATURE (K) 500

THE CP DEPARTURE IS 19.0925235 (J / MOL K)
THE REAL GAS CP IS 131.209961 (J / MOL K)

NEXT TEMPERATURE ? (Y / N) N
NEXT PRESSURE ? (Y / N) N
NEXT COMPOUND ? (Y / N) N
```

ONE OF THE INDEPENDENT VARIABLES IS TEMPERATURE

The following four cases must be discussed:

$$Tv, \quad Th, \quad Tu, \quad Ts$$

Case *Tv*

First of all the state of aggregation of the substance defined by variables Tv must be decided. For this purpose a good estimate of the molar volume of the saturated liquid, v_{sat} can be obtained from the following empirical relation:

$$v_{\text{sat}} = v_c z_c^{(1 - T_r)^{0.2857}}$$

This relation does not result from an equation of state; its only role here is to help to decide the state of aggregation of the body. The molar volume of saturated vapor is derived from the equation of state:

$$v^\circ = \frac{\mathbf{R}Tz[T, P^\circ[T]]}{P^\circ}$$

When v_{sat} and v° are known, the following can be said of the state of the system:

$v \geq v^\circ$	Vapor
$v^\circ > v > v_{sat}$	Two phases
$v \leq v_{sat}$	Liquid

For the calculation of the compressibility coefficient:

1. The quadratic virial equation yields an explicit expression,

$$z = 1 + \frac{B_r}{\phi_r}$$

2. The explicit form of the Redlich–Kwong equation for the pressure is

$$P = \frac{\mathbf{R}T}{v - b} - \frac{a}{T^{0.5}v(v + b)}$$

$$z = \frac{Pv}{\mathbf{R}T}$$

3. The Lee-Kesler equation yields an explicit expression for the compressibility coefficient,

$$z = 1 + \frac{B_r}{\phi_r} + \frac{C_r}{\phi_r^2} + \frac{D_r}{\phi_r^5} + \frac{c_4}{T_r^3\phi_r^2}\left(\beta + \frac{\gamma}{\phi_r^2}\right)\exp\left(-\frac{\gamma}{\phi_r^2}\right)$$

When the compressibility coefficient is known, pressure is calculated by substitution:

$$P = \frac{\mathbf{R}Tz[T, \phi]}{v}$$

Now all further calculation is reduced to the basic case *PT*.

Cases *Th*, *Tu*, and *Ts*

These three cases are discussed together.

Let us rewrite the equations for enthalpy, internal energy, and entropy in the following forms:

$$h - (h^*[T] + \Delta h) = 0$$

$$u - (h^*[T] + \Delta h - z\mathbf{R}T) = 0$$

$$s - \left(s^*[T] + \Delta s - \mathbf{R}\ln\frac{P}{P_u}\right) = 0$$

where

$$\Delta h = \begin{cases} \Delta h[T, P, z[T, P]] \\ \text{or} \\ \Delta h[T, \phi, z[T, \phi]] \end{cases}$$

$$\Delta s = \begin{cases} \Delta s[T, P, z[T, P]] \\ \text{or} \\ \Delta s[T, \phi, z[T, \phi]] \end{cases}$$

In all the three cases the problem is whether ϕ or P, and indirectly z, must be known for the calculation of Δh and Δs. Thus, depending on whether h, u, or s is the independent variable, we form, from the relevant equation and from an appropriate equation of state, a system of equations with P or ϕ and z as variables. After solving the system of equations, the calculation of the properties yet unknown reduces to the PT or ϕT case.

Algorithm and Program: Calculation of Pressure and Volume from Temperature and Enthalpy

Write a program for the calculation of the molar volume of a real gas if the temperature and the molar enthalpy are known. Use the Peng–Robinson equation of state.

Algorithm.

1. Read the sequential number of the desired compound in the data file.
2. Call the database-reader routine.
3. Read the actual temperature and enthalpy.
4. Calculate the constants in the Peng–Robinson equation.
5. Calculate a pressure range bracketing the solution (start with $\phi \leq P \leq 4P_c$).
6. Find the pressure interval containing the solution by calculating the enthalpy from the actual temperature.
7. Tighten the interval by repeating steps 5 and 6 three times.
8. Calculate the volume and compressibility factor from the actual temperature and the calculated pressure.
9. Print the results.

Program HTEMP. A possible realization:

```
]PR#0
]LOAD HTEMP
```

```
]LIST

 10  REM :CALCULATION OF PRESSURE
      AND VOLUME FROM TEMPERATURE
      AND ENTHALPY
 20  DIM A(15)
 30  PRINT : PRINT
 40  INPUT "ENTER SEQUENTIAL NUMBER
      OF COMPOUND IN DATAFILE
      "; IS
 50  GOSUB 590
 60  INPUT "ENTER THE ENTHALPY (J /
      MOL) "; HI
 70  INPUT "ENTER THE TEMPERATURE
      (K) "; T
 80  DEF FN F(Z) = Z ∧ 3 - (1 - B
      ) * Z ∧ 2 + (A - 3 * B ∧ 2 -
      2 * B) * Z - (A * B - B ∧ 2 -
      B ∧ 3)
 90  DEF FN FD(Z) = 3 * Z ∧ 2 -2
      * (1 - B) * Z + (A - 3 * B ∧
      2 - 2 * B)
100  R = 8.3144
110  TC = A(3)
120  PC = A(2)
130  TR = T / TC
140  OM = A(4)
150  M = 0.37464 + 1.54226 * OM -
      0.26992 * OM ∧ 2
160  AL = 0.45724 * R * R * TC * T
      C / PC * (1 + M * (1 - SQR
      (TR))) ∧ 2
170  BL = 0.0778 * R * TC / PC
180  PL = 0:PU = 3 * PC
190  FOR J = 1 TO 4
200  FOR P = PL TO PU STEP (PU -
      PL) / 10
210  A = AL * P / R / R / T / T
220  B = BL * P / R / T
230  Z1 = 6
240  I = 0
250  H = FN F(Z1) / FN FD(Z1)
260  I = I + 1
270  IF I > 50 THEN PRINT : PRINT
      "ITERATION NOT CONVERGENT": GOTO
      340
280  Z2 = Z1 - H
290  IF ABS (Z2 - Z1) < 1E - 5 THEN
      GOTO 320
```

```
300  Z1 = Z2
310  GOTO 250
320  DH = R * ((T * (Z2 - 1) - 2.0
      8 * TC * (1 + M) * (1 + M *
      (1 - SQR (TR))) * LOG ((Z2
      + 2.414 * B) / (Z2 - 0.414 *
      B))))
330  H0 = ((((A( 12) * T + A(11)) *
      T + A(10)) * T + A(9)) * T +
      A(8)) * T + A(7)
340  H = DH + H0
350  TK = TR: IF TR > 1 THEN TK =
      0.999999
360  IF P = PL AND J = 1 AND HI >
      H THEN PRINT : PRINT "TOO
      HIGH INPUT ENTHALPY": GOTO 470
370  IF P = PL AND J = 1 AND TR <
      1 AND HI < H - A(6) * ((1 -
      TK) / (1 - A(5) / TC)) ∧ 0.3
      8 THEN PRINT : PRINT "LIQUID
      STATE ": GOTO 470
380  IF H < HI THEN GOTO 410
390  IF P = PU AND H > HI THEN PRINT
      : PRINT "NO SOLUTION IN THE
      PRESSURE RANGE
      "; PL; "-"; PU: GOTO 470
400  NEXT P
410  PL = P - (PU - PL) / 10: PU =
      P
420  NEXT J
430  PRINT : PRINT
440  PRINT "THE PRESSURE "; P; " (P
      A)"
450  PRINT "THE MOLAR VOLUME "Z2
      * R * T / P; " (M3 / MOL)"
460  PRINT "THE COMPRESSIBILITY
      FACTOR "; Z2
470  PRINT : PRINT
480  INPUT "NEXT TEMPERATURE ? (Y
      /N) "; A$
490  PRINT
500  IF A$ = "Y" THEN GOTO 70
510  INPUT "NEXT ENTHALPY ? (Y / N)
      "; A$
520  PRINT
530  IF A$ = "Y" THEN GOTO 60
540  PRINT
550  INPUT "NEXT COMPOUND ? (Y / N)
      "; A$
```

```
560  PRINT
570  IF A$ = "Y" THEN GOTO 40
580  END
590  D$ = CHR$ (4)
600  PRINT D$; "OPEN PHDAT"
610  PRINT D$; "POSITION PHDAT, R";
      IS - 1
620  PRINT D$; "READ PHDAT"
630  INPUT A(1), A(2), A(3), A(4), A(5),
      A(6), A(7), A(8), A(9), A(10),
      A(11), A(12), A(13), A(14), A(15)
640  PRINT D$; "CLOSE PHDAT"
650  PRINT D$; "IN#0"
660  RETURN
```

Example. Run the program for methane at temperature 400 K and molar enthalpy −53134 J/mole, and for propane at temperature 400 K and molar enthalpy −69068 J/mole.

```
]RUN

ENTER THE SEQUENTIAL NUMBER OF COMPOUND IN DATA FILE 1
ENTER THE ENTHALPY (J / MOL) -53134
ENTER THE TEMPERATURE (K) 400

THE PRESSURE 995023.26 (PA)
THE MOLAR VOLUME 3.31848793E-03 (M3 / MOL)
THE COMPRESSIBILITY FACTOR .992847555

NEXT TEMPERATURE ? (Y / N) N
NEXT ENTHALPY ? (Y / N) N
NEXT COMPOUND ? (Y / N) Y

ENTER SEQUENTIAL NUMBER OF COMPOUND IN DATA FILE 3
ENTER THE ENTHALPY (J / MOL) -69068
ENTER THE TEMPERATURE (K) 400

THE PRESSURE 9999426.15 (PA)
THE MOLAR VOLUME 1.36758421E-04 (M3 / MOL)
THE COMPRESSIBILITY FACTOR .411185935

NEXT TEMPERATURE ? (Y / N) N
NEXT ENTHALPY ? (Y / N) N
NEXT COMPOUND ? (Y / N) N
```

Remark. If the desired enthalpy is larger than the ideal-gas enthalpy at the actual temperature, then the problem has no solution.

ONE OF THE INDEPENDENT VARIABLES IS PRESSURE

The following four cases must be discussed:

$$pv, \quad Ph, \quad Pu, \quad Ps$$

Case *Pv*

The unknown variable is temperature T, and this can be calculated with an appropriate equation of state. Inconvenience arises from the fact that the equations of state are isotherm equations, and their parameters too include the temperature. Therefore, a system of two equations is solved. One of these equations is an equation of state; the other is

$$v - \frac{\mathbf{R}Tz[P,T]}{P} = 0$$

in this case with P as parameter.

The two variables of the system are T and z.

After the solving the system of equations, the calculation of h, u, and s is reduced to the basic case PT.

Cases *Ph*, *Pu*, and *Ps*

These three cases are discussed together. However, first the case Ph is shown in Figure 9.4. The enthalpy correction is given as a function of reduced temperature in terms of a set of reduced temperature parameters:

$$\frac{h - h^*[T]}{\mathbf{R}T_c} = \frac{\Delta h}{\mathbf{R}T_c} = \frac{\Delta h[T_r, P_r]}{\mathbf{R}T_c}$$

We plot in the figure the reduced isobar P_r relevant to the given case.

When T is known, we have $(h - h^*)/\mathbf{R}T_c$ immediately from the figure: the ordinate of the intersection of the given reduced isotherm with the reduced isobar is the enthalpy correction sought. However, now we do not know which is the isotherm in question. We know on the other hand that $h^*[T_r]$ increases monotonically with pressure, and hence, $\Delta h = h - h^*[T_r]$ decreases monotonically with increasing pressure. The curve of Δh has been drawn on the plot, and it can be seen that Δh decreases also with increasing reduced pressure. The intersection of the curve with the reduced isobar lies along that isotherm for which

$$\Delta h = h - h^*[T_r] = \Delta h[T_r, P_r]$$

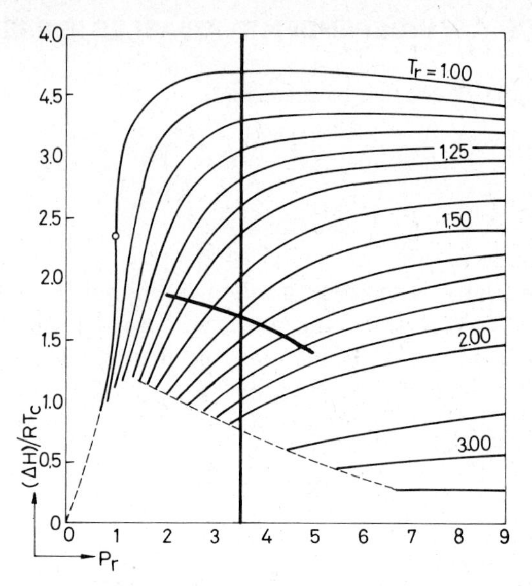

FIGURE 9.4. Isothermal enthalpy correction.

In the course of a calculation based on the equation of state, the two-variable function $\Delta h[P_r T_r]$ is not directly available, but we have the relation

$$\Delta h[T, P, z[T, P]] \quad \text{or} \quad \Delta h[T_r, \phi_r, z[T_r, \phi_r]]$$

The solution in the case *Ps* can be found by an identical argument, shown in Figure 9.5. Here too, the criterion of the solution is

$$\Delta s = s - s^*[T_r, P_r] = \Delta s[T_r, P_r]$$

In the course of a calculation based on the equation of state, the enthalpy correction or entropy correction is not available in the form of a two-variable function. The two corrections, in the form of an algebraic expression containing z, are at our disposal for the basic case. Therefore, in general, we start for our numerical solution from the equations for h, u, or s rewritten with zero on the right (as in the cases *Th*, *Tu*, and *Ts*). The other equation will be the appropriate equation of state for the calculation of z. This is needed because the parameters of the equations for h, u, and s are functions of the unknown temperature:

$$z = z[T, P, \pi[T]]$$

The solution of the system of equations yields the values of T and z in the given state. The calculation of the two variables yet unknown reduces to that in the basic case *PT*.

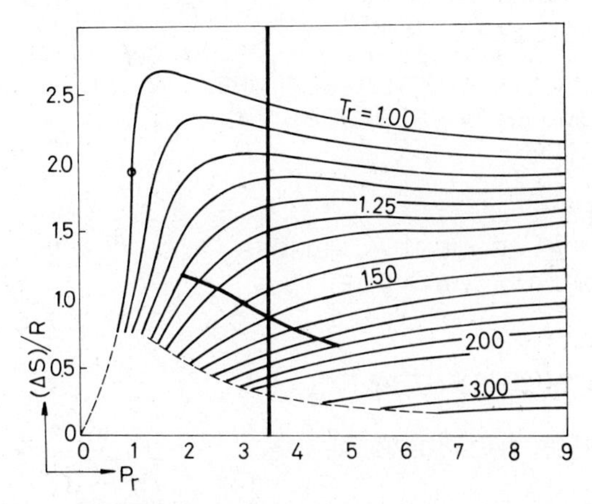

FIGURE 9.5. Isothermal entropy correction.

Algorithm and Program: Calculation of Temperature and Volume from Pressure and Enthalpy

Write a program for the calculation of the temperature and molar volume of a real gas if the pressure and the molar enthalpy are known. Use the Peng–Robinson equation of state.

Algorithm.

1. Read the sequential number of the components in the datafile.
2. Call the database-reader routine.
3. Read the actual pressure and enthalpy.
4. Calculate the constants in the Peng–Robinson equation.
5. Calculate a temperature range bracketing the solution (starting with $T_b - 20 \leq T \leq 5T_c$).
6. Find a new temperature interval for the solution by calculating the enthalpy from the actual pressure.
7. Tighten the interval by repeating steps 5 and 6 three times.
8. Calculate the volume and the compressibility factor from the actual pressure and from the calculated temperature.
9. Print the results.

Program HPRES. A possible realization:

```
]PR#0
]LOAD HPRES
```

```
]LIST

 10  REM :CALCULATION OF TEMPERATURE
      AND VOLUME FROM PRESSURE
      AND ENTHALPY
 20  DIM A(15)
 30  PRINT : PRINT
 40  INPUT "ENTER SEQUENTIAL NUMBER
      OF COMPOUND IN DATA FILE
      "; IS
 50  GOSUB 600
 60  INPUT "ENTER THE ENTHALPY (J /
      MOL) "; HI
 70  INPUT "ENTER THE PRESSURE (PA
      ) "; P
 80  DEF FN F(Z) = Z ∧ 3 - (1 - B
      ) * Z ∧ 2 + (A - 3 * B ∧ 2 -
      2 * B) * Z - (A * B - B ∧ 2 -
      B ∧ 3)
 90  DEF FN FD(Z) = 3 * Z ∧ 2 - 2
      * (1 - B) * Z + (A - 3 * B ∧
      2 - 2 * B)
100  R = 8.3144
110  TC = A(3)
120  PC = A(2)
130  OM = A(4)
140  M = 0.37464 + 1.54226 * OM -
      0.26992 * OM ∧ 2
150  AL = 0.45724 * R * R * TC * T
      C / PC * (1 + M * (1 - SQR
      (TR))) ∧ 2
160  BL = 0.0778 * R * TC / PC
170  TL = A(5) - 20:TU = 5 * TC
180  FOR J = 1 TO 4
190  FOR T = TU TO TL STEP - (TU
      - TL) / 10
200  TR = T / TC
210  A = AL * P / R / R / T / T
220  B = BL * P / R / T
230  Z1 = 6
240  I = 0
250  H = FN F(Z1) / FN FD(Z1)
260  I = I + 1
270  IF I > 50 THEN PRINT : PRINT
      "ITERATION NOT CONVERGENT": GOTO
      350
280  Z2 = Z1 - H
290  IF ABS (Z2 - Z1) < 1E - 5 THEN
      GOTO 320
300  Z1 = Z2
```

```
310  GOTO 250
320  IF Z2 < 0.4 THEN Z2 = 0.4
330  DH = R * ((T * (Z2 - 1) - 2.0
      8 * TC * (1 + M) * (1 + M *
      (1 - SQR (TR))) * LOG ((Z2
      + 2.414 * B) / (Z2 - 0.414 *
      B))))
340  H0 = ((((A(12) * T + A(11)) *
      T + A(10)) * T + A(9)) * T +
      A(8)) * T + A(7)
350  H = DH + H0
360  TK = TR: IF TR > 1 THEN TK =
      0.999999
370  IF T = TU AND J = 1 AND HI >
      H THEN PRINT : PRINT "TOO
      HIGH INPUT ENTHALPY": GOTO 480
380  IF H < HI THEN GOTO 420
390  IF J = 1 AND TR < 1 AND HI <
      H - A(6) * ((1 - TK) / (1 -
      A(5) / TC)) ∧ 0.38 THEN PRINT
      : PRINT "LIQUID STATE": GOTO
      480
400  IF T = TL AND H > HI AND J =
      1 THEN PRINT : PRINT "NO
      SOLUTION IN THE TEMPERATURE
      RANGE "; TL; "-"; TU: GOTO 480
410  NEXT T
420  TL = T: TU = T + (TU - TL) /10
430  NEXT J
440  PRINT : PRINT
450  PRINT "THE TEMPERATURE "; T;
      " (K)"
460  PRINT "THE MOLAR VOLUME "; Z2
      * R * T / P; " (M3 / MOL)"
470  PRINT "THE COMPRESSIBILITY
      FACTOR "; Z2
480  PRINT : PRINT
490  INPUT "NEXT PRESSURE ? (Y / N)
      "; A$
500  PRINT
510  IF A$ = "Y" THEN GOTO 70
520  INPUT "NEXT ENTHALPY ? (Y / N)
      "; A$
530  PRINT
540  IF A$ = "Y" THEN GOTO 60
550  PRINT
560  INPUT "NEXT COMPOUND ? (Y / N)
      "; A$
570  PRINT
580  IF A$ = "Y" THEN GOTO 40
```

```
590  END
600  D$ = CHR$ (4)
610  PRINT D$; "OPEN PHDAT"
620  PRINT D$; "POSITION PHDAT, R";
      IS - 1
630  PRINT D$; "READ PHDAT"
640  INPUT A(1), A(2), A(3), A(4), A(5),
      A(6), A(7), A(8), A(9), A(10),
      A(11), A(12), A(13), A(14), A(15)
650  PRINT D$; "CLOSE PHDAT"
660  PRINT D$; "IN#0"
670  RETURN
```

Example. Run the program for methane at pressure 10^6 Pa and enthalpy $-53{,}134$ J/mole, and for propane at pressure 10^7 Pa and enthalpy $-69{,}068$ J/mole.

```
]RUN

ENTER SEQUENTIAL NUMBER OF COMPOUND IN DATA FILE 1
ENTER THE ENTHALPY (J / MOL) -53134
ENTER THE PRESSURE (PA) 1E6

THE TEMPERATURE 399.431436 (K)
THE MOLAR VOLUME 3.20293693E - 03 (M3 / MOL)
THE COMPRESSIBILITY FACTOR .964440038

NEXT PRESSURE ? (Y / N) N
NEXT ENTHALPY ? (Y / N) N
NEXT COMPOUND ? (Y / N) Y

ENTER SEQUENTIAL NUMBER OF COMPOUND IN DATA FILE 3
ENTER THE ENTHALPY (J / MOL) -69068
ENTER THE PRESSURE (PA) 1E7

THE TEMPERATURE 345.4783 (K)
THE MOLAR VOLUME 2.14799458E - 04 (M3 / MOL)
THE COMPRESSIBILITY FACTOR .747793168

NEXT PRESSURE ? (Y / N) N
NEXT ENTHALPY ? (Y / N) N
NEXT COMPOUND ? (Y / N) N
```

ONE OF THE INDEPENDENT VARIABLES IS THE MOLAR VOLUME

The following three cases have to be discussed:

$$vh, \quad vu, \quad vs$$

The cases Tv and Pv have been already dealt with.

These three cases are discussed together. In each we start from the form of equations h, u, or s with zero on the right. Each of them contains two unknowns: T and z. Thus, a further equation is needed, which will be the appropriate equation of state for the calculation of z. This is necessary because the parameters of our starting equations, too, depend on the unknown temperature. The solution of the system of equations yields the values of T and z in the given state.

When these values are known, the calculation of the two variables yet unknown is reduced to that in the case PT.

Algorithm and Program: Calculation of Temperature and Pressure from Volume and Enthalpy

Write a program for the calculation of the temperature and pressure of a real gas if the molar volume and molar enthalpy are known. Use the Peng–Robinson equation of state.

Algorithm.

1. Read the sequential number of the component in the datafile.
2. Call the database-reader routine.
3. Read the actual molar enthalpy and molar volume.
4. Calculate the Peng–Robinson constants.
5. Calculate a temperature range bracketing the solution (starting with $T_b - 20 \leq T \leq 5T_c$).
6. Find the temperature interval of the solution by calculating the pressure from the actual volume.
7. Tighten the interval by repeating steps 5 and 6 three times.
8. Print the results.

Program HVOL. A possible realization:

```
]PR#0
]LOAD HVOL
]LIST

 10  REM :CALCULATION TEMPERATURE
      AND PRESSURE FROM VOLUME AND
      ENTHALPY
 20  DIM A(15)
 30  PRINT : PRINT
 40  INPUT "ENTER SEQUENTIAL NUMBE
      R OF COMPOUND IN DATA FILE
      "; IS
 50  GOSUB 500
 60  INPUT "ENTER THE ENTHALPY (J /
```

```
    MOL) "; HI
 70 INPUT "ENTER THE MOLAR VOLUME
    (M3 / MOL) "; V
 80 DEF FN F(Z) = Z ∧ 3 - (1 - B
    ) * Z ∧ 2 + (A - 3 * B ∧ 2 -
    2 * B) * Z - (A * B - B ∧ 2 -
    B ∧ 3)
 90 DEF FN FD(Z) = 3 * Z ∧ 2 - 2
    * (1 - B) * Z + (A - 3 * B ∧
    2 - 2 * B)
100 R = 8.3144
110 TC = A(3)
120 PC = A(2)
130 OM = A(4)
140 M = 0.37464 + 1.54226 * OM -
    0.26992 * OM ∧ 2
150 AL = 0.45724 * R * R * TC * T
    C / PC * (1 + M * (1 - SQR
    (TR))) ∧ 2
160 BL = 0.0778 * R * TC / PC
170 TL = A(5) - 20:TU = 5 * TC
180 FOR J = 1 TO 4
190 FOR T = TU TO TL STEP - (TU
    - TL) / 10
200 TR = T / TC
210 P = R * T / (V - BL) - AL / (
    V * (V + BL) + BL * (V - BL)
    )
220 Z2 = P * V / R / T
230 DH = R * ((T * (Z2 - 1) - 2.0
    8 * TC * (1 + M) * (1 + M *
    (1 - SQR (TR))) * LOG ((Z2
    + 2.414 * B) / (Z2 - 0.414 *
    B))))
240 H0 = ((((A( 12) * T + A(11)) *
    T + A(10)) * T + A(9)) * T +
    A(8)) * T + A(7)
250 H = DH + HO
260 TK = TR: IF TR > 1 THEN TK =
    0.999999
270 IF T = TU AND J = 1 AND HI >
    H THEN PRINT : PRINT "TOO
    HIGH INPUT ENTHALPY": GOTO 380
280 IF H < HI THEN GOTO 320
290 IF J = 1 AND TR < 1 AND HI <
    H - A(6) * ((1 - TK) / (1 -
    A(5) / TC)) ∧ 0.38 THEN PRINT
    : PRINT "LIQUID STATE": GOTO
    380
300 IF T = TL AND H > HI AND J =
```

```
     1 THEN PRINT : PRINT "NO
     SOLUTION IN THE TEMPERATURE
     RANGE "; TL; "-"; TU: GOTO 380
310  NEXT T
320  TL = T:TU = T + (TU - TL) /1
     0
330  NEXT J
340  PRINT : PRINT
350  PRINT "THE TEMPERATURE "; T;
     "(K)"
360  PRINT "THE PRESSURE "; P;
     " (PA)"
370  PRINT "THE COMPRESSIBILITY
     FACTOR "; Z2
380  PRINT : PRINT
390  INPUT "NEXT VOLUME ? (Y / N)
     "; A$
400  PRINT
410  IF A$ = "Y" THEN GOTO 70
420  INPUT "NEXT ENTHALPY ? (Y / N)
     "; A$
430  PRINT
440  IF A$ = "Y" THEN GOTO 60
450  PRINT
460  INPUT "NEXT COMPOUND ? (Y / N)
     "; A$
470  PRINT
480  IF A$ = "Y" THEN GOTO 40
490  END
500  D$ = CHR$ (4)
510  PRINT D$; "OPEN PHDAT"
520  PRINT D$; "POSITION PHDAT, R";
     IS - 1
530  PRINT D$; "READ PHDAT"
540  INPUT A(1), A(2), A(3), A(4), A(5),
     A(6), A(7), A(8), A(9), A(10),
     A(11),A(12),A(13),A(14),A(15)
550  PRINT D$; "CLOSE PHDAT"
560  PRINT D$; "IN#0"
570  RETURN
580  IF J = 1 AND TR < 1 AND HI <
     H - A(6) * ((1 - TK) / (1 -
     A(5) / TC)) ∧ 0.38 THEN PRINT
     : PRINT "LIQUID
     STATE": GOTO 380
```

Example. Run the program for methane at molar volume $-53{,}134$ J/mole and molar volume 3.3184×10^{-3} m^3/mole, and for propane at molar enthalpy $-69{,}068$ J/mole and molar volume 1.367×10^{-4} m^3/mole.

```
]RUN

ENTER SEQUENTIAL NUMBER OF COMPOUND IN DATA FILE 1
ENTER THE ENTHALPY (J / MOL) -53134
ENTER THE MOLAR VOLUME (M3 / MOL) 3.3184E - 3

THE TEMPERATURE 399.431436 (K)
THE PRESSURE 966416.94 (PA)
THE COMPRESSIBILITY FACTOR .965650819

NEXT VOLUME ? (Y / N) N
NEXT ENTHALPY ? (Y / N) N
NEXT COMPOUND ? (Y / N) Y

ENTER SEQUENTIAL NUMBER OF COMPOUND IN DATA FILE 3
ENTER THE ENTHALPY (J / MOL) -69068
ENTER THE MOLAR VOLUME (M3 / MOL) 1.367E - 4

THE TEMPERATURE 278.10119 (K)
THE PRESSURE 11117731.5 (PA)
THE COMPRESSIBILITY FACTOR .657280779

NEXT VOLUME ? (Y / N) N
NEXT ENTHALPY ? (Y / N) N
NEXT COMPOUND ? (Y / N) N
```

THE INDEPENDENT VARIABLES ARE *h*, *u*, OR *s*

The following three cases have to be discussed:

$$hu, \quad hs, \quad us$$

Case *hu*

Let us write the equation for the difference between h and u with zero on the right:

$$(h - u) - \mathbf{R}Tz[T, P] = 0$$

The equation for enthalpy, written similarly, is

$$h - (h^*[T] + \Delta h[T, P, z[T, P]]) = 0$$

These two equations must be solved simultaneously. The unknowns are P and T, and indirectly z, which can be calculated in the way described for the case PT at each step in the calculation.

Cases *hs* and *us*

The system of equations consisting of equations for h and s, or for u and s, must be set up and solved. The unknown variables are P and T, and indirectly z, which can be calculated in the way described for the case PT at each step in the calculation.

Let us recall here Figures 9.4 and 9.5. It was shown in these how to find, in the case hP or sP, the isotherm T belonging to the given state. Namely, the enthalpy correction is calculated for each isotherm by means of the enthalpy polynomial:

$$\frac{\Delta h}{\mathbf{R}T_c} = \frac{h - h^*[T]}{\mathbf{R}T_c}$$

The geometric locus of these points gives $\Delta h[T_r]$, but on the other hand it also gives $\Delta h[P_r]$, that is, we obtain that function $P_r[T_r]$ along which the enthalpy correction is

$$\Delta h[T_r, P_r] = h - h^*[T]$$

This $P_r[T_r]$ curve is drawn in Figure 9.5. The point on the curve having the ordinate $(s - s^*)/\mathbf{R}$ gives (P_r, T_r), that is, the pair of values (P, T) to which the prescribed enthalpy and the prescribed entropy correspond.

SUMMARY OF SINGLE-PHASE CASES

For the sake of a more concise discussion let us introduce the following notation for the variables:

$$T \to T$$

$$P, v \to Y$$

$$h, u, s \to X$$

1. In the *YT* case the task is divided into two steps: in the first step the compressibility coefficient is calculated (Table 9.2), in the second step the variables X (Table 9.1).

2. In the *XT* case the following equation must be solved:

$$F[Y] = \left(X - \left(X^*[T] + \Delta X[T, Y, z[T, Y]]\right)\right) = 0$$

where Y_0 is the root of the equation. When this is known, all other calculations can be reduced to those in Table 9.1. Here, and in all the other cases, $z = z[T, Y, \pi[T]]$.

TABLE 9.2
Calculation of Compressibility Factor

EOS	PT Case	vT Case
Virial EOS	$z = 1 + \frac{B[T]P}{\mathbf{R}T}$	$\phi_r = \frac{vP_c}{\mathbf{R}T_c}$ $z = 1 + \frac{B_v[T_r]}{\phi_r}$
Cubic EOS		
a. Redlich–Kwong	$z^3 - z^2 + (A^2P - B^2P^2 - BP)z - ABP^2 = 0$	$P = \frac{\mathbf{R}T}{v - b} - \frac{a}{T^{0.5}v(v + b)}$
b. Peng–Robinson	$z^3 - (1 - B)z^2 + (A - 3B^2 - 2B)z - (AB - B^2 - B^3) = 0$	$P = \frac{\mathbf{R}T}{v - b} - \frac{a}{v(v + B) + b(v - b)}$ $z = \frac{Pv}{\mathbf{R}T}$
Lee–Kesler EOS	$\frac{P_r\phi_r}{T_r} - \left(1 + \frac{B_v}{\phi_r} + \frac{C}{\phi_r^2} + \frac{D}{\phi_r^5} + \frac{C_4}{Tr^3\phi_R^2 r}\left(B + \frac{\gamma}{\phi_r^2}\right)\exp\left(B + \frac{\gamma}{\phi_r^2}\right)\right) = 0$ $z = \frac{P_r\phi_r}{T_r}$ $v = \frac{z\mathbf{R}T}{P}$	$z = 1 + \frac{B_v}{\phi_r} + \frac{C}{\phi_r^2} + \frac{D}{\phi_r^5} + \frac{C_4}{T_r^3\phi_r^2}\left(B + \frac{\gamma}{\phi_r^2}\right)\exp\left(B + \frac{\gamma}{\phi_r^2}\right)$ $P = \frac{z\mathbf{R}T}{v}$

3. In the XY case the following equation must be solved:

$$G[T] = (X - (X^*[T] + \Delta X[T, Y, z[T, Y, \pi[T]])) = 0$$

where T_0 is the root of the equation. The inconvenience of the calculation arises from the fact that the parameters $\pi[T]$ of the equation of state are functions of the temperature (Table 9.1). When T_0 is known, all further calculations are reduced to those in Table 9.1.

4. In the Y_1Y_2 case—that is, in the Pv case—the following equation must be solved:

$$G[T] = \left(Y_1 - \frac{\mathbf{R}Tz[T, Y_2, \pi[T]]}{Y_2} \right) = 0$$

where T_0 is the root of the equation. When this and the compressibility coefficient z are known, all further calculations are reduced to those in Table 9.1.

5. In the X_1X_2 case the following system of equations must be solved for T_0 and Y_0:

$$F[Y] = (X_1 - (X_1^*[T] + \Delta X_1[T, Y, z[T, Y, \pi[T]]])) = 0$$

$$G[T] = (X_2 - (X_2^*[T] + \Delta X_2[T, Y, z[T, Y, \pi[T]]])) = 0$$

The steps in the solution make use of the fact that T and Y cannot change independently, for thermodynamic reasons. Let us assume that $T = T_1$. Now the parameters of the first equation can be calculated, so that we have an equation in one unknown variable, and we seek its root $Y_0[T_1]$ corresponding to T_1, as in the XT case. The calculated values of T_1 and $Y_0[T_1]$ are substituted in the second equation. The temperature is varied until

$$G[T] = (X_2 - (X_2^*[T_0] + \Delta X_2[T_0, Y_0[T_0], z[T_0, Y_0[T_0], \pi[T_0]]])) = 0$$

10

PHASE EQUILIBRIUM IN ONE-COMPONENT SYSTEMS

The phases of a one-component body are open toward one another, that is, both energy transfer and substance transfer are possible between them. For both phases the Gibbs–Duhem equation is valid:

$$s^{\iota}\, dT - v^{\iota}\, dP + d\mu = 0 \qquad \iota = 1, 2$$

The condition of equilibrium between the two phases is the identity of the temperature, pressure, and chemical potential of each component of the contacting phases.

Before beginning the discussion of phase equilibrium, we must deal with the chemical potential. It will be remembered that the chemical potential is equal to the molar free enthalpy, but nothing has been said yet of its explicit functional form.

ISOTHERMAL CHANGE OF MOLAR FREE ENTHALPY WITH PRESSURE

The partial derivative of the free enthalpy with respect to pressure is

$$\left(\frac{\partial g}{\partial P}\right)_T = v$$

Starting from this, we can say that the free enthalpy of one mole of matter changes at constant temperature from g_1 to g_2, when the pressure changes from P_1 to P_2. The change is

$$\Delta g = g[P_2, T] - g[P_1, T] = \int_{g_1}^{g_2} dg = \int_{P_1}^{P_2} v[P, T]\, dP$$

The indicated integration can be performed if the function $v[P, T]$ is known for the specific case. It is known at any rate for the ideal gas, and thus in that case

$$\int_{P_1}^{P_2} v^*[P,T]\,dP = \int_{P_1}^{P_2} \frac{\mathbf{R}T}{v}\,dP = \mathbf{R}T\int_{\ln P_1}^{\ln P_2} d\left(\ln\frac{P}{P_u}\right)$$

where P_u is the standard pressure of the gas. (Though the limits are eliminated in the integration, we wish to stress that the argument of the logarithm cannot in fact be a quantity with dimension. It is P/P_u that is dimensionless.) After integration we have for the change in isothermal free molar enthalpy

$$g^*[P_2,T] - g^*[P_1,T] = \mathbf{R}T\ln\frac{P_2}{P_1}$$

This relation is not true for real gases. The isothermal change in molar free enthalpy for real gases is

$$\Delta g = g[P_2,T] - g[P_1,T]$$

Let us separate this difference into two parts. Let the gas first go from initial pressure P_1 to the standard state of pressure P_u, and from there, in a second step, to the final state P_2:

$$\Delta g = \Delta g_{P_u,P_1} + \Delta g_{P_2,P_u}$$

The following notation is introduced:

$$\Delta g_{P_2,P_u} = g[P_2,T] - g[P_u,T] \equiv \mathbf{R}T\ln a_2$$

$$\Delta g_{P_1,P_u} \equiv g[P_1,T] - g[P_u,T] \equiv \mathbf{R}T\ln a_1$$

Here a new dimensionless quantity, the *activity* a, has been defined. The molar free enthalpy of the gas at temperature T and pressure P, relative to the gas at standard pressure P_u, is

$$g[P,T] = g[P_u,T] + \mathbf{R}T\ln a$$

Here $g[P_u,T]$, the constant of integration, is the standard free enthalpy. This is a function of T only, since P_u is the standard pressure, and thus by convention constant.

What can be said of this integration constant? Let us add to both sides of the equation the quantity

$$-\mathbf{R}T\ln\frac{P}{P_u}$$

We have then the following expression:

$$g[P,T] - \mathbf{R}T \ln \frac{P}{P_u} = g[P_u,T] + \mathbf{R}T \ln \frac{aP_u}{P}$$

Since any real gas behaves at sufficiently low pressure as an ideal gas

$$\lim_{P\to 0} a = \frac{P}{P_u}$$

and then it is also true that

$$\lim_{P\to 0} \frac{aP_u}{P} = 1$$

If the pressure approaches zero, then the second term on the right side of the equation above becomes zero and

$$g[P_u,T] = \lim_{P\to 0} \left(g[P,T] - \mathbf{R}T \ln \frac{P}{P_u} \right)$$

This means that knowing the free enthalpy of the gas as a function of pressure, and extrapolating it to pressure $P = 0$, we can obtain the sought constant of integration $g[P_u, T]$, the standard free enthalpy. Accordingly, the standard free enthalpy is the free enthalpy of the gas at standard pressure P_u in the hypothetical ideal gas state.

Let us return now to our initial question, how large is the change in isothermal molar free enthalpy between pressures P_1 and P_2:

$$\Delta g = g[P_2,T] - g[P_1,T]$$

Let us write the previous equation for both states and take their difference:

$$g[P_2,T] - \mathbf{R}T \ln \frac{P_2}{P_u} = g[P_u,T] + \mathbf{R}T \ln \frac{a_2 P_u}{P_2}$$

$$g[P_1,T] - \mathbf{R}T \ln \frac{P_1}{P_u} = g[P_u,T] + \mathbf{R}T \ln \frac{a_1 P_u}{P_1}$$

$$g[P_2,T] - g[P_1,T] - \mathbf{R}T \ln \frac{P_2}{P_1} = \mathbf{R}T \ln \frac{a_2 P_u}{a_1 P_u} - \mathbf{R}T \ln \frac{P_2}{P_1}$$

Accordingly, the difference in molar free enthalpy is

$$\Delta g = \mathbf{R}T \ln \frac{a_2}{a_1}$$

FUGACITY

The quantity aP_u is given the special name *fugacity*:

$$f \equiv aP_u$$

Fugacity is not dimensionless, but has the dimensions of pressure. Its importance lies in the fact that it can be directly calculated from (P, v) data or equations of state. Namely,

$$g[P,T] - g[P_u,T] = \mathbf{R}T \ln \frac{f}{P_u}$$

where (to mention it again) P_u is the standard pressure, and hence constant. From this

$$\ln \frac{f}{P_u} = \frac{g[P,T] - g[P_u,T]}{\mathbf{R}T}$$

Let us differentiate with respect to pressure. On both sides the second term is zero. Using the known thermodynamic relation $(\partial g/\partial P)_T = v$, we have

$$\left(\frac{\partial \ln f}{\partial P}\right)_T = \frac{1}{\mathbf{R}T}\left(\frac{\partial g[P,T]}{\partial P}\right)_T = \frac{v}{\mathbf{R}T}$$

After rearrangement we have

$$d \ln f = \frac{v}{\mathbf{R}T}\, dP$$

Subtracting $d \ln P$ from both sides of the equation, we have

$$d \ln f - d \ln P = d \ln \frac{f}{P} \equiv d \ln \nu = \frac{v}{\mathbf{R}T}\, dP - d \ln P$$

that is,

$$d \ln \nu = \left(\frac{v}{\mathbf{R}T} - \frac{1}{P}\right) dP$$

where $\nu_f \equiv f/P$ is the fugacity coefficient.

After factoring out $-1/\mathbf{R}T$ and integrating between 0 and P we have

$$\ln \nu = -\frac{1}{\mathbf{R}T}\int_0^P \left(\frac{\mathbf{R}T}{P} - v\right) dP$$

$$= -\frac{1}{\mathbf{R}T}\int_0^P (v^* - v)\, dP$$

With decreasing pressure a real gas approaches an ideal one:

$$\lim_{P \to 0} v = v^*$$

Accordingly, the fugacity coefficient approaches unity with decreasing pressure.

To think over the meaning of this, let us carry out the following thought experiment: Consider one mole of gas at temperature T and pressure P, and assume that intermolecular forces do not act between the molecules of the gas. Thus the case is that of an ideal gas, the volume of which is $v^* = \mathbf{R}T/P$.

Let us now "switch on" intermolecular forces acting between the molecules. At unchanged temperature and pressure, the volume is changed by the intermolecular forces, which perform work in the process. The molecules acquire potential energy, and since we have an isothermal, isobaric case, the relevant potential is the free enthalpy. The molar free enthalpy of the real gas, relative to that of the ideal gas, is

$$\Delta g = g - g^* = \mathbf{R}T \ln \frac{f}{P}$$

The quantity $v^* - v$ can be readily measured or calculated from the equation of state as a function of pressure. Numerical integration can be conveniently performed (Figure 10.1). After substitution in the equation of state, integration can be carried out analytically. For example, from the Redlich–Kwong equation, we obtain

$$\ln \frac{f}{P} = (z - 1) - \ln(z - BP) - \frac{A^2}{B} \ln\left(1 + \frac{BP}{z}\right)$$

From the Peng–Robinson equation,

$$\ln \frac{f}{P} = (z - 1) - \ln(z - B) - \frac{A}{2\sqrt{2}\,B} \ln \frac{z + 2.414B}{z - 0.414B}$$

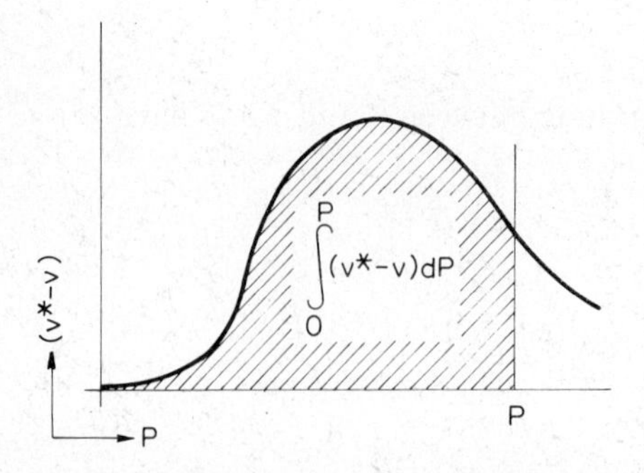

FIGURE 10.1. Fugacity.

and from the Lee–Kesler equation,

$$\ln \frac{f}{P} = (z - 1) - \ln z + \frac{B_v}{\phi_r} + \frac{C_r}{2\phi_r^2} + \frac{D_r}{5\phi_r^5} + E$$

where the denotation is the same as used in the preceding chapter (Table 9.1).

DEGREES OF FREEDOM OF TWO-PHASE SYSTEMS

The phase rule states that systems of this kind have one degree of freedom. Hence, by fixing the temperature its pressure is determined. Because it belongs to an equilibrium state, this pressure is called the equilibrium vapor pressure.

Here two questions arise:

1. How does the equilibrium vapor pressure change with temperature?
2. Does the phase transition at equilibrium obey an energetic relation, and if so, what kind?

THE CLAUSIUS – CLAPEYRON EQUATION

In conjunction with the first question the derivative of the function $P°[T]$ is of interest. The derivative contains the equilibrium temperature and pressure. Just these two variables are contained in the Gibbs–Duhem equation, which is now written separately for the two open phases of the system:

$$d\mu = -s^{\alpha}\, dT + v^{\alpha}\, dP$$

$$d\mu = -s^{\beta}\, dT + v^{\beta}\, dP$$

μ, P, and T are not indexed according to phase, since these have to be identical because of the known conditions of equilibrium. Thus, the difference of the two equations is

$$-(s^{\alpha} - s^{\beta})\, dT + (v^{\alpha} - v^{\beta})\, dP° = 0$$

and hence the required derivative is

$$\frac{dP°}{dT} = \frac{s^{\alpha} - s^{\beta}}{v^{\alpha} - v^{\beta}} = \frac{\Delta s}{\Delta v}$$

where Δs = change in molar entropy associated with the phase transformation,

Δv = change in molar volume associated with the phase transformation.

The change in entropy of the phase transformation cannot be directly measured; however, the change in enthalpy of phase transformation can be measured calorimetrically. The relation between these two quantities is

$$\Delta s = \frac{\Delta h}{T}$$

Inserting this relation in the expression for dP°/dT, the following equation is obtained:

$$\frac{dP^\circ}{dT} = \frac{\Delta h}{T\Delta v}$$

This is the Clausius–Clapeyron equation; it is notable in that all the quantities involved can be directly measured. The equation answers both questions raised at the beginning of this section.

Equilibrium between Liquid and Vapor Phases

The Clausius–Clapeyron equation can be directly applied to the case of equilibrium between liquid and vapor phases. Here our equation takes the following form:

$$\frac{dP^\circ}{dT} = \frac{\Delta h^{\alpha\beta}}{T(v^\alpha - v^\beta)}$$

where P° = equilibrium vapor pressure at temperature T,
$\Delta h^{\alpha\beta}$ = molar enthalpy of evaporation at temperature T and pressure P°,
v^α = molar volume of the vapor phase,
v^β = molar volume of the liquid phase.

Thus, a relation is established between the temperature, the equilibrium vapor pressure, the difference of the molar volumes of the phases, and the enthalpy of evaporation. Since $\Delta h^{\alpha\beta}$ and $v^\alpha - v^\beta$ are in all cases positive, the Clausius–Clapeyron equation shows that the equilibrium vapor pressure increases with temperature. The equation allows us to go from the enthalpy of evaporation to the equilibrium vapor pressure, and conversely, to calculate the enthalpy of evaporation from data on the equilibrium vapor pressure.

Conditions for Phase Equilibrium

The three conditions for phase equilibrium are

$$T^\alpha = T^\beta$$

$$P^\alpha = P^\beta$$

$$\mu^\alpha = \mu^\beta$$

The chemical potential is equal to the molar free enthalpy. The fugacity has been defined in terms of the molar free enthalpy:

$$d\mu = dg = \mathbf{R}T\,d\ln\frac{f}{P_u}$$

so that we can say that the three conditions for equilibrium of open phases are

$$T^\alpha = T^\beta$$

$$P^\alpha = P^\beta$$

$$f^\alpha = f^\beta$$

If the temperature is fixed as independent variable

$$T = T^\alpha = T^\beta$$

then no degree of freedom is left to the system: the equilibrium vapor pressure and the equilibrium chemical potential are determined by the equation of state, corresponding to the substance class and to the pressure–temperature region. The roots z^α and z^β of the equation of state satisfy the following system of equations:

$$P^\circ = F[T, z^\alpha] = F[T, z^\beta]$$

$$\ln\frac{f}{P^\circ} = G[T, z^\alpha] = G[T, z^\beta]$$

Here G is the arithmetic expression obtained from the equation of state for the logarithm of the fugacity coefficient.

CALCULATION OF EQUILIBRIUM VAPOR PRESSURE

1. Assuming that at not too high pressure and temperature equilibrium vapor behaves as an ideal gas, we can write

$$v^\alpha = \frac{\mathbf{R}T}{P^\circ}$$

The fugacity coefficient $f/P = 1$, that is, the condition of the equality of fugacities is trivially satisfied. At the same time,

$$v^\alpha \gg v^\beta$$

—that is, the molar volume of the liquid is negligible compared to that of the vapor. The Clausius–Clapeyron equation can be written then in the following form:

$$\frac{dP^\circ}{dT} = \frac{\Delta h^{\alpha\beta}}{T(v^\alpha - v^\beta)} = \frac{P^\circ \Delta^{\alpha\beta}}{\mathbf{R}T^2}$$

Integrating after the separation of the variables and assuming that $\Delta h^{\alpha\beta} =$ const, we obtain for equilibrium vapor pressure

$$P^\circ = \exp\left(-\frac{\Delta h^{\alpha\beta}}{\mathbf{R}T} \right) + \text{const}$$

Thus the equilibrium vapor pressure increases exponentially with temperature. The condition $\Delta h^{\alpha\beta} =$ const, however, can be maintained only in a very narrow temperature range. It was shown in Chapter 3, in the graph of $\ln P^\circ$ versus $1/T$, that the relation is actually not linear in the interval from $T_r = 0.6$ to T_c.

2. In the case of the Redlich–Kwong equation of state, the compressibility coefficients for the vapor and the liquid phases are obtained as the roots of the following equation:

$$z^3 - z^2 + (AP - BP - B^2P^2)z - ABP^2 = 0$$

The fugacity coefficient is given by

$$\ln \frac{f}{P} = (z - 1) - \ln(z - BP) - \frac{A^2}{B} \ln\left(1 + \frac{BP}{z}\right)$$

If the pressure is equal to the equilibrium vapor pressure, then the condition of the equality of fugacities is satisfied:

$$\ln \frac{f^\alpha}{P^\circ} - \ln \frac{f^\beta}{P^\circ} = 0$$

Practically, this solution does not yield a result of satisfactory accuracy, because the Redlich–Kwong equation is not sufficiently sensitive to the density of the saturated liquid.

3. The following empirical formula, well applicable for nonpolar substances, was fitted to vapor-pressure data calculated with the Lee–Kesler equation of state:

$$\ln P_r = 5.92714 - 6.0968T_r^{-1} - 1.228862 \ln T_r + 0.169347T_r^6$$

$$+\omega(15.2518 - 15.6875T_r^{-1} - 13.4721 \ln T_r + 0.43577T_r^6)$$

where P_r° = reduced equilibrium vapor pressure,
T_r = reduced temperature,
ω = omega parameter.

The equation of state yields for the fugacity coefficient

$$\ln \frac{f}{P} = (z - 1) - \ln z + \frac{B_v}{\phi_r} + \frac{C_r}{2\phi_r^2} + \frac{D_r}{5\phi_r^5} + E$$

The wide range of applicability of the Lee–Kesler equation of state can be attributed to the fact that in fitting the parameters of the equation the condition

$$f^\alpha = f^\beta$$

has also been taken into consideration.

Boiling Point

The *boiling point* is that temperature at which vapor and liquid phases are in equilibrium at the prescribed pressure. The normal boiling point T_{nbp} corresponds to atmospheric pressure (101,325 Pa).

Enthalpy of Evaporation

The enthalpy-versus-temperature diagram of a one-component body can be seen in Figure 10.2. This shows also the enthalpy of the given substance, calculated for the ideal-gas state. An isobar, corresponding to the equilibrium vapor pressure at temperature T is also indicated. In the figure

h^* = molar enthalpy of the ideal gas,
h^α = molar enthalpy of the saturated vapor,
h^β = molar enthalpy of the saturated liquid,
$\Delta h^{\alpha\beta}$ = molar enthalpy of evaporation.

On what is the enthalpy of evaporation expended? It follows from the definition of enthalpy that

$$\Delta h^{\alpha\beta} = \Delta u^{\alpha\beta} + P^\circ \Delta v^{\alpha\beta}$$

This means that the enthalpy of evaporation has two components: $\Delta u^{\alpha\beta}$, the

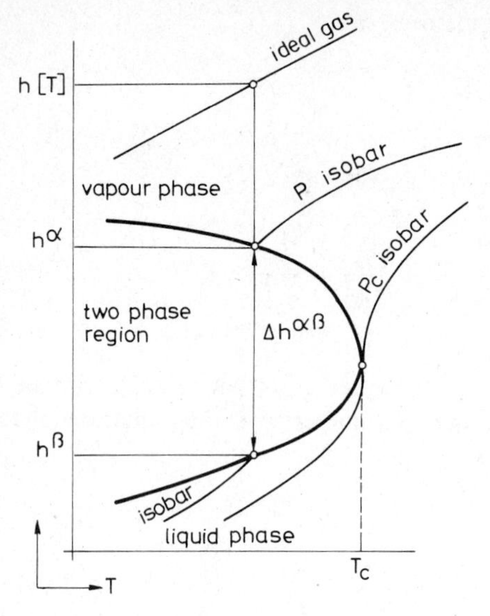

FIGURE 10.2. Two-phase region of one-component system: (H, T) diagram.

molar internal energy of evaporation, and $P^\circ\,\Delta v^{\alpha\beta}$, the volumetric work done by the expanding substance on the surroundings at pressure P°.

The internal energy of evaporation acts against the intermolecular potential energy. While this latter keeps the molecules in the vicinity of one another and provides for the existence of the liquid state, the internal energy of evaporation increases the mean kinetic energy of the molecules by just as much as is needed to remove them from one another. Thereby the liquid phase diminishes, and the molecules fill up the volume, so that the pressure established is in equilibrium with the external pressure.

This model of evaporation can be made quantitative by substituting a suitable equation of state into the equation for the isothermal enthalpy correction:

1. In accordance with critical remarks concerning the Redlich–Kwong equation, we show by way of example that for saturated vapor

$$h^\alpha - h^* = \mathbf{R}T\left((1 - z^\alpha) + \frac{3A_{\mathrm{RK}}^2}{2B_{\mathrm{RK}}}\ln\left(1 + \frac{B_{\mathrm{RK}}P^\circ}{z^\alpha}\right)\right)$$

and for saturated liquid

$$h^\beta - h^* = \mathbf{R}T\left((1 - z^\beta) + \frac{3A_{\mathrm{RK}}^2}{2B_{\mathrm{RK}}}\ln\left(1 + \frac{B_{\mathrm{RK}}P^\circ}{z^\beta}\right)\right)$$

The enthalpy of evaporation is then

$$\Delta h^{\alpha\beta} = \mathbf{R}T\left((z^{\alpha} - z^{\beta}) - \frac{3A_{\mathrm{RK}}^{2}}{2B_{\mathrm{RK}}} \ln\left(\frac{1 + \dfrac{B_{\mathrm{RK}}P^{\circ}}{z^{\alpha}}}{1 + \dfrac{B_{\mathrm{RK}}P^{\circ}}{z^{\beta}}}\right)\right)$$

2. Inserting the pressure formula satisfying the Lee–Kesler equation of state into the form of the Clausius–Clapeyron equation solved for $\Delta h^{\alpha\beta}$, we have

$$\frac{\Delta h^{\alpha\beta}}{\mathbf{R}T_{\mathrm{c}}(z^{\alpha} - z^{\beta})} = 6.09648 - 1.28862T_{\mathrm{r}} + 1.016T_{\mathrm{r}}^{7}$$

$$+\omega(15.6875 - 13.4721T_{\mathrm{r}} + 2.615T_{\mathrm{r}}^{7})$$

3. An excellent empirical formula for the enthalpy of evaporation was suggested by Watson:

$$\Delta h^{\alpha\beta} = \Delta h_{\mathrm{nbp}}^{\alpha\beta}\left(\frac{1 - T_{\mathrm{r}}}{1 - T_{\mathrm{r,nbp}}}\right)^{0.38}$$

where $T_{\mathrm{r,nbp}}$ = reduced temperature at normal boiling point,
$\Delta h_{\mathrm{nbp}}^{\alpha\beta}$ = enthalpy of evaporation at normal boiling point.

These are two empirical, substance-specific parameters.

THE TWO-PHASE RANGE

The one-component homogeneous (gas or liquid) phase has two degrees of freedom. In the (P, T) plane the boundary of the two phases is the vapor-pressure curve (Figure 9.1). Along this curve two phases can coexist; at a temperature higher than the critical, the liquid phase does not exist. It is customary to call the rarefied phase below the critical temperature *vapor*, and above it *gas*.

Conditions for the state of aggregation of one-component body are summarized in Figure 10.3. The figure shows the envelope curve of the two-phase region in six different coordinate systems. The same nine kinds of states are represented in all six drawings. At pressure P_1 there are four:

1. Liquid below its boiling point.
2. Saturated liquid at the boiling point.
3. Saturated vapor at the boiling point.
4. Vapor above its boiling point.

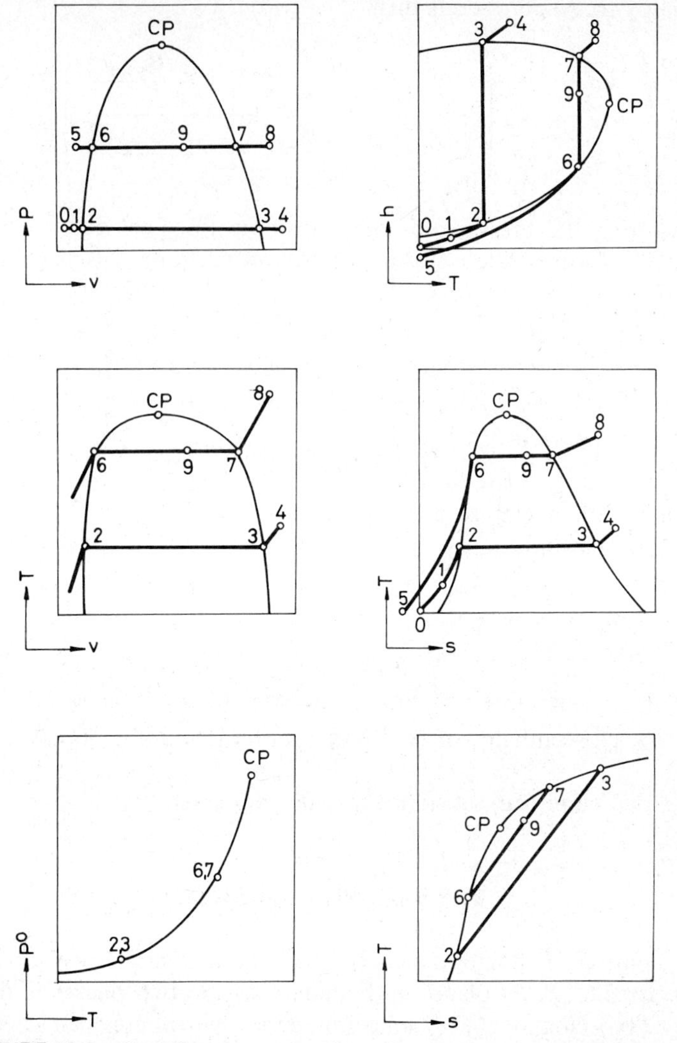

FIGURE 10.3. Six different representations of a two-phase one-component system.

At pressure P_2 there are five:

5. Liquid below its boiling point.
6. Saturated liquid at the boiling point.
7. Saturated vapor at the boiling point.
8. Vapor above the boiling point.
9. Vapor and liquid at the boiling point.

The state in the two-phase equilibrium range is characterized by one of the six properties taken into consideration so far, together with the vapor ratio ε.

We have the following six cases:

$$T_{\text{bp}}\varepsilon, \quad P^\circ\varepsilon, \quad v\varepsilon, \quad h\varepsilon, \quad u\varepsilon, \quad s\varepsilon$$

Using this concise notation, we can say that the fundamental equation of the two-phase equilibrium system is the following:

$$X - X_{\text{sat}}^{\alpha}\varepsilon + X_{\text{sat}}^{\beta}(1 - \varepsilon)$$

where X can mean any of the four molar quantities v, u, h, s. The property X of the saturated vapor or of the saturated liquid is a function of the boiling point only:

$$X_{\text{sat}}^{\alpha} = X^{*}\left[T_{\text{bp}}\right] + \Delta X\left[T_{\text{bp}}, P^\circ\left[T_{\text{bp}}\right], z^{\alpha}\left[T_{\text{bp}}, P^\circ\left[T_{\text{bp}}\right], \pi\left[T_{\text{bp}}\right]\right]\right]$$

$$X_{\text{sat}}^{\beta} = X^{*}\left[T_{\text{bp}}\right] + \Delta X\left[T_{\text{bp}}, P^\circ\left[T_{\text{bp}}\right], z^{\beta}\left[T_{\text{bp}}, P^\circ\left[T_{\text{bp}}\right], \pi\left[T_{\text{bp}}\right]\right]\right]$$

The cases of v_{sat}^{α} and v_{sat}^{β} have been already discussed in conjunction with the equations of state.

Cases $T_{\text{bp}}\varepsilon$ and $P^\circ\varepsilon$

The calculation of these two cases does not need hard thinking. Either of T_{bp} and P° can be calculated from the other. When both are known, z_{sat}^{α} and z_{sat}^{β} can be calculated from the equation of state. In a third step, X_{sat}^{α} and X_{sat}^{β} can be calculated with the arithmetic expressions of Table 9.1. The fourth step is substitution into the fundamental equation of the two-phase equilibrium system.

Cases $v\varepsilon$, $h\varepsilon$, $u\varepsilon$, and $s\varepsilon$

In these four cases (let X now also mean v), that temperature T_{bp} must be found at which the prescribed values of X and ε are established. This temperature also determines the equilibrium vapor pressure. Thus the following nonlinear equation must be solved:

$$X - \left(\varepsilon X_{\text{sat}}^{\alpha}[T_0] + (1 - \varepsilon)\, X_{\text{sat}}^{\beta}[T_0]\right) = 0$$

where the root of the equation is $T_0 = T_{\text{bp}}$. When T_{bp}, P°, and z_{sat}^{g} and z_{sat}^{1} are known, the unknown properties can be calculated in the way described above.

We have finished with cases involving a prescribed vapor ratio. The discussion is continued with cases of unknown vapor ratio, where the question is whether the substance is in a homogeneous phase or in the two-phase region.

Cases *Pv*, *Ph*, *Pu*, and *Ps*

In these cases pressure is one of the state variables. It can be seen from Figure 10.2 that if $P > P_c$, then only one phase can exist. If $P < P_c$, then starting along a given isobar from a point above the critical temperature in the direction of decreasing temperatures, the following can be said about the state of aggregation:

Vapor	if $X \geq X^{\alpha}_{\text{sat}}$
Two-phase	if $X^{\alpha}_{\text{sat}} > X > X^{\beta}_{\text{sat}}$
Liquid	if $X \leq X^{\beta}_{\text{sat}}$

It can be decided on the basis of these criteria whether we have a one- or a two-phase case. We have discussed the one-phase case; in the two-phase case the equilibrium pressure $P = P°$ determines the boiling point T_{bp}, and when z^{α}_{sat} and z^{β}_{sat} are known, the values of X^{α}_{sat} and X^{β}_{sat} can be determined. Then the vapor ratio is

$$\varepsilon = \frac{X - X^{\beta}_{\text{sat}}}{X^{\alpha}_{\text{sat}} - X^{\beta}_{\text{sat}}}$$

When the vapor ratio is known, all the remaining unknown properties can be calculated with the fundamental equation of two-phase equilibrium systems.

Cases *hv*, *uv*, *sv*, *hu*, *hs*, and *us*

These six cases are discussed together. They are not of equal practical importance, and a given case may require a special technique because of calculation difficulties. Nevertheless, ordinarily we may start by a two-phase system. Taking a temperature value somewhat lower than the critical temperature, vapor ratios $\varepsilon_1[T]$ and $\varepsilon_2[T]$ are calculated for properties X_1 and X_2, and then, decreasing the temperature in suitable steps this calculation is repeated a few times. If $\varepsilon_1[T]$ and $\varepsilon_2[T]$ approach one another with decreasing temperature, then the substance is actually in the two-phase range, and in the state sought if $\varepsilon_1 = \varepsilon_2$. When the vapor ratio, T_{bp}, and $P°$ are known, all further properties can be calculated in the way described.

If $\varepsilon_1[T]$ and $\varepsilon_2[T]$ move apart with decreasing temperature, then the solution is to be found in the single-phase region.

11

MIXTURES

A homogeneous phase containing two or more components is called a *mixture*. In studying the thermodynamics of mixtures, one naturally wishes to derive the properties of a mixture of given composition from the properties of the pure components forming it. This means that the mixture is considered as the final state of a process, in the initial state of which the components are present in the pure state, and in the same quantity as in the mixture, and at the temperature and pressure of the mixture. This state is chosen as the reference state. Other choices are possible, but in any case it is meaningful to discuss a thermodynamic property of a mixture only in relation to whatever reference state has been taken as the initial state.

Attention is called to a further circumstance. It was shown in the discussion of one-component bodies that classical thermodynamics gives completely general relations between the variables, and these relations are made concrete by taking into consideration the chemistry of the molecules forming the body; that is, the work of intermolecular and intramolecular forces is expressed by suitable constitutive equations, and substituted into the general relations. In this chapter on mixtures the procedure will be the same: general relationships of the thermodynamics of mixtures will be elaborated, and then the chemistry of the types of molecules forming the mixture will be taken into account. In this respect the shape of the molecules and intermolecular forces play an important role.

To start with, let us assume that the molecules of the components forming the mixture are rigid spheres of near-identical size, and that no attractive force acts between them. With the notation used in Chapter 1 this can be formulated in the following way:

$$\varepsilon[r_{AA}] = \begin{cases} \infty & \text{if} \quad r_{AA} < \sigma_A \\ 0 & \text{if} \quad r_{AA} \geq \sigma_A \end{cases}$$

$$\varepsilon[r_{BB}] = \begin{cases} \infty & \text{if} \quad r_{BB} < \sigma_B \\ 0 & \text{if} \quad r_{BB} \geq \sigma_B \end{cases}$$

$$\varepsilon[r_{AB}] = \begin{cases} \infty & \text{if} \quad r_{AB} < \frac{1}{2}(\sigma_A + \sigma_B) \\ 0 & \text{if} \quad r_{AB} \geq \frac{1}{2}(\sigma_A + \sigma_B) \end{cases}$$

where $\varepsilon[\cdot]$ = pair potential,
r_{AA} = actual distance of the two A molecules,
r_{BB} = actual distance of the two B molecules,
r_{AB} = actual distance of one A molecule and one B molecule,
σ_A = diameter of an A molecule.
σ_B = diameter of a B molecule.

These three pair potentials imply that substance A in the pure state, substance B in the pure state, and also the mixture of these two substances behave as ideal gases.

CONCEPT OF THE IDEAL MIXTURE AND ITS THREE MODELS

The Kinetic Model of the Ideal Mixture

The basis of the first model is the kinetic theory of gases. Consider a vessel immersed in a heat bath of volume V. The vessel is divided by two superposed, movable, semipermeable walls into two parts. In one part of the space there are only molecules of type A, in the other only molecules of type B, their numbers being n_A and n_B, respectively (Figure 11.1a).

In the initial state the two parts are in equilibrium at the two semipermeable walls:

$$P_A = \frac{n_A \mathbf{R} T}{V_A} = P_B = \frac{n_B \mathbf{R} T}{V_B}$$

and naturally

$$V_A + V_B = V$$

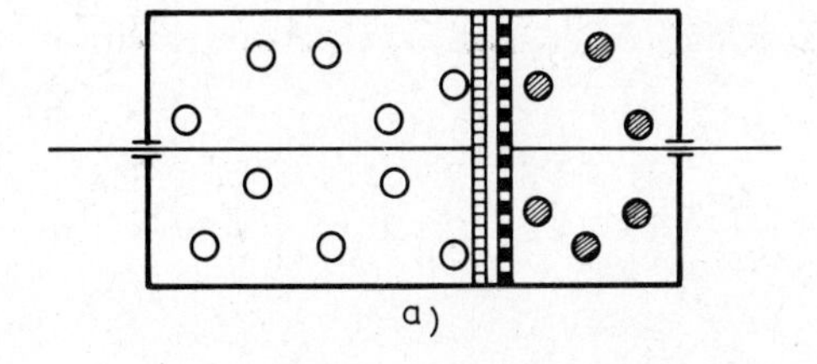

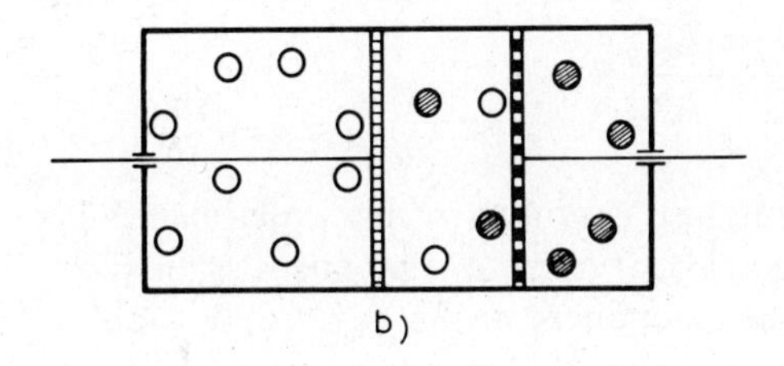

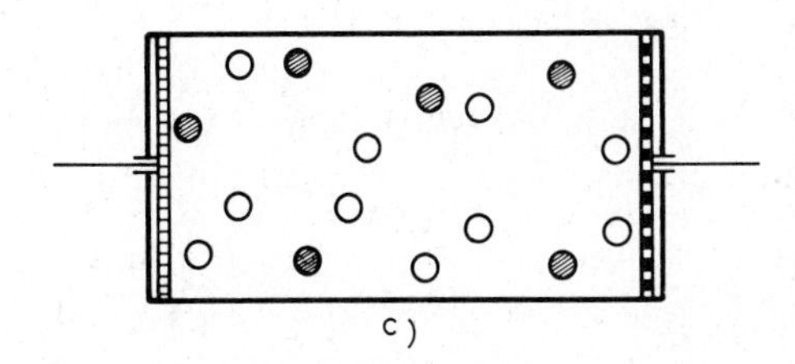

FIGURE 11.1. Mixing: independent expansion of components; (a) components in pure state, (b) displacement of semipermeable pistons by expanding components, (c) complete mixing.

The semipermeable wall initially in contact with molecules of type A permits only the passing of A molecules; the other wall, only of B molecules. This transfer actually occurs, as the concentration of molecules of either kind is zero on the far side of the semipermeable wall, and as we know, molecules flow from a place of higher chemical potential (concentration) to a place of lower chemical potential.

As a result of flow through the two semipermeable walls the numbers of A molecules and of B molecules decrease in the respective volumes, while in the space enclosed by the two movable (and indeed moving) semipermeable walls, the number of both types of molecules increases: a mixture is formed by a spontaneous process. During this process mechanical equilibrium is maintained:

$$P_A = P = P_B$$

where P is the pressure established in the enclosed volume.

Let Δn_A and Δn_B be the numbers of molecules that have penetrated into the enclosed volume. Then

$$P_A = \frac{n_A - \Delta n_A}{V_A - \Delta V_A}\mathbf{R}T$$

$$P_B = \frac{n_B - \Delta n_B}{V_B - \Delta V_B}\mathbf{R}T$$

$$P = \frac{\Delta n_A + \Delta n_B}{\Delta V_A + \Delta V_B}\mathbf{R}T$$

How long does the formation of mixture continue in this experiment? As long as the chemical potential (concentration) of pure component A or pure component B is higher than that of the component of the mixture. But this is always the case, because (Figure 11.1*b*)

$$\frac{n_A}{V_A} = \frac{\Delta n_A}{\Delta V_A} > \frac{\Delta n_A}{\Delta V_A + \Delta V_B}$$

Accordingly, the formation of mixture is terminated when there is no more pure A or B component. Then the volume enclosed by the semipermeable walls is

$$\Delta V_A + \Delta V_B = V$$

Observe that both components are in the same volume. This is just the essence of a mixture. Moreover, this common volume is larger than in the reference state. Another way of looking at this is that on mixing the components expand into a common volume (Figure 11.1*c*).

In the final state the pressure of the mixture is:

$$P = \frac{n_A + n_B}{V}\mathbf{R}T$$

This can be written in the following form:

$$P = P_A + P_B \frac{n_A}{V}\mathbf{R}T + \frac{n_B}{V}\mathbf{R}T$$

where P_A = partial pressure of A molecules,
P_B = partial pressure of B molecules,
n_A = amount of component A in volume V,
n_B = amount of component B in volume V.

Changing over to the general multicomponent case, and dividing the equation for the partial pressure by the equation for the total pressure, we have after reduction

$$\frac{P_i}{P} = \frac{n_i}{\sum n_i} = y_i$$

where y_i is the mole fraction of component i. Thus

$$P_i = y_i P$$

The partial pressure of component i is proportional to its mole fraction. This is Dalton's law.

Let us define now the partial volume. In the reference state of the components, the pressure of the mixture is P, its temperature T, and the amounts of the components $n_1, n_2, \ldots, n_i, \ldots, n_k$. The volume of a single pure component in the reference state is

$$V_i = n_i \frac{\mathbf{R}T}{P}$$

The molar volume of component i from the equation of state of the ideal gas is

$$v_i = \frac{\mathbf{R}T}{P}$$

The contribution of component i to the total volume is

$$V_i = n_i v_i$$

Hence, the volume of the mixture is

$$V = \sum_i V_i = \sum n_i v_i$$

Dividing this relationship into the preceding one, we have

$$\frac{V_i}{V} = -\frac{n_i v_i}{\sum\limits_i n_i v_i} = \frac{n_i}{\sum\limits_i n_i} = y_i$$

Cancellation of the molar volume v_i is permitted, because it is independent of the nature of the ideal gas at given pressure P and temperature T. From this follows Amagat's law, according to which the partial volume of component i is proportional to its mole fraction:

$$V_i = y_i V$$

With entropy the case is different. In the reference state, that is, in the pure state, the molar volume of component i is v_i, while in the mixture it is Σv_i, as the total volume is available to the molecules. Thus, on mixing, component i expands, as shown by the thought experiment of mixture formation at the beginning of the chapter. The partial derivative of molar entropy with respect to volume, according to the Maxwell relation, is

$$\left(\frac{\partial s}{\partial v}\right)_T = \left(\frac{\partial P}{\partial T}\right)_v$$

Let us apply the Maxwell relation to an ideal gas:

$$\left(\frac{\partial s}{\partial v}\right)_T = \frac{P}{T} = \frac{\mathbf{R}}{v}$$

whence

$$ds = \frac{\mathbf{R}}{v}\, dv = \mathbf{R}\, d \ln v$$

Let us apply the expression to component i. After integration,

$$\bar{s}_i - s_i = \mathbf{R} \ln \frac{\sum v_i}{v_i} = -\mathbf{R} \ln y_i$$

The partial molar entropy of component i in the mixture is

$$\bar{s}_i = s_i - R \ln y_i$$

where s_i = molar entropy of component i in the reference state,
$\bar{s}_i$ = partial molar entropy of component i in the mixture,
y_i = mole fraction of component i in the mixture.

Thus, the partial molar entropy of component i is higher than the molar entropy of the pure substance in the reference state.

The molar entropy of the mixture is

$$s = \sum_i y_i \bar{s}_i = \sum_i y_i (s_i - \mathbf{R} \ln y_i)$$

$$= \sum_i y_i s_i - \mathbf{R} \sum_i (y_i \ln y_i)$$

Remember that s is the molar entropy of the mixture at given composition, $\Sigma y_i s_i$ is the linear combination of the molar entropy of the pure components, and the difference

$$\Delta s^{(\text{id})} = -\mathbf{R}\sum_i (y_i \ln y_i)$$

is the molar entropy of mixing of ideal-gas components.

Algorithm and Program: Graphical Representation of Ideal Gas Entropy of Mixing

Write a program for the graphical representation of the ideal gas entropy of mixing at different concentrations.

Program GRMIXENTR. A possible realization:

```
]LOAD GRMIXENTR,D1
]PR#0
]LIST

 10  REM :GRAPHICAL REPRESENTATION
      OF ENTROPY OF MIXING IN TWO
      -COMPONENT SYSTEM
 20  REM : FOR IDEAL GAS
 30  HPLOT 0,0
 40  HGR
 50  HPLOT 10,0 TO 10,150
 60  HPLOT 10,150 TO 200,150
 70  HPLOT 200,150 TO 200,0
 80  HPLOT 10,150
 90  FOR I = 1 TO 49
100  X1 = I * 1 / 50
110  X2 = 1 - X1
120  SM = -8.3144 * (X1 * LOG (
      X1) + X2 * LOG (X2))
130  X = 10 + 190 * X1
140  Y = 150 - 20 * SM
150  HPLOT TO X,Y
160  NEXT I
170  HPLOT TO 200,150
180  INPUT "END ? (Y) "; A$
190  TEXT
200  END
```

Remark. Run the program, watch the monitor and explain the result.

Lattice Model of the Ideal Mixture

Let us pass over to the discussion of the second model, and consider for this purpose a lattice, with molecules of types A and B arranged at the lattice points. Mixing consists exclusively in a change in the steric arrangement of the molecules, but this is not accompanied by a change in the energy state of the molecules.

Let the mixture be two-component, the numbers of molecules being n_A and n_B, respectively, with a single molecule arranged at each lattice point, and no lattice points unoccupied. The question is now, of what does the process of mixing consist in this model? In the initial state molecules of type A are sited in space at neighboring lattice points, leaving no free lattice points between. Molecules of type B are similarly arranged in the other half of the lattice. Now mixing consists in the exchange of one A and one B molecule. By these changes of place a multitude of arrangements can be produced in the lattice. Each possible arrangement represents a microstate of the system. Now the formulation of the question is, how many microstates are possible in the initial and in the final state (Figure 11.2).

Since each microstate is of equal probability, the Boltzmann relationship will be used:

$$S_{\text{initial}} = \mathbf{k} \ln W_{\text{initial}}$$

$$S_{\text{final}} = \mathbf{k} \ln W_{\text{final}}$$

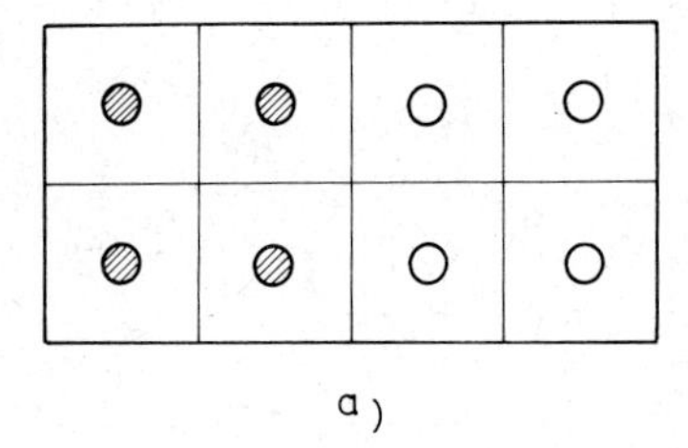

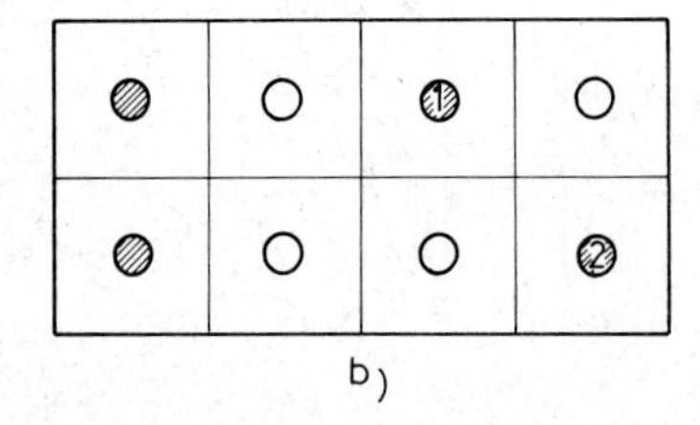

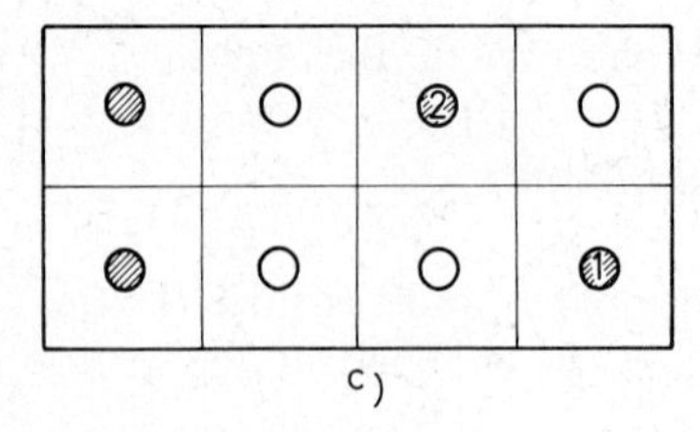

FIGURE 11.2. Mixing: exchange of molecules in a lattice; (a) two types of molecules, (b) exchange of two pairs of molecules, (c) exchange of two identical molecules.

In the initial state a single arrangement is possible:

$$W_{\text{initial}} = 1$$

In the final state each configuration is of equal probability, so we seek the number of possible configurations. There are $n_A + n_B$ molecules in the lattice, and the number of possible configurations would be $(n_A + n_B)!$ if molecules of identical type were distinguished (Figure 11.2*b*). However, since the exchange of two A molecules in some arrangement (Figure 11.2*c*) evidently does not produce a new microstate, $(n_A + n_B)!$ is an overestimate of the possible arrangements. The overestimation factor because of molecules of type A is $n_A!$; because of molecules of type B, $n_B!$. Hence, the number of possible arrangements is

$$W_{\text{final}} = \frac{(n_A + n_B)!}{n_A! n_B!}$$

Hence the mixing entropy is

$$\Delta S = S_{\text{final}} - S_{\text{initial}} = \mathbf{k} \ln W_{\text{final}} - \mathbf{k} \ln W_{\text{initial}}$$

$$= \mathbf{k} \ln \frac{(n_A + n_B)!}{n_A! n_B!}$$

$$= \mathbf{k}(\ln (n_A + n_B)! - \ln n_A! - \ln n_B!)$$

Let us use Stirling's formula to eliminate the factorials:

$$\Delta S = -\mathbf{k}(n_A + n_B)\ln(n_A + n_B) - n_A \ln n_A - n_B \ln n_B$$

$$= -\mathbf{k}\left(n_A \ln \frac{n_A}{n_A + n_B} + n_B \ln \frac{n_B}{n_A + n_B}\right)$$

Normalizing on both sides, molar entropy of mixing is

$$\Delta s^{(\text{id})} = -\mathbf{R}(x_A \ln x_A + x_B \ln x_B)$$

The same result is obtained as in the case of the first model.

Algorithm and Program: Entropy of Mixing in a Crystal

Write a program for the calculation of the entropy of mixing in an idealized crystal. The model: There are two crystals; at the beginning, the first is made of A molecules and the second of B molecules. Molecules are exchanged between the crystals. This change produces entropy, and the problem is to calculate this entropy production.

Algorithm.

1. Read the initial number of A molecules in crystal 1.
2. Read the initial number of B molecules in crystal 2.
3. Exchange one A molecule with one B molecule, and calculate the possible ways of doing so—in other words, calculate the number of microstates (see Chapter 1, program NODIST).
4. Repeat step 3 until in one of the crystals every molecule has been replaced.
5. Find the total number of microstates in this process.
6. Calculate the entropy of mixing,

$$\Delta S_{\text{mix}} = \mathbf{k} \ln W$$

7. Print the results.

Program ENTRMIX. A possible realization:

```
]PR#0
]LOAD ENTRMIX
]LIST

 10  REM :CALCULATION OF ENTROPY OF
      MIXING ON THE BASIS OF
      STATISTICAL MECHANICS
 20  REM :CONSIDER TWO IDEALIZED
      CRYSTALS; AT THE BEGINNING IN
      THE LEFT CRYSTAL THERE ARE
      N1 ATOMS OF TYPE A,
 30  REM :IN THE RIGHT CRYSTAL N2
      ATOMS OF TYPE B. THE QUESTIONS:
      HOW MANY POSSIBLE STATES,
      WHICH IS THE MOST
      PROBABLE DISTRIBUTION OF ATOMS,
      HOW MUCH IS THE ENTROPY
      OF MIXING ?
 40   PRINT : PRINT
 50  SP = 0
 60  SM = 0
 70  PRINT "ENTER NUMBER OF 'A' ATOMS"
 80  INPUT "IN THE LEFT CRYSTAL
      ";N1
 90  PRINT
100  PRINT "ENTER NUMBER OF 'B' ATOMS"
110  INPUT "IN THE RIGHT CRYSTAL
```

```
        ";N2
120  K = N1
130  IF N2 < K THEN K = N2
140  PRINT : PRINT
150  PRINT " ATOMS LEFT I ATOMS
      RIGHT I POS. STATES"
160  PRINT " A B I A
      B I NO."
170  PRINT "------------I--------
      ------I-----------"
180  FOR I = 0 TO K
190  LA = N1 - I
200  LB = I
210  RB = N2 - I
220  RA = I
230  N = N1:M = 1
240  GOSUB 500
250  PS = OM
260  N = N2
270  GOSUB 500
280  PS = PS * OM
290  SP = SP + PS
300  PRINT TAB( 3);LA; TAB( 9);L
      B; TAB( 13);"I"; TAB( 18);RA
      : TAB( 24);RB; TAB( 28);"I";
      TAB( 32);PS
310  NEXT I
320  PRINT "------------I--------
      ------I-----------"
330  PRINT : PRINT
340  PRINT "THE PRODUCTION OF ENTROPY"
350  K = 1.380662E - 23
360  PRINT "DURING THE MIXING IS
      "; K * LOG (SP)
370  PRINT : PRINT
380  PRINT "NUMBER OF ARRANGEMENTS
      "; SP
390  PRINT : PRINT
400  INPUT "NEXT CALCULATION ? (Y
      /N) "; A$
410  IF A$ = "Y" THEN GOTO 40
420  END
430  REM :CALCULATION OF FACTORIALS
440  F = 1
450  IF M = 0 THEN GOTO 490
460  FOR II = 1 TO M
470  F = F * II
480  NEXT II
```

```
490  RETURN
500  REM :CALCULATION OF NUMBER
      OF POSSIBLE STATES. KNOWN:
      NUMBER OF ATOMS IN PLANE (N),
      NUMBER OF MOVED ATOMS (M)
510  OM = 1
520  IF N = M THEN GOTO 590
530  IF M = 0 THEN GOTO 590
540  GOSUB 430
550  FOR JJ = N TO N - M + 1 STEP
      - 1
560  OM = OM * JJ
570  NEXT JJ
580  OM = OM / F
590  RETURN
```

Example. Run the program for the number of A molecules equal to 4, 6, 10 and for the number of B molecules equal to 4, 6, 10, 15.

```
]RUN

ENTER NUMBER OF 'A' ATOMS
IN THE LEFT CRYSTAL 4

ENTER NUMBER OF 'B' ATOMS
IN THE RIGHT CRYSTAL 4
```

ATOMS LEFT		ATOMS RIGHT		POS. STATES
A	B	A	B	NO.
4	0	0	4	1
3	1	1	3	16
2	2	2	2	36
1	3	3	1	16
0	4	4	0	1

```
THE PRODUCTION OF ENTROPY
DURING THE MIXING IS 5.86573595E - 23

NUMBER OF ARRANGEMENTS 70

NEXT CALCULATION ? (Y / N) Y

ENTER NUMBER OF 'A' ATOMS
IN THE LEFT CRYSTAL 6

ENTER NUMBER OF 'B' ATOMS
IN THE RIGHT CRYSTAL 6
```

ATOMS LEFT		ATOMS RIGHT		POS. STATES
A	B	A	B	NO.
6	0	0	6	1
5	1	1	5	36
4	2	2	4	225
3	3	3	3	400
2	4	4	2	225
1	5	5	1	36
0	6	6	0	1

THE PRODUCTION OF ENTROPY
DURING THE MIXING IS 9.42814328E − 23

NUMBER OF ARRANGEMENTS 924

NEXT CALCULATION ? (Y / N) Y

ENTER NUMBER OF 'A' ATOMS
IN THE LEFT CRYSTAL 10

ENTER NUMBER OF 'B' ATOMS
IN THE RIGHT CRYSTAL 10

ATOMS LEFT		ATOMS RIGHT		POS. STATES
A	B	A	B	NO.
10	0	0	10	1
9	1	1	9	100
8	2	2	8	2025
7	3	3	7	14400
6	4	4	6	44100
5	5	5	5	63504
4	6	6	4	44100
3	7	7	3	14400
2	8	8	2	2025
1	9	9	1	100
0	10	10	0	1

THE PRODUCTION OF ENTROPY
DURING THE MIXING IS 1.67429999E − 22

NUMBER OF ARRANGEMENTS 184756

NEXT CALCULATION ? (Y / N) Y

ENTER NUMBER OF 'A' ATOMS
IN THE LEFT CRYSTAL 10

ENTER NUMBER OF 'B' ATOMS
IN THE RIGHT CRYSTAL 15

ATOMS LEFT A	B	ATOMS RIGHT A	B	POS. STATES NO.
10	0	0	15	1
9	1	1	14	150
8	2	2	13	4725
7	3	3	12	54600
6	4	4	11	286650
5	5	5	10	756756
4	6	6	9	1051050
3	7	7	8	772200
2	8	8	7	289575
1	9	9	6	50050
0	10	10	5	3003

THE PRODUCTION OF ENTROPY
DURING THE MIXING IS 2.07098213E − 22

NUMBER OF ARRANGEMENTS 3268760

NEXT CALCULATION ? (Y / N) N

Remark. You can see that the entropy of mixing is a symmetric function.
The result becomes more readable if you multiply the entropy value by 10^{25} and take the integer part.

Model of the Ideal Mixture in Statistical Mechanics

The third model is offered by statistical mechanics. It has been shown that the relationship between entropy and partition function is

$$s = \mathbf{k} \ln Q + \frac{u}{T}$$

where Q = partition function, and
u = molar internal energy, which does not change in our case (that of an ideal gas) during mixing.

Let us pass over to the molecular partition function:

$$Q = \left(\frac{qve}{\mathbf{N}}\right)^{\mathbf{N}}$$

$$\ln Q = \mathbf{N}\left(\ln \frac{qv}{\mathbf{N}} + 1\right)$$

Thus we have for the molar entropy

$$s = \mathbf{kN}\left(\ln \frac{qv}{\mathbf{N}} + 1\right) + \frac{u}{T}$$

The entropy in the initial state is the sum of the entropies of the separate substances A and B:

$$S_{\text{initial}} = \mathbf{k}n_A\left(\ln \frac{q_A v_A}{n_A} + 1\right) + \frac{u_A}{T} + \mathbf{k}n_B\left(\ln \frac{q_B v_B}{n_B} + 1\right) + \frac{u_B}{T}$$

The entropy in the final state is

$$S_{\text{final}} = \mathbf{k}n_A\left(\ln \frac{q_A(v_A + v_B)}{n_A} + 1\right) + \frac{u_A}{T} + \mathbf{k}n_B\left(\ln \frac{q_B(v_A + v_B)}{n_B} + 1\right) + \frac{u_B}{T}$$

The entropy of mixing is

$$\begin{aligned} S_{\text{final}} - S_{\text{initial}} &= \mathbf{k}n_A \ln(v_A + v_B) + \mathbf{k}n_B \ln(v_A + v_B) \\ &\quad - \mathbf{k}n_A \ln v_A - \mathbf{k}n_B \ln v_B \\ &= -\mathbf{k}\left(n_A \ln \frac{v_A}{v_A + v_B} + n_B \ln \frac{v_B}{v_A + v_B}\right) \end{aligned}$$

In the ideal mixture of ideal-gas components the volume fraction is equal to the mole fraction. If we normalize on both sides, then we have for the molar entropy of mixing, in accordance with the two previous models,

$$\Delta s^{(\text{id})} = -\mathbf{R}(x_A \ln x_A + x_B \ln x_B)$$

Generalized for multicomponent mixtures, this becomes

$$\Delta s^{(\text{id})} = -\mathbf{R}\sum_i (x_i \ln x_I)$$

FREE ENTHALPY OF MIXING

Three mixing models have been discussed. The same phenomenon was treated with three methods, developed earlier, and naturally, identical results were obtained. Studying the mixing of molecules with no interaction between them, we found that mixing in an isothermal–isobaric spontaneous process is accompanied by an increase in entropy. What conclusions can be drawn from this about the free enthalpy of mixing? We have for this quantity

$$\Delta g^{(\text{id})} = \Delta h^{(\text{id})} - T\Delta s^{(\text{id})}$$

As our case excludes energetic interaction, $\Delta h^{(id)} = 0$, so that the free enthalpy of mixing is negative in this case; that is, mixing is accompanied by a decrease in free enthalpy:

$$\begin{aligned}\Delta g^{(id)} &= -T\Delta s^{(id)} \\ &= \mathbf{R}T\sum_i (x_i \ln x_i)\end{aligned}$$

Algorithm and Program: Graphical Representation of Ideal-Gas Free Energy of Mixing

Write a program for plotting the free energy of mixing when the mixing fluids are ideal gases.

Algorithm.

1. Read the temperature.
2. Calculate the ideal-gas free energy of mixing at a given concentration, and plot it.
3. Repeat step 2.

Program GRMIXFREEENERGY. A possible realization:

```
]PR#0
]LOAD GRMIXFREEENERGY
]LIST

 10  REM GRAPHICAL REPRESENTATION
      OF FREE ENERGY OF MIXING IN
      TWO-COMPONENT SYSTEM
 20  REM :OF IDEAL GASES
 30  HPLOT 0,0
 40  HGR
 50  HPLOT 10,0 TO 10,150
 60  HPLOT 10,0 TO 200,0
 70  HPLOT 200,150 TO 200,0
 80  HPLOT 10,0
 90  INPUT "ENTER TEMPERATURE (K)
      "; T
100  FOR I = 1 TO 49
110  X1 = I * 1 / 50
120  X2 = 1 - X1
130  GM = 8.3144 * T * (X1 * LOG
      (X1) + X2 * LOG (X2))
140  X = 10 + 190 * X1
150  Y = - 0.05 * GM
160  HPLOT TO X,Y
```

```
170  NEXT I
180  HPLOT TO 200,0
190  INPUT "NEXT TEMPERATURE ? (Y
      /N) "; A$
200  IF A$ = "Y" THEN GOTO 80
210  TEXT
220  END
```

Example. Run the program for temperatures 100, 200, 500 K.

```
]RUN
ENTER TEMPERATURE (K) 100
NEXT TEMPERATURE ? (Y / N) Y
ENTER TEMPERATURE (K) 200
NEXT TEMPERATURE ? (Y / N) Y
ENTER TEMPERATURE (K) 500
NEXT TEMPERATURE ? (Y / N) N
```

Remark. Watch the screen and explain the curves.

Algorithm and Program: Graphical Representation of Free Energy of a Mixture of Ideal Gases

Write a program for plotting the free energy of a mixture of two ideal gases.

Algorithm.

1. Read the temperature.
2. Read the free energies of the components at the actual temperature.
3. Calculate the free energy of mixing at various concentrations (see program GRMIXFREEENERGY).
4. Calculate and plot the free energy of the system,

$$g^{\circ}_{\text{mixture}}(T) = X_A g^{\circ}_A(T) + X_B g^{\circ}_B(T) + \Delta g^{(\text{id})}_m(T, X_A, X_B)$$

   where X_A, X_B = molar fractions of components A and B (their sum must be 1),
   g° = ideal-gas free energy,
   $\Delta g^{(\text{id})}_m$ = free energy of perfect mixing.

5. Repeat step 4.

Program GRFREEENERGY. A possible realization:

```
]PR#0
]LOAD GRFREEENERGY
]LIST
```

```
 10  REM :GRAPHICAL REPRESENTATION
      OF FREE-ENERGY CHANGE ON
      MIXING OF IDEAL GASES
 30  HPLOT 0,0
 40  HGR
 50  HPLOT 10,0 TO 10,150
 60  HPLOT 10,150 TO 200,150
 70  HPLOT 200,150 TO 200,0
 80  HPLOT 10,10
 90  INPUT "ENTER TEMPERATURE (K)
      "; T
100  PRINT "ENTER FREE ENERGY OF
      COMPONENT 'A' "
110  INPUT "AT TEMPERATURE 'T' (J
      /MOL) "; GA
120  PRINT "ENTER FREE ENERGY OF
      COMPONENT 'B' "
130  INPUT "AT TEMPERATURE 'T' (J
      /MOL) "; GB
140  FOR I = 1 TO 49
150  X1 = I * 1 / 50
160  X2 = 1 - X1
170  GM = 8.3144 * T * (X1 * LOG
      (X1) + X2 * LOG (X2))
180  G = GB * X1 + GA * X2 + GM
190  X = 10 + 190 * X1
200  Y = (1 - G / GA) * 160 + 10
210  HPLOT TO X,Y
220  NEXT I
230  HPLOT TO 200,(1 - GB / GA) *
      160 + 10
240  INPUT "NEXT TEMPERATURE ? (Y
      /N) "; A$
250  IF A$ = "Y" THEN GOTO 80
260  TEXT
270  END
```

Example. Run the program for components with molar free energies 3000 and 2000 J/mole, at temperatures 300 and 200 K.

```
]PR#0
]RUN
ENTER TEMPERATURE (K) 300
ENTER FREE ENERGY OF COMPONENT 'A'
AT TEMPERATURE 'T' (J / MOL) 3000
ENTER FREE ENERGY OF COMPONENT 'B'
AT TEMPERATURE 'T' (J / MOL) 2000
```

```
NEXT TEMPERATURE ? (Y / N) Y
ENTER TEMPERATURE (K) 200
ENTER FREE ENERGY OF COMPONENT 'A'
AT TEMPERATURE 'T' (J / MOL) 3000
ENTER FREE ENERGY OF COMPONENT 'B'
AT TEMPERATURE 'T' (J / MOL) 2000
NEXT TEMPERATURE ? (Y / N) N
```

Remark. Watch the screen and explain the curves.

Algorithm and Program: Ideal Mixture of Liquids

Write a program for calculating the molar volume of an ideal mixture.

Algorithm.

1. Read the actual temperature and pressure.
2. Read the datafile sequential numbers of the compounds.
3. Call database-reader routine.
4. Calculate the liquid molar volumes of the components (see program ESLIQU in Chapter 3).
5. Read the composition of the mixture.
6. Calculate molar volume of the mixture:

$$v_{mix} = x_1 v_1 + (1 - x_1) v_2$$

7. Print the results.

Program VLEIDID. A possible realization:

```
]PR#0
]LOAD VLEIDID
]LIST

 10  REM :CALCULATION OF MIXTURE
      MOLAR VOLUME
 20  REM :SUPPOSING THE MIXTURE TO
      BE IDEAL
 30  PRINT : PRINT
 40  DIM A(15)
 50  PRINT
 60  INPUT "ENTER TEMPERATURE (K)
      "; T
 70  PRINT
 80  INPUT "ENTER PRESSURE (PA) ";
      P
```

```
 90  PRINT
100  PRINT "ENTER THE SEQUENTIAL
      NUMBER"
110  INPUT "OF FIRST COMPONENT
      "; IS
120  GOSUB 410
130  TC = A(3)
140  PC = A(2)
150  OM = A(4)
160  GOSUB 520
170  V1 = V
180  PRINT
190  PRINT "ENTER THE SEQUENTIAL
      NUMBER"
200  INPUT "OF SECOND COMPONENT
      "; IS
210  GOSUB 410
220  TC = A(3)
230  PC = A(2)
235  OM = A(4)
240  GOSUB 520
250  V2 = V
260  PRINT
270  PRINT "ENTER MOLECULAR FRACTION "
280  INPUT "OF FIRST COMPONENT "
      ;X1
290  VM = X1 * V1 + (1 - X1) * V2
300  PRINT : PRINT
310  PRINT "THE MOLAR VOLUME OF"
320  PRINT "THE MIXTURE "; VM; " M
      3 / MOL"
330  PRINT
340  INPUT "NEXT COMPOSITION ? (Y
      /N) "; A$
350  IF A$ = "Y" THEN GOTO 260
360  INPUT "NEXT COMPOUNDS ? (Y / N
      ) "; A$
370  IF A$ = "Y" THEN GOTO 90
380  INPUT "NEXT TEMPERATURE OR P
      RESSURE ? (Y / N) "; A$
390  IF A$ = "Y" THEN GOTO 50
400  END
410  REM :READING OF SPECIFIED
      RECORD OF PHDAT FILE INTO ARRAY
      A(15)
420  REM :THE SEQUENTIAL NUMBER OF
      THE COMPOUND IN THE CALLING
      PROGRAM HAS TO BE "IS"
430  REM :ARRAY A HAS TO BE DECLA
      RED IN THE CALLING PROGRAM
```

```
440  D$ = CHR$ (4)
450  PRINT D$; "OPEN PHDAT"
460  PRINT D$; "POSITION PHDAT, R";
      IS - 1
470  PRINT D$; "READ PHDAT"
480  INPUT A(1), A(2), A(3), A(4), A(5),
      A(6), A(7), A(8), A(9), A(10),
      A(11), A(12), A(13), A(14), A(15)
490  PRINT D$; "CLOSE PHDAT"
500  PRINT D$; "IN#0"
510  RETURN
520  DEF FN Z0(PR) = (.46407 - 0
      .73221 * TR + 0.45256 * TR ∧
      2) * PR - (0.00871 - 0.02939
      * TR + 0.02775 * TR ∧ 2) *
      PR ∧ 2
530  DEF FN Z1(PR) = - (0.02676
      + 0.28376 * TR - 0.2834 * T
      R ∧ 2) * PR - (0.10209 - 0.3
      2504 * TR + 0.25376 * TR ∧ 2
      ) * PR ∧ + (0.00919 - 0.03
      016 * TR - 0.02485 * TR ∧ 2)
      * PR ∧ 3
540  TR = T / TC
550  PR = P / PC
560  Z = FN Z0(PR) + OM * FN Z1(
      PR)
570  V = Z * 8.3144 * T / P
580  RETURN
```

Example. Run the program for a mixture of chloroform and 1,1-dichloroethane at temperatures 250 and 288.2 K, at pressure 101,325 Pa, and at mole fractions 0.5, 0.6, and 0.7.

```
]RUN

ENTER TEMPERATURE (K) 250

ENTER PRESSURE (PA) 101325

ENTER THE SEQUENTIAL NUMBER
OF FIRST COMPONENT 10

ENTER THE SEQUENTIAL NUMBER
OF SECOND COMPONENT 11

ENTER MOLECULAR FRACTION
OF SECOND COMPONENT 0.5

THE MOLAR VOLUME OF
THE MIXTURE 7.8003406E - 05 M3 / MOL
```

```
NEXT COMPOSITION ? (Y / N) Y

ENTER MOLECULAR FRACTION
OF FIRST COMPONENT 0.6

THE MOLAR VOLUME OF
THE MIXTURE 7.76390022E - 05 M3 / MOL

NEXT COMPOSITION ? (Y / N) Y

ENTER MOLECULAR FRACTION
OF FIRST COMPONENT 0.7

THE MOLAR VOLUME OF
THE MIXTURE 7.72745985E - 05 M3 / MOL

NEXT COMPOSITION ? (Y / N) N
NEXT COMPOUNDS ? (Y / N) N
NEXT TEMPERATURE OR PRESSURE ? (Y / N) Y

ENTER TEMPERATURE (K) 288.2

ENTER PRESSURE (PA) 101325

ENTER THE SEQUENTIAL NUMBER
OF FIRST COMPONENT 10

ENTER THE SEQUENTIAL NUMBER
OF SECOND COMPONENT 11

ENTER MOLECULAR FRACTION
OF FIRST COMPONENT 0.5

THE MOLAR VOLUME OF
THE MIXTURE 8.10233204E - 05 M3 / MOL

NEXT COMPOSITION ? (Y / N) Y

ENTER MOLECULAR FRACTION
OF FIRST COMPONENT 0.6

THE MOLAR VOLUME OF
THE MIXTURE 8.0650388E - 05 M3 / MOL

NEXT COMPOSITION ? (Y / N) Y

ENTER MOLECULAR FRACTION
OF FIRST COMPONENT 0.7

THE MOLAR VOLUME OF
THE MIXTURE 8.02774556E - 05 M3 / MOL

NEXT COMPOSITION ? (Y / N) N
NEXT COMPOUNDS ? (Y / N) N
NEXT TEMPERATURE OR PRESSURE ? (Y / N) N
```

Remark. Explain the results.

THERMODYNAMICS OF REAL MIXTURES

So far we have discussed *ideal* mixtures, the component molecules of which are rigid spheres of near-identical size, which do not attract one another. These are rather special conditions. Our task is now to deal with the thermodynamics of real mixtures.

In the discussion of the extensive properties of real mixtures, it is convenient to treat the contributions of the individual components separately. This is to be understood in the following way: Let M be the actual value of an arbitrary extensive property (volume, enthalpy, entropy, free enthalpy) of one mole of mixture at given temperature, pressure, and composition:

1. Let us choose the contributions of the individual components in the following way (Figure 11.3):

$$M = \sum x_i \overline{M}_i$$

where $\overline{M}_i$ is the partial molar M of component i, and generally depends on the composition.

2. Let M_i denote the molar value of the property M for the ith component in the pure state at the temperature and pressure of the mixture (i.e., in the reference state). Then

$$M = \sum_i x_i M_i + \sum_i x_i \Delta M_i$$

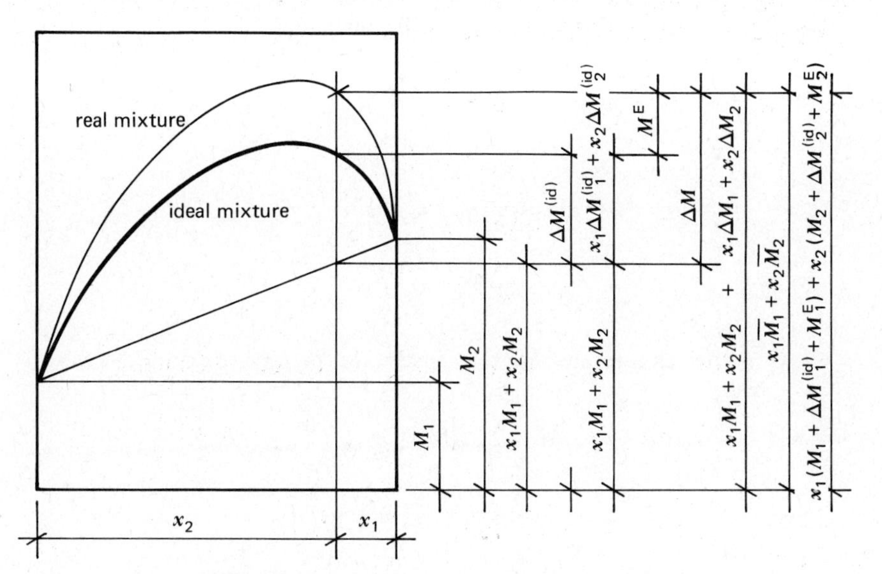

FIGURE 11.3. Ideal mixture as a reference state.

Here ΔM_i is the partial molar mixing M of component i, which is just the difference between the partial molar $\tilde{M}_i$ and the M_i for the pure state:

$$\Delta M_i = \overline{M}_i - M_i$$

3. For an ideal mixture,

$$M^{(\mathrm{id})} = \sum_i x_i \overline{M}_i^{(\mathrm{id})}$$

$$= \sum_i x_i M_i + \sum x_i \Delta M_i^{(\mathrm{id})}$$

Here $\overline{M}_i^{(\mathrm{id})}$ is the partial molar M of component i in the ideal mixture:

$$\overline{M}_i^{(\mathrm{id})} = M_i + \Delta M_i^{(\mathrm{id})}$$

where $\Delta M_i^{(\mathrm{id})}$ is the partial molar mixing M of component i in the ideal mixture. This is just the difference between the value of partial molar for the mixture (i.e., $\overline{M}_i^{(\mathrm{id})}$) and the molar M value of the substance in the reference state (i.e., M_i):

$$\Delta M_i^{(\mathrm{id})} = \overline{M}_i^{(\mathrm{id})} - M_i$$

The partial molar mixing $\Delta M^{(\mathrm{id})}$ can be calculated on the basis of the model of ideal mixtures for any arbitrary property, as it does not depend on the individual properties of the component. In particular,

$$\Delta s_i^{(\mathrm{id})} = -\mathbf{R} \ln x_i$$

$$\Delta g_i^{(\mathrm{id})} = \mathbf{R}T \ln x_i$$

$$\Delta h_i^{(\mathrm{id})} = 0$$

$$\Delta v_i^{(\mathrm{id})} = 0$$

4. Consider as intermediate reference state the ideal mixture. Now

$$M = \sum_i x_i \overline{M}_i$$

$$= \sum_i x_i \left(\overline{M}_i^{(\mathrm{id})} + M_i^{\mathrm{E}} \right)$$

Here M_i^{E} is the excess M of component i in the real mixture, that is, the difference from the intermediate reference state:

$$M_i^{\mathrm{E}} = \overline{M}_i - \overline{M}_i^{(\mathrm{id})}$$

Thus, in a real mixture the partial molar $\overline{M}_i$ of component i has the following form:

$$\overline{M}_i = M_i + \Delta M_i^{(\mathrm{id})} + M_i^{\mathrm{E}}$$

Of the three terms:

1. The first is the value of a property of the pure substance, independent of composition.
2. The second term follows from the model of the ideal mixture, and does not contain component-specific parameters.
3. The third can be calculated with the model of real mixture. It depends on the composition of the mixture, and contains component-specific parameters.

We will return later to these real mixture models.

Partial Molar Properties

Let M be the value of an arbitrary extensive property in one mole of mixture. If the amount of the mixture is n moles, then

$$nM = M[P, T, n_1, n_2, \ldots, n_k]$$

where P = pressure,
T = temperature,
n_i = amount of the ith component,
k = number of components.

The total amount of the mixture is

$$n = \sum_i n_i$$

Differentiating the previous expression, we have

$$d(nM) = \left(\frac{\partial nM}{\partial P}\right)_{T,n} dP + \left(\frac{\partial nM}{\partial T}\right)_{P,n} dT + \sum_{i=1}^{k} \left(\frac{\partial nM}{\partial n_i}\right) dn_i$$

The first term on the right represents the change in quantity (nM) with pressure at constant temperature and composition, the second term represents the change in quantity with temperature at constant pressure and composition, and terms under the summation sign represent the change in quantity with amount of the ith component, at constant temperature and pressure and otherwise constant composition. The partial derivative under the summation sign is the partial molar quantity $\overline{M}_i$ of the ith component:

$$\overline{M}_i = \left(\frac{\partial nM}{\partial n_i} \right)_{T,P,n_j}$$

This is an intensive quantity, being the derivative of a homogeneous linear function.

Integrating the previous equation for the case of constant temperature and pressure, we have according to Euler's theorem

$$nM = \sum_{i=1}^{k} \overline{M}_i n_i$$

Differentiating this, and comparing it with the preceding form of $d(nM)$, a special form of the Gibbs–Duhem equation is obtained:

$$\sum_{i=1}^{k} n_i \, d\overline{M}_i = 0$$

This can be alternatively written in terms of mole fractions in the following form:

$$\sum_{i=1}^{k} x_i \, d\overline{M}_i$$

Integrated, this becomes

$$M = \sum_{i=1}^{k} x_i \overline{M}_i$$

A partial molar quantity cannot be directly measured. However, the volume of the mixture can be measured dilatometrically, and the change of enthalpy calorimetrically. Restricting ourselves again to the two-component case, we will show how a partial molar quantity belonging to a given composition can be calculated from experimental data.

In a binary mixture,

$$\Delta M = M - \sum_{i=1}^{2} M_i x_i = x_1(\overline{M}_1 - M_1) + x_2(\overline{M}_2 - M_2)$$

$$= (1 - x_2)(\overline{M}_1 - M_1) + x_2(\overline{M}_2 - M_2)$$

Figure 11.4 shows the change in ΔM with x_2. Differentiating this relationship with respect to x_2,

$$\frac{\partial \Delta M}{\partial x_2} = (1 - x_2)\left(\frac{\partial \overline{M}_1}{\partial x_2}\right) + x_2\left(\frac{\partial \overline{M}_2}{\partial x_2}\right) - (\overline{M}_1 - M_1) - (\overline{M}_2 - M_2)$$

The sum of the first two terms on the right is zero according to the Gibbs–Duhem equation deduced in the foregoing:

$$\sum_{i=1}^{2} x_i \, d\overline{M}_i = (1 - x_2)\left(\frac{\partial \overline{M}_1}{\partial x_2}\right) + x_2\left(\frac{\partial \overline{M}_2}{\partial x_2}\right) = 0$$

Substituting the last term on the right from the previous equation for a partial molar mixing property, we have

$$\overline{M}_2 - M_2 = \frac{\Delta M}{x_2} - \frac{(1 - x_2)(\overline{M}_1 - M_1)}{x_2}$$

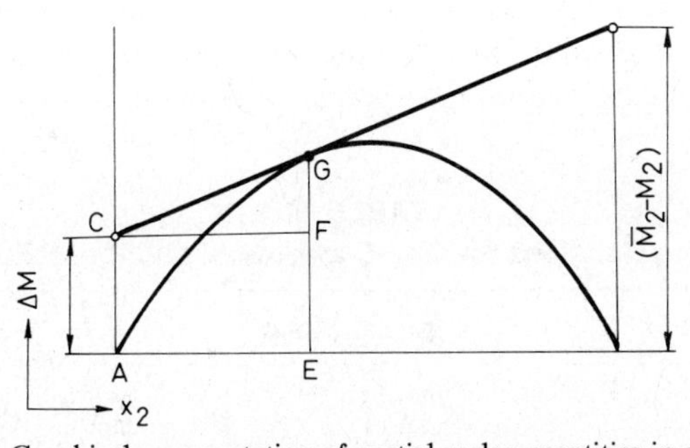

FIGURE 11.4. Graphical representation of partial molar quantities in a binary mixture.

$$\overline{CA} = \overline{FE} = (\overline{M}_1 - M_2)$$

$$\overline{GE} = \overline{GF} + \overline{FE} = \overline{GF} + \overline{CA} = \Delta M[x_2]$$

$$\overline{GF}/\overline{FC} = (\partial \Delta M/\partial x_2)x_2$$

Substituting this, we have

$$\frac{\partial \Delta M}{\partial x_2} = \frac{\Delta M}{x_2} - \frac{(1 - x_2)(\overline{M}_1 - M_1)}{x_2} - (\overline{M}_1 - M_1)$$

$$= \frac{\Delta M}{x_2} - (\overline{M}_1 - M_1)\left(\frac{1 - x_2}{x_2} + 1\right)$$

$$= \frac{\Delta M}{x_2} - \frac{\overline{M}_1 - M_1}{x_2}$$

We multiply through by x_2 and reduce the equation. The relationship sought for the partial molar quantity of component 1 is the following:

$$\overline{M}_1 = M_1 + \Delta M[x_2] + x_2\left(\frac{\partial \Delta M}{\partial x_2}\right)_{x_2}$$

where M_1 = molar M of component 1 in the pure state at the temperature and pressure of the mixture,
$\Delta M[x_2]$ = mixing property at a given value of the mole fraction x_2,
$(\partial \Delta M/\partial x_2)_{x_2}$ = slope of the tangent to the curve of ΔM versus x_2 at a given value of the mole fraction x_2.

It can be seen from the figure that the intercept of the tangent is $\overline{M}_1 - M_1$.

On the basis of similar considerations, the intercept of the tangent for component 2 is $\overline{M}_2 - M_2$.

The elaboration of the thermodynamics of homogeneous mixtures is a fairly simple task if we take into consideration the following theorem: Each thermodynamic relationship valid for one-component homogeneous phases maintains its validity in the case of multicomponent phases of constant composition, if

TABLE 11.1
Corresponding Relations for One-Component and for Mixture Properties

Pure Component	Partial Molar	Molar Excess
$v_i = \left(\frac{\partial g_i}{\partial P}\right)_T$	$\bar{v}_i = \left(\frac{\partial \bar{g}_i}{\partial P}\right)_{T,x}$	$v^{\mathrm{E}} = \left(\frac{\partial g^{\mathrm{E}}}{\partial P}\right)_{T,x}$
$h_i = -\mathbf{R}T^2\frac{\partial(g_i/\mathbf{R}T)}{\partial T}$	$\bar{h}_i = -\mathbf{R}T^2\frac{\partial(\bar{g}_i/\mathbf{R}T)}{\partial T}$	$h^{\mathrm{E}} = -\mathbf{R}T^2\frac{\partial(g^{\mathrm{E}}/\mathbf{R}T)}{\partial T}$
$s_i = -\left(\frac{\partial g_i}{\partial T}\right)_P$	$\bar{s}_i = -\left(\frac{\partial \bar{g}_i}{\partial T}\right)_{P,x}$	$s^{\mathrm{E}} = -\left(\frac{\partial g^{\mathrm{E}}}{\partial T}\right)_{P,x}$

TABLE 11.2
Mixing Properties

$$\Delta M = \sum_i x_i(\overline{M}_i - M_i)$$

$$\Delta g = \sum_i x_i(\overline{g}_i - g_i)$$

$$\Delta v = \sum_i x_i(\overline{v}_i - v_i) = \frac{1}{P}\sum_i x_i\left(\frac{\partial(\overline{g}_i - g_i)}{\partial \ln P}\right)_{T,x}$$

$$\Delta h = \sum_i x_i(\overline{h}_i - h_i) = -\sum_i x_i\left(\frac{\partial(\overline{g}_i - g_i)/\mathbf{R}T}{\partial \ln T}\right)_{P,x}$$

$$\Delta s = \sum_i x_i(\overline{s}_i - s_i) = \sum_i x_i\left(\frac{\partial(\overline{g}_i - g_i)}{\partial T}\right)_{P,x}$$

the appropriate variables are substituted for the respective:

1. Mixture molar variables.
2. Partial molar variables.
3. Excess variables.

The relationships valid for one-component homogeneous phases were deduced in Chapter 6.

Thus, for example, we have the formulas in Table 11.1. By the substitution of these relationships, the molar free enthalpy, volume, enthalpy, and entropy of mixing can be written in the form shown in Table 11.2.

Excess Properties

It is remarkable that all the mixing properties could be written as a function of $\overline{g}_i - g_i$. The question now is whether we can say something about the difference between the partial molar free enthalpy and the molar free enthalpy of the pure substance.

It was shown in Chapter 10 that in the case of one-component homogeneous phases the isothermal change in molar free enthalpy can be expressed in terms of the activity:

$$dg_i = \mathbf{R}T\,d \ln a_i$$

This relationship is also taken for mixture component i:

$$d\overline{g}_i = \mathbf{R}T\,d \ln \overline{a}_i$$

where $\bar{a}_i$ is the activity of component i in the mixture. However, in an ideal mixture,

$$dg_i^{(\mathrm{id})} = \mathbf{R}T\, d \ln x_i$$

—that is, in an ideal mixture the activity of component i is equal to its mole fraction.

The difference between the actual partial molar free enthalpy and its value for an ideal mixture is the *molar excess free enthalpy*,

$$g_i^{\mathrm{E}} = \mathbf{R}T \ln a_i - \mathbf{R}T \ln x_i$$

$$= \mathbf{R}T \ln \frac{a_i}{x_i}$$

$$\equiv \mathbf{R}T \ln \gamma_i$$

where γ_i is the activity coefficient of component i in the real mixture. The activity coefficient can be also interpreted according to the foregoing as the correction factor for the mole fraction:

$$\bar{a}_i = \gamma_i x_i$$

Hence the molar excess free enthalpy of the mixture is

$$g^{\mathrm{E}} = \mathbf{R}T \sum_i x_i \ln \gamma_i$$

Knowing the expression for the molar free enthalpy of the mixture, the (logarithm of the) activity coefficient can be calculated with the following relationship:

$$\ln \gamma_i = \frac{1}{\mathbf{R}T}\left(\frac{\partial g^{\mathrm{E}}}{\partial x_i}\right)$$

If we deal with real-gas mixtures, the fugacity must be used instead of the activity. Since the relationship between the activity and fugacity for the pure substance is

$$a_i P_u = f_i$$

the partial fugacity of a component of the mixture is

$$\bar{a}_i P_u = \bar{f}_i$$

Thus, in a real-gas mixture, the molar excess free enthalpy of component i is

$$g_i^{\mathrm{E}} = \mathbf{R}T \ln \frac{\bar{f}_i}{P_{\mathrm{u}}} - \mathbf{R}T \ln \frac{P_i}{P_{\mathrm{u}}}$$

$$= \mathbf{R}T \ln \frac{\bar{f}_i}{P_i}$$

$$= \mathbf{R}T \ln \frac{\bar{f}_i}{y_i P}$$

where g_i^{E} = molar excess free enthalpy of component i,
$\bar{f}_i$ = partial fugacity of component i in the mixture,
P_{u} = standard pressure,
P_i = partial pressure of component i in the mixture,
P = total pressure of the mixture,
y_i = mole fraction of component i in the mixture.

In the case of liquid mixtures the activity coefficients γ_i must be known for practical calculations. These can be obtained if we are able to estimate, on the basis of an adequate model of liquid structure, the actual value of the excess partial molar free enthalpy g_i^{E}. This requires special considerations.

In the case of gas mixtures special considerations are not necessary, because the equation of state contains sufficient information for the calculation of the fugacity of real mixtures.

Algorithm and Program: Comparison of Fugacity Coefficient of a Pure Compound and a Mixture Component

Write a program for the determination of the partial fugacity and pure-component fugacity.

Algorithm.

1. Read the sequential numbers of the mixture components.
2. Read the temperature and pressure of the mixture.
3. Read the partial pressure of the selected component in the mixture.
4. Read the composition of the two-component mixture.
5. Read the pure-component physical properties from the datafile.
6. Calculate the pseudo-one-component properties of the mixture.
7. Calculate the compressibility factor and parameters a and b of the pure first component.

8. Calculate a and b for the second component.
9. Calculate the pseudo-one-component a and b for the Redlich–Kwong equation of state.
10. Calculate the compressibility factor of the mixture.
11. Calculate the partial fugacity of the first component with the equation

$$\ln \phi_{\text{first}} = \frac{b_{\text{first}}}{b_{\text{mixture}}}(z_{\text{mixture}} - 1) - \ln(z_{\text{mixture}} - z_{\text{mixture}} h_{\text{mixture}})$$

$$+ \frac{a_{\text{mixture}}}{b_{\text{mixture}} \mathbf{R} T^{3/2}} \left(\frac{b_{\text{first}}}{b_{\text{mixture}}} - \frac{2\Sigma_k y_k a_k}{a_{\text{mixture}}} \right) \ln(1 + h_{\text{mixture}})$$

$$h = \frac{bP}{z}$$

12. Print the results.

Program PARTIALFUGACITY,D2. A possible realization:

```
]LOAD PARTIALFUGACITY,D2
]LIST

 10  REM :CALCULATION OF THE FUGACITY
      COEFFICIENT OF A PURE COMPO
      NENT AND THE PARTIAL FUGACITY
      OF THE SAME COMPONENT IN
      MIXTURE
 20  DIM A(15)
 30  PRINT : PRINT
 40  INPUT "ENTER SEQ. NO. OF FIRST
      COMPONENT "; C1
 50  PRINT
 60  INPUT "ENTER SEQ. NO. OF SECO
      ND COMPONENT "; C2
 70  QJ = 0
 80  PRINT
 90  INPUT "ENTER TEMPERATURE (K)
      "; T
100  PRINT
110  INPUT "ENTER PRESSURE OF MIX
     TURE (PA) "; P
120  PRINT
130  PRINT "ENTER PARTIAL PRESSURE "
140  INPUT "OF FIRST COMPONENT (P
     A) "; PM
150  PRINT
160  PRINT "ENTER MOLE FRACTION"
```

```
170  INPUT "OF FIRST COMPONENT ";
     Y1
180  GOSUB 700
190  TC = T1; PC = P1
200  GOSUB 500
210  AH = AL
220  BH = BL
230  HH = H2
240  ZH - Z2
250  FH = EXP (ZH - 1 - LOG (ZH -
      ZH * HH) - AH * LOG (1 + HH
      ) / BH / R / T ∧ (3 / 2))
260  TC = T2:PC = P2
270  GOSUB 500
280  BL = Y1 * BH + (1 - Y1) * BL
290  AS = AL
300  TC = TP:PC = PP
310  GOSUB 500
320  AL = Y1 * (1 - Y1) * AL + Y1 *
      Y1 * AH + (1 - Y1) ∧ 2 * AS
330  GOSUB 580
340  AM = AL
350  BM = BL
360  ZM = Z2
370  HM = H2
380  HK = Y1 * AH + (1 - Y1) * 0.4
      278 * R * R * TP ∧ 2.5 / PP
390  FM = EXP (BH / BM * (ZM - 1)
      - LOG (ZM - ZM * HM) + AM *
      (BH / BM - 2 * HK / AM) * LOG
      (1 + HM) / BM / R / T ∧ (3 /
      2))
400  PRINT : PRINT
410  PRINT "FUGACITY COEFFICIENT
     OF FIRST COMPONENT"
420  PRINT "WHEN IT IS PURE: "; FH
430  PRINT
440  PRINT "PARTIAL FUGACITY COEF
     FICIENT OF FIRST"
450  PRINT "COMPONENT IN MIXTURE:
      "; FM
460  PRINT : PRINT
470  INPUT "NEXT CALCULATION (Y / N
     ) "; A$
480  IF A$ = "Y" THEN GOTO 40
490  END
500  REM :CALCULATION OF COMPRESS
     IBILITY
510  REM :FACTOR BY REDLICH-KWONG
```

```
520  REM :EQUATION OF STATE
530  R = 8.3144
540  CA = 0.4278
550  CB = 0.0867
560  AL = CA * R * R * TC ∧ 2.5 /
      PC
570  BL = CB * R * TC / PC
580  B = BL / R / T
590  A2 = AL / R / R / T ∧ 2.5
600  Z1 = 1
610  1 = 0
620  H2 = B * P / Z1
630  I = I + 1
640  IF 8 > 150 THEN PRINT : PRINT
     "ITERATION NOT CONVERGENT ":
      GOTO 690
650  Z2 = 1 / (1 - H2) - A2 * H2 /
      B / (1 + H2)
660  IF ABS (Z2 - Z1) < 1E - 5 THEN
      GOTO 690
670  Z1 = Z2
680  GOTO 620
690  RETURN
700  REM :CALCULATION OF PSEUDO-
      ONE-COMPONENT PROPERTIES FROM
      COMPOSITION AND PURE COMPON
      ENT DATA FOR BINARY MIXTURES
710  REM :USING LEE-KESLER MIXING
      RULES
720  REM :COMPOSITION IS IN 'Y1',
      SEQUENTIAL NUMBERS OF COMPOU
     NDS IN MIXTURE ARE IN 'C1' A
     ND IN 'C2'
730  IF QJ = 1 THEN GOTO 810
740  QJ = 1
750  IS = C1
760  GOSUB 870
770  T1 = A(3):P1 = A(2):V1 = A(14
     ):O1 = A(4)
780  IS = C2
790  GOSUB 870
800  T2 = A(3):P2 = A(2):V2 = A(14
     ):O2 = A(4)
810  VP = 1 / 8 * (Y1 * (1 - Y1) *
      (V1 ∧ (1 / 3) + V2 ∧ (1 / 3)
      ) ∧ 3 + Y1 * Y1 * (2 * V1 ∧
      (1 / 3)) ∧ 3 + (1 - Y1) * Y1
      * (V2 ∧ (1 / 3) + V1 ∧ (1 /
      3)) ∧ 3 + (1 - Y1) ∧ 2 * (2 *
      V2 ∧ (1 / 3)) ∧ 3)
```

```
820  TP = 1 / 8 / VP * (Y1 * (1 -
      Y1) * (V1 ∧ (1 / 3) + V2 ∧ (
      1 / 3)) ∧ 3 * SQR (T1 * T2)
      + Y1 * Y1 * (2 * V1 ∧ (1 /
      3)) ∧ 3 * T1 + (1 - Y1) * Y1
      * (V2 ∧ (1 / 3) + V1 ∧ (1 /
      3)) ∧ 3 * SQR (T2 * T1) + (
      1 - Y1) ∧ 2 * (2 * V2 ∧ (1 /
      3)) ∧ 3 * T2)
830  OP = (Y1 * O1 + (1 - Y1) * O2
      )
840  R = 8.3144
850  PP = (0.2905 - 0.085 * OP) *
      R * TP / VP
860  RETURN
870  REM :READING OF SPECIFIED
     RECORD OF PHDAT FILE INTO AR
     RAY A(15)
880  REM :ARRAY A HAS TO BE DECLA
     RED IN THE CALLING PROGRAM
890  D$ = CHR$ (4)
900  PRINT D$; "OPEN PHDAT, D1"
910  PRINT D$; "POSITION PHDAT, R";
     IS - 1
920  PRINT D$; "READ PHDAT"
930  INPUT A(1), A(2), A(3), A(4), A(5),
      A(6), A(7), A(8), A(9), A(10),
      A(11), A(12), A(13), A(14), A(15)
940  PRINT D$; "CLOSE PHDAT"
950  PRINT D$; "IN#0"
960  RETURN
```

Example. Run the program for a hydrogen–propane mixture. The molar fraction of the hydrogen is 0.341. The temperature is 400 K. The pressure of the mixture is 3386788 Pa; the partial pressure of the hydrogen is 1356336 Pa.

```
]RUN

ENTER SEQ. NO. OF FIRST COMPONENT 17

ENTER SEQ. NO. OF SECOND COMPONENT 3

ENTER TEMPERATURE (K) 400

ENTER PRESSURE OF MIXTURE (PA) 3386788

ENTER PARTIAL PRESSURE
OF FIRST COMPONENT (PA) 1356336

ENTER MOLE FRACTION
OF FIRST COMPONENT 0.341
```

FUGACITY COEFFICIENT OF FIRST COMPONENT
WHEN IT IS PURE: 1.01667233

PARTIAL FUGACITY COEFFICIENT OF FIRST
COMPONENT IN MIXTURE: 1.01502147

NEXT CALCULATION (Y / N) N

Remark. Explain the results.

Gas Mixtures

In the case of gas mixtures the Lewis–Randall rule can be used for the calculation of the fugacity of the component in the mixture:

$$\bar{f}_i = y_i f_i$$

where $\bar{f}_i$ = fugacity of the ith component in the mixture at temperature T and pressure P, if its molar fraction is y_i,

f_i = fugacity of the pure component i at the same temperature T and pressure P.

Note that this extended definition of ideal mixtures is consistent with the definition given at the beginning of this chapter.

The molar free enthalpy of the pure substance is

$$dg_i = \mathbf{R}T\,d\ln\frac{f_i}{P_\mathrm{u}}$$

and the partial molar free enthalpy

$$d\bar{g}_i = \mathbf{R}T\,d\ln\frac{\bar{f}_i}{P_\mathrm{u}}$$

whence

$$\bar{g}_i - g_i = \mathbf{R}T\ln y_i$$

while in an ideal mixture

$$\bar{g}_i^{(id)} - g_i = \mathbf{R}T\ln y_i$$

where $\bar{g}_i$ = partial molar free enthalpy of the ith component in the real mixture,

g_i = molar free enthalpy of the pure ith component at identical temperature and pressure,

$\bar{g}_i^{(\mathrm{id})}$ = partial molar free enthalpy in the ideal mixture.

Thus, it emerges that a real-gas mixture for which the Lewis–Randall rule is valid behaves as an ideal mixture.

According to the fugacity rule, two (or more) components behaving in the pure state as real gases may form an ideal mixture. Let these two components be A and B. There is attraction between A and A, between A and B, and between B and B molecules; let us assume that the corresponding pair potentials are all of Lennard-Jones type. The two-component mixture will be ideal if the parameters of the A–B pair potential differ only slightly from those of the A–A and B–B pair potentials.

It is usual to take for the parameters of the pair potentials

$$\sigma_{AB} = \tfrac{1}{2}(\sigma_{AA} + \sigma_{BB})$$

$$\varepsilon_{0(AB)} = \sqrt{\varepsilon_{0(AA)}\varepsilon_{0(BB)}}$$

Macro Properties of Ideal Mixtures

What is the change in volume when real-gas components form an ideal mixture?

The respective Maxwell relations between $\bar{g}_i$, P, and v_i have been shown, for the case of constant composition y and temperature T, to be

$$v_i^{(\mathrm{id})} = \left(\frac{\partial \bar{g}_i^{(\mathrm{id})}}{\partial P}\right)_{T,y} \qquad v_i = \left(\frac{\partial g_i}{\partial P}\right)_T$$

The difference of the two is

$$v_i^{(\mathrm{id})} - v_i = \left(\frac{\partial\left(\bar{g}_i^{(\mathrm{id})} - g_i\right)}{\partial P}\right)_{T,y}$$

$$= \left(\frac{\partial(\mathbf{R}T \ln y_i)}{\partial P}\right)_{T,y} = 0$$

because the derivative of a constant is zero. This, however, means in the case of interest to us that the isothermal–isobaric formation of an ideal mixture from pure components is not accompanied by a change in volume:

$$v^{(\mathrm{id})} = \sum_i y_i v_i$$

Using the appropriate Maxwell relations, it can be shown in a similar way that with the exception of nf, ng, and ns the change in the other extensive variables is also zero when an ideal mixture is formed from real-gas components.

Let us now inspect one of the exceptions. What change in entropy accompanies mixture formation in this case? The difference between the partial molar entropy and the molar entropy for component i is

$$\left(\bar{s}_i^{(\text{id})} - s_i\right) = -\left(\frac{\partial\left(g_i^{(\text{id})} - g_i\right)}{\partial T}\right)_{P,y}$$

$$= -\left(\frac{\partial(\mathbf{R}T \ln y_i)}{\partial T}\right)_{P,y}$$

$$= -\mathbf{R} \ln y_i$$

Thus the molar of entropy mixing of a mixture formed from real-gas components is the same as the entropy of mixing of the ideal mixture:

$$\Delta s = \Delta s^{(\text{id})} = -\mathbf{R}\sum_i y_i \ln y_i$$

Solubility of Gases in Liquid

It was said in the foregoing that the concept of the ideal mixture can be extended to all those mixtures for which the following linear relationship is true:

$$\bar{f}_i = y_i f_i$$

This is valid for the whole composition region, even in the case $y_i = 1$. Therefore

$$\lim_{y_i \to 1}\left(\frac{\bar{f}_i}{y_i}\right) = f_i$$

where $\bar{f}_i$ = the partial fugacity of component i in the mixture

f_i = the fugacity of the pure component at the temperature and pressure of the mixture.

This relationship can be applied without any difficulty to an ideal mixture of real gases. Furthermore, the thermodynamic definition is independent of the state of aggregation and is valid also for liquid mixtures:

$$\bar{f}_i = x_i f_i$$

However, here two difficulties arise.

1. There are cases in which component i does not exist in the pure state at the temperature and pressure of the mixture. This difficulty can be evaded by defining an extrapolated hypothetical state, in which the fugacity is

$$f_i^{(\text{hyp})}[T, P]$$

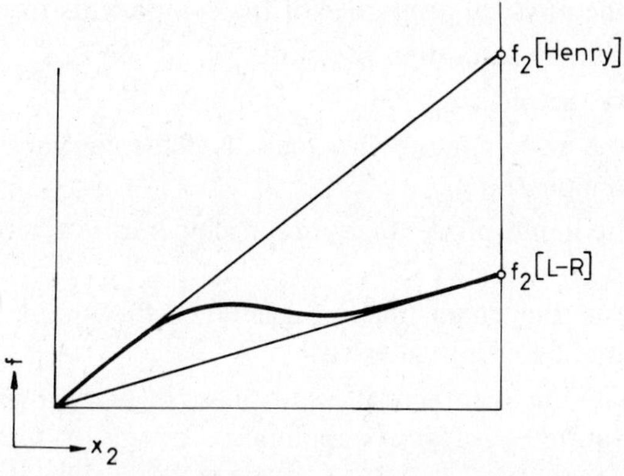

FIGURE 11.5. Henry region and Lewis–Randall region in a binary mixture.

2. The other difficulty is met if the partial fugacity of component i is not a linear function of the mole fraction over the whole composition range. Evidently, in this case the application of relationships relevant to the ideal mixture is acceptable only in that range where linearity is satisfactorily approximated. Two such regions exist. The region $x_i \to 1$ has already been mentioned; we have yet to examine $x_i \to 0$. The former is called the Lewis–Randall region, the latter the Henry region (Figure 11.5). In the Henry region the partial fugacity of component i in the mixture is proportional to its mole fraction:

$$\bar{f}_i = x_i f_i^{(\text{hyp})}[T, P] = x_i k_i[\text{Henry}]$$

Here k_i[Henry] is called Henry's constant; it is physically a fugacity and thus has the dimensions of pressure, and it equals the fugacity of component i at the temperature of the mixture in a standard state defined in the following way:

$$\lim_{x_i \to 0} \frac{\bar{f}_i}{x_i} = f_i^{(\text{hyp})}[T, P] = k_i[\text{Henry}]$$

Algorithm and Program: Calculation of Hypothetical Vapor-Phase Fugacity Coefficient

Write a program for the calculation of the pure-component vapor-phase fugacity coefficient at a temperature and pressure where the pure component is in the liquid phase.

Algorithm.

1. Read the sequential number of the component.
2. Read the temperature and pressure.

3. Read the physical properties of the components from the datafile.
4. Calculate the saturation pressure.
5. Identify the phase.
6. If in the vapor phase, then calculate the fugacity coefficient at the given temperature.
7. If in the liquid phase, then find a temperature where the vapor phase exists.
8. Calculate the vapor-phase fugacity coefficient at two temperatures where the liquid phase exists.
9. Calculate the hypothetical vapor-phase fugacity coefficient at the given temperature by linear extrapolation.
10. Print the results.

Program HYPOTHETICALFUGACITY,D2. A possible realization:

```
]LOAD HYPOTHETICALFUGACITY,D2
]LIST

 10  REM :CALCULATION OF THE FUGAC
      ITY COEFFICIENT IN HYPOTHETIC
      AL VAPOR STATE
 20  DIM A(15)
 30  PRINT : PRINT
 40  INPUT "ENTER SEQ. NO. OF THE
      COMPONENT "; C1
 50  PRINT
 60  INPUT "ENTER PRESSURE (PA) ";
      P
 70  PRINT
 80  PRINT "ENTER TEMPERATURE (K) "
 90  PRINT "WHERE YOU WOULD LIKE
      TO COMPUTE"
100  INPUT "THE FUGACITY COEFFICIENT"; T
110  PRINT : PRINT
120  IS = C1
130  GOSUB 640
140  TC = A(3):PC = A(2)
150  TK = T
160  GOSUB 740
170  IF PS > P THEN GOTO 200
180  T = T + 5
190  GOTO 160
200  IF T < > TK THEN GOTO 270
210  GOSUB 440
220  F = EXP (Z2 - 1 - LOG (Z2 -
      Z2 * H2) - AL * LOG (1 + H2
      ) / BL / R / T ^ (3 / 2))
```

```
230  PRINT "REAL GAS PHASE AT TEM
     PERATURE "; TK
240  PRINT : PRINT "THE FUGACITY
     COEFFICIENT "; F
250  PRINT : PRINT
260  GOTO410
270  GOSUB 440
280  F1 = EXP (Z2 - 1 - LOG (Z2 -
      Z2 * H2) - AL * LOG (1 + H2
      ) / BL / R / T ∧ (3 / 2))
290  T1 = T
300  T = T + 10
310  T2 = T
320  GOSUB 440
330  F2 = EXP (Z2 - 1 - LOG (Z2 -
      Z2 * H2) - AL * LOG (1 + H2
      ) / BL / R / T ∧ (3 / 2))
340  F = (F2 - F1) / (T2 - T1) * (
      TK - T1) + F1
350  PRINT : PRINT
360  PRINT "LIQUID PHASE AT TEMPE
      RATURE "; TK
370  PRINT
380  PRINT "THE HYPOTHETICAL GAS-
      PHASE"
390  PRINT "FUGACITY COEFFICIENT:
      "; F
400  PRINT
410  INPUT "NEXT CALCULATION ? (Y
      / N) "; A$
420  IF A$ = "Y" THEN GOTO 30
430  END
440  REM :CALCULATION OF COMPRESS
      IBILITY
450  FACTOR BY REDLICH-KWONG
460  REM :EQUATION OF STATE
470  R = 8.3144
480  CA - 0.4278
490  CB = 0.867
500  AL = CA * R * R * TC ∧ 2.5 /
      PC
510  BL = CB * R * TC / PC
520  B = BL / R / T
530  A2 = AL / R / R / T ∧ 2.5
540  Z1 = 1
550  I = 0
560  H2 = B * P / Z1
570  I = I + 1
580  IF I –150 THEN PRINT : PRINT
```

```
         "ITERATION NOT CONVERGENT ":
         GOTO 630
590   Z2 = 1 / (1 − H2) − A2 * H2 /
         B / (1 + H2)
600   IF ABS (Z2 − Z1) -1E − 5 THEN
         GOTO 630
610   Z1 = Z2
620   GOTO 560
630   RETURN
640   REM :READING OF SPECIFIED
      RECORD OF PHDAT FILE INTO AR
      RAY A(15)
650   REM :ARRAY A HAS TO BE DECLA
      RED IN THE CALLING PROGRAM
660   D$ = CHR$ (4)
670   PRINT D$; "OPEN PHDAT, D1"
680   PRINT D$; "POSITION PHDAT, R";
      IS − 1
690   PRINT D$; "READ PHDAT"
700   INPUT A(1), A(2), A(3), A(4), A(5),
         A(6), A(7), A(8), A(9), A(10),
         A(11), A(12), A(13), A(14), A(15)
710   PRINT D$; "CLOSE PHDAT"
720   PRINT D$; "IN#0"
730   RETURN
740   PC = A(2)
750   TC = A(3)
760   OM = A(4)
770   TB = A(5)
780   TR = T / TC
790   PS = 5.92714 − 6.09648 / TR −
         1.28862 * LOG (TR) + 0.1693
         47 * TR ∧ 6 + OM * (15.2518 −
         15.6875 / TR − 13.4712 * LOG
         (TR) + 0.43577 * TR ∧ 6)
800   PS = EXP (PS) * PC
810   RETURN
```

Example. Run the program for butane. Calculate the hypothetical vapor fugacity coefficient at temperature 374 K and pressure 20×10^5 Pa. Repeat the calculation at temperature 390 K and pressure 20.265×10^5 Pa.

```
]RUN

NTER SEQ. NO. OF THE COMPONENT 4

ENTER PRESSURE (PA) 20E5

ENTER TEMPERATURE (K)
WHERE YOU WOULD LIKE TO COMPUTE
THE FUGACITY COEFFICIENT 374
```

```
LIQUID PHASE AT TEMPERATURE 374

THE HYPOTHETICAL GAS-PHASE
FUGACITY COEFFICIENT: .748202201

NEXT CALCULATION ? (Y / N) Y

ENTER SEQ. NO. OF THE COMPONENT 4

ENTER PRESSURE (PA) 20.265E5

ENTER TEMPERATURE (K)
WHERE YOU WOULD LIKE TO COMPUTE
THE FUGACITY COEFFICIENT 390

REAL GAS PHASE AT TEMPERATURE 390

THE FUGACITY COEFFICIENT .771657576
NEXT CALCULATION ? (Y / N) N
```

Remark. Explain the results.

Liquid Mixtures

In the case of real liquid mixtures: for the interpretation and explication of experimental evidence, special mixture models containing interaction parameters are needed.

In the laboratory, volume effects manifested during mixing can be directly measured with a dilatometer, and thermal effects with a calorimeter. Indirectly (from vapor–liquid or liquid–liquid equilibrium measurements), the molar excess free enthalpy of mixing can also be determined. This means that terms in the defining equation of the molar excess free enthalpy of mixing can be separately evaluated by a suitable series of experiments:

$$g^{E} = u^{E} + Pv^{E} - Ts^{E}$$

The excess of entropy mixing, of particular importance for model development, also becomes accessible, since g^E, u^E, and v^E can be separately measured.

On the right-hand side of the equation, energetic interaction between the molecules is comprised in the term u^E; nonideality because of the shape and arrangement of the molecules, in the terms Pv^E and Ts^E.

Figure 11.6 shows four cases. In each of these h^E and Ts^E are of the same order of magnitude. In case *a* the deviation of g^E is negative, in the other cases positive. The g^E-versus-x curves are of quasiparabolic shape, but not necessarily symmetric.

If the energy effect and the entropy effect compensate one another, the mixture behaves as a quasi-ideal mixture (Figure 11.6*c*). There are also cases where s^E is negative (Figure 11.6*d*), that is, the mixing entropy is lower than in the ideal mixture (but of course positive). This means that orientation disorder is higher in the pure component than in the mixture.

CALCULATION OF THERMODYNAMIC EXCESS QUANTITIES

There are several possibilities for the description of thermodynamic excess quantities, but none of the theoretical or empirical expressions elaborated so far can be put to general use. The following must be considered for each problem when choosing from the existing models:

1. Simplicity of the mathematical form of the equation for further handling.
2. Data available for the determination of the parameters of the equation.
3. In the case of multicomponent mixtures, the possibility of estimation from data on binary mixtures.
4. Restrictions on the accuracy of the relationship.

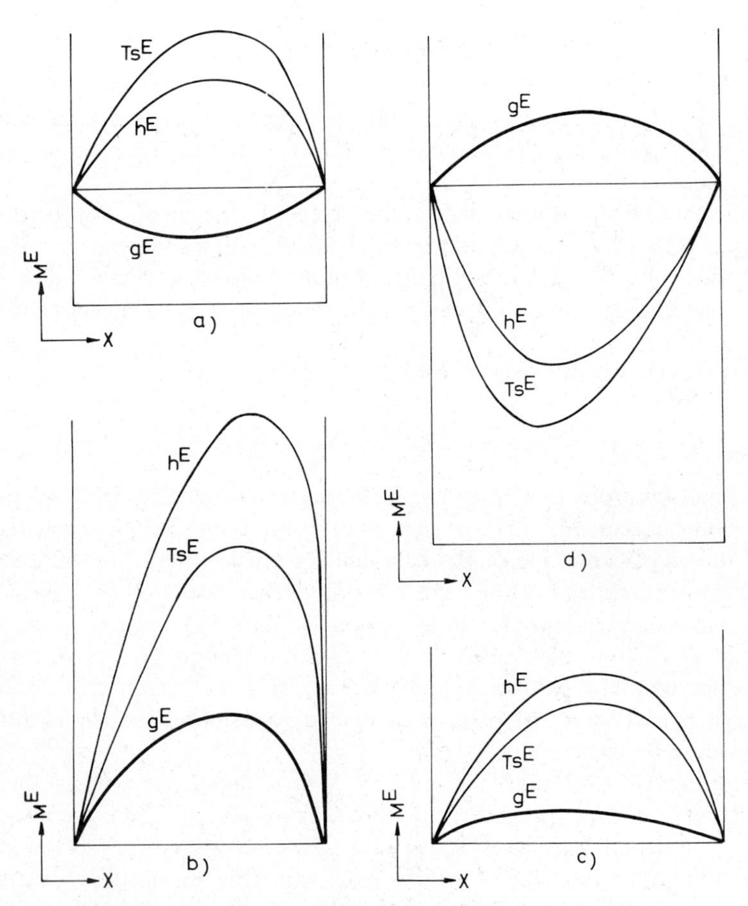

FIGURE 11.6. Excess quantities for four different binary mixtures.

5. Intelligibility of the physical content of the parameters.
6. Unusual nonidealities, as for example segregation.

Programs WRVLEDAT and REVLEDAT

In the following programs for calculating activity coefficients of components of binary mixtures, some special files are used. The name of each file begins with the letter D. The content of these files is the following:

1. First record:

T, NO, S1, S2

where T = temperature of measurements (K),
NO = number of measurements stored in the file, (the number of records of the file is NO + 1),
S1 = sequential number of the first component of the mixture in file PHDAT,
S2 = sequential number of the second component of the mixture in file PHDAT.

2. Second to (NO + 1)th

X, G1, G2, GE

where X = molar fraction of the first component in the mixture,
G1 = activity coefficient of the first component in the mixture at concentration $(X, (1 - X))$,
G2 = activity coefficient of the second component in the mixture at concentration $(X, (1 - X))$,
GE = $\Delta G_{mix}/(\mathbf{R}TX(1 - X))$.

For the creation of such files use program WRVLEDAT.

Listing of Program WRVLEDAT.

```
]LOAD WRVLEDAT
]LIST

 10  REM :EVALUATION OF DATAFILES
      CONTAINING VLE DATA
 20  REM :THE DATA ARE X,GAMMA1,GA
      MMA2,GE / R / T / X1 / X2
 30  REM :IN THE FIRST RECORD THER
```

```
       E ARE THE DATA:T,NO. OF
       MEASUREMENTS, SEQUENTIAL NUMB
       ER OF COMPOUNDS IN THE TWO-
       COMPONENT MIXTURE
  40   PRINT : PRINT
  50   INPUT "ENTER THE NAME OF THE
       FILE "; A$
  60   D$ =  CHR$ (4)
  70   PRINT D$; "OPEN "; A$
  80   PRINT D$; "CLOSE "; A$
  90   PRINT : PRINT
 100   INPUT "ENTER TEMPERATURE (K)
       "; T
 110   INPUT "ENTER NO. OF POINTS ";
       NO
 120   INPUT "ENTER SEQ. NO. 1 "; S1
 130   INPUT "ENTER SEQ. NO. 2 "; S2
 140   PRINT D$; "APPEND "; A$
 150   PRINT D$; "WRITE"; A$
 160   PRINT T; ";"; NO; ","; S1; ","; S2
 170   PRINT D$; "PR#0"
 180   FOR I  =  1 TO NO
 190   INPUT "X =  "; X
 200   INPUT "G1 = "; G1
 210   INPUT "G2 =  "; G2
 220   INPUT "GE / (RTX1X2) =  "; GE
 230   PRINT D$; "APPEND "; A$
 240   PRINT D$; "WRITE "; A$
 250   PRINT X; ","; G1; ","; G2; ","; GE
 260   PRINT D$; "PR#0"
 270   PRINT
 280   NEXT I
 290   PRINT D$; "CLOSE "; A$
 300   INPUT "NEXT DATAFILE ? (Y / N)
       "; C$
 310   IF C$ ="Y" THEN GOTO 10
 320   END
```

For reading the content of the files use program REVLEDAT.

Listing of Program REVLEDAT.

```
]LOAD REVLEDAT
]LIST

 10   INPUT "ENTER INDENT "; KK
 20   INPUT "ENTER THE FILENAME "; A
      $
 30   D$ =  CHR$ (4)
```

```
 40  PRINT D$; "OPEN "; A$
 50  PRINT D$; "READ "; A$
 60  INPUT T,NO,S1,S2
 70  PRINT
 80  PRINT D$; "PR#"; KK
 90  PRINT T; TAB(10); NO: TAB(15);
      S1; TAB(25); S2
100  PRINT
110  FOR I = 1 TO NO
120  PRINT D$; "READ "; A$
130  INPUT X,G1,G2,GE
140  PRINT D$; "PR#"; KK
150  PRINT X; TAB(7); G1; TAB(20
     );G2; TAB(32);GE
160  NEXT I
170  INPUT "NEXT READING ? (Y / N)
     "; C$
180  IF C$ = "Y" THEN GOTO 20
190  END
```

Sample run:

```
]RUN
ENTER INDENT 1
ENTER THE FILENAME D16

288.2 21 10 11

        0       1.0120955    .99999988    0
        .05     1.0112247    1.0000105    0
        .1      1.0103512    1.0000868    .011967
        .15     1.0094719    1.0002003    .012318
        .2      1.0086126    1.0003901    .012434
        .25     1.0077658    1.0006361    .012678
        .3      1.006938     1.0009336    .012866
        .35     1.0061378    1.0013208    .012996
        .4      1.0053549    1.001792     .013194
        .45     1.0046062    1.0023403    .013386
        .5      1.0038996    1.002986     .013559
        .55     1.0032368    1.0037184    .013756
        .6      1.002615     1.0045576    .013938
        .65     1.0020504    1.0055075    .014117
        .7      1.001543     1.0065746    .014311
        .75     1.0010891    1.0077553    .014511
        .8      1.0007153    1.009067     .014664
        .85     1.0004044    1.0105267    .014868
        .9      1.0001793    1.0121078    .015024
        .95     1.0000467    1.0138636    .015175
        1       1            1.0157557    .015438

NEXT READING ? (Y / N) Y
```

ENTER THE FILENAME D17

288.2 21 10 12

0	1.406744	1	0
.05	1.2997332	1.0019522	0
.1	1.2217941	1.006918	.315166
.15	1.1641464	1.0138483	.291709
.2	1.1210852	1.0219421	.270674
.25	1.0887146	1.0306358	.251559
.3	1.0643358	1.0394917	.234186
.35	1.0459967	1.0481968	.21832
.4	1.032295	1.0565004	.203807
.45	1.0220909	1.0642385	.190461
.5	1.0146437	1.07129	.178198
.55	1.0093012	1.0775404	.16691
.6	1.0055676	1.082921	.156459
.65	1.0030622	1.08741	.146746
.7	1.0014668	1.0909977	.137746
.75	1.0005293	1.093648	.129387
.8	1.0000629	1.0953951	.121554
.85	.99989909	1.096241	.114283
.9	.99989206	1.0961943	.107499
.95	.99996018	1.0953178	.101035
1	1	1.0936069	.095102

NEXT READING ? (Y / N) Y
ENTER THE FILENAME D18

288.2 21 10 7

0	1.4235525	1	0
.05	1.3992939	1.0004263	0
.1	1.3746901	1.0018654	.362409
.15	1.3497715	1.0044985	.372465
.2	1.3246069	1.0085325	.383036
.25	1.2992411	1.0141993	.394132
.3	1.2737408	1.0218887	.405701
.35	1.2482071	1.0319328	.4181
.4	1.2227173	1.0447998	.431179
.45	1.1973982	1.0610924	.444977
.5	1.1723795	1.0816078	.45962
.55	1.1478243	1.107275	.475276
.6	1.1239262	1.1392832	.491969
.65	1.1009111	1.1793356	.509732
.7	1.079072	1.2295933	.528794
.75	1.0587378	1.2929735	.549289
.8	1.0403538	1.3735924	.571272
.85	1.0244627	1.4773512	.594976
.9	1.0117731	1.6127377	.620642
.95	1.0032034	1.7927256	.6485
1	1	2.0373554	.6678866

NEXT READING ? (Y / N) Y

ENTER THE FILENAME D8

288.2 21 16 15

0	3.5082064	1	0
.05	2.7270794	1.0062943	0
.1	2.227457	1.0227671	1.920766
.15	1.8919964	1.046752	1.850583
.2	1.6580324	1.076354	1.798448
.25	1.4899511	1.1101427	1.760706
.3	1.3664885	1.1470671	1.734791
.35	1.2742119	1.186228	1.718906
.4	1.2044344	1.2268858	1.711893
.45	1.1512699	1.2684183	1.713055
.5	1.1106167	1.3102589	1.721858
.55	1.0795612	1.3518972	1.738101
.6	1.055933	1.3928719	1.761851
.65	1.0381536	1.4327688	1.793233
.7	1.0249634	1.4711962	1.832767
.75	1.0154381	1.5078096	1.88108
.8	1.0087652	1.5422544	1.939179
.85	1.0043678	1.5742617	2.008399
.9	1.0017071	1.6035547	2.090454
.95	1.0003614	1.6298809	2.188083
1	1	1.6530495	2.304569

NEXT READING ? (Y / N) Y
ENTER THE FILENAME D5

288.2 21 13 10

0	9.4291315	1	0
.05	5.506855	1.0131187	2.057786
.1	3.7581615	1.044486	1.907517
.15	1.78292	1.08745548	1.78292
.2	2.2722769	1.1387806	1.676854
.25	1.9139776	1.1966953	1.58486
.3	1.6695538	1.2601662	1.504009
.35	1.4959164	1.3283806	1.431842
.4	1.368906	1.4008455	1.366929
.45	1.2740431	1.4770489	1.307972
.5	1.2021799	1.5565376	1.254013
.55	1.1473141	1.6388645	1.204354
.6	1.1053104	1.7235126	1.15834
.65	1.0732708	1.8099556	1.115526
.7	1.0490694	1.8975296	1.07545
.75	1.031126	1.9856224	1.037854
.8	1.0182209	2.0734797	1.002468
.85	1.0093775	2.1602602	.969004
.9	1.0038042	2.2451353	.937206
.95	1.0008342	2.40345	.897685
1	1	2.73585	.765421

NEXT READING ? (Y / N) Y

ENTER THE FILENAME D4

288.2 21 14 10

0	6.743893	1	0
.05	4.0228348	1.0124054	0
.1	2.8577223	1.0403118	1.71294
.15	2.2429457	1.0765753	1.562911
.2	1.8750753	1.118	1.443175
.25	1.635962	1.1629429	1.344393
.3	1.4713974	1.2104912	1.26094
.35	1.353425	1.2600517	1.18925
.4	1.2662849	1.3112431	1.126763
.45	1.2005463	1.3638325	1.071612
.5	1.150198	1.4176092	1.022536
.55	1.1112576	1.4724874	.978446
.6	1.0810556	1.5283613	.938587
.65	1.0575752	1.5851669	.902422999
.7	1.0395002	1.6428967	.869255
.75	1.0257359	1.7015295	.838907
.8	1.0155315	1.7610674	.810869
.85	1.0082722	1.8215466	.784969
.9	1.0034857	1.8829918	.760927
.95	1.0008211	1.9454479	.738454
1	1	2.008997	.717399

NEXT READING ? (Y / N) Y
ENTER THE FILENAME D2

298.2 21 22 23

0	1.4762	1	9.53E − 07
.05	1.4428	1.0006	.018861
.1	1.4098	1.0024	.0365
.15	1.3772	1.0058	.0529
.2	1.3451	1.0109	.0679
.25	1.3136	1.0179	.0814
.3	1.2826	1.0271	.09341
.35	1.2524	1.039	.103
.4	1.2231	1.0539	.11206
.45	1.1946	1.0724	.118
.5	1.1672	1.0952	.122
.55	1.1411	1.123	.124
.6	1.1163	1.157	.124
.65	1.0931	1.1983	.121
.7	1.0716	1.2487	.115
.75	1.0522	1.3105	.105
.8	1.0352	1.3865	.0929
.85	1.0209	1.4806	.0764
.9	1.0098	1.5982	.0556
.95	1.0026	1.7469	.0303
1	1	1.9374	0

NEXT READING ? (Y / N) N

Calculation of Excess Free Enthalpy of Mixing from the Equation of State: The van Laar Equation

This theory starts from the assumptions that:

1. The molecules are rigid spheres.
2. The molecules are of near-identical size, as in the model of the ideal mixture.
3. However, attractive forces act between them.

It follows from the first two conditions that:

1. Mixing is not accompanied by change in volume: $v^E = 0$.
2. The excess entropy of mixing is zero: $s^E = 0$.

It follows from these two conditions that the molar excess free enthalpy of mixing is equal to the internal energy of mixing:

$$g^E = u^E$$

This energy must be calculated, and van Laar postulated that this is indeed possible with an equation of state taking into consideration the attraction between the molecules. To that end he constructed the following cycle (Figure 11.7):

a. The two pure components in the liquid state isothermally evaporate at pressure P into a space of arbitrarily low pressure. Here both components behave as ideal gases.

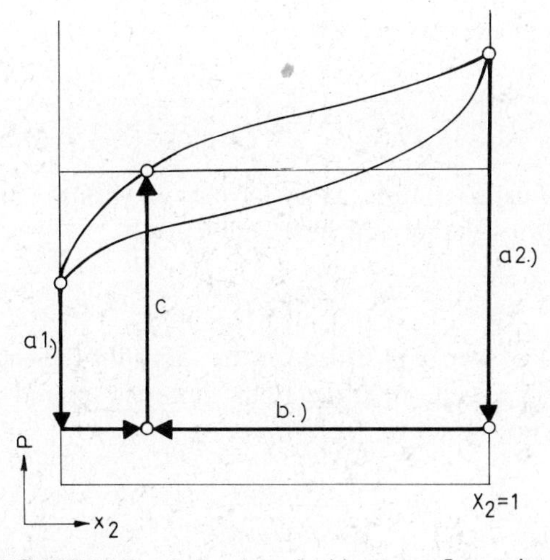

FIGURE 11.7. Formation of a binary van Laar mixture.

b. In the low-pressure space the pure components mix, and an ideal mixture is formed.

c. The mixture is compressed to pressure P in an isothermal process. In this way the liquid mixture in the prescribed state is produced.

The change in energy of the cycle—that is, the energy of mixing sought—is

$$u^{\mathrm{E}} = \Delta u_{\mathrm{a}} + \Delta u_{\mathrm{b}} + \Delta u_{\mathrm{c}}$$

For the calculation of step **a** the following known thermodynamic relationship is used:

$$\left(\frac{\partial u}{\partial v}\right)_T = T\left(\frac{\partial P}{\partial T}\right)_v - P$$

For the equation of state needed to perform concrete calculations, van Laar used the van der Waals equation. In this case

$$\frac{\partial u}{\partial v} = \frac{a}{v^2}$$

The mole fractions in the mixture are x_1, x_2. The change in internal energy per mole of component is

$$x_1 \Delta u_{\mathrm{a}1} = \int_{v_1^{(1)}}^{\infty} \frac{a_1 x_1}{v^2}\, dv = \frac{a_1 x_1}{v_1^{(1)}}$$

$$x_2 \Delta u_{\mathrm{a}2} = \int_{v_2^{(1)}}^{\infty} \frac{a_2 x_2}{v^2}\, dv = \frac{a_2 x_2}{v_2^{(1)}}$$

Sufficiently far from the critical point, the liquid molar volume can be replaced by the constant b of the van der Waals equation, so that in step **a** the change in internal energy of the mixture is

$$\Delta u_{\mathrm{a}} = \frac{a_1 x_1}{b_1} + \frac{a_2 x_2}{b_2}$$

Step **b** does not cause trouble, as we have to do with the mixing of ideal-gas components, and thus the mixing is thermoneutral:

$$\Delta u_{\mathrm{b}} = 0$$

The calculation of step **c** is based on the fact that the relationship used in step **a** is valid also for the mixture, if the latter is treated as a pseudo-one-component substance. Then we have

$$\Delta u_{\mathrm{c}} = -\frac{a_{\mathrm{mixture}}}{b_{\mathrm{mixture}}} = -\frac{a_{12}}{b_{12}}$$

But now we must say something about the derivation of the parameters a_{mixture} and b_{mixture}. van Laar used the following mixing rules:

$$\sqrt{a_{12}} = x_1\sqrt{a_1} + x_2\sqrt{a_2}$$

$$b_{12} = x_1 b_1 + x_2 b_2$$

The first rule states that the parameter a_{12}, characteristic of the interaction, is to be treated as a geometric mean; the second rule states that mixing proceeds without change in volume. After substitution and rearrangement, we have for the molar excess free enthalpy of mixing

$$g^{\mathrm{E}} = \frac{x_1 x_2 b_1 b_2}{x_1 b_1 + x_2 b_2}\left(\frac{\sqrt{a_1}}{b_1} + \frac{\sqrt{a_2}}{b_2}\right)^2$$

where the following notation is introduced:

$$A_{12} = \frac{b_1}{\mathbf{R}T}\left(\frac{\sqrt{a_1}}{b_1} - \frac{\sqrt{a_2}}{b_2}\right)^2$$

$$A_{21} = \frac{b_2}{\mathbf{R}T}\left(\frac{\sqrt{a_2}}{b_1} - \frac{\sqrt{a_2}}{b_2}\right)^2$$

Algebraic transformation yields the following expression for the molar excess free enthalpy of mixing:

$$\frac{x_1 x_2}{g^{\mathrm{E}}} = \frac{1}{\mathbf{R}T}\left(\frac{x_1}{A_{21}} + \frac{x_2}{A_{12}}\right)$$

For the activity coefficients we obtain

$$\ln \gamma_1 = \frac{A_{12} x_2^2}{\left(\dfrac{A_{12}}{A_{21}} x_1 + x_2\right)^2}$$

$$\ln \gamma_2 = \frac{A_{21} x_1^2}{\left(x_1 + \dfrac{a_{21}}{A_{12}} x_2\right)^2}$$

The conclusions are:

1. The logarithm of the activity coefficient is inversely proportional to temperature. This follows from the fact that $s^{\mathrm{E}} = 0$.

2. The activity coefficient is always larger than one; the theory interprets only mixtures of positive deviation. This is a consequence of the mixing rule, because

$$a_{12} > x_1 a_1 + x_2 a_2$$

If the parameter a of the mixing rule were also a linear combination, then the activity coefficient would be $\gamma_1 = \gamma_2 = 1$.

It follows from the van der Waals equation of state that P_c is proportional to the quotient $\sqrt{a}/b$. A further consequence of this is that the nonideality increases with the difference between the critical pressures of the two components. A mixture of components of identical critical pressure should be ideal. Regrettably this is not the case. This inadequacy in the predictions does not arise from van Laar's argument or from the simplifications used by him, but from the limitations of the van der Waals equation of state and the mixing rules just used. Remember that in the introduction of the concept of a pseudo-one-component phase in Chapter 8, other empirical mixing rules were recommended.

These critical remarks are theoretically rather grave, but experience shows that the van Laar equation can nevertheless be used effectively in a lot of cases. Therefore, it is best to consider A_{12} and A_{21} as parameters without physical content, to be fitted to experimental data. In this case, however, the use of another form of function is also justified, and no objection can be raised against introducing another variable instead of the mole fraction.

Algorithm and Program: Calculation of the Activity Coefficient with the van Laar Equation

Write a program for the calculation of the activity coefficient with the van Laar equation. Estimate the parameters of the equation from the measured activity coefficients at $x = 0$ and $x = 1$. Compare the calculated activity coefficients with the measured ones.

Algorithm.

1. Read the name of the datafile containing the datafile sequential numbers of the compounds of the mixture, the temperature, and the measured activity coefficients.
2. Read the first record of the measurement file, which contains sequential numbers, the temperature, and the numbers of measurements.
3. Read the second and the last record of the measurement file (γ_1 at $x_1 = 0$ and γ_2 at $x = 1$).
4. Calculate the parameters in the van Laar equation.
5. Calculate the activity coefficients with the van Laar equation.

6. Print the calculated and measured data.

Program VLEVANLAAR. A possible realization:

```
]PR#1
]LOAD VLENVANLAAR
]LIST

 10  REM :CALCULATION OF PARAMETER
      S OF VAN LAAR EQUATION FROM
      ACTIVITY COEFFICIENTS AT X1 =
      0 AND X1 = 1
 20  REM :COMPARISON OF CALCULATED
      WITH MEASURED RESULTS.
      THE MEASURED DATA RE IN FILE
      D18, BUT YOU CAN GIVE ANOT
      HER FILE NAME INSTEAD.
 30  PRINT : PRINT
 40  PRINT "ENTER THE NAME OF THE
      FILE CONTAINING"
 50  INPUT "MEASURED DATA (EQ. D17)
      "; A$
 60  PRINT
 70  N = 1
 80  GOSUB 700
 90  PRINT "SEQ. NO. OF COMPONENTS
      OF MIXTURE "; C1; TAB(39); C2
100  PRINT
110  N = 2
120  GOSUB 700
130  A1 = LOG (G1)
140  N = NO + 1
150  GOSUB 700
160  A2 = LOG (G2)
170  DE = 0
180  PRINT " X GAMMA1(MES) GA
     MMA1(CALC) DEV(%)"
190  PRINT "=======================
      ===================="
200  FOR N = 2 TO NO + 1 STEP 2
210  GOSUB 700
220  GC = EXP (A1 * (1 - X) ∧ 2 /
      (A1 / A2 * X + 1 - X) ∧ 2)
230  DE = (G1 - GC) ∧ 2 + DE
240  DD = (G1 - GC) / G1 * 100
250  D% = DD * 100
260  DD = D% / 100
270  PRINT X; TAB(6); G1; TAB(20
      ); GC; TAB(34);DD
```

```
280  NEXT N
290  PRINT
300  PRINT "======================
     ===================== "
310  DE = SQR (DE) / (NO - 1)
320  PRINT
330  INVERSE
340  INPUT "TO CONTINUE PRESS
     'RETURN' "; Q$
350  PRINT
360  NORMAL
370  PRINT "STANDARD DEVIATION OF
      GAMMA1"
380  PRINT "CALC. BY VAN LAAR EQ
     . "; DE
390  PRINT
400  PRINT "CONSTANTS OF VAN LAAR
      EQUATION"
410  PRINT "A12, A21 ARE "; A1; TAB(
      25);A2
420  PRINT
430  INVERSE
440  INPUT "TO CONTINUE PRESS
     'RETURN' "; Q$
450  PRINT
460  NORMAL
470  DE = 0
480  PRINT : PRINT
490  PRINT "X GAMMA2(MES) GA
     MMA2(CALC) DEV(%)"
500  PRINT "======================
     ===================== "
510  FOR N = 2 TO NO + 1 STEP 2
520  GOSUB 700
530  GC = EXP (A2 * X ∧ 2 / (A2 /
      A1 * (1 - X) + X) ∧ 2)
540  DE = (G2 - GC) ∧ 2 + DE
550  DD = (G2 - GC) / G2 * 100
560  D% = DD * 100
570  DD = D% / 100
580  PRINT X; TAB(6);G2; TAB(20
     );GC; TAB(34);DD
590  NEXT N
600  PRINT
610  PRINT "======================
     ===================== "
620  DE = SQR (DE) / (NO - 1)
630  PRINT
640  PRINT "STANDARD DEVIATION OF
      GAMMA2"
```

```
650  PRINT "CALC. BY VAN LAAR EQ
     . "; DE
660  PRINT
670  INPUT "NEXT CALCULATION ? (Y
     / N) "; C$
680  IF C$ = "Y" THEN GOTO 30
690  END
700  REM :READING ROUTINE FOR VLE
      DATAFILES. THE NAMES OF THE
      FILES HAVE TO BE IN VARIABLE
      "A$". THE ROUTINE READS T,
      NO, C1, AND C2 WHEN N = 1. IF N
      > 1 AND  < NO, READS THE CORRESP
      ONDING X, G1, G2, AND GE / R / T /
      X1 / X2
710  D$ =  CHR$ (4)
720  PRINT D$; "OPEN "; A$
730  PRINT D$; "POSITION "; A$; ", R"
     ;N - 1
740  PRINT D$;"READ "; A$
750  IF N  =  1 THEN INPUT T,NO,C1
     ,C2
760  IF N  < >  1 THEN INPUT X,G1
     ,G2,GE
770  PRINT D$; "CLOSE "; A$
780  PRINT D$; "IN#0"
790  RETURN
```

Example. Run the program for datafile D18, which contains the measured data on a chloroform–heptane mixture. Repeat the calculation for datafile D17, containing the measured data on chloroform–1,2-dichloroethane.

```
]RUN
ENTER THE NAME OF THE FILE CONTAINING
MEASURED DATA (EQ. D17) D17

SEQ. NO. OF COMPONENTS OF MIXTURE 10 12
```

X	GAMMA1(MES)	GAMMA1(CALC)	DEV(%)
0	1.406744	1.406744	−.01
.1	1.2217941	1.1833569	3.14
.2	1.1210852	1.09355143	2.45
.3	1.0643358	1.0503982	1.3
.4	1.032295	1.02756594	.45
.5	1.0146437	1.01483562	−.02
.6	1.0055676	1.00758386	−.21
.7	1.0014668	1.00348868	−.21
.8	1.0000629	1.00129232	−.13
.9	.99989206	1.00027352	−.04
1	1	1	0

TO CONTINUE PRESS 'RETURN'

STANDARD DEVIATION OF GAMMA1
CALC. BY VAN LAAR EQ. 2.48092461E − 03

CONSTANTS OF VAN LAAR EQUATION
A12,A21 ARE .341277815 .0894813166

TO CONTINUE PRESS 'RETURN'

X	GAMMA2(MES)	GAMMA2(CALC)	DEV(%)
0	1	1	0
.1	1.006918	1.00795866	−.11
.2	1.0219421	1.02154662	.03
.3	1.0394917	1.0350443	.42
.4	1.0565004	1.04717322	.88
.5	1.07129	1.05777408	1.26
.6	1.082921	1.06698255	1.47
.7	1.0909977	1.07499499	1.46
.8	1.0953951	1.08199959	1.22
.9	1.0961943	1.08815863	.73
1	1.0936069	1.0936069	−.01

STANDARD DEVIATION OF GAMMA2
CALC. BY VAN LAAR EQ. 1.61618973E − 03

NEXT CALCULATION ? (Y / N) Y

ENTER THE NAME OF THE FILE CONTAINING
MEASURED DATA (EQ. D17) D18

SEQ. NO. OF COMPONENTS OF MIXTURE 10 7

X	GAMMA1(MES)	GAMMA1(CALC)	DEV(%)
0	1.4235525	1.4235525	0
.1	1.3746901	1.37329106	.1
.2	1.3246069	1.32247167	.16
.3	1.2737408	1.27143485	.18
.4	1.2227173	1.22066721	.16
.5	1.1723795	1.17086919	.12
.6	1.1239262	1.12306503	.07
.7	1.079072	1.07879016	.02
.8	1.0403538	1.04043113	−.01
.9	1.0117731	1.01188941	−.02
2	1	1	0

TO CONTINUE PRESS 'RETURN'

STANDARD DEVIATION OF GAMMA1
CALC. BY VAN LAAR EQ. 2.18851809E − 04

```
CONSTANTS OF VAN LAAR EQUATION
A12,A21 ARE .353155508 .711652595

TO CONTINUE PRESS 'RETURN'
```

X	GAMMA2(MES)	GAMMA2(CALC)	DEV(%)
0	1	1	0
.1	1.0018654	1.00194528	−.01
.2	1.0085325	1.00870658	−.02
.3	1.0218887	1.02213001	−.03
.4	1.0447998	1.04495935	−.02
.5	1.0816078	1.08142668	.01
.6	1.1392832	1.13836069	.08
.7	1.2295933	1.22740751	.17
.8	1.3735924	1.3698502	.27
.9	1.6127377	1.60815591	.28
1	2.0373554	2.0373554	0

```
STANDARD DEVIATION OF GAMMA2
CALC. BY VAN LAAR EQ. 3.19291193E − 04

NEXT CALCULATION ? (Y / N) N
```

Remark. Explain the results.

Calculation of Excess Free Enthalpy of Mixing from Experimental Data: Wohl's Treatment

The model of the ideal mixture involves the two assumptions that the molecules are spherical and that they are of substantially identical size. As long as these two assumptions are true, the mole fraction and volume fraction are equal. Otherwise, the mole fraction and volume fraction are not equal. This has two consequences:

1. The configuration entropy differs from that of the ideal mixture.
2. The number of neighbors of a given molecule becomes important, since interaction is possible between the molecules.

Let us use for the calculation of the excess free enthalpy of mixing the following relationship:

$$\frac{g^{\mathrm{E}}}{\mathbf{R}T} \cdot \frac{1}{\Sigma_i q_i x_i} = \sum_{ij} \phi_i \phi_j a_{ij} + \sum_{ijk} \phi_i \phi_j \phi_k a_{ijk} + \sum_{ijkl} \phi_i \phi_j \phi_k \phi_l a_{ijkl}$$

where the parameter q_i is the *effective mole volume* of component i, proportional to the number of neighboring molecules in contact with a molecule of i.

The effective mole volume presumably increases with the size of the molecule, but need not be strictly proportional to it. The variable x_i is the mole fraction, and the variable ϕ_i is the effective volume fraction:

$$\phi_i = \frac{q_i x_i}{\Sigma q_i x_i}$$

Actually the quotient of the effective molar volumes is needed, because

$$\phi_1 = \frac{x_1}{x_1 + \dfrac{q_2}{q_1}x_2 + \dfrac{q_3}{q_1}x_3 + \cdots}$$

$$\phi_2 = \frac{x_2}{x_1 + \dfrac{q_2}{q_1}x_2 + \dfrac{q_3}{q_1}x_3 + \cdots}$$

Finally, the Wohl equation contains the empirical constants a_{ij}, a_{ijk}, and so on. These measure the interactions of the groups of molecules *ij*, *ijk*, and so on. Each group must contain at least two types of molecules, so that, for example, $a_{ii} = 0$, but $a_{iij} \neq 0$. An investigation of the equation shows that each new term takes into account interactions which were not taken into account by the preceding terms. (The equation is similar in this respect to the virial equation for gases.) Equations of higher degree than four are not used in practice. Some approximate physical content can be assigned to the parameters of the equation, and thus Wohl's representation can be systematically extended to multicomponent systems.

On the other hand, *a priori* assumptions may be made with respect to the ratio of effective molar volumes, and distortions arising from this can then be compensated for in the estimation of the anyway empirical coefficients. The coefficients are mixture-specific; there is no way to transfer them to other systems.

For a binary mixture the second-degree form of the Wohl equation is the following:

$$\frac{g^{\mathrm{E}}}{\mathbf{R}T} = (q_1 x_1 + q_2 x_2)(A_0 \phi_1 \phi_2)$$

where $A_0 = a_{ij}$. The reduction

$$\frac{q_2}{q_1} = 1$$

is due to Margules. Using the definition of volume fraction, we obtain for this case that the volume fraction is equal to the molar fraction:

$$\phi_i = \frac{x_i}{x_1 + (q_2/q_1)x_2} = \frac{x_i}{x_1 + x_2} = x_i$$

whence the Margules equation of second degree:

$$\frac{g^E}{\mathbf{R}T} = (x_1 + x_2)(A_0 x_1 x_2)$$

Dividing both sides by the product $x_1 x_2$ gives

$$\frac{g^E}{\mathbf{R}T x_1 x_2} = A_0$$

This means that the quantity $g^E/\mathbf{R}Tx_1x_2$ is constant over the whole concentration range. Binary mixtures meeting this criterion are called simple regular mixtures. The Margules equation of second degree gives for the activity coefficients

$$\ln \gamma_1 = \frac{1}{\mathbf{R}T}\frac{\partial g^E}{\partial x_1} = \frac{A_0}{\mathbf{R}T} x_2^2$$

$$\ln \gamma_1 = \frac{1}{\mathbf{R}T}\frac{\partial g^E}{\partial x_2} = \frac{A_0}{\mathbf{R}T} x_1^2$$

For binary mixtures the Wohl equation of third degree has the following form:

$$\frac{g^E}{\mathbf{R}T} = q_1 x_1 - q_2 x_2$$

since

$$a_{12} = a_{21}$$

$$a_{111} = a_{121} = a_{211}$$

$$a_{122} = a_{212} = a_{221}$$

New constants are introduced:

$$A_{12} = q_1(2a_{12} + 3a_{112})$$

$$A_{21} = q_2(2a_{12} + 3a_{122})$$

Substituting these, we have for molar excess free enthalpy of mixing

$$\frac{g^{\mathrm{E}}}{\mathbf{R}T} = A_{12}x_1\phi_2^2 + A_{21}x_2\phi_1^2$$

First of all, using the definition of volume fraction, we rewrite the Wohl equation of third degree:

$$\frac{g^{\mathrm{E}}}{\mathbf{R}T} = A_{12}x_1\left(\frac{x_2}{x_1 + (q_2/q_1)x_2}\right)^2 + A_{21}x_2\left(\frac{x_1}{x_1 + (q_2/q_1)x_2}\right)^2$$

From this we obtain, by the substitution $q_2/q_1 = A_{21}/A_{12}$, the van Laar equation:

$$\frac{\mathbf{R}Tx_1x_2}{g^{\mathrm{E}}} = \left(\frac{x_1}{A_{21}} + \frac{x_2}{A_{12}}\right)$$

The Margules equation of third degree is obtained by the substitution $q_2/q_1 = 1$. Now the volume fractions become equal to the molar fractions:

$$\frac{g^{\mathrm{E}}}{\mathbf{R}T} = A_{12}x_1x_2^2 + A_{21}x_2x_1^2$$

Dividing both sides by the product x_1x_2 gives

$$\frac{g^{\mathrm{E}}}{\mathbf{R}Tx_1x_2} = A_{12}x_2 + A_{21}x_1$$

This means that the quantity $g^{\mathrm{E}}/\mathbf{R}Tx_1x_2$ is a linear function of the variable x_1, characteristic of the composition. If this does not lead to the desired results in the processing of experimental data, further terms can be introduced on the right-hand side. The greater is the deviation from the straight line yielded by the Margules equation, and the more asymmetric this deviation becomes, the greater is the need to introduce new terms and parameters on the right-hand side. Thus, for example, we have:

1. The three-parameter Margules equation

$$\frac{g^{\mathrm{E}}}{\mathbf{R}Tx_1x_2} = A_{12}x_2 + A_{21}x_1 + Cx_1x_2$$

2. The four-parameter equation

$$\frac{g^{\mathrm{E}}}{\mathbf{R}Tx_1x_2} = A_{12}x_2 + A_{21}x_1 - \frac{\alpha_{12}\alpha_{21}x_1x_2}{\alpha_{12}x_1 + \alpha_{21}x_2}$$

3. The five-parameter equation

$$\frac{g^{\mathrm{E}}}{\mathbf{R}Tx_1x_2} = A_{12}x_2 + A_{21}x_1 - \frac{\alpha_{12}\alpha_{21}x_1x_2}{\alpha_{12}x_1 + \alpha_{21}x_2 + \eta x_1x_2}$$

The activity coefficients are calculated on the basis of the relationship

$$\frac{1}{\mathbf{R}T}\frac{\partial g^{\mathrm{E}}}{\partial x_i} = \ln\gamma_i$$

In the case of van Laar's equation the relations already known are obtained:

$$\ln\gamma_1 = A_{12}\phi_2^2 = \frac{A_{12}x_2^2}{((A_{12}/A_{21})x_1 + x_1)^2}$$

$$\ln\gamma_2 = A_{21}\phi_1^2 = \frac{A_{21}x_1^2}{(x_1 + (A_{21}/A_{12})x_1)^2}$$

The Margules equation gives for the activity coefficients

$$\ln\gamma_1 = x_2^2(A_{12} + 2(A_{21} - A_{12})x_1)$$

$$= (2A_{21} - A_{12})x_2^2 + 2(A_{12} - A_{21})x_2^3$$

$$\ln\gamma_2 = x_1^2(A_{21} + 2(A_{12} - A_{21})x_2)$$

$$= (2A_{12} - A_{21})x_1^2 + 2(A_{21} - A_{12})x_1^3$$

The van Laar and Margules equations cannot be well used in cases with strong asymmetry. The limit of applicability for both is

$$\frac{1}{2} < \frac{A_{12}}{A_{21}} < 2$$

Algorithm and Program: Calculation of the Activity Coefficient with the Margules Equation

Write a program for the calculation of the activity coefficient with the Margules equation. Estimate the parameters of the equation from the measured activity coefficients at $x = 0$ and $x = 1$. Compare the calculated activity coefficients with the measured ones.

Algorithm.

1. Read the name of the datafile containing the datafile sequential numbers of the compounds of the mixture, the temperature, and the measured activity coefficients.
2. Read the first record of the measurement file, which contains the sequential numbers, the temperature, and the number of measurements.
3. Read the second and the last record of the measurement file (γ_1 at $x = 0$ and γ_2 at $x = 1$).
4. Calculate the Margules-equation parameters.
5. Calculate the activity coefficients with the Margules equation.
6. Print the calculated and measured data.

Program VLEMARGULES. A possible realization:

```
]LOAD VLEMARGULES
]LIST

 10  REM :CALCULATION OF PARAMETER
      S OF MARGULES EQUATION FROM
      ACTIVITY COEFFICIENTS AT X1 =
      0 AND X1 = 1
 20  REM :COMPARISON OF CALCULATED
      WITH MEASURED RESULTS.
      THE MEASURED DATA ARE IN FILE
      D17, BUT YOU CAN GIVE ANOTHER
      FILENAME INSTEAD.
 30  PRINT : PRINT
 40  PRINT "ENTER THE NAME OF THE
      FILE CONTAINING"
 50  INPUT "MEASURED DATA (EQ. D17)
      "; A$
 60  PRINT
 70  N = 1
 80  GOSUB 700
 90  PRINT "SEQ. NO. OF COMPONENTS
      OF MIXTURE "; C1; TAB(39);C2
100  PRINT
110  N = 2
120  GOSUB 700
130  A1 = LOG (G1)
140  N = NO + 1
150  GOSUB 700
160  A2 = LOG (G2)
170  DE = 0
180  PRINT " X GAMMA1(MES) GA
      MMA1(CALC) DEV(%)"
```

```
190  PRINT "======================
     ======================"
200  FOR N = 2 TO NO + 1 STEP 2
210  GOSUB 700
220  GC = EXP ((1 - X) ∧ 2 * (A1 +
     2 * (A2 - A1) * X))
230  DE = (G1 - GC) ∧ 2 + DE
240  DD = (G1 - GC) / G1 * 100
250  D% = DD * 100
260  DD = D% / 100
270  PRINT X; TAB( 6);G1; TAB( 20
     );GC; TAB( 34);DD
280  NEXT N
290  PRINT
300  PRINT "======================
     ======================"
310  DE = SQR (DE) / (NO - 1)
320  PRINT
330  INVERSE
340  INPUT "TO CONTINUE PRESS
     'RETURN' "; Q$
350  PRINT
360  NORMAL
370  PRINT "STANDARD DEVIATION OF
     GAMMA1"
380  PRINT "CALC. BY MARGULES EQ.
     "; DE
390  PRINT
400  PRINT "CONSTANTS OF MARGULES
     EQUATION"
410  PRINT "A12,A21 ARE "; A1; TAB(
     25); A2
420  PRINT
430  INVERSE
440  INPUT "TO CONTINUE PRESS
     'RETURN' "; Q$
450  PRINT
460  NORMAL
470  DE = 0
480  PRINT : PRINT
490  PRINT " X GAMMA2(MES) GA
     MMA2(CALC) DEV(%)"
500  PRINT "======================
     ======================"
510  FOR N = 2 NO + 1 STEP 2
520  GOSUB 700
530  GC = EXP (X ∧ 2 * (A2 + 2 *
     (A1 - A2) * (1 - X)))
540  DE = (G2 - GC) ∧ 2 + DE
```

```
550  DD = (G2 - GC) / G2 * 100
560  D% = DD * 100
570  DD = D% / 100
580  PRINT X; TAB(6); G2; TAB(20);
      GC; TAB(34); DD
590  NEXT N
600  PRINT
610  PRINT "======================
      ===================="
620  DE = SQR (DE) / (NO - 1)
630  PRINT
640  PRINT "STANDARD DEVIATION OF
      GAMMA2"
650  PRINT "CALC. BY MARGULES EQ.
      "; DE
660  PRINT
670  INPUT "NEXT CALCULATION ? (Y
      / N) "; C$
680  IF C$ = "Y" THEN GOTO 30
690  END
700  REM :READING ROUTINE FOR VLE
      DATAFILES. THE NAME OF THE
      FILE HAS TO BE IN VARIABLE
      "A$". THE ROUTINE READS T,
      NO, C1 AND C2 WHEN N = 1. IF N
      > 1 AND < NO, READS THE CORRESP
      ONDING X, G1, G2, AND GE / R / T /
      X1 / X2
710  D$ = CHR$ (4)
720  PRINT D$; "OPEN "; A$
730  PRINT D$; "POSITION "; A$; ", R"
      ;N - 1
740  PRINT D$; "READ "; A$
750  IF N = 1 THEN INPUT T,NO,C1
      ,C2
760  IF N < > 1 THEN INPUT X,G1
      ,G2,GE
770  PRINT D$; "CLOSE "; A$
780  PRINT D$; "IN#0"
790  RETURN
```

Example. Run the program for datafile D17, which contains the measured data on a chloroform–1,2-dichloroethane mixture.

```
]RUN

ENTER THE NAME OF THE FILE CONTAINING
MEASURED DATA (EQ. D17) D17
```

SEQ. NO. OF COMPONENTS OF MIXTURE 10 12

X	GAMMA1(MES)	GAMMA1(CALC)	DEV(%)
0	1.406744	1.406744	−.01
.1	1.2217941	1.26572363	−3.6
.2	1.1210852	1.16644178	−4.05
.3	1.0643358	1.09767901	−3.14
.4	1.032295	1.05163135	−1.88
.5	1.0146437	1.02262242	−.79
.6	1.0055676	1.00627916	−.08
.7	1.0014668	.998989156	.24
.8	1.0000629	.99753917	.25
.9	99989206	.998881068	.1
1	1	1	0

TO CONTINUE PRESS 'RETURN'

STANDARD DEVIATION OF GAMMA1
CALC. BY MARGULES EQ. 3.72504453E − 03

CONSTANTS OF MARGULES EQUATION
A12,A21 ARE .341277815 .0894813166

TO CONTINUE PRESS 'RETURN'

X	GAMMA2(MES)	GAMMA2(CALC)	DEV(%)
0	1	1	0
.1	1.006918	1.0054419	.14
.2	1.0219421	1.01988944	.2
.3	1.0394917	1.04058149	−.11
.4	1.0565004	1.06466686	−.78
.5	1.07129	1.08906492	−1.66
.6	1.082921	1.1104115	−2.54
.7	1.0909977	1.12510236	−3.13
.8	1.0953951	1.12944679	−3.11
.9	1.0961943	1.11993528	−2.17
1	1.0936069	1.0936069	−.01

STANDARD DEVIATION OF GAMMA2
CALC. BY MARGULES EQ. 3.17499033E − 03

NEXT CALCULATION ? (Y / N) N

Remark. Explain the results.

Calculation of Excess Free Enthalpy of Mixing from Experimental Data: The Redlich – Kister Mode of Treatment

Instead of the general form of this equation, first the expression for two-component, and then the one for three-component mixtures will be shown. For two-component mixtures,

$$\frac{g^{\mathrm{E}}}{\mathbf{R}T} = x_1x_2\left(A_{12} + B_{12}(x_1 - x_2) + C_{12}(x_1 - x_2)^2 + D_{12}(x_1 - x_2)^3 + \cdots\right)$$

It is easy to see the relation of this equation to the equations discussed so far.

For three-component mixtures the Redlich–Kister equation is

$$\begin{aligned}\frac{g^{\mathrm{E}}}{\mathbf{R}T} = {} & x_1x_2\left(A_{12} + B_{12}(x_1 - x_2) + C_{12}(x_1 - x_2)^2 + D_{12}(x_1 - x_2)^3 + \cdots\right)\\ & + x_1x_3\left(A_{13} + B_{13}(x_1 - x_3) + C_{13}(x_1 - x_3)^2\right.\\ & \left. + D_{13}(x_1 - x_3)^3 + \cdots\right)\\ & + x_2x_3\left(A_{23} + B_{23}(x_2 - x_3) + C_{23}(x_2 - x_3)^2\right.\\ & \left. + D_{23}(x_2 - x_3)^3 + \cdots\right)\\ & + x_1x_2x_3\left(A_{123} + B_{120}(x_1 - x_2) + B_{130}(x_1 - x_3)\right.\\ & \left. + B_{230}(x_2 - x_3) + \cdots\right)\end{aligned}$$

The practical question is, to what number of terms is it worthwhile to go in the fitting of experimental results. In the following only binary mixtures will be discussed.

We now show the relationship of the Redlich–Kister equation with the van Laar and Margules equations in the case of binary mixtures, for which the Redlich–Kister equation can be written in the following form:

$$\frac{g^{\mathrm{E}}}{\mathbf{R}Tx_1x_2} = A + B(x_1 - x_2) + C(x_1 - x_2)^2$$

Four cases are discussed:

1. $A = 0$, $B = 0$, $C = 0$. The excess free enthalpy is zero; the mixture is ideal.

2. $A = A_0$, $B = 0$, $C = 0$. Then

$$\frac{g^{\mathrm{E}}}{\mathbf{R}Tx_1x_2} = A_0$$

This criterion is satisfied by simple, regular mixtures.

3. $A \neq 0$, $B \neq 0$, $C = 0$. Then

$$\frac{g^{E}}{\mathbf{R}Tx_1x_2} = A + B(x_1 - x_2)$$

Let us multiply the parameter A by $x_1 + x_2$:

$$\frac{g^{E}}{\mathbf{R}Tx_1x_2} = (x_1 + x_2)A + B(x_1 - x_2)$$

Rearrange with respect to x_1 and x_2:

$$\frac{g^{E}}{\mathbf{R}Tx_1x_2} = (A + B)x_1 + (A - B)x_2$$

Introduce the parameters $A + B = A_{21}$ and the $A - B = A_{12}$:

$$\frac{g^{E}}{\mathbf{R}Tx_1x_2} = A_{21}x_1 + A_{12}x_2$$

This is identical with the Margules equation.

4. $A' \neq 0$, $B' \neq 0$, $C = 0$. Now we take the relationship

$$\frac{\mathbf{R}Tx_1x_2}{g^{E}} = A' + B'(x_1 - x_2)$$

The parameter A' is multiplied by $x_1 + x_2$:

$$\frac{\mathbf{R}Tx_1x_2}{g^{E}} = A'(x_1 + x_2) + B'(x_1 - x_2)$$

Rearrangement with respect to x_1 and x_2 gives

$$\frac{\mathbf{R}Tx_1x_2}{g^{E}} = (A' + B')x_1 + (A' - B')x_2$$

Introduce the following notation:

$$A' + B' = \frac{1}{A_{21}}$$

$$A' - B' = \frac{1}{A_{12}}$$

Then we have

$$\frac{\mathbf{R}Tx_1x_2}{g^{\mathrm{E}}} = \frac{x_1}{A_{21}} + \frac{x_2}{A_{12}}$$

This is the van Laar equation.

MOLECULAR STRUCTURE AND THERMODYNAMIC EXCESS PROPERTIES OF MIXTURES

The molar excess free enthalpy has no meaning without a definition of the standard state to which it is referred. The reference state of the thermodynamic excess quantities used here is the ideal solution. This choice is arbitrary, but usually convenient in its consequences, though in some cases another choice will be more so. It results in our having to use mole fraction as the variable for characterizing the composition, and in some cases, such as polymer and electrolyte solutions, it is more convenient to use another composition variable. The effective volume fraction is such a variable in the Wohl representation.

The excess free enthalpy consists of two terms:

$$g^{\mathrm{E}} = h^{\mathrm{E}} - Ts^{\mathrm{E}}$$

If we have some model for g^{E}, it can be calculated directly, as was done in the foregoing, but it will be a gain in physical understanding to formulate separate models for h^{E} and s^{E}, because these are physically fundamental quantities, related to molecular properties, while g^{E} is only their combination. The excess enthalpy is primarily connected with energetic interaction between the molecules, while the excess entropy primarily depends on the structure of the mixture, that is, on the steric arrangement of the molecules. (This leads to the concepts of random arrangement and segregation.)

Regrettably, excess enthalpy and excess entropy are not independent of one another. The Gibbs–Helmholtz equation gives the relation between them:

$$h^{\mathrm{E}} = -\mathbf{R}T^2\left(\frac{\partial(g^{\mathrm{E}}/\mathbf{R}T)}{\partial T}\right)_{P,x}$$

Thus, even if we construct independent models, their interdependence must be taken into account. Nevertheless, fundamentally we have to do with physically different phenomena. Historically speaking, authors making proposals for g^{E} have usually thought in terms of h^{E} or of s^{E}.

A strictly theoretical equation, taking into account the nature of both h^{E} and s^{E}, is at present beyond our powers, though many efforts have been made to deduce expressions, primarily for practical purposes, that have some theoretical basis. In general, all the theories of liquid mixtures existing today are of rather limited validity. Nevertheless, for practical purposes they are indispensable, and relationships deduced from even rather special theories are preferable to the purely empirical formulas discussed above. We therefore attempt now to summarize and coordinate the mixture models most often used.

Three classes of models will be distinguished, according to the form and size of the molecules:

1. Spherical molecules of equal diameter.
2. Molecules of unequal size.
3. Molecules consisting of segments.

In all three classes our fundamental model is the lattice. The basis of distinction is the number of lattice points occupied by the molecule.

In the first class all the molecules occupy a single lattice point, without regard to the type of the molecule. The form of the molecules of the second subclass is specified by the number of lattice points occupied by the molecule. Molecules forming the mixture occupy according to their types various numbers of lattice points, but aside from this, the structure of the molecule does not play a role in the model.

In the third class too, types of molecules are characterized by the number of lattice points occupied by them; molecules are considered as consisting of segments, and each segment occupies one lattice point. Moreover, a further parameter is used for the characterization of the molecules: the so-called number of contacts.

In the first class, the "natural" composition variable of the mixture is the mole fraction—for example, in the binary case,

$$x_i = \frac{n_i}{n_1 + n_2}$$

In the second class the volume fraction plays a role:

$$\phi_i = \frac{x_i v_i}{x_1 v_1 + x_2 v_2}$$

where v_i is the molar volume of the component i at the temperature and pressure of the mixture in the pure state. In the third class a further fraction is involved: the so-called surface fraction:

$$\Theta_i = \frac{x_i q_i}{x_1 q_1 + x_2 q_2}$$

Its nature will be explained later in detail.

From the point of view of interaction between the molecules four classes are distinguished:

1. No interaction; in solutions of this kind, nonideality arises only from the form of the molecules.
2. Mild interaction, which does not produce local concentration differences.
3. Strong interaction, producing local concentration differences, with complete miscibility over the whole concentration range.
4. Very strong interaction, producing local changes of concentration, leading to segregation in part of the concentration range.

The discussion will be continued according to this two-dimensional classification. For each model it will be specified:

1. What aspects of molecular form and size are taken into consideration.
2. What energetic actions between the molecules are taken into consideration.

Thermoneutral Mixtures of Spherical Molecules of Identical Size

Here we only refer back to what has already been said on the basis of the lattice model. This model leads to an ideal mixture, and cannot be used when the molecules are not of identical size, even if the mixing is thermoneutral. For completeness we repeat that the molar free enthalpy of mixing is

$$\Delta g^{(\mathrm{id})} = \mathbf{R}T(x_1 \ln x_1 + x_2 \ln x_2)$$

$$g^{\mathrm{E(id)}} = 0$$

Thermoneutral Mixtures of Molecules of Different Size

Molecules of different size can also be arranged in the lattice. We suppose that molecule 1 occupies r_1 adjacent lattice points, while molecule 2 occupies r_2 lattice points. Let us define the two volume fractions

$$\phi_1 = \frac{x_1 r_1}{x_1 r_1 + x_2 r_2}$$

$$\phi_2 = \frac{x_2 r_2}{x_1 r_1 + x_2 r_2}$$

The combinatorial task is the same as in the case of the ideal mixture, with the difference that now we have to arrange the quantities $n_1 r_1$ and $n_2 r_2$ instead of

n_1 and n_2. Thus, the molar mixing entropy is

$$\Delta s = -\mathbf{R}(x_1 \ln \phi_1 + x_2 \ln \phi_2)$$

and the excess molar entropy of mixing is

$$s^{\mathrm{E}} = -\mathbf{R}(x_1 \ln \phi_1 + x_2 \ln \phi_2) - \mathbf{R}(x_1 \ln x_1 + x_2 \ln x_2)$$

$$= -\mathbf{R}\left(x_1 \ln \frac{\phi_1}{x_1} - x_2 \ln \frac{\phi_2}{x_2} \right)$$

The excess molar free enthalpy of mixing is

$$g^{\mathrm{E}} = \mathbf{R}T\left(x_1 \ln \frac{\phi_1}{x_1} + x_2 \ln \frac{\phi_2}{x_2} \right)$$

This relationship is called the Flory–Huggins equation.

Thermoneutral Mixtures of Multisegment Molecules

A molecule consisting of characteristic atomic groups is called a multisegment molecule. For example, pentane is a multisegment molecule the segments of which are in order (using the nomenclature of organic chemistry): methyl–methylene–methylene–methylene–methyl (see Figure 11.12*a* below). For such multisegment molecules the lattice model can be used under the assumption that one segment falls on each lattice point, and all the components of the mixture consist of segments occupying the same volume.

In a mixture of multisegment molecules it must be taken into consideration not only that a molecule is surrounded by z other molecules, but also that contact occurs between the segments. If the molecule has r segments, and makes on average zq contacts, then we may take the resulting nonideality into consideration by means of a variable defined on the analogy of mole fraction and volume fraction. Let us call this variable the *surface fraction*:

$$\Theta_1 = \frac{x_1 z q_1}{x_1 z q_1 + x_2 z q_2} = \frac{x_1 q_1}{x_1 q_1 + x_2 q_2}$$

$$\Theta_2 = \frac{x_2 q_2}{x_1 q_1 + x_2 q_2}$$

Two molecules participate in a single contact; thus, the number of contacts per molecule is $\frac{1}{2} z q$, and each contact contributes to the nonideality. The contribution of a single contact to the molar excess free enthalpy is

$$\mathbf{R}T \ln \frac{\Theta_1}{\phi_1}$$

Thus the contribution of $\frac{1}{2}zq_1$ contacts is

$$\frac{g_1^{\mathrm{E}}}{\mathbf{R}T} = \sum_{1}^{\frac{1}{2}zq_1} \ln \frac{\Theta_1}{\phi_1} = \ln\left(\frac{\Theta_1}{\phi_1}\right)^{\frac{1}{2}zq_1} = \tfrac{1}{2}zq_1 \ln \frac{\Theta_1}{\phi_1}$$

In the case of a binary mixture the excess molar free enthalpy has the following form:

$$g^{\mathrm{E}} = \mathbf{R}T\left(x_1 \ln \frac{\phi_1}{x_1} + x_2 \ln \frac{\phi_2}{x_2} + \tfrac{1}{2}x_1 zq_1 \ln \frac{\Theta_2}{\phi_1} + \tfrac{1}{2}x_2 zq_2 \ln \frac{\Theta_2}{\phi_2}\right)$$

Here we have a purely combinatorial effect and its resulting configuration entropy, which is not manifested in isolation, but in combination with the energetic result of the interaction between the segments. The separation of the two effects into additive terms is an artifice of model formulation. Thus we must later return to this additive energetic term.

Summary of Results on Thermoneutral Mixtures

Since for a thermoneutral mixture

$$h^{\mathrm{E}} = 0$$

the Gibbs–Duhem relation has the following form:

$$h^{\mathrm{E}} = -\mathbf{R}T^2 \sum_i \left(\frac{\partial \ln \gamma_i}{\partial T}\right)_{x_1 P} = 0$$

It is evident that in a thermoneutral mixture the activity coefficient is independent of temperature:

$$\left(\frac{\partial \ln \gamma_i}{\partial T}\right)_{x_1 P} = 0$$

THE REGULAR MIXTURE

We started at the beginning of this chapter from the lattice model of liquid mixtures, and defined as ideal that mixture in which the component molecules are rigid spheres of near identical size, and do not attract one another. Let us now omit this third condition, and consider a two-component mixture in which the numbers of molecules are n_1 and n_2 respectively. Let the values of the pair potentials in the equilibrium state of the mixture be ε_{11}, ε_{22}, and ε_{12}. The coordination number is z, which means that each molecule has z immediate neighbors. If a molecule of component 1 surrounded by like molecules is exchanged with a molecule of component 2 surrounded by like molecules, so that the two molecules are placed in foreign surroundings, then, per molecule, the value of the resulting isothermal–isobaric energy change produced in this way, that is, the change in free enthalpy, is

$$w = z\left(\varepsilon_{12} - \tfrac{1}{2}\varepsilon_{11} - \tfrac{1}{2}\varepsilon_{22}\right)$$

Let us call w the *exchange energy* (Figure 11.8).

For the lattice energy of the mixture the following can be written:

$$E_{\text{lattice}} = \tfrac{1}{2}zn_{11}\varepsilon_{11} + \tfrac{1}{2}zn_{22}\varepsilon_{22} + n_{12}\left(\varepsilon_{12} - \tfrac{1}{2}\varepsilon_{11} - \tfrac{1}{2}\varepsilon_{22}\right)$$

That is, the lattice energy results from the energies of 1–1, 2–2, and 1–2 molecule pairs. For the numbers of molecule pairs the following stoichiometric relation is valid:

$$zn_1 = 2n_{11} + n_{12}$$

$$zn_2 = 2n_{22} + n_{12}$$

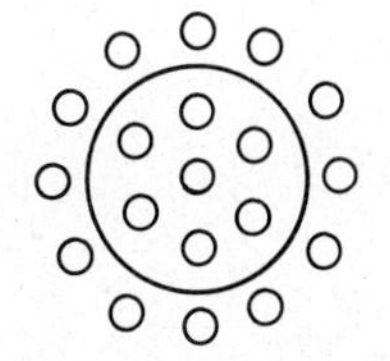
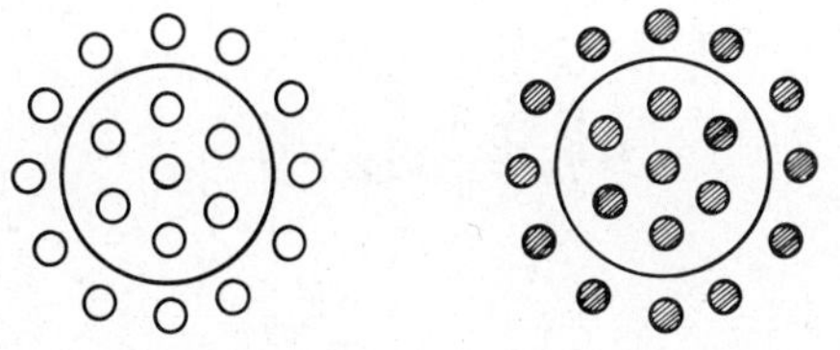
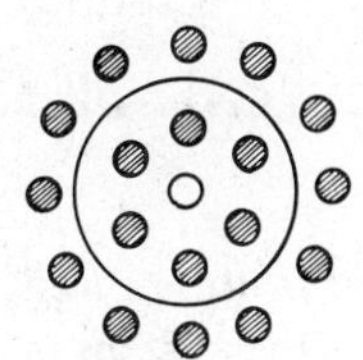
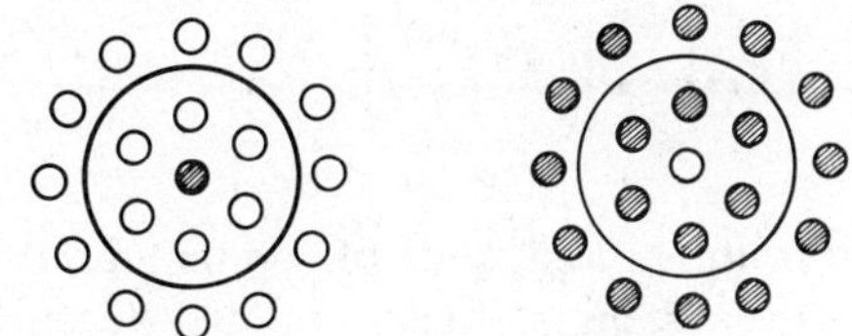

FIGURE 11.8. Exchange of two different molecules.

By the definition of exchange energy,

$$E_{\text{lattice}} = E_{11} + E_{22} + \frac{1}{z} n_{12} w$$

The molar free energy of the mixture, according to statistical thermodynamics, is

$$f = -\mathbf{k}T \ln Q$$

What does the partition function mean in this case? Clearly, here too it is the sum of the Boltzmann factors:

$$Q = \sum_{n_{12}} g[n_1, n_2, n_{12}] \exp\left(-\frac{E_{11} + E_{22} + \frac{1}{z} n_{12} w}{\mathbf{k}T}\right)$$

The partition function can be factored:

$$Q = \exp\left(-\frac{E_{11} + E_{22}}{\mathbf{k}T}\right) \cdot \sum g[n_1, n_2, n_{12}] \exp\left(-\frac{n_{12} w}{z\mathbf{k}T}\right)$$

where n_{12} is the number of pairs which can be arranged at a lattice point. However, several arrangements are possible at a given n_{12} value, their number being $g[n_1, n_2, n_{12}]$. This means that energy level $n_{12} w / z\mathbf{k}T$ is g-fold degenerate. (It is customary to call the statistical weight g in the thermodynamics of mixtures the *combinatorial factor*.)

Accordingly, the free energy is

$$f = -\mathbf{k}T \ln Q$$

$$= E_{11} + E_{22} - \mathbf{k}T \ln\left(\sum_{n_{12}} g[n_1, n_2, n_{12}] \exp\left(-\frac{n_{12} w}{z\mathbf{k}T}\right)\right)$$

whence the molar free energy of mixing is

$$\Delta f = f - E_{11} - E_{12}$$

$$= -\mathbf{k}T \ln\left(\sum_{n_{12}} g[n_1, n_2, n_{12}] \exp\left(-\frac{n_{12} w}{z\mathbf{k}T}\right)\right)$$

The difficulty of the calculation is that in the present state of combinatorial analysis we do not know how degenerate the system is at given n_{12}. However, we can give the sum of the combinatorial factors:

$$\sum_{n_{12}} g[n_1, n_2, n_{12}] = \frac{(n_1 + n_2)!}{n_1! n_2!}$$

Let us substitute this expression into the equation for the free energy of mixing:

$$\Delta f = -\mathbf{k}T \ln\left(\frac{(n_1 + n_2)!}{n_1! n_2!} \sum_{n_{12}} \exp\left(-\frac{n_{12}w}{z\mathbf{k}T}\right)\right)$$

Applying the technique usual in statistical mechanics, let us take instead of the summation its largest term,

$$\Delta f = -\mathbf{k}T \ln\left(\frac{(n_1 + n_2)!}{n_1! n_2!} \exp\left(-\frac{n_{12}^* w}{z\mathbf{k}T}\right)\right)$$

where n_{12}^* is the most probable value of the number of 1–2 pairs. This is evaluated by the following argument. The probability that a molecule of substance 1 or substance 2 occupies a selected lattice point is

$$\frac{n_1}{n_1 + n_2} \quad \text{or} \quad \frac{n_2}{n_1 + n_2}$$

respectively. The probability that a 1–2 pair of molecules occupies two selected adjacent lattice points is

$$\frac{2n_1 n_2}{(n_1 n_2)^2}$$

The factor 2 arises from the exchangeability of 1 and 2.

The number of all the pairs in the mixture is

$$\tfrac{1}{2}z(n_1 + n_2)$$

Let us multiply the probability of the formation of a 1–2 molecule pair by the number of all pairs:

$$n_{12}^* = \tfrac{1}{2}z(n_1 + n_2)\frac{2n_1 n_2}{(n_1 + n_2)^2}$$

$$= \frac{2n_1 n_2}{n_1 + n_2}$$

This is the most probable value of the number of 1–2 pairs in a random arrangement of the molecules. Let us substitute this value into the expression of free energy of mixing:

$$\Delta f = -\mathbf{k}T \ln\left(\frac{(n_1 + n_2)!}{n_1! n_2!} \exp\left(-\frac{n_1 n_2 w}{\mathbf{k}T(n_1 + n_2)}\right)\right)$$

$$= -\mathbf{k}T \ln \frac{(n_1 + n_2)!}{n_1! n_2!} + \frac{n_1 n_2 w}{n_1 + n_2}$$

Using Stirling's formula, the factorials can be eliminated:

$$\Delta f = \mathbf{k}T(n_1 \ln x_1 + n_2 \ln x_2) + \frac{n_1 n_2 w}{n_1 + n_2}$$

Since $n_1 + n_2 = \mathbf{N}$ and $W = w\mathbf{N}$, the molar free energy of mixing is

$$\Delta f = \mathbf{R}T(x_1 \ln x_1 + x_2 \ln x_2) + x_1 x_2 W$$

The first term on the right is the free energy of mixing of the ideal mixture. Thus, the molar excess free energy—or the molar excess free enthalpy, which is the same in the case of liquid mixtures—is

$$f^{\mathrm{E}} = g^{\mathrm{E}} = x_1 x_2 W$$

We note the following:

1. This is identical with the Margules equation of second degree,

$$g^{\mathrm{E}} = x_1 x_2 \mathbf{R}TA_0$$

The parameter A_0 in that equation can thus be identified:

$$A_0 = \frac{W}{\mathbf{R}T}$$

2. Here we have to do with a mixture of positive deviation. The pair potential ε_{12} is to a good approximation

$$\varepsilon_{12} = \sqrt{\varepsilon_{11}\varepsilon_{22}}$$

and the geometric mean is smaller than the arithmetic mean:

$$\varepsilon_{12} < \tfrac{1}{2}(\varepsilon_{11} + \varepsilon_{22})$$

Since all the pair potentials are negative,

$$\varepsilon_{11}, \varepsilon_{22}, \varepsilon_{12} < 0$$

Thus w is always positive. The condition for an ideal mixture is $w = 0$.

3. The activity coefficients are

$$\ln \gamma_1 = \frac{w}{\mathrm{k}T} x_2^2 = \frac{W}{\mathbf{R}T} x_2^2$$

$$\ln \gamma_2 = \frac{w}{\mathrm{k}T} x_2^2 = \frac{W}{\mathbf{R}T} x_1^2$$

The question now is, what can be said about the molar exchange energy.

Regular Mixtures of Molecules of Different Size

The lattice model is used again, but we assume that molecules of type 1 occupy r_1, and molecule of type 2 occupy r_2, lattice points. Moreover, let us assume that the mixture is formed successive pairs of molecules leaving their pure-component surroundings and changing places. The energetic consequence of this is expressed by the exchange energy, which has been defined already in conjunction with the theory of regular mixtures of spherical molecules of identical size.

By the same argument as there, but taking into consideration the different space requirements of the molecules through the lattice parameters, we have for the molar excess free enthalpy

$$g^{\mathrm{E}} = \frac{(x_1 r_1)(x_2 r_2)}{x_1 r_1 + x_2 r_2} W$$

where r_1, r_2 give the number of lattice points occupied by a molecule of type 1 or 2, whether in the pure state or in a mixture.

In this equation there are two cardinal problems:

1. The lattice parameters must be expressed somehow in terms of macro quantities.
2. A suitable method must be found for the estimation of the molar exchange energy W.

As concerns the first problem, it is rather obvious that the quotient of lattice parameters is equal to the quotient of molar volumes:

$$\frac{r_1}{r_2} = \frac{v_1}{v_2}$$

Substituting this into the equation, we have

$$g^{\mathrm{E}} = \frac{(v_1 x_1)(v_2 x_2)}{v_1 x_1 + v_2 x_2} W$$

Let us introduce the volume fractions:

$$\phi_1 = \frac{v_1 x_1}{v_1 x_1 + v_2 x_2}$$

$$\phi_2 = \frac{v_2 x_2}{v_1 x_1 + v_2 x_2}$$

and rewrite the expression for the excess free enthalpy of mixing in terms of them:

$$g^{\mathrm{E}} = \phi_1 \phi_2 (v_1 x_1 + v_2 x_2) W$$

It remains to estimate the molar exchange energy.

Estimation of Molar Exchange Energy in the Theory of Regular Mixtures

Let us define the cohesive energy in the following way:

$$c = \frac{\Delta u}{v}$$

where Δu = molar internal energy of evaporation from the saturated liquid state to the perfect-gas state at the temperature of the mixture,
v = molar volume of the saturated liquid.

If one mole of type-1 molecules evaporates from the surrounding of type-1 molecules, then

$$c_{11} = \frac{\Delta u_1}{v_1}$$

Similarly, if one mole of type-2 molecules evaporates from the surrounding of type-2 molecules,

$$c_{22} = \frac{\Delta u_2}{v_2}$$

However, if one mole of type-1 molecules evaporates from the surrounding of type-2 molecules, or vice versa, then the cohesive energy is the geometrical mean of the cohesive energies of the pure components:

$$c_{12} = c_{21} = \sqrt{c_{11}c_{22}}$$

We can write

$$\left(\sqrt{c_{11}} - \sqrt{c_{22}}\right)^2 = c_{11} + c_{22} - 2\sqrt{c_{11}c_{22}}$$

We recall that the definition of exchange energy is

$$w = z\left(\varepsilon_{12} - \tfrac{1}{2}\varepsilon_{11} - \tfrac{1}{2}\varepsilon_{22}\right)$$

$$= z\left(\sqrt{\varepsilon_{11}\varepsilon_{22}} - \tfrac{1}{2}\varepsilon_{11} - \tfrac{1}{2}\varepsilon_{22}\right)$$

Here ε_{11} is the energy needed to remove a single type-1 molecule from the surrounding of type-1 molecules to infinite distance—that is, to evaporate it. On the other hand, c_{11} is the energy needed to evaporate one mole of type-1 molecules from volume v_1. Hence in the theory of regular mixtures the molar exchange energy is

$$W = \left|c_{11} + c_{22} - 2\sqrt{c_{11}c_{22}}\right|$$

Let us define the solubility parameters:

$$\delta_1 \equiv \sqrt{c_{11}} = \sqrt{\frac{\Delta u_1}{v_1}}$$

$$\delta_2 \equiv \sqrt{c_{22}} = \sqrt{\frac{\Delta u_2}{v_2}}$$

With this notation the molar excess free enthalpy of mixing can be written in the following way:

$$g^{\mathrm{E}} = \phi_1\phi_2(\delta_1 - \delta_2)^2(x_1v_1 + x_2v_2)$$

It follows from this that the activity coefficients are

$$\ln\gamma_1 = \frac{1}{\mathbf{R}T}v_1\phi_2^2(\delta_1 - \delta_2)^2$$

$$\ln\gamma_2 = \frac{1}{\mathbf{R}T}v_2\phi_1^2(\delta_1 - \delta_2)^2$$

Let us introduce into these equations the following notation:

$$A_{12} = \frac{v_1}{\mathbf{R}T}(\delta_1 - \delta_2)^2$$

$$A_{21} = \frac{v_2}{\mathbf{R}T}(\delta_1 - \delta_2)^2$$

Then, changing over to mole fractions, the activity coefficients are seen to be

$$\ln \gamma_1 = A_{12}\phi_2^2 = \frac{A_{12}x_2^2}{((A_{12}/A_{21})x_1 + x_2)^2}$$

$$\ln \gamma_2 = A_{21}\phi_1^2 = \frac{A_{21}x_1^2}{(x_1 + (A_{21}/A_{12})x_2)^2}$$

This shows that a relation of van Laar type, with all its disadvantages, is inherent in the model. On the other hand, it has the advantage that A_{12} and A_{21} can be calculated from known macro properties in principle.

The dependence of the activity coefficient on temperature must be investigated. It was shown for the regular mixture that

$$s^{\mathrm{E}} = 0$$

so that

$$\frac{\partial s^{\mathrm{E}}}{\partial x_1} = 0$$

The excess free enthalpy of mixing is

$$g^{\mathrm{E}} = h^{\mathrm{E}}$$

According to known relationships

$$\frac{\partial g^{\mathrm{E}}}{\partial T} = -s^{\mathrm{E}} = 0 \quad \text{and} \quad \left(\frac{\partial g^{\mathrm{E}}}{\partial x_1}\right) = \mathbf{R}T \ln \gamma_1$$

whence

$$\left(\frac{\partial^2 g^{\mathrm{E}}}{\partial T\, \partial x_1}\right) = 0 \quad \text{and} \quad \frac{\partial^2 g^{\mathrm{E}}}{\partial x_1\, \partial T} = \frac{\partial(\mathbf{R}T \ln \gamma_1)}{\partial T} = 0$$

This means that in a regular mixture the activity coefficient is inversely proportional to absolute temperature:

$$\ln \gamma_1 \propto \frac{1}{T}$$

Algorithm Program: Calculation of the Activity Coefficient with the Regular-Mixture Theory

Write a program for the calculation of activity coefficients using the regular-mixture theory. This treatment allows the calculation of the activity coefficient from thermodynamic data alone. Compare the calculated results with measured data.

Algorithm.

1. Read the name of the datafile containing the datafile sequential numbers of the compounds of the mixture, the temperature, and the measured activity coefficients.
2. Read the actual pressure.
3. Read the first record of the measurement file, containing the sequential numbers, the temperature, and the number of measurements.
4. Calculate ΔH_{vap}, the enthalpy of vaporization at the actual temperature, by means of the Watson equation for the two components.
5. Calculate the liquid molar volumes of the two components.
6. Calculate the solubility parameters δ of the two components.
7. Calculate the volume fractions ϕ of the two components in the mixture.
8. Calculate the activity coefficients γ of the two components in the mixture.
9. Find the percentage deviations between the calculated activity coefficients and the measured ones; print the values and the deviations.

Program VLEREGULAR. A possible realization:

```
]LOAD VLEREGULAR
]LIST

 10  REM :CALCULATION OF ACTIVITY
     COEFFICIENT BY REGULAR-MIXTURE
     THEORY. THE PROGRAM CALC
     ULATES THE SOLUBILITY PARAME
     TERS TOO.

 20  REM :COMPARISON OF CALCULA
     TED RESULTS WITH MEASURED ONES.
     THE MEASURED DATA ARE IN
```

```
       FILE D2 , BUT YOU CAN GIVE
       ANOTHER FILE NAME INSTEAD.
 30  DIM A(15)
 40  R = 8.3144
 50  PRINT : PRINT
 60  PRINT "ENTER THE NAME OF THE
       FILE CONTAINING"
 70  INPUT "MEASURED DATA (EQ. D2)
       "; A$
 80  PRINT
 90  INPUT "ENTER PRESSURE (PA)";
       P
100  PRINT : PRINT
110  N = 1
120  GOSUB 710
130  PRINT "SEQ. NO. OF COMPONENT
       S OF MIXTURE "; C1; TAB(39);
       C2
140  PRINT
150  GOSUB 810
160  DE = 0
170  PRINT " X GAMMA1(MEAS) GA
       MMA1(CALC) DEV(%)"
180  PRINT "======================
       ====================="
190  FOR N = 2 TO NO + 1 STEP 2
200  GOSUB 710
210  F2 = (1 - X) * V2 / (X * V1 +
       (1 - X) * V2)
220  GC = EXP (V1 * F2 * (DI - D2
       ) ∧ 2 / R / T)
230  DE = (G1 - GC) ∧ 2 + DE
240  DD = (G1 - GC) / G1 * 100
250  D% = DD * 100
260  DD = D% / 100
270  PRINT X; TAB(6); G1; TAB(20
       ); GC; TAB(34); DD
280  NEXT N
290  PRINT
300  PRINT "======================
       ====================="
310  DE = SQR (DE) / (NO - 1)
320  PRINT
330  INVERSE
340  INPUT "TO CONTINUE PRESS
       'RETURN' "; Q$
350  PRINT
360  NORMAL
```

```
370  PRINT "STANDARD DEVIATION OF
      GAMMA1"
380  PRINT "CALC. REGULAR SOLUTION
      TH. "; DE
390  PRINT
400  PRINT "SOLUBILITY PARAMETERS
      OF COMPOUNDS"
410  PRINT "DELTA1 DELTA2 "; D1; TAB(
      28); D2
420  PRINT
430  INVERSE
440  INPUT "TO CONTINUE PRESS
      'RETURN' "; Q$
450  PRINT
460  NORMAL
470  DE = 0
480  PRINT : PRINT
490  PRINT " X GAMMA2(MAS) GA
      MMA2(CALC) DEV(%)"
500  PRINT "======================
      ======================"
510  FOR N = 2 TO NO + 1 STEP 2
520  GOSUB 710
530  F1 = X * V1 / (X * V1 + (1 -
      X) * V2)
540  GC = EXP (V2 * F1 * (D1 - D2
      ) ∧ 2 / R / T)
550  DE = (G2 - GC) ∧ 2 + DE
560  DD = (G2 - GC) / G2 * 100
570  D% = DD * 100
580  DD = D% / 100
590  PRINT X; TAB( 6);G2; TAB( 20
      ); GC; TAB(34); DD
600  NEXT N
610  PRINT
620  PRINT "======================
      ======================"
630  DE = SQR (DE) / (NO - 1)
640  PRINT
650  PRINT "STANDARD DEVIATION OF
      GAMMA2"
660  PRINT "CALC. BY REGULAR SOLU
      TION TH. "; DE
670  PRINT
680  INPUT "NEXT CALCULATION ? (Y
      /N) "; C$
690  IF C$ = "Y" THEN GOTO 50
700  END
```

```
 710  REM :READING ROUTINE FOR VLE
       DATAFILES. THE NAME OF THE
       FILE HAVE TO BE IN VARIABLE
       "A$". THE ROUTINE READS T,
       NO, C1 AND C2 WHEN N = 1. IF N
       > 1 AND  < NO READS THE CORRESP
       ONDING X, G1, G2 AND GE / R / T /
       X1 / X2
 720  D$ = CHR$ (4)
 730  PRINT D$; "OPEN"; A$
 740  PRINT D$; "POSITION"; A$; ,R"
      ; N - 1
 750  PRINT D$; "READ "; A$
 760  IF N = 1 THEN INPUT T,NO,C1
      ,C2
 770  IF N < > 1 THEN INPUT X,G1
      ,G2,GE
 780  PRINT D$; "CLOSE "; A$
 790  PRINT D$; "IN#0"
 800  RETURN
 810  REM :CALCULATION OF SOLUBILITY
       PARAMETERS AND MOLAR VOLUMES
 820  IS = C1
 830  GOSUB 1070
 840  TC = A(3)
 850  PC = A(2)
 860  OM = A(4)
 870  TB = A(5)
 880  DH = A(6)
 890  GOSUB 1250
 900  GOSUB 1180
 910  V1 = V
 920  H1 = HV
 930  D1 = SQR ((H1 - R * T) / V1)
 940  IS = C2
 950  GOSUB 1070
 960  TC = A(3)
 970  PC = A(2)
 980  OM = A(4)
 990  TB = A(5)
1000  DH = A(6)
1010  GOSUB 1250
1020  GOSUB 1180
1030  V2 = V
1040  H2 = HV
1050  D2 = SQR ((H2 - R * T) / V2)
1060  RETURN
1070  REM :READING OF SPECIFIED
       RECORD OF PHDAT FILE INTO
```

```
      ARRAY A(15)
1080  REM :THE SEQUENTIAL NUMBER
      OF THE COMPOUND IN THE CALLI
      NG PROGRAM HAS TO BE "IS"
1090  REM :ARRAY A HAS TO BE DECL
      ARED IN THE CALLING PROGRAM
1100  D$ = CHR$ (4)
1110  PRINT D$; "OPEN PHDAT"
1120  PRINT D$; "POSITION PHDAT, R"
      ;IS - 1
1130  PRINT D$; "READ PHDAT"
1140  INPUT A(1), A(2), A(3), A(4), A(5),
      A(6), A(7), A(8), A(9), A(10),
      A(11), A(12), A(13), A(14), A(15)
1150  PRINT D$; "CLOSE PHDAT"
1160  PRINT D$; "IN#0"
1170  RETURN
1180  DEF FN Z0(PR) = (.46407 -
      0.73221 * TR + 0.45256 * TR ∧
      2) * PR - (0.00871 - 0.02939
      * TR + 0.02775 * TR ∧ 2) *
      PR ∧ 2
1190  DEF FN Z1(PR) = - (0.0267
      6 + 0.28376 * TR - 0.2834 *
      TR ∧ 2) * PR - (0.10209 - 0.
      32504 * TR + 0.25376 * TR ∧
      2) * PR ∧ 2 + (0.00919 - 0.0
      3016 * TR - 0.02485 * TR ∧ 2
     ) * PR ∧ 3
1200  TR = T / TC
1210  PR = P / PC
1220  Z = FN Z0(PR) + OM * FN Z1
      (PR)
1230  V = Z * 8.3144 * T / P
1240  RETURN
1250  REM :CALCULATION OF ENTHALP
      Y OF VAPORIZATION BY WATSON
      EQN.
1260  HV = DH * ((T - TC) / (TB -
      TC)) ∧ 0.38
1270  RETURN
```

Example. Run the program for measurement file D1, which contains 20 measured values of γ_1 and γ_2 for the benzene–2, 4-dimethylpentene mixture. Let the pressure be 101,325 Pa.

```
]RUN
ENTER THE NAME OF THE FILE CONTAINING
```

MEASURED DATA (EQ. D2) D2
ENTER PRESSURE (PA) 101325
SEQ. NO. OF COMPONENTS OF MIXTURE 22 23

X	GAMMA1(MEAS)	GAMMA1(CALC)	DEV(%)
0	1.4762	1.27150205	13.86
.1	1.4098	1.25350901	11.08
.2	1.3451	1.23410482	8.25
.3	1.2826	1.21311935	5.41
.4	1.2231	1.1903546	2.67
.5	1.1672	1.16557896	.13
.6	1.1163	1.13851994	−2
.7	1.0716	1.10885529	−3.48
.8	1.0352	1.07620164	−3.97
.9	1.0098	1.04010039	−3.01
1	1	1	0

TO CONTINUE PRESS 'RETURN'

STANDARD DEVIATION OF GAMMA1
CALC. REGULAR SOLUTION TH. .0149190679

SOLUBILITY PARAMETERS OF COMPOUNDS
DELTA1 DELTA2 17106.5316 14428.2441

TO CONTINUE PRESS 'RETURN'

X	GAMMA2(MEAS)	GAMMA2(CALC)	DEV(%)
0	1	1	0
.1	1.0024	1.02542299	−2.3
.2	1.0109	1.05399369	−4.27
.3	1.0271	1.0863222	−5.77
.4	1.0539	1.12318399	−6.58
.5	1.0952	1.16557896	−6.43
.6	1.157	1.21481729	−5
.7	1.2487	1.27264723	−1.92
.8	1.3865	1.34145032	3.24
.9	1.5982	1.42454883	10.86
1	1.9374	1.52670661	21.19

STANDARD DEVIATION OF GAMMA2
CALC. BY REGULAR SOLUTION TH. .0234743174

NEXT CALCULATION ? (Y / N) N

Remark. Explain the results.

LOCAL INHOMOGENEITY

An assumption in the Flory–Huggins theory of thermoneutral mixtures is that mixing proceeds without change in volume: $V^{\mathrm{E}} = 0$. This assumption is based on the assumption that the molecules are arranged at random in the mixture. This is the definition of the concept of homogeneous phase in phenomenological thermodynamics. According to the molecular concept the random arrangement of the molecules is not a basic condition; indeed, it is natural to expect that the different sizes and forms of the molecules, as well as the stronger or weaker attractions acting between them or their segments, will interfere with their completely random arrangement. This physical fact must be taken into account by a satisfactory theory of mixtures. Such *local inhomogeneity* is a qualitative notion, the quantitative description of which is also needed. This leads us to the concept of *local composition*. Figure 11.9 illustrates this idea. The domain of the central molecule, that is, the region containing its immediate neighbors, is enclosed in a heavy circle. The local composition and local mole fraction are interpreted as the values in this domain. In a larger vicinity fifteen molecules of type 1 and fifteen of type 2 are to be found. Thus, the stoichiometric mole fraction is $x_1 = x_2 = 0.5$. In the domain, however, there are nine molecules around the central molecule (Figure 11.9*a*). The local mole fraction of type 2 around type 1 can be found by counting:

$$x_{21} = \frac{\text{number of molecules of type 2 around central molecule of type 1}}{\text{total number molecules around central molecule of type 1}} = \frac{5}{8}$$

$$x_{11} = \frac{\text{number of molecules of type 1 around central molecule of type 1}}{\text{total number molecules around central molecule of type 1}} = \frac{3}{8}$$

The sum of the local mole fractions around the central molecule of type 1 is

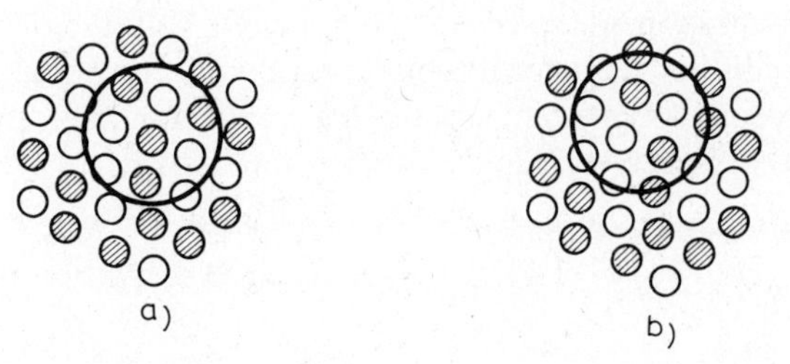

FIGURE 11.9. Local inhomogeneity: domain of central molecule (*a*) of type 1; (*b*) of type 2.

thus

$$x_{21} + x_{11} = 1$$

Similarly, for a central molecule of type 2, from Figure 11.9 *b* we find

$$x_{12} = \frac{4}{4+5} = \frac{4}{9}$$

$$x_{22} = \frac{5}{4+5} = \frac{5}{9}$$

The sum of local mole fractions in the domain of the central molecule of type 2 is thus

$$x_{12} + x_{22} = 1$$

From the point of view of the thermodynamic behavior of molecules of types 1 and 2, local concentrations are decisive. Therefore, the model becomes usable if we formulate constitutive equations for the local mole fractions. Three equations of this kind will be described.

THE NRTL EQUATION

NRTL stands for "nonrandom two-liquid." The background of the equation is the following model: In a binary mixture each molecule of component 1 is surrounded by components 1 and 2, and vice versa. Consequently the behavior of both components is determined by their local and not by their stoichiometric concentrations. "Two-liquid" applied to this theory means simply that instead of two components two supercomponents ("liquids") are considered, each of which can be characterized by its local concentration. The mixture of these two supercomponents replaces the real two-component mixture.

The molar excess free enthalpy of the mixture is then

$$g^{E} = x_1\bar{g}_1 + x_2\bar{g}_2 - x_1 g_{11} - x_2 g_{22}$$

where $\bar{g}_1$, $\bar{g}_2$ = partial molar free enthalpies of supercomponents 1 and 2 in the mixture, namely

$$\bar{g}_1 = x_{11}g_{11} + x_{21}g_{21}$$

$$\bar{g}_2 = x_{12}g_{12} + x_{22}g_{22}$$

x_{11}, x_{21} = local mole fractions of pure components 1 and 2 in supercomponent (1),
x_{12}, x_{22} = local molar fractions of the pure components in supercomponent (2),
g_{11} = molar free enthalpy of pure component 1,
g_{22} = molar free enthalpy of pure component 2,
g_{12} = molar free enthalpy of component 1 in supercomponent (2),
g_{21} = molar free enthalpy of component 2 in supercomponent (1).

Substituting $\bar{g}_1$ and $\bar{g}_2$ in the equation, we have for the excess molar free enthalpy of the mixture

$$g^{\mathrm{E}} = x_1(x_{11}g_{11} + x_{21}g_{21}) + x_2(x_{12}g_{12} + x_{22}g_{22}) - x_1g_{11} - x_2g_{22}$$

The sum of the local molar fractions is 1, and thus

$$x_{11} = 1 - x_{21}$$

$$x_{22} = 1 - x_{12}$$

On substitution we have the following simplified form of the equation for the molar excess free enthalpy:

$$g^{\mathrm{E}} = x_1x_{21}(g_{21} - g_{11}) + x_2x_{12}(g_{12} - g_{22})$$

According to Renon and Prausnitz, characteristic local molar fractions are obtained from the stoichiometric molar fractions weighted by the Boltzmann factors:

$$x_{21} = \frac{x_2\exp(-\alpha g_{21}/\mathbf{R}T)}{x_1\exp(-\alpha g_{11}/\mathbf{R}T) + x_2\exp(-\alpha g_{22}/\mathbf{R}T)}$$

$$x_{12} = \frac{x_1\exp(-\alpha g_{12}/\mathbf{R}T)}{x_1\exp(-\alpha g_{11}/\mathbf{R}T) + x_2\exp(-\alpha g_{22}/\mathbf{R}T)}$$

where α is the parameter of nonrandom arrangement. Substitution yields for the equation for the molar excess free enthalpy

$$\frac{g^{\mathrm{E}}}{\mathbf{R}T} = x_1 \frac{g_{21} - g_{11}}{\mathbf{R}T} \cdot \frac{x_2 \exp(-\alpha(g_{21} - g_{11})/\mathbf{R}T)}{x_1 + x_2 \exp(-\alpha(g_{21} - g_{11})/\mathbf{R}T)}$$

$$+ x_2 \frac{g_{12} - g_{22}}{\mathbf{R}T} \cdot \frac{x_1 \exp(-\alpha(g_{12} - g_{22})/\mathbf{R}T)}{x_2 + x_1 \exp(-\alpha(g_{12} - g_{22})\mathbf{R}T)}$$

Let us introduce the following notation:

$$\tau_{21} = \frac{g_{21} - g_{11}}{\mathbf{R}T} \qquad A_{21} = \exp(-\alpha\tau_{21})$$

$$\tau_{12} = \frac{g_{12} - g_{22}}{\mathbf{R}T} \qquad A_{12} = \exp(-\alpha\tau_{12})$$

Then we have for the equation of molar excess free enthalpy

$$\frac{g^{\mathrm{E}}}{\mathbf{R}T} = x_1 x_2 \left(\frac{\tau_{21} A_{21}}{x_1 + x_2 A_{21}} + \frac{\tau_{12} A_{12}}{x_2 + x_1 A_{12}} \right)$$

The activity coefficients are

$$\ln \gamma_1 = x_2^2 \left(\frac{\tau_{21} A_{21}^2}{(x_1 + x_2 A_{21})^2} + \frac{\tau_{12} A_{12}^2}{(x_2 + x_1 A_{12})^2} \right)$$

$$\ln \gamma_2 = x_1^2 \left(\frac{\tau_{12} A_{12}^2}{(x_2 + x_1 A_{12})^2} + \frac{\tau_{21} A_{21}^2}{(x_1 + x_2 A_{21})^2} \right)$$

If $\alpha = 0$, then we have a random distribution, and the NRTL equation passes over to the following equation:

$$g^{\mathrm{E}} = x_1 x_2 (2g_{12} - g_{11} - g_{22})$$

This form of function is known from the theory of regular mixtures, and the quantity in parentheses can be identified as the molar exchange energy.

Algorithm and Program: Calculation of the Activity Coefficient with the NRTL Equation

Write a program for the calculation of the activity coefficient with the NRTL equation. Compare the calculated results with measured data.

Algorithm.

1. Read the name of the datafile containing the datafile sequential numbers of the compounds of the mixture, the temperature, and the measured activity coefficients.
2. Read the first record of the measurement file, which contains the sequential numbers, the temperature, and the number of measurements.
3. Read the parameter of the NRTL equation: $\tau_{12}, \tau_{21}, \alpha$.
4. Calculate the constants A_{12} and A_{21}.
5. Calculate γ_1 and γ_2 with the NRTL equation.
6. Find the percentage deviations from the measured data, and print the results.

Program VLENRTL. A possible realization:

```
]LOAD VLENRTL
]LIST
 10  REM :COMPARISON OF CALCULATE
      D AND MEASURED ACTIVITY COEFF
      ICIENTS FOR "NRTL" EQUATION
 20  REM :MEASURED ACTIVITY COEFFIC
      IENTS ARE IN FILE "DS",
      BUT YOU CAN GIVE ANOTHER
      FILENAME INSTEAD.
 30  PRINT : PRINT
 40  PRINT "ENTER THE NAME OF THE
      FILE CONTAINING"
 50  INPUT "MEASURED DATA (EQ. D8)
      "; A$
 60  PRINT
 70  N = 1
 80  GOSUB 700
 90  PRINT "SEQ. NO. OF COMPONENTS
      OF MIXTURE "; C1; TAB(39))C2
100  PRINT
110  INPUT "ENTER 'TAU12' (NRTL
      EQN.) "; T1
115  INPUT "ENTER 'TAU21'
      "; T2
120  INPUT "ENTER 'ALFA' (NRTL
      EQN.) "; AL
130  PRINT : PRINT
135  A1 = EXP ( - AL * T1)
140  A2 = EXP ( - AL * T2)
150  PRINT "NRTL CONSTANT A12 = ";
      A1
```

```
155  PRINT "NRTL CONSTANT A21 = ";
      A2
160  PRINT : PRINT
170  DE = 0
180  PRINT " X GAMMA1(MEAS) GA
      MMA1(CALC) DEV(%)"
190  PRINT "=======================
      ===================== "
200  FOR N = 2 TO N0 + 1 STEP 2
210  GOSUB 700
220  GC = EXP ((1 - X) ∧ 2 * (T2 *
      A2 ∧ 2 / (X + (1 - X) * A2) ∧
      2 + T1 * A1 ∧ 2 / ((1 - X) +
      X * A1) ∧ 2))
230  DE = (G1 - GC) ∧ 2 + DE
240  DD = (G1 - GC) / G1 * 100
250  D% = DD * 100
260  DD = D% / 100
270  PRINT X; TAB( 6);G1; TAB( 20
     );GC; TAB( 34);DD
280  NEXT N
290  PRINT
300  PRINT "=======================
      ====================="
310  DE = SQR (DE) / (NO - 1)
320  PRINT
330  INVERSE
340  INPUT "TO CONTINUE PRESS
      'RETURN' "; Q$
350  PRINT
360  NORMAL
370  PRINT "STANDARD DEVIATION OF
      GAMMA1"
380  PRINT "CALC. BY NRTL EQ.
      ((; DE
390  PRINT
420  PRINT
430  INVERSE
440  INPUT "TO CONTINUE PRESS 'RE
      TURN' "; Q$
450  PRINT
460  NORMAL
470  DE = 0
480  PRINT : PRINT
490  PRINT " X GAMMA2(MEAS) GA
      MMA2(CALC) DEV(%)"
500  PRINT "=======================
      ===================== "
510  FOR N = 2 TO NO + 1 STEP 2
```

```
520  GOSUB 700
530  GC = EXP (X ∧ 2 * (T1 * A1 ∧
      2 / ((1 - X) + X * A1) ∧ 2 +
      T2 * A2 ∧ 2 / (X + (1 - X) *
      A2) ∧ 2))
540  DE = (G2 - GC) ∧ 2 + DE
550  DD = (G2 - GC) / G2 * 100
560  D% = DD * 10
570  DD = D% / 10
580  PRINT X; TAB(6); G2; TAB(20);
      GC; TAB(34); DD
590  NEXT N
600  PRINT
610  PRINT "========================
      ===================="
620  DE = SQR (DE) / (NO - 1)
630  PRINT
640  PRINT "STANDARD DEVIATION OF
      GAMMA2"
650  PRINT "CALC. BY NRTL EQ.
      "; DE
660  PRINT
670  INPUT "NEXT CALCULATION ? (Y
      /N) "; C$
680  IF C$ = "Y" THEN GOTO 30
690  END
700  REM :READING ROUTINE FOR VLE
      DATAFILES. THE NAME OF THE
      FILE HAS TO BE IN VARIABLE
      "A$". THE ROUTINE READS T,
      NO, C1 AND C2 WHEN N = 1, IF N
      > 1 AND < NO READS THE CORRESP
      ONDING X, G1, G2 AND GE / R / T /
      X1 / X2
710  D$ = CHR$ (4)
720  PRINT D$; "OPEN"; A$
730  PRINT D$; "POSITION"; A$; "R"
      ; N - 1
740  PRINT D$; "READ "; A$
750  IF N = 1 THEN INPUT T,NO,C1
      ; C2
760  IF N < > 1 THEN INPUT X,G1,
      G2,GE
770  PRINT D$; "CLOSE"; A$
780  PRINT D$; "IN#0"
790  RETURN
```

Example. Run the program for datafile **D8**, containing 21 measured values of γ_1 and γ_2 for acetic acid–water mixtures. The NRTL parameters of this mixture are

$$\tau_{12} = -0.245147$$

$$\tau_{21} = 1.55749$$

$$\alpha = 0.44506$$

```
]RUN
ENTER THE NAME OF THE FILE CONTAINING
MEASURED DATA (EQ. D8) D8
SEQ. NO. OF COMPONENTS OF MIXTURE 16  15
ENTER 'TAU12' (NRTL EQN.) -0.245147
ENTER 'TAU21' 1.55749
ENTER 'ALFA' (NRTL EQN.) 0.44506
NRTL CONSTANT A12 =  1.11527959
NRTL CONSTANT A21 =  .499985341
```

X	GAMMA1(MEAS)	GAMMA1(CALC)	DEV(%)
0	3.5082064	3.49930458	.25
.1	2.227457	2.22824416	−.04
.2	1.6580324	1.65822656	−.02
.3	1.3664885	1.3661223	.02
.4	1.2044344	1.20412228	.02
.5	1.1106167	1.11059825	0
.6	1.055933	1.05618044	−.03
.7	1.0249634	1.02531806	−.04
.8	1.0087652	1.00904368	−.03
.9	1.0017071	1.00181283	−.02
1	1	1	0

```
TO CONTINUE PRESS 'RETURN'

STANDARD DEVIATION OF GAMMA1
CALC. BY NRTL  EQ. 4.48350613E - 04
TO CONTINUE PRESS 'RETURN'
```

X	GAMMA1(MEAS)	GAMMA1(CALC)	DEV(%)
0	1	1	0
.1	1.0227671	1.00994061	1.2
.2	1.076354	1.03211417	4.1
.3	1.1470671	1.05897533	7.6
.4	1.2268858	1.08605944	11.4
.5	1.3102589	1.11059825	15.2
.6	1.3928719	1.13086503	18.8
.7	1.4711962	1.1458271	22.1
.8	1.5422544	1.15494014	25.1
.9	1.6035547	1.15801407	27.7
1	1.6530495	1.15511819	30.1

```
STANDARD DEVIATION OF GAMMA2
CALC. BY NRTL EQ. .0458406293

NEXT CALCULATION ? (Y / N) N
```

Remark. We got the parameters τ and α as a result of a nonlinear parameter estimation applied to the measured data. Explain the results.

Extension of the NRTL Equation

The NRTL equation can be extended to multicomponent systems:

$$\frac{g^{\mathrm{E}}}{\mathbf{R}T} = \sum_{i=1}^{k} x_i \frac{\sum_{j=1}^{k} \tau_{ji} x_j}{\sum_{l=1}^{k} A_{li} x_l}$$

In this equation

$$\tau_{ji} = \frac{g_{ji} - g_{ii}}{\mathbf{R}T} \qquad g_{ij} = g_{ji}$$

$$A_{ji} = \exp(-\alpha_{ji}\tau_{ji}) \qquad \alpha_{ji} = \alpha_{ij}$$

It is remarkable that in multicomponent mixtures too, only the binary coefficients are relevant, that is, triple and higher-order interactions need not be taken into consideration.

THE WILSON EQUATION

The description of a thermoneutral mixture of molecules of nonidentical size led to the Flory–Huggins equation

$$\frac{g^{\mathrm{E}}}{\mathbf{R}T} = x_1 \ln \frac{\phi_1}{x_1} + x_2 \ln \frac{\phi_2}{x_2}$$

where the volume fractions, on the basis of the lattice model, are

$$\phi_1 = \frac{r_1 x_1}{r_1 x_1 + r_2 x_2}$$

$$\phi_2 = \frac{r_2 x_2}{r_1 x_1 + r_2 x_2}$$

Here no kind of local inhomogeneity is concerned; the lattice parameters are eliminated by means of the suggestive assumption that their ratio is equal to the ratio of molar volumes:

$$\frac{r_1}{r_2} = \frac{v_1}{v_2}$$

If we wish now to introduce local inhomogeneity into the model, this is done by taking for the ratio of lattice parameters in the volume fractions, instead of the quotient of molar volumes, their quotient weighted by the quotient of the Boltzmann factors:

$$\frac{r_1}{r_2} = \frac{v_1 \exp(-g_{12}/\mathbf{R}T)}{v_2 \exp(-g_{22}/\mathbf{R}T)} = \frac{v_1}{v_2} \exp\left(-\frac{g_{12} - g_{22}}{\mathbf{R}T}\right) = A_{12}$$

$$\frac{r_2}{r_1} = \frac{v_2}{v_1} \exp\left(-\frac{g_{21} - g_{11}}{\mathbf{R}T}\right) = A_{21}$$

where A_{12} and A_{21} are parameters

g_{12} = energy needed for the removal of one mole of type-1 molecules from a surrounding of type-2 molecules,

g_{22} = energy needed for the removal of one mole of type-2 molecules from the liquid phase,

g_{21} = energy needed for the removal of one mole of type-2 molecules from a surrounding of type-1 molecules,

g_{11} = energy needed for the removal of one mole of type-1 molecules from the liquid phase.

Our explanation concerning the physical nature of g_{ij} is not exact, as we have no notion from what physical data parameters the A_{12} and A_{21} are to be calculated. In practice these parameters are obtained by a fit of Wilson's equation to experimental data. This is achieved by the following procedure. Let us substitute into the volume fractions the parameters

$$\phi_1 = \frac{x_1}{x_1 + \dfrac{r_2}{r_1} x_2} = \frac{x_1}{x_1 + \dfrac{v_2}{v_1} \exp\left(-\dfrac{g_{21} - g_{11}}{\mathbf{R}T}\right)} = \frac{x_1}{x_1 + A_{21} x_2}$$

$$\phi_2 = \frac{x_2}{A_{12} x_1 + x_2}$$

whence

$$\frac{x_1}{\phi_1} = x_1 + A_{21} x_2$$

$$\frac{x_2}{\phi_2} = A_{12} x_1 + x_2$$

We substitute these expressions into the Flory–Huggins equation:

$$g^{E} = -x_1 \ln(x_1 + A_{21}x_2) - x_2 \ln(A_{12}x_1 + x_2)$$

What kind of model does this substitution imply? The Flory–Huggins equation takes into consideration the combinatorial effect on the formation of a mixture consisting of molecules of different sizes, with the implicit assumption that the mixture is also locally homogeneous in the domain of every molecule. What we do now is dispense with that assumption. The development of local inhomogeneity is qualitatively explained in terms of the energy difference $g_{ij} - g_{jj}$, and quantitatively on the basis of the intuitive idea of Wilson, by providing the mole fraction with a correction factor.

From the Wilson equation we have for the activity coefficients

$$\ln \gamma_1 = -\ln(x_1 + A_{21}x_2) + x_2 \left(\frac{A_{21}}{x_1 + A_{21}x_2} - \frac{A_{12}}{x_2 + A_{12}x_1} \right)$$

$$\ln \gamma_2 = -\ln(x_2 + A_{12}x_1) + x_1 \left(\frac{A_{12}}{x_2 + A_{12}x_1} - \frac{A_{21}}{x_1 + A_{21}x_2} \right)$$

Extension of the Wilson Equation

The Wilson equation can be extended to multicomponent mixtures:

$$\frac{g^{E}}{RT} = -\sum_{i=1}^{k} x_i \ln\left(\sum_{j=1}^{k} A_{ji} x_j \right)$$

It is remarkable that in a multicomponent mixture too, only the binary coefficients are relevant, that is, triple and higher-order interactions need not be taken into consideration.

Algorithm and Program: Calculation of the Activity Coefficient with the Wilson Equation

Write a program for the calculation of the activity coefficient with the Wilson equation. Make a comparison with measured data.

Algorithm.

1. Read the name of the datafile containing the datafile sequential numbers of the compounds of the mixture, the temperature, and the measured activity coefficients.
2. Read the first record of the measurement file, containing the sequential numbers, the temperature, and the number of measurements.

3. Read the second and the last record of the measurement file (γ_1 at $x = 0$ and γ_2 at $x = 1$).
4. Calculate the constants A_{12} and A_{21} of the Wilson equation.
5. Calculate γ_1 and γ_2 with the Wilson equation.
6. Calculate the percentage deviation from the measured data, and print of the results.

Program VLEWILSON. A possible realization:

```
]LOAD VLEWILSON
]LIST

 10  REM :CALCULATION OF PARAMET
      ERS OF WILSON EQUATION FRO
      M ACTIVITY COEFFICIENTS AT X
      1 = 0 AND X1 = 1
 20  REM :COMPARISON OF CALCULATED
      RESULTS WITH MEASURED ONES.
      THE MEASURED DATA ARE IN FILE
      D5 , BUT YOU CAN GIVE ANO
      THER FILE NAME INSTEAD.
 30  PRINT : PRINT
 40  PRINT "ENTER THE NAME OF THE
      FILE CONTAINING"
 50  INPUT "MEASURED DATA (EQ. D5)
      "; A$
 60  PRINT
 70  N = 1
 80  GOSUB 820
 90  PRINT "SEQ. NO. OF COMPONENTS
      OF MIXTURE "; C1; TAB(39); C2
100  PRINT
110  N = 2
120  GOSUB 820
130  V1 = G1
140  N = NO + 1
150  GOSUB 820
160   G1 = V1
170   V1 = 1000:V2 = 1000
180  A2 = 0.9
190  L1 = 1 - A2 - LOG (G1)
200  A1 = EXP (L1)
210  L2 = 1 - AL - LOG (G2)
220  A2 = EXP (L2)
230  IF ABS (V1 - A1) + ABS (V2
      - A2) < 1E - 5 THEN GOTO 2
      70
240  V1 = A1
```

```
250  V2 = A2
260  GOTO 190
270  PRINT
280  PRINT : PRINT
290  DE = 0
300  PRINT " X GAMMA1(MEAS) GA
      MMA1(CALC) DEV(%)"
310  PRINT "========================
      ===================="
320  FOR N = 2 TO NO + 1 STEP 2
330  GOSUB 820
340  GC = EXP ( - LOG (X + (1 -
      X) * A1) + (1 - X) * (A1 / (
      X + A1 * (1 - X)) - A2 / ((1
      - X) + A2 * X)))
350  DE = (G1 - GC) ^ 2 + DE
360  DD = (G1 - GC) / G1 * 100
370  D% = DD * 100
380  DD = D% / 100
390  PRINT X; TAB(6); G1; TAB(20
     ); GC; TAB(34); DD
400  NEXT N
410  PRINT
420  PRINT "========================
      ===================="
430  DE = SQR (DE) / (NO - 1)
440  PRINT
450  INVERSE
460  INPUT "TO CONTINUE PRESS
     'RETURN' "; Q$
470  PRINT
480  NORMAL
490  PRINT "STANDARD DEVIATION OF
      GAMMA1"
500  PRINT "CALC. BY WILSON EQ.
     "; DE
510  PRINT
520  PRINT"CONSTANTS OF WILSON
      EQUATION "
530  PRINT "A12,A21 ARE "; A1; TAB(
      25); A2
540  PRINT
550  INVERSE
560  INPUT "TO CONTINUE PRESS
     'RETURN' "; Q$
570  PRINT
580  NORMAL
590  DE = 0
600  PRINT : PRINT
```

```
610  PRINT " X GAMMA2(MEAS) GAMMA2
      (CALC) DEV(%)"
620  PRINT "=========================
      ====================== "
630  FOR N = 2 TO NO + 1 STEP 2
640  GOSUB 820
650  GC = EXP ( - LOG (1 - X + X
      * A2) - X * (A1 / (X + A1 *
      (1 - X)) - A2 / ((1 - X) + A
      2 * X)))
660  DE = (G2 - 0 GC) ∧ 2 + DE
670  DD = (G2 - GC) / G2 * 100
680  D% = DD * 100
690  DD = D% / 100
700  PRINT X; TAB(6);G2; TAB(20
      ); GC; TAB(34); DD
710  NEXT N
720  PRINT
730  PRINT "=========================
      ====================== "
740  DE = SQR (DE) / (NO - 1)
750  PRINT
760  PRINT "STANDARD DEVIATION OF
      GAMMA2"
770  PRINT "CALC. BY WILSON EQ.
      "; DE
780  PRINT
790  INPUT "NEXT CALCULATION ? (Y
      /N) "; C$
800  IF C$ = "Y" THEN GOTO 30
810  END
820  REM :READING ROUTINE FOR VLE
      DATAFILES, THE NAME OF THE
      FILE HAS TO BE IN VARIABLE
      "A$". THE ROUTINE READS T,
      NO, C1, AND C2 WHEN N = 1. IF N
      > 1 AND < NO, READS THE CORRESP
      ONDING X, G1, G2 AND GE / R / T /
      X1 / X1
830  D$ = CHR$ (4)
840  PRINT D$;"OPEN "; A$
850  PRINT D$;"POSITION "; A$; ",R"
      ; N - 1
860  PRINT D$; "READ "; A$
870  IF N = 1 THEN INPUT T,NO,C1
      ,C2
880  IF N < > 1 THEN INPUT X,G1
      ,G2,GE
890  PRINT D$; "CLOSE "; A$
```

```
900  PRINT D$; "IN#0"
910  RETURN
```

Example. Run the program for datafile D5, which contains 20 measured values of γ_1 and γ_2 for methanol–chloroform mixtures.

```
]RUN

ENTER THE NAME OF THE FILE CONTAINING
MEASURED DATA (EQ. D5) D5

SEQ. NO. OF COMPONENTS OF MIXTURE 13 10
```

X	GAMMA1(MEAS)	GAMMA1(CALC)	DEV(%)
0	9.4291315	9.4291502	−.01
.1	3.7581615	3.62472664	3.55
.2	2.2722769	2.2699705	.1
.3	1.6695538	1.70944819	−2.39
.4	1.368906	1.41638921	−3.47
.5	1.2021799	1.2441243	−3.49
.6	1.1053104	1.13692325	−2.87
.7	1.0490694	1.06934876	−1.94
.8	1.0182209	1.02833177	−1
.9	1.0038042	1.00661878	−.29
1	1	1	0

```
TO CONTINUE PRESS 'RETURN'

STANDARD DEVIATION OF GAMMA1
CALC. BY WILSON EQ. 7.89564884E − 03

CONSTANTS OF WILSON EQUATION
A12,A21 ARE .119362121 .8817874

TO CONTINUE PRESS 'RETURN'
```

X	GAMMA2(MEAS)	GAMMA2(CALC)	DEV(%)
0	1	1	0
.1	1.044486	1.04454441	−.01
.2	1.1387806	1.13175	.61
.3	1.2601662	1.24235688	1.41
.4	1.4008455	1.37358479	1.94
.5	1.5565376	1.52635326	1.93
.6	1.7235126	1.7031298	1.18
.7	1.8975296	1.9074294	−.53
.8	2.0734797	2.14374084	−3.39
.9	2.2451353	2.4176155	−7.69
1	2.73585	2.73585	0

STANDARD DEVIATION OF GAMMA2
CALC. BY WILSON EQ. 9.64628917E − 03

NEXT CALCULATION ? (Y / N) N

Remark. Explain the results.

Now we make a small detour, and call attention to technical literature describing several variations of the Wilson equation and of the NRTL equation. These have not in fact found favor, and both the Wilson equation and the NRTL equation are most popular in their original form. Nevertheless, the technique for the modification of the Wilson equation will be shown.

The One-Parameter Form of the Wilson Equation

Wilson's equation contains two adjustable parameters, g_{12} and g_{21}, the physical content of which has been already discussed. Recall, however, the theory of regular mixtures. In that theory the density of the internal energy of evaporation was called the cohesion energy, and it was assumed that if one mole of type-1 molecules evaporates from the surrounding of type-2 molecules, or one mole of type-2 molecules evaporates from the surrounding of type-1 molecules, then the molar cohesion energy can be taken as the geometrical mean of the cohesion energies of the pure components: $c_{12} = c_{21}$.

On the analogy of this, let us postulate than in the Wilson equation $g_{12} = g_{21}$. This actually means that the same energy is needed when one mole of type-1 molecules leaves the surrounding of type-2 molecules, or when one mole of type-2 molecules leaves the surrounding of type-1 molecules. Assuming this, the Wilson equation takes the following form:

$$\frac{g^{\mathrm{E}}}{\mathbf{R}T} = x_1 \ln\left(x_1 + x_2 \frac{v_2}{v_1} \exp\left(-\frac{g_{12} - g_{11}}{\mathbf{R}T}\right)\right) - x_2 \ln\left(x_2 + x_1 \frac{v_1}{v_2} \exp\left(-\frac{g_{12} - g_{22}}{\mathbf{R}T}\right)\right)$$

Let us introduce the following notation:

$$A_{12} = 1 - \frac{v_1}{v_2} \exp\left(-\frac{g_{12} - g_{11}}{\mathbf{R}T}\right)$$

$$A_{21} = 1 - \frac{v_2}{v_1} \exp\left(-\frac{g_{12} - g_{22}}{\mathbf{R}T}\right)$$

Then the one-parameter Wilson equation is

$$\frac{g^E}{\mathbf{R}T} = -x_1 \ln(1 - A_{21}x_1) - x_2 \ln(1 - A_{12}x_2)$$

where g_{ii} is the internal energy of evaporation of pure component i at the temperature and pressure of the system. Thus, there remains a single parameter to be fitted: g_{12}.

By partial differentiation we have for the activity coefficient

$$\ln \gamma_1 = -\ln(1 - A_{21}x_2) + x_2\left(\frac{x_2 A_{12}}{1 - A_{12}x_1} - \frac{x_1 A_{12}}{1 - A_{21}x_2}\right)$$

$$\ln \gamma_2 = -\ln(1 - A_{12}x_1) + x_1\left(\frac{x_2 A_{12}}{1 - A_{12}x_1} - \frac{x_1 A_{12}}{1 - A_{21}x_2}\right)$$

Extrapolated to infinite dilution, the terminal activity coefficients are

$$\ln \gamma_1^{\infty} = \ln(1 - A_{21}) + A_{12}$$

$$\ln \gamma_2^{\infty} = \ln(1 - A_{12}) + A_{21}$$

If we calculate from the experimental γ_1^{∞} and γ_2^{∞} the values of A_{12} and A_{21} and from them the values of g_{12} and g_{21}, then the latter will differ somewhat from the original values, but the calculation can be continued iteratively to find that value g_{12} which reflects the experimental data fairly well under the assumption $g_{12} - g_{21} = 0$.

The Enthalpic Wilson Equation

In the theory of regular mixtures of molecules of different volumes, the following expression was obtained for the excess free enthalpy of mixing:

$$g^{\mathrm{E}} = \frac{(r_1 x_1)(r_2 x_2)}{r_1 x_1 + r_2 x_2} W$$

If energetic interaction between the molecules produces local inhomogeneity, then the equation above must be modified in two ways:

1. The molar exchange energy cannot be calculated as prescribed by the theory of regular mixtures, but must be calculated according to Wilson's assumption with respect to the quotient r_1/r_2.
2. Wilson's expression for the local mole fractions must be substituted into the molar exchange energy.

As concerns the first modification, we start from Wilson's assumption that

$$\frac{r_1}{r_2} = \frac{v_1}{v_2} \exp\left(-\frac{g_{21} - g_{22}}{\mathbf{R}T}\right) = A_{21}$$

with the further assumption, made in conjunction with the one-parameter

Wilson equation, that $g_{12} = g_{21}$. Let us take logarithms and form the product $A_{12}A_{21}$:

$$\frac{2g_{12} - g_{22} - g_{11}}{\mathbf{R}T} = -\ln(A_{12}A_{21})$$

On the left of the equation we have molar exchange energy:

$$W = -\mathbf{R}T\ln(A_{12}A_{21})$$

Concerning the second modification, Wilson's substitution must be introduced into the exchange energy:

$$\frac{x_1}{\phi_1} = x_1 + A_{21}x_2 \qquad \frac{x_2}{\phi_2} = x_2 + A_{12}x_1$$

With the further special assumption that one of the lattice constants $r_2 = 1$, the excess free enthalpy of mixing finally appears as

$$g^{\mathrm{E}} = -\frac{x_1}{x_1 + A_{21}x_2} \cdot \frac{x_2}{x_2 + A_{12}x_1}\ln(A_{12}A_{21})$$

This is the enthalpic Wilson equation.

For the activity coefficients we have

$$\ln\gamma_1 = -\frac{x_2\ln(A_{12}A_{21})}{(x_1 + A_{21}x_2)(x_2 + A_{12}x_1)}$$

$$\times\left(1 + x_1\left(1 + \frac{1}{x_1 + A_{21}x_2} - \frac{A_{12}}{x_2 + A_{12}x_1}\right)\right)$$

$$\ln\gamma_2 = -\frac{x_1\ln(A_{12}A_{21})}{(x_1 - A_{21}x_2)(x_2 + A_{12}x_1)}$$

$$\times\left(1 + x_2\left(1 - \frac{A_{21}}{x_1 + A_{21}x_2} - \frac{1}{x_2 + A_{12}x_1}\right)\right)$$

The Orye Equation

The Orye equation takes into consideration in the excess free enthalpy the excess enthalpy arising from energetic interaction, and the excess entropy arising from incompletely random arrangement for geometrical reasons. Actually, Orye's equation is the sum of the Wilson equation and the enthalpic

Wilson equation:

$$\frac{g^{\mathrm{E}}}{\mathbf{R}T} = -x_1 \ln(x_1 + A_{21}x_2) - x_2 \ln(A_{12}x_1 + x_2) - \frac{x_1 x_2 \ln(A_{12}A_{21})}{(x_1 + A_{21}x_2)(x_2 + A_{12}x_1)}$$

The importance of the Orye equation is that it fits experimental data better than its parents, the Wilson equation and the enthalpic Wilson equation. The fitting could be further improved by taking instead of the sum of the two parent equations a general linear combination of them. This would naturally mean the introduction of a third parameter.

With this, the discussion of the family of Wilson equations is terminated. The topic has been handled with restraint: several further equations belonging to this family have not been discussed. It is easy to create such variants, but in practice they have not found wide application. Nor has the Orye equation. However, there is one point worth mentioning: the Orye equation, at least in intention, takes into account the two possible causes of nonideality: the combinatorial and the energetic.

THE UNIQUAC EQUATION

The starting idea of the UNIQUAC equation is that the equation for the excess free enthalpy consists of additive terms that can be treated separately:

$$g^{\mathrm{E}} = h^{\mathrm{E}} - Ts^{\mathrm{E}}$$

The second term on the right is the measure of order arising from the arrangement of multisegmented molecules in the mixture, and the first term is the manifestation of the energetic interaction between the molecules. In the deduction of the UNIQUAC equation the two effects are separated by analyzing the excess free enthalpy into a combinatorial and a residual term:

$$g^{\mathrm{E}} = g^{\mathrm{E}}(\text{combinatorial}) + g^{\mathrm{E}}(\text{residual})$$

Combinatorial Excess Free Enthalpy

Let us discuss first the combinatorial contribution. g^{E}(combinatorial) involves the contribution due to the arrangement of multisegmented molecules in the mixture, in the calculation of which all kinds of interactions between the molecules are left out of consideration. An evaluation resting on the lattice model can be performed with the apparatus of statistical mechanics, under the assumption that each possible microstate is of identical energy. This deduction will not be repeated here. We recall that in the subsection "Thermoneutral Mixtures of Multisegment Molecules" of the section "Molecular Structure and Thermodynamic Excess Properties of Mixtures," we got by heuristical deduction to the final result that combinatorial excess free enthalpy is

$$\frac{g^{\mathrm{E}}(\text{combinatorial})}{\mathbf{R}T} = x_1 \ln \frac{\phi_1}{x_1} + x_2 \ln \frac{\phi_2}{x_2} + \tfrac{1}{2} z q_1 x_1 \ln \frac{\Theta_1}{\phi_1} + \tfrac{1}{2} z q_2 x_2 \ln \frac{\Theta_2}{\phi_2}$$

where the average surface fraction is

$$\Theta_i = \frac{x_i q_i}{\sum x_i q_i}$$

The combinatorial excess free enthalpy involves surrounding-independent parameters, but none are adjustable. This is the crucial point of the separation into combinatorial and residual contributions, because all the effects not included in the combinatorial contribution are taken into consideration in the residual contribution, incorporating them in parameters to be fitted anyway.

For example, consider a two-component mixture, containing six molecules each of type 1 and of type 2. Let these be arranged in a two-dimensional lattice, and their sizes be respectively $r_1 = 3$ and $r_2 = 2$. In this two-dimensional lattice the coordination number z is 4, which means that each lattice point has four immediate neighbors. What is the average surface fraction of the components?

Let us count the number of contacts of molecules of type 1 and type 2 in this lattice. We obtain

$$q_1 z = 8$$

$$q_2 z = 6$$

and the quantities q_i are thereby determined.

Somewhat more generally, consider a molecule of type 1, occupying r_1 lattice points. Alternatively we can say that a molecule of type 1 consists of r_1 segments. The segments are of (nearly) identical size, but they differ with respect to the magnitude of their external contact surface. For example, in *n*-pentane the area of each of the two terminal methyl groups is larger than that of a methylene group (Figure 11.10*a*). On the other hand, the central carbon atom of *iso*-pentane has no external contact surface at all (Figure 11.10*b*)

The number of the external contacts of a single molecule of type 1 is zq_1, where q_1 is a quantity proportional to the external surface area of the molecule. The lattice parameters of a molecule of type 1 are r_1 and q_1.

The following relation exists between q, z and r (for molecules containing no ring):

$$\tfrac{1}{2} z (r - q) = r - 1$$

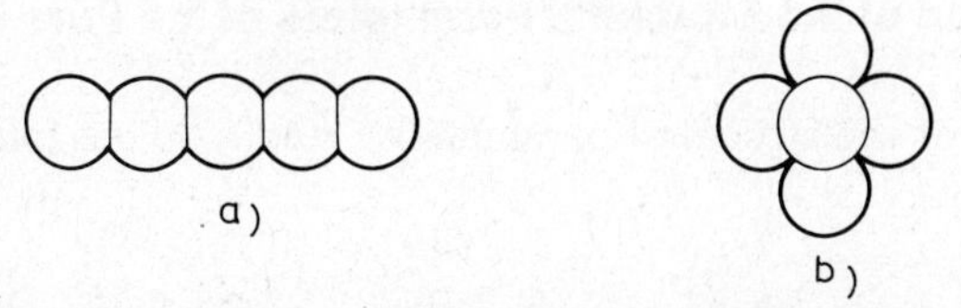

FIGURE 11.10. Molecule and segments: (a) n-pentane, (b) *iso*-pentane.

In our example,

$$\tfrac{1}{2}z(r_1 - q_1) = r_1 - 1 \qquad \tfrac{1}{2}4(3 - 2) = 3 - 1 = 2$$

$$\tfrac{1}{2}z(r_2 - q_2) = r_2 - 1 \qquad \tfrac{1}{2}4(2 - 1.5) = 2 - 1 = 1$$

The average volume fraction is

$$\phi_i = \frac{r_i x_i}{\sum_i r_i x_i}$$

The mole fractions in our case are $x_1 = x_2 = 0.5$; the volume fractions,

$$\phi_1 = \frac{r_1 x_1}{r_1 x_1 + r_2 x_2} = \frac{3 \times 0.5}{3 \times 0.5 + 2 \times 0.5} = \frac{3}{5}$$

$$\phi_2 = \frac{r_2 x_2}{r_1 x_1 + r_2 x_2} = \frac{2 \times 0.5}{3 \times 0.5 + 2 \times 0.5} = \frac{2}{5}$$

The average surface fraction can be expressed in terms of the quantity $q_i z$:

$$\Theta_i = \frac{(q_i z) x_i}{\sum_i (q_i z) x_i}$$

The average surface fractions are

$$\Theta_1 = \frac{8 \times 0.5}{8 \times 0.5 + 6 \times 0.5} = \frac{4}{7}$$

$$\Theta_2 = \frac{6 \times 0.5}{8 \times 0.5 + 6 \times 0.5} = \frac{3}{7}$$

The average volume fractions and average surface fractions differ from one another, and are suitable for the description of different effects in the model of mixtures.

Calculation of the Structural Parameters of the Pure Component

The definition of the structural parameters r and q is the following:

$$r_i = \frac{V_{wi}}{V_{ws}}$$

$$q_i = \frac{A_{wi}}{A_{ws}}$$

where V_{wi} and A_{wi} are the van der Waals volume and surface of a segment of type i, known from the work of Bondi (1968), and V_{ws} and A_{ws} the van der Waals volume and surface of the standard segment. The selection of the standard segment is a question of convention.

Let methylene be the standard segment in a polymethylene chain. The volume of the standard segment is 15.17 cm^3 $mole^{-1}$, and the area of the standard segment 2.5×10^9 cm^2 $mole^{-1}$. Substituting these values into the defining equation of the structural parameters, we have

$$r_i = \frac{V_{wi}}{15.17}$$

$$q_i = \frac{A_{wi}}{2.5 \times 10^9}$$

We will return to the calculation of V_{wi} and A_{wi} in connection with the UNIFAC method.

Residual Excess Free Enthalpy

In our discussion we have not assumed till now an energetic interaction between the molecules or segments. Thus, what was said is relevant to athermal mixtures. In the conceptual experiment performed, multisegment molecules were placed in a lattice, so that one segment fell on each lattice point. The question was exclusively one of arrangement, and interaction between molecules or segments was ruled out. In the first part of the conceptual experiment only the part of the free enthalpy arising from combinatorial effects was taken into consideration.

In the present second part of the conceptual experiment, attractive–repulsive forces between the molecules or segments will be taken into account. These forces displace the molecules from their initial arrangement. This gives a new additive contribution to the excess free enthalpy, additive because it does not rearrange the molecules, but only "distorts" the existing arrangement.

Thus, the UNIQUAC equation needs further elaboration. We seek the residual excess free enthalpy.

What is the situation in the equilibrium state established after allowing for repulsive–attractive forces? Consider that these forces decay within a short distance, and act only between adjacent molecules, or rather between some segment of one molecule and the segment of another molecule in direct contact with the first. The effects between various segments are naturally different, so that u_{ij}, expressing the effect between molecules i and j, represents some kind of average over the segments. Let us introduce the notation $g_{ij} = \frac{1}{2}zu_{ij}$ for the energy needed to evaporate into vacuum one mole of type-j molecules from the surrounding of type-i molecules, taking into consideration that z type-i molecules surround one type-j molecule.

The displacement of the molecules with respect to their arrangement with no field of force means moreover that local surface fractions are relevant instead of average surface fractions in the development of the residual excess free enthalpy. This statement is of value if some constitutive relationship, linking the local surface fraction to the average surface fraction, is available.

Recall that Wilson introduced local mole fractions into the athermal Flory–Huggins model. His intuitive assumption was that mole fractions weighted with the appropriate Boltzmann factor can account for the excess free enthalpy. (The Boltzmann factor includes g_{ij}.) Wilson concluded that the substitution

$$\frac{x_1}{\phi_1} = x_1 + A_{21}x_2 \qquad \frac{x_2}{\phi_2} = x_2 + A_{12}x_1$$

must be made in the Flory–Huggins equation, where

$$A_{21} = \frac{v_2}{v_1}\exp\left(-\frac{g_{21} - g_{11}}{\mathbf{R}T}\right)$$

On the anlogy of this, we suggest substituting for the fraction in the equation of the combinatorial excess free enthalpy the following expression:

$$\frac{\Theta_1}{\phi_1} = \Theta_1 + \Theta_2 A_{21}$$

This means that the residual excess free enthalpy can be written in the following form:

$$\frac{g^{\mathrm{E}}(\text{residual})}{\mathbf{R}T} = q_1 x_1 \ln(\Theta_1 + \Theta_2 A_{21}) + q_2 x_2 \ln(\Theta_2 + \Theta_1 A_{12})$$

where A_{21} and A_{12} are adjustable parameters.

The Complete UNIQUAC Equation

The complete UNIQUAC equation consists of the combinatorial and residual contributions:

$$\frac{g^{\mathrm{E}}}{\mathbf{R}T} = x_1 \ln \frac{\phi_1}{x_1} + x_2 \ln \frac{\phi_2}{x_2} + \frac{z}{2}\left(q_1 x_1 \ln \frac{\Theta_1}{\phi_1} + q_2 x_2 \ln \frac{\Theta_2}{\phi_2}\right)$$

$$+ q_1 x_1 \ln(\Theta_1 + \Theta_2 A_{21}) + q_2 x_2 \ln(\Theta_2 + \Theta_1 A_{12})$$

This equation contains only two adjustable parameters, A_{21} and A_{12}.

The activity coefficients yielded by the complete UNIQUAC equation are

$$\ln \gamma_1 = \ln \frac{\phi_1}{x_1} + \frac{z_2}{2} q_1 \ln \frac{\Theta_1}{\phi_1} + \phi_2\left(l_1 + \frac{r_1}{r_2} l_2\right) - q_1 \ln(\Theta_1 + \theta_2 A_{21})$$

$$+ \Theta_2 q_2 \left(\frac{A_{21}}{\Theta_1 + \Theta_2 A_{21}} - \frac{A_{12}}{\Theta_2 + \Theta_1 A_{12}}\right)$$

and $\ln \gamma_2$ obtained by exchange of the indices.

The UNIQUAC equation can be extended to multicomponent mixtures. Its great merit in this case is that only binary interactions are involved: an increase in component number does not necessitate the introduction of supplementary parameters. The binary interaction parameters can be transferred to any arbitrary system in which the two given components occur.

The combinatorial free-enthalpy contribution in the UNIQUAC equation does not contain adjustable parameters; only a knowledge of the geometric data (group volume, group area) on the segments of the molecule is needed for its calculation. However, the residual contribution contains the adjustable parameters A_{12} and A_{21}. The question presents itself, whether it is possible to express these parameters in terms of the same sets which are used for the calculation of the combinatorial term. It must be taken into consideration that this time the two adjustable parameters are related to a pair of components, not to a single component. As we know of these parameters that they can be applied to any arbitrary surrounding, the above question must be modified; we ask instead whether an arbitrary solution can be regarded as a solution of all the groups from which the molecules forming the solution are built up. Giving a positive answer to this question, it can be said that the essence of nonideality is not the interaction of the molecules, but the interaction of the groups.

The UNIFAC method, which represents the application of the solution-of-functional-groups concept to the UNIQUAC theory, rests on this consideration.

THE UNIFAC EQUATION

The aim of the solution-of-functional-groups concept is to make possible the use of existing vapor–liquid equilibrium data for the estimation of equilibrium systems for which no direct experimental data are available. In order to achieve this, activity coefficients obtained experimentally are reduced in a suitable way, to obtain parameters characteristic of group-pair interactions, which are expected to be independent of the surroundings. These parameters are used for the estimation of the activity coefficients of other systems not yet studied, presuming of course that these systems contain only the group pairs studied. "Group" in this context means suitably selected part of the molecular structure, such as for example $—CH_3$, $—COCH_2—$, and $—CH_2Cl$. In any case the use of a suitable segmentation is a key issue in the success of the solution-of-functional-groups concept. For example, methylethylketone is not $(CH_3)(CO)(CH_2)(CH_3)$, but $(CH_3CO)(CH_2)(CH_3)$. This reflects that the (CH_3) group beside the (CO) group behaves differently than that beside the (CH_2) group. This distinction is justified from the point of view of chemical reactivity and phase-equilibrium behavior alike.

The postulates of the solution-of-functional-groups concept are as follows.

In accordance with the UNIQUAC theory, the activity coefficient can be expressed as the sum of two terms: the combinatorial part follows fundamentally from the size and shape of the molecules, and the residual part fundamentally from the energetic interaction:

$$\ln \gamma_i = \ln \gamma_i^{(c)} + \ln \gamma^{(r)}$$

The combinatorial part can be calculated when the structual parameters of the participating groups are known. These are the volumes R_k and the surface areas Q_k.

The combinatorial activity coefficient is given by the UNIQUAC theory:

$$\ln \gamma_i^{(c)} = \ln \frac{\phi_i}{x_i} + \frac{z}{2} q_i \ln \frac{\Theta_i}{\phi_i} + l_i - \frac{\phi_i}{x_i} \sum_j x_j l_j$$

where

$$l_i = \frac{z}{2}(r_i - q_i) - (r_i - 1)$$

the coordination number being by convention

$$z = 10$$

the surface fraction

$$\Theta_i = \frac{q_i x_i}{\sum_j q_j x_j}$$

and the volume fraction

$$\phi_i = \frac{r_i x i}{\sum_j r_j x_j}$$

The structural parameters of the molecule are

$$r_i = \sum_k \nu_k^{(i)} R_k$$

$$q_i = \sum_k \nu_k^{(i)} Q_k$$

where k = ordinal index of the groups,
$\nu_k^{(i)}$ = number of groups k in molecule i,
$R_k = V_{wk}/15.17$,
$Q_k = A_{wk}/2.5 \times 10^9$.

These are the structural parameters for group k. The structural parameters of the molecule, r_i and q_i, have been already defined in connection with the UNIQUAC equation. What is essential for our purpose now is that structural parameters pertinent to the groups have been published in tabular form by the authors of UNIFAC, and the structural parameters of the molecules can be calculated from these data.

The combinatorial activity coefficient contains only structural parameters and is independent of temperature.

The residual term of the UNIQUAC equation is

$$\ln \gamma_i^{(r)} = q_i \left(1 - \ln\left(\sum_j \Theta_j A_{ji} \right) - \sum_j \frac{\Theta_j A_{ji}}{\sum_k \Theta_k A_{ki}} \right)$$

where

$$A_{ji} = \exp\left(- \frac{g_{ji} - g_{ii}}{\mathbf{R}T} \right)$$

If we wish to generate the residual activity coefficient of the component additively from the residual activity coefficients of the groups, then a function $\ln \Gamma_k$ of similar form, assignable to each structural group, is needed:

$$\ln \Gamma_k = Q_k \left(1 - \ln\left(\sum_m \Theta_m \psi_{mk} \right) - \sum_m \frac{\Theta_m \psi_{km}}{\sum_n \Theta_n \psi_{nm}} \right)$$

where m and n are the ordinal indices of the groups. Here the surface fraction of group m is

$$\Theta_m = \frac{\Theta_m X_m}{\sum_n \Theta_n X_n}$$

where the group fraction of group m is

$$X_m = \frac{\sum_j \nu_m^{(i)} x_i}{\sum_i \sum_n \nu_n^{(i)} x_i}$$

and $\nu_m^{(i)}$ is the stoichiometric index of group m in molecule i. This equation also contains the temperature-dependent quantity

$$\psi_{nm} = \exp\left(-\frac{g_{nm} - g_{mm}}{\mathbf{R}T}\right) = \exp\left(-\frac{a_{nm}}{T}\right)$$

which in turn contains the parameter a_{nm} of group interaction. On the other hand

$$\psi_{mn} = \exp\left(-\frac{g_{mn} - g_{nn}}{\mathbf{R}T}\right) = \exp\left(-\frac{a_{mn}}{T}\right)$$

Thus it is evident that $a_{nm} \neq a_{mn}$. Each pair of groups has two interaction parameters, but three-group and higher-order interaction parameters are not needed even in multicomponent mixtures. The parameters were estimated from measured equilibrium data. The parameter a_{nm} of group interaction is independent of temperature and has the dimensions of temperature.

In the logarithmic form the group activity coefficients are additive:

$$\ln \gamma_i^{(r)} = \sum_k \nu_k^{(i)} \left(\ln \Gamma_k - \ln \Gamma_k^{(i)}\right)$$

where Γ_k = residual activity coefficient of group k in the solution,
$\Gamma_k^{(i)}$ = residual activity coefficient of group k in the reference solution, containing only type-i molecules,
$\nu_k^{(i)}$ = number of groups k in a type-i molecule.

Now the following must be considered. The activity coefficient of component i was defined so that when $x_i = 1$ then $\gamma_i = 1$, that is, the activity coefficient of pure component i is 1. Thus, the activity coefficient of group k cannot be 1 in the pure component, and neither is it independent of the component in which it is present. Consider, for example, that in pure methylethylketone the group fraction of the CH_3CO group is

$$\frac{1(CH_3CO)}{1(CH_3CO) + 1(CH_2) + 1(CH_3)} = \frac{1}{3}$$

while in pure acetone it is

$$\frac{1(CH_3CO)}{1(CH_3CO) + 1(CH_3)} = \frac{1}{2}$$

The individual group contributions, in any surrounding containing $1, 2, \ldots, N$ groups, depend only on the group concentrations and on the temperature:

$$\left.\begin{matrix} \Gamma_k \\ \Gamma_k^{(i)} \end{matrix}\right\} = \mathrm{F}[x_1, x_2, \ldots, x_N, T]$$

that is, the same function is valid for Γ_k and for Γ_k^i.

Algorithm and Program: Calculation of the Activity Coefficient with the UNIFAC Method

Write a program to calculate the activity coefficient for binary mixtures by the UNIFAC group-contribution method.

Algorithm.

1. Read the next data:
   (a) The number of different functional subgroups in the first molecule an integer between 1 and 64, according to the UNIFAC subgroup definitions).
   (b) The number of groups of type I in the first molecule.
   (c) The number of different functional (sub)groups in the second molecule.
   (d) The UNIFAC codes of groups of type I in the second molecule.
   (e) The number of groups of type I in the second molecule.
   (f) The group size R and surface area Q of each group of type I (according to the UNIFAC table).
   (g) The interaction parameter between I and J subgroups (according to the UNIFAC table). Subgroups within the same main group have group interaction parameters equal to zero.
2. Read the actual temperature.
3. Read the composition of the binary mixture.

4. Calculate the combinatorial activity coefficient.
5. Calculate the residual activity coefficient of group I in the solution.
6. Calculate the residual activity coefficient of group I in a reference solution.
7. Print the results.

Program UNIFAC. A possible realization:

```
]LOAD UNIFAC,D2
]LIST

 10  REM :CALCULATION OF ACTIVITY
      COEFFICIENT BY UNIFAC METHOD
 20  REM :NUMBER OF COMPONENTS = 2;
      MAX. NO. OF GROUPS = 8
 30  REM :N CONTAINS THE CODES OF
      GROUPS, THE NUMBER OF GROUPS
      OF KIND "I" IN MOLECULES 1 AND 2
 40  REM :P CONTAINS THE GROUP INT
      ERACTION PARAMETERS
 50  REM :R AND Q ARE THE MOLECULAR
      VOLUMES AND SURFACE AREAS
 60  DIM N(3,8),P(8,8),Q(8),R(8),Y
      (3,2)G(4,8),A$(10)
 70  FOR I = 1 TO 8
 80  Q(I) = R(I) = N(1,I) = N(2,I) =
      N(3,I) = 0
 90  FOR J = 1 TO 8
100  P(I,J) = 0
110  NEXT J
120  NEXT I
130  INPUT "NO. OF DIFFERENT GROUPS
      IN COMPONENT 1. "; N0
140  FOR I = 1 TO N0
150  PRINT "UNIFAC CODE OF "; I
      "GROUP OF 1. COMPONENT"
160  INPUT " "; N(1,I)
170  PRINT "NO. OF "; N(1,I);
      "GROUP IN 1. COMPONENT"
180  INPUT " "; N(2,I)
190  NEXT I
200  INPUT "NO. OF DIFFERENT
      GROUPS IN COMPONENT 2. "; N1
210  FOR J = 1 TO N1
220  PRINT "UNIFAC CODE OF "; J;
      "GROUP OF 2. COMPONENT"
230  INPUT " "; N2
240  FOR I = 1 TO N0
```

```
250  IF N(1,I) = N2 THEN GOTO 30
     0
260  NEXT I
270  N0 = N0 + 1
280  I = N0
290  N(1,I) = N2
300  PRINT "NO. OF "; (1,I);
     "GROUP IN 2. COMPONENT"
310  INPUT " "; N(3,I)
320  NEXT J
330  FOR I = 1 TO N0
340  PRINT "PARAMETERS R AND Q OF
     "; N(1,I); " GROUP"
350  INPUT R(I),Q(I)
360  NEXT I
370  FOR I = 1 TO N0
380  FOR J = 1 TO N0
390  IF I = J THEN GOTO 420
400  PRINT "INTERACTION PARAMETER
      BETWEEN "; N(1,I); "-" N(1,J);
      "K"
410  INPUT " "; P(I,J)
420  NEXT J
430  NEXT I
440  INPUT "ENTER TEMPERATURE (K)";
      T
450  PRINT
460  PRINT TAB(20); "INPUT DATA"
470  PRINT : PRINT : PRINT
480  PRINT "GROUP CODES:
     ";
490  FOR I = 1 TO N0
500  PRINT N(1,I); SPC(3);
510  NEXT I
520  PRINT
530  PRINT : PRINT
540  FOR I = 1 TO N0
550  FOR J = 1 TO N0
560  P(I,J) = EXP ( - P(I,J) / T)
570  NEXT J
580  NEXT I
590  Y(1,1) = 0
600  Y(1,2) = 0
610  Y(2,1) = 0
620  Y(2,2) = 0
630  FOR I = 1 TO N0
640  Y(1,1) = Y(1,1) + N(2,I) * R(I)
650  Y(1,2) = Y(1,2) + N(3,I) * R(I)
660  Y(2,1) = Y(2,1) + N(2,I) * Q(I)
```

```
670  Y(2,2) = Y(2,2) + N(3,I) * Q(I)
680  NEXT I
690  Y(3,1) = 5 * (Y(1,1) - Y(2,1)
      ) - (Y(1,1) - 1)
700  Y(3,2) = 5 * (Y(1,2) - Y(2,2)
      ) - (Y(1,2) - 1)
710  PRINT : PRINT
720  INPUT "ENTER MOLAR FRACTION
      OF FIRST COMPONENT "; X1
730  X2 = 1 - X1
740  REM :CALCULATION OF ACTIVITY
      COEFFICIENTS
750  REM :CALCULATION OF COMBINA
      TORIAL PART
760  Q = Y(2,1) * X1 + Y(2,2) * X2
770  R = Y(1,1) * X1 + Y(1,2) * X2
780  L = Y(3,1) * X1 + Y(3,2) * X2
790  G1 = LOG (Y(1,1) / R) + 5 *
      LOG (Y(2,1) / Q * R / Y(1,1
      )) * Y(2,1) + Y(3,1) - Y(1,1
      ) / R * L
800  G2 = LOG (Y(1,2) / R) + 5 *
      LOG (Y(2,2) / Q * R / Y(1,2
      )) * Y(2,2) + Y(3,2) - Y(1,2
      ) / R * L
810  GOSUB 1160
820  FOR I = 1 TO N0
830  G1 = G1 + N(2,I) * G(4,I)
840  G2 = G2 + N(3,I) * G(4,I)
850  NEXT I
860  X1 = 1
870  X2 = 0
880  GOSUB 1160
890  FOR I = 1 TO N0
900  G1 = G1 - N(2,I) * G(4,I)
910  NEXT I
920  X1 = 0
930  X2 = 1
940  GOSUB 1160
950  FOR I = 1 TO N0
960  G2 = G2 - N(3,I) * G(4,I)
970  NEXT I
980  G1 = EXP (G1)
990  G2 = EXP (G2)
1000 PRINT : PRINT
1010 PRINT "ACTIVITY COEFFICIENT
      OF FIRST COMPONENT "; G1
1020 PRINT
1030 PRINT "ACTIVITY COEFFICIENT
```

```
        OF SECOND COMPONENT "; G2
1040  INPUT "NEXT COMPOSITION? (Y
        /N) "; B$
1050  IF B$ = "Y" THEN GOTO 710
1060  FOR I = 1 TO N0
1070  FOR J = 1 TO N0
1080  P(I,J) = - T * LOG (P(I,J))
1090  NEXT J
1100  NEXT I
1110  INPUT "NEXT TEMPERATURE ?
        (Y / N) "; B$
1120  IF B$ = "Y" THEN GOTO 440
1130  INPUT "NEXT MIXTURE ? (Y / N)
        "; B$
1140  IF B$ = "Y" THEN GOTO 70
1150  END
1160  REM :SUBROUTINE FOR CALCULA
        TION OF RESIDUAL PART
1170  S0 = 0
1180  S1 = 0
1190  FOR I = 1 TO N0
1200  G(1,I) = X1 * N(2,I) + X2 *
        N(3,I)
1210  S0 = S0 + G(1,I)
1220  NEXT I
1230  FOR I = 1 TO N0
1240  G(1,I) = G(1,I) / S0
1250  G(2,I) = G(1,I) * Q(I)
1260  S1 = S1 + G(2,I)
1270  NEXT I
1280  FOR I = 1 TO N0
1290  G(2,I) = G(2,I) / S1
1300  G(3,I) = 0
1310  NEXT I
1320  FOR I = 1 TO N0
1330  FOR J = 1 TO N0
1340  G(3,I) + G(3,I) + G(2,J) * P
        (J,I)
1350  NEXT J
1360  NEXT I
1370  FOR I = 1 TO N0
1380  G = 1 - LOG (G(3,I))
1390  FOR J = 1 TO N0
1400  G = G - G(2,J) * P(I,J) / G(
        3,J)
1410  NEXT J
1420  G(4,I) = Q(I) * G
1430  NEXT I
1440  RETURN
```

Example. Run the program for a heptane–acetone mixture at temperature 353 K and 0.5–0.5 mole-fraction composition.

```
]RUN
NO. OF DIFFERENT GROUPS IN COMPONENT 1. 2
UNIFAC CODE OF 1 GROUP OF 1. COMPONENT
 1
NO. OF 1 GROUP IN 1. COMPONENT
 2
UNIFAC CODE OF 2 GROUP OF 1. COMPONENT
 2
NO. OF 2 GROUP IN 1. COMPONENT
 5
NO. OF DIFFERENT GROUPS IN COMPONENT 2. 2
UNIFAC CODE OF 1 GROUP OF 2. COMPONENT
 1
NO. OF 1 GROUP IN 2. COMPONENT
 1
UNIFAC CODE OF 2 GROUP OF 2. COMPONENT
 22
NO. OF 22 GROUP IN 2. COMPONENT
 1
PARAMETERS R AND Q OF 1 GROUP
?0.9011
??0.848
PARAMETERS R AND Q OF 2 GROUP
?0.6744
??0.54
PARAMETERS R AND OF 22 GROUP
?1.6722
??1.488
INTERACTION PARAMETER BETWEEN 1-2,K
 0
INTERACTION PARAMETER BETWEEN 1-22,K
 476.4
INTERACTION PARAMETER BETWEEN 2-1,K
 0
INTERACTION PARAMETER BETWEEN 2-22,K
 476.4
INTERACTION PARAMETER BETWEEN 22-1,K
 26.76
INTERACTION PARAMETER BETWEEN 22-2,K
 26.76
ENTER TEMPERATURE (K) 353

 INPUT DATA

GROUP CODES: 1 2 22

ENTER MOLAR FRACTION OF FIRST COMPONENT 0.5
```

```
ACTIVITY COEFFICIENT OF FIRST COMPONENT 1.42006972
ACTIVITY COEFFICIENT OF SECOND COMPONENT 1.49979979

NEXT COMPOSITION? (Y / N) Y

ENTER MOLAR FRACTION OF FIRST COMPONENT 0.1

ACTIVITY COEFFICIENT OF FIRST COMPONENT 3.63172568
ACTIVITY COEFFICIENT OF SECOND COMPONENT 1.02012104

NEXT COMPOSITION? (Y / N) N
NEXT TEMPERATURE ? (Y / N) Y
ENTER TEMPERATURE (K) 360

 INPUT DATA

GROUP CODES: 1 2 22

ENTER MOLAR FRACTION OF FIRST COMPONENT 0.5

ACTIVITY COEFFICIENT OF FIRST COMPONENT 1.40811711
ACTIVITY COEFFICIENT OF SECOND COMPONENT 1.49072003

NEXT COMPOSITION? (Y / N) N
NEXT TEMPERATURE ? (Y / N) N
NEXT MIXTURE ? (Y / N) N
```

COMPARISON OF DIFFERENT EQUATIONS

We have finished the description of models that are used with good effect in practice for the representation of mixtures. These models are more or less

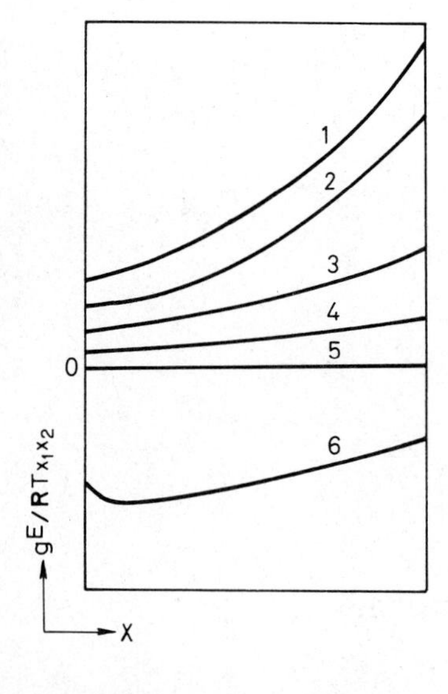

FIGURE 11.11. Binary mixtures: 1, chloroform–methanol; 2, chloroform–ethanol; 3, chloroform–heptane; 4, chloroform–(1,2)-dichloroethane; 5, chloroform–(1,1)-dichloroethane; 6, chloroform–acetone.

demanding computationally, and there is a practical problem of achieving the description having the prescribed accuracy with the model requiring the least calculation. One group of models contain parameters which can be calculated from measured data; the other group are based on a physical picture and can predict experimental input.

Figure 11.11 shows $g^E/\mathbf{R}Tx_1x_2$ for six binary mixtures, plotted as a function of x_1. One component of each mixture is chloroform. One of the six mixtures is ideal (chloroform–1,1-dichloroethane); four mixtures show positive, and one negative, deviation. The 1,2 isomer of dichloroethane forms with chloroform a rather moderately positive-deviation mixture. A somewhat greater positive deviation is displayed by the heptane–chloroform mixture. Chloroform with ethanol, and still more with methanol, shows strong positive deviation and asymmetry. The mixture of chloroform with acetone is the only one with negative deviation.

The last mixture is also among the six mixtures in Figure 11.12. One component of each of the other five mixtures is also acetone. Of these, the water–acetone mixture also displays strongly negative deviation. It is interesting that the ethanol–acetone mixture is ideal, and the mixtures of acetone with both methanol and ethanol are symmetric, while the acetone–pentane and acetone–carbon disulfide mixtures show a positive deviation.

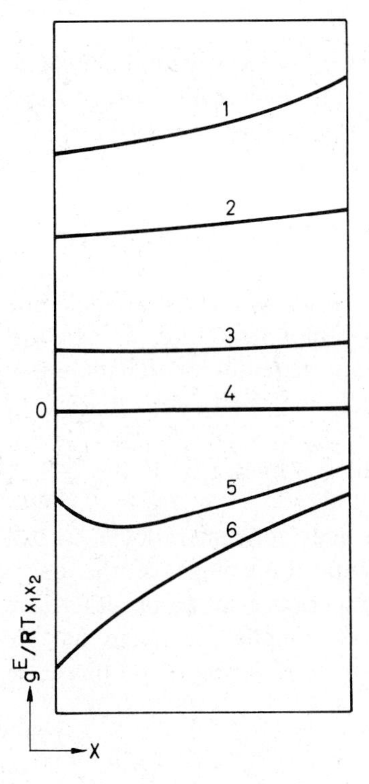

FIGURE 11.12. $g^E/\mathbf{R}Tx_1x_2$ versus x_1 for binary mixtures: 1, acetone–carbon disulfide; 2, acetone–pentane; 3, methanol–acetone; 4, ethanol–acetone; 5, acetone–chloroform; 6, water–acetone.

Presuming that the form of the function $g^E/\mathbf{R}Tx_1x_2$ is known from experimental data or from some prediction, the procedure is the following with mixtures of positive deviation. If the relationship:

1. Is a straight line, use the Margules equation.
2. Is mildly concave, use the van Laar equation.
3. Is strongly concave, but without an extremum, use the Wilson equation.
4. Has an extremum, use the NRTL equation.

For mixtures with negative deviation the use of the NRTL equation is recommended.

With these two- or three-parameter equations parameter estimation is not completely free of problems, as the parameters show strong statistical correlation. This does not cause difficulties in all cases, but has the consequence that the parameters are not independent of the surrounding, even if this would follow from the deterministic model, as for example from the model of regular mixtures, or in the cases of the Wilson and the NRTL equations.

In the absence of experimental data, recourse must be had to a model that can to predict the properties of the mixture when the physical properties of its pure components are known. Two such models are recommended:

1. For mixtures of nonpolar or slightly polar components, the model of regular mixtures can be used.
2. If strong nonideality is to be expected, the UNIFAC method should be used. This method must be used also if strong nonideality occurs in one part of the concentration range in the formation of two liquid phases.

TWO-COMPONENT SYSTEMS OF LIMITED SOLUBILITY

The mixture models discussed above (NRTL, Wilson, UNIQUAC) took into account the energetic interaction between the components and, as a consequence of this, the development of local inhomogeneities in the molecular domains of both components. The extent of inhomogeneity depends on the chemical nature of the given component pair, and can vary greatly. This was particularly mentioned in conjunction with the parameter α of the NRTL equation.

Inhomogeneity can even be so strong that it does not remain local, so that inhomogeneity appears–at least in a certain composition range–on the macro level. This is the phenomenon of *segregation*: Experience shows that there are liquid pairs that are not miscible over the whole composition range, but are more or less soluble in one another, and form two equilibrium liquid phases in contact.

This means, from a theoretical point of view, that the free enthalpy of a given amount of substance, distributed between the two phases in equilibrium, is lower than it would be in a hypothetical homogeneous phase, containing the same amount of substance. In this case, the free-enthalpy function of an adequate model describing the properties of the homogeneous phase mixture displays two local minima. The two-phase region is formed along the straight line which is a double tangent of the free enthalpy curve: if the composition of the whole two-component system is given by the mole fraction x_1, then the mole fraction of component 1 will be x_{11} in one phase, and x_{12} in the other, so that $x_{11} < x_1 < x_{12}$, and the quantitative ratio of the two phases is obtained by linear combination (Figure 11.13)

The two-phase region between x_{11} and x_{12} can be divided into three parts. In the regions between x_{11} and x_B and between x_C and x_{12}, the system is metastable, while in the region between x_B and x_C it is unstable. By this we understand that a homogeneous solution of mole fraction x_1 belonging to the metastable region can be prepared at a temperature higher than that of the investigation, and the temperature can then be adjusted by slow cooling without segregation. The same cannot be achieved in the unstable range. The segregation of the metastable phase is brought about by simple mechanical action (shaking).

No classical thermodynamic explanation can be given for this process as indeed classical thermodynamics does not say anything about the path between two (equilibrium) states of a system; it only predicts the difference between properties in these two states. Nevertheless, we can reason in the following way. The segregation of a mixture in a metastable state is accompanied by a decrease in free enthalpy. As a process, segregation begins with the development of local inhomogeneity in the homogeneous phase, and thus the phase, maintaining for the time being its macroscopic homogeneity, is locally separated into supercomponents, somewhat as discussed in the NRTL model. The molar free enthalpy of one of the components is higher, and that of the other lower, than the molar free enthalpy of the homogeneous phase, and it is certain

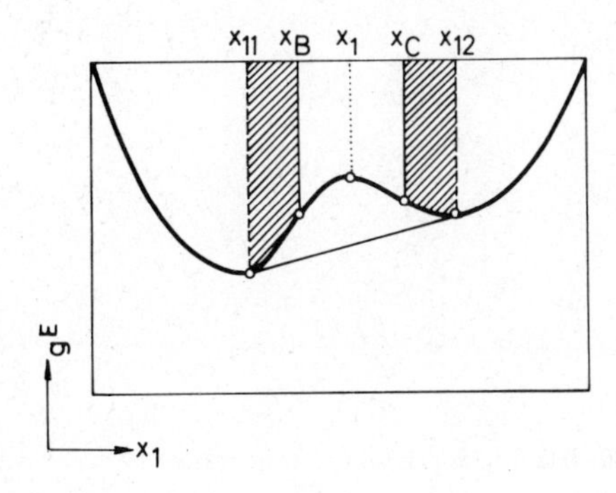

FIGURE 11.13. Segregation in a two-component liquid.

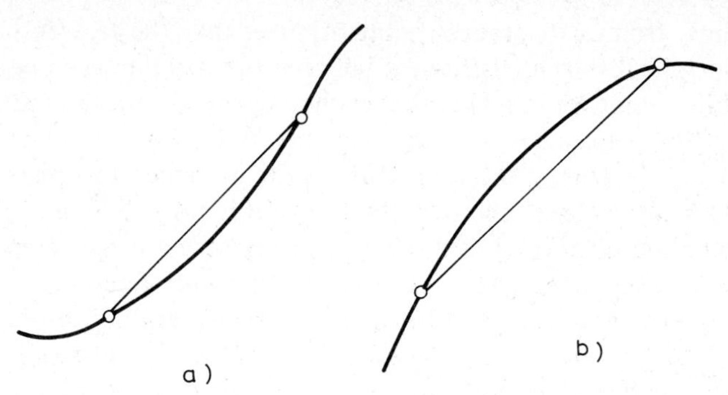

FIGURE 11.14. Segregation: (a) metastable range, (b) unstable range.

that their linear combination in the metastable region is higher than that of the corresponding homogeneous phase (Figure 11.14). Therefore, segregation does not start spontaneously in the metastable region, so that an external mechanical intervention is needed. On the other hand, in the unstable region the linear combination of the free enthalpies of the two supercomponents is smaller than the free enthalpy of the corresponding homogeneous phase, and thus segregation begins spontaneously.

The curve of g versus x is plotted again in Figure 11.15, marking the noteworthy points on the curves as characterized by the first, second and third derivatives:

Minimum	A	$dg/dx = 0$	$d^2g/dx^2 > 0$
Inflection	B	$d^2g/dx^2 = 0$	$d^3g/dx^3 < 0$
Inflection	C	$d^2g/dx^2 = 0$	$d^3g/dx^3 > 0$
Minimum	D	$dg/dx = 0$	$d^2g/dx^2 > 0$

If the temperature changes in the appropriate direction, then the inflection points approach one another, and the slope of the tangents at the inflec-

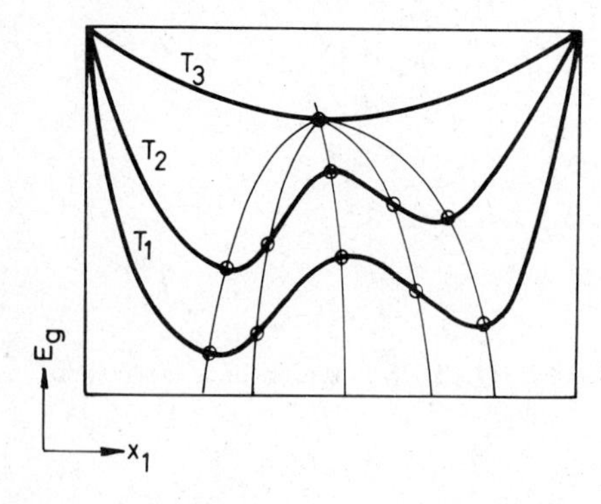

FIGURE 11.15. Effect of temperature on segregation.

tion points decreases. Finally, the two minimum points and the two inflection points merge into a single critical point, at which the curve has a horizontal tangent; the critical point is the minimum of the free-enthalpy function corresponding to the critical solution temperature. Accordingly, the criteria for the critical point are

$$A \text{ and } D: \quad \frac{dg}{dx} = 0$$

$$B \text{ and } C: \quad \frac{d^2g}{dx^2} = 0$$

$$B \text{ and } C: \quad \frac{d^3g}{dx^3} = 0$$

$$\frac{d^4g}{dx^4} > 0$$

the last criterion following from the condition that at a minimum the first nonvanishing derivative must be positive.

These very analytical considerations make possible the discussion of a given mixture model with regard to its suitability for describing the mixture properties of substances of limited solubility.

The molar free enthalpy of the mixture is

$$g = x_1 g_1 + x_2 g_2 + \mathbf{R}T(x_1 \ln x_1 + x_2 \ln x_2) + g^{\mathrm{E}}$$

where g^{E} is the molar excess free enthalpy, which is described by a suitable model. Thus, criteria for the critical mixing temperature are

$$\frac{d^2g}{dx_1^2} = \frac{\mathbf{R}T_c}{x_1 x_2} + \frac{d^2g^{\mathrm{E}}}{dx_1^2} = 0$$

$$\frac{d^3g}{dx_1^3} = \frac{\mathbf{R}T_c(x_1 - x_2)}{x_1^2 x_2^2} + \frac{d^3g^{\mathrm{E}}}{dx_1^3} = 0$$

$$\frac{d^4g}{dx_1^4} = 2\mathbf{R}T_c \frac{1 - 3x_1 x_2}{x_1^2 x_2^3} + \frac{d^4g^{\mathrm{E}}}{dx_1^4} > 0$$

To decide whether a mixture model is suitable for the description of segregation, the validity of these criteria for the g^{E} function following from the model must be investigated:

1. In the case of an ideal mixture, $g^{\mathrm{E}} = 0$ and $d^2g^{\mathrm{E}}/dx_1^2 = 0$. Owing to this, none of the criteria are met. An ideal mixture is a solution over the whole concentration range.

2. If the mixture is symmetric, that is,

$$g^{\mathrm{E}} = A_0(x_1 x_2)$$

then

$$\frac{d^2 g^{\mathrm{E}}}{dx_1^2} = -2A_0 \qquad \frac{d^3 g^{\mathrm{E}}}{dx_1^3} = 0$$

It follows from this that

$$\frac{\mathbf{R}T_{\mathrm{c}}}{x_1 x_2} = -2A_0$$

On the other hand:

$$\frac{\mathbf{R}T_{\mathrm{c}}(x_1 - x_2)}{x_1^2 x_2^2} = 0$$

which means that at the critical point mole fractions are equal:

$$x_1 = x_2 = x_{\mathrm{c}}$$

The third criterion is also fulfilled, because

$$\frac{2\mathbf{R}T_{\mathrm{c}}}{x_1^2 x_3^3} - \frac{3x_1 x_2}{x_1^2 x_2^3} > 0$$

This means that a symmetric mixture may exist in two phases over the whole concentration range. Segregation occurs if

$$A_0 > 2\mathbf{R}T_{\mathrm{c}}$$

3. The Wilson equation in its original form is not suitable for the description of segregation. If a third parameter is introduced into it, however, the resulting variant is found suitable:

$$\frac{g^{\mathrm{E}}}{\mathbf{R}T} = -C\left(x_1 \ln(x_1 + A_{21}x_2) + x_2 \ln(x_2 + A_{12}x_1)\right)$$

The value of the parameter C falls within the range from 1.5 to 5.

4. The NRTL equation can be applied also in its original form to mixtures of limited solubility.

5. The UNIQUAC equation and the UNIFAC method can be used for mixtures of limited solubility.

Algorithm and Program: Determination of Metastable Region

Write a program for the determination of the metastable region when the molar free-enthalpy function of the mixture is known.

Algorithm.

1. Calculate the coordinates of the inflection point.
2. Calculate the coordinates of points for which the tangent is the same.
3. Print the results.

Program METASTABLE,D2. A possible realization:

```
]LOAD METASTABLE,D2
]LIST

 10  REM :CALCULATION OF BOUNDS OF
      METASTABLE STATES
 20  REM :THE CHANGE OF FREE ENERGY
      WITH COMPOSITION IS GIVEN IN
      POLYNOMIAL FORM
 30  K = 0.2
 40  RESTORE
 50  READ A,B,C,D,E
 60  DATA 1.5,- 13,59.3333,- 88,42.6
      667
 70  DEF FN G(X) = A + B * X + C *
      X * X + D * X * X * X + E *
      X * X * X * X
 80  DEF FN DG(X) = B + 2 * C * X
      + 3 * D * X * X + 4 * E * X
      * X * X
 90  DEF FN BB(X) = FN G(X) - X *
      FN DG(X)
100  LI = ( - 6 * D - SQR (36 * D
      ^ 2 - 48 * E * 2 * C)) / 24
      / E
110  UI = ( - 6 * D + SQR (36 * D
      ^ 2 - 48 * E * 2 * C)) / 24
      / E
120  LL = LI
130  FOR XX = 0 TO LL STEP 0.02
140  M = FN DG(XX)
150  B1 = FN BB(XX)
160  FOR X = UI TO 1 STEP 0.02
170  M2 = FN DG(X)
180  B2 = FN BB(X)
190  IF ABS (M2 - M) + ABS (B2 -
      B1) < K THEN GOTO 240
```

```
200  NEXT X
210  NEXT XX
220  PRINT : PRINT "NO SOLUTION"
230  GOTO 280
240  PRINT : PRINT
250  PRINT "THE FIRST METASTABLE
      RANGE IS"
260  PRIN "X1  =  "; XX; "X2  =  "; LI
270  PRINT : PRINT
280  PRINT "THE SECOND METASTABLE
      RANGE IS"
290  PRINT "X3 =  "; UI;" X4 =  "; X
300  END
```

Example. In the program the following function describes the function $g(x)$ at a given temperature and pressure:

$$g(x) = a + bx + cx^2 + dx^3 + ex^4$$

$$a = 1.5 \quad b = -13 \quad c = 59.33 \quad d = -88 \quad e = 42.6$$

```
]RUN

THE FIRST METASTABLE RANGE IS
X1  =  .2 X2  =  .330967609

THE SECOND METASTABLE RANGE IS
X3  =  .700281585 X4  =  .840281585
```

12

VAPOR – LIQUID EQUILIBRIA

We recall that the two formulations of the phase-equilibrium criterion are:

1. The appropriate potential function has a minimum.
2. The chemical potentials of a given species are identical in the phases in contact.

We are going to deal with the second criterion first. Properly speaking, there aren't any theoretical difficulties; it is only some practical–technical information we need to know. Hence, this chapter will start with the discussion of binary mixtures as representing the simplest case. Meanwhile, we will get acquainted with the terminology that is also needed for the general discussion of multicomponent systems.

After this, the problems of calculating vapor–liquid equilibria will be dealt with.

Finally, we will scrutinize the whole subject once again to demonstrate how the phase equilibria can be calculated by minimization of the free energy.

VAPOR – LIQUID EQUILIBRIUM FOR BINARY MIXTURES

The number of degrees of freedom of such a system, given by the Gibbs phase rule, is 2. Accordingly one of the pairs (x, T) or (x, P) or (x, y) is usually chosen as intensive variables, where x and y are mole fractions of the more volatile component (of lower boiling point in the pure state) in liquid respectively in gas phase.

Vapor–liquid equilibrium systems will be discussed not in general, but according to special combinations as follows:

Gas Phase	Liquid Phase
Mixture of ideal gases	Ideal mixture
Mixture of real gases	Ideal mixture
Mixture of ideal gases	Nonideal mixture
Mixture of real gases	Nonideal mixture

Mixture of Ideal Gases – Ideal Liquid Mixture

For mixtures composed of ideal gaseous species, the chemical potential of any species can be given as

$$\mu_j[T, P, y_i] = \mu_j[T, P_u] + \mathbf{R}T \ln \frac{P}{P_u} + \mathbf{R}T \ln y_j$$

The chemical potential of any component in an ideal liquid solution is

$$\mu_j^{(1)} = \mu_j^\circ + \mathbf{R}T \ln x_j$$

At equilibrium, the chemical potentials of the species in question are identical in the coexisting phases:

$$\mu_j[T, P, y_i] = \mu_j^{(1)}$$

and thus

$$\mu_j[T, P_u] + \mathbf{R}T \ln \frac{P}{P_u} + \mathbf{R}T \ln y_j = \mu_j^\circ + \mathbf{R}T \ln x_j$$

This relationship holds for any composition, even for a pure component, so that

$$x_j = y_j = 1$$

Consequently, the total pressure is equal to the equilibrium vapor pressure ($P = P^\circ$), and the standard potential can be written as

$$\mu_j^\circ = \mu_j[T, P_u] + \mathbf{R}T \ln \frac{P^\circ}{P_u}$$

Let us substitute this into the previous equation:

$$\mu_j[T, P_u] + \mathbf{R}T \ln \frac{P}{P_u} + \mathbf{R}T \ln y_j = \mu_j[T, P_u] + \mathbf{R}T \ln \frac{P^\circ}{P_u} + \mathbf{R}T \ln x_j$$

After reduction and making use of Dalton's law, we have

$$\overline{P}_j = Py_j = P_j^\circ x_j$$

Thus the partial pressure of a given component in the vapor phase is proportional to its mole fraction in the liquid phase, and the proportionality factor is the vapor pressure of the pure component. The relationship is named

$$\overline{P}_1 = P_1^\circ x_1$$

$$\overline{P}_2 = P_2^\circ x_2$$

The sum of the partial pressures gives the total pressure:

$$P = \overline{P}_1 + \overline{P}_2 = x_1(P_1^\circ - P_2^\circ) + P_2^\circ$$

That is, the total pressure of the system is a linear function of the composition at any prescribed temperature (Figure 12.1). In its validity range, Raoult's law enables us to calculate the composition and total pressure of the equilibrium vapor phase at any specified temperature, using the liquid compositions.

In practice, it may also be important to calculate the equilibrium vapor composition at specified total pressure using the liquid composition. However, this would necessitate knowing the equilibrium temperature. In accordance with the Raoult's law,

$$y_j = \frac{x_j P_j^\circ[T]}{P}$$

where P_j° depends only on temperature. In addition, the following restriction must be satisfied:

$$\Sigma y_j = 1$$

Since x_j and P on the right-hand side of the former equation are specified, the value of P° can be found. In other words, the temperature T is adjusted to satisfy the restriction $\Sigma y_j = 1$.

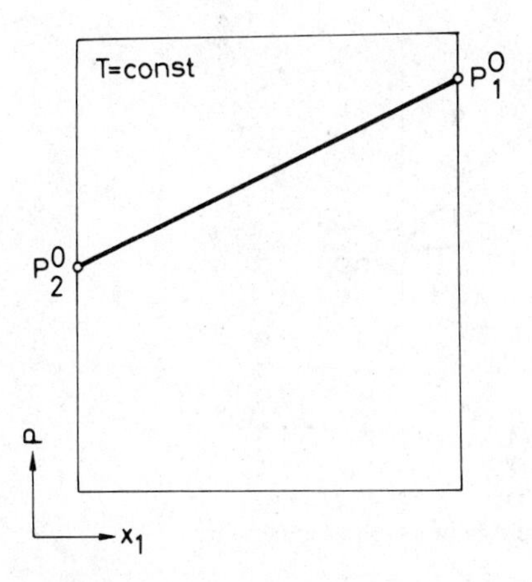

FIGURE 12.1. Effect of composition on total pressure of a binary system.

On the other hand, if the equilibrium liquid composition is to be calculated at given total pressure P using the vapor composition, then the following relationship may be applied:

$$x_j = \frac{y_i P}{P_j^{\circ}[T]}$$

The temperature has to be found such that

$$\sum_j x_j = 1$$

Thus, it can be seen that, at constant total pressure, the mixture temperature at equilibrium depends on the composition, that is,

$$T = T_{\mathrm{b}}[x]$$

$$T = T_{\mathrm{d}}[y]$$

This means that, at given temperature and total pressure, the composition of the vapor and the liquid phases at equilibrium are well defined (Figure 12.2).

The functions T_{d} and T_{b} are named the dew-point and bubble-point curves, respectively, since above the dew point only the vapor phase can exist, and below the bubble point, only the liquid phase. Experiment shows that the total pressure P increases, the dew-point and bubble-point curves move toward higher temperatures and approach each other (Figure 12.3). In the critical region, there are some special phenomena which will not be discussed here.

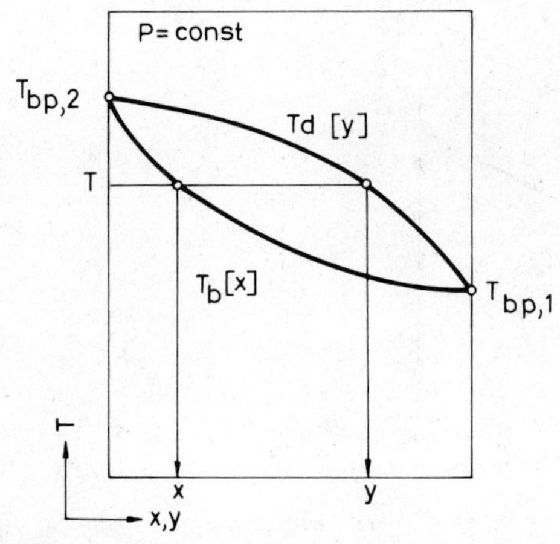

FIGURE 12.2. Dew-point temperature and bubble-point temperature.

The iterative calculations implied in the foregoing are inconvenient. A reasonably good approximation can be made in the following simple way. For binary mixtures one may apply the Raoult's law:

$$y_1 P = x_1 P_1^\circ$$

$$y_2 P = x_2 P_2^\circ$$

Let us divide the first equation by the second one:

$$\frac{y_1}{y_2} = \frac{x_1 P_1^\circ}{x_2 P_2^\circ}$$

and introduce the following notation:

$$\alpha^\circ \equiv \frac{P_1^\circ}{P_2^\circ} = \frac{y_1 x_2}{y_2 x_1}$$

Thus α° represents the ratio of the equilibrium vapor pressures. According to experiment this quotient usually depends weakly on the temperature, so that it can be taken as constant over the whole composition range. Let us define the

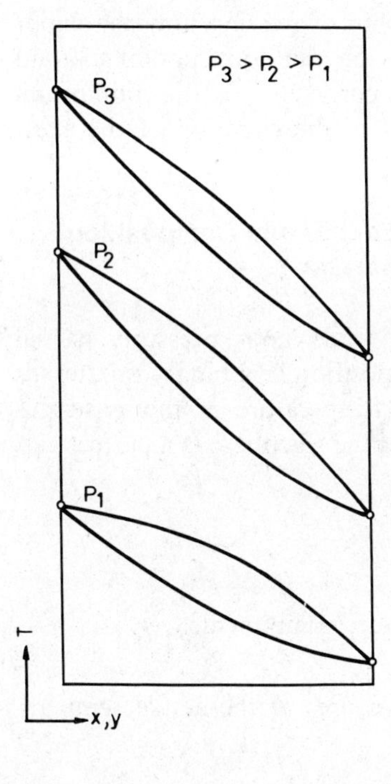

FIGURE 12.3. Effect of pressure on dew-point and bubble-point curves.

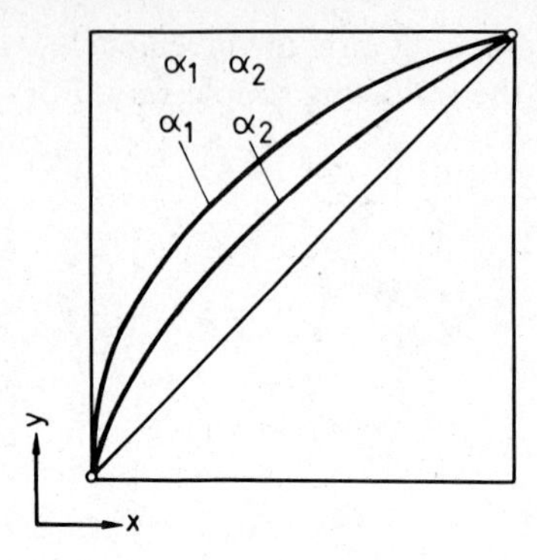

FIGURE 12.4. Relative volatility as parameter of the equilibrium curve.

additional symbols

$$x_1 = x$$

$$y_1 = y$$

$$x_2 = 1 - x_1$$

$$y_2 = 1 - y_1$$

After substituting and rearranging,

$$y = \frac{\alpha^\circ x}{(\alpha^\circ - 1)x + 1}$$

In this manner, we have arrived at a relationship expressing how the vapor mole fraction of the light component depends on the corresponding liquid mole fraction at constant pressure P, with α° as parameter. So the equilibrium composition can easily be calculated by means of the equation of the equilibrium curve (Figure 12.4).

Algorithm and Program: Calculation of Equilibrium Composition for Ideal Liquid and Ideal Gas

Write a program for the calculation of the total vapor pressure, partial pressures of components, and vapor-phase composition in a binary mixture as a function of liquid composition. The actual temperature of the system is known. The liquid phase is an ideal mixture, and the gas phase is a perfect gas.

Algorithm.

1. Read the actual temperature.
2. Read the sequential numbers of the mixture components.
3. Call the property-datafile reader routine.
4. Calculate the pure-component vapor pressures at the actual temperature.

5. Calculate the partial pressures of the components in kilopascals.
6. Calculate the total pressure of the system.
7. Calculate the composition of the vapor phase.
8. Print the results.

Program DISTIDID. A possible realization:

```
]PR#1
]LOAD DISTIDID,D1
]LIST

 10  REM :CALCULATION OF VLE COMPO
      SITIONS AND PARTIAL AND TOTAL
      PRESSURE
 20  REM :SUPPOSING THE LIQUID AND
      GAS TO BE IDEAL
 30  PRINT : PRINT
 40  DIM A(15)
 50  PRINT
 60  INPUT "ENTER TEMPERATURE (K)
      "; T
 70  PRINT
 80  PRINT "ENTER THE SEQUENTIAL
      NUMBER"
 90  INPUT "OF FIRST COMPONENT
      "; IS
100  GOSUB 460
110  TC = A(3)
120  PC = A(2)
130  OM = A(4)
140  GOSUB 570
150  S1 = PS
160  PRINT
170  PRINT "ENTER THE SEQUENTIAL
      NUMBER"
180  INPUT "OF SECOND COMPONENT
      "; IS
190  GOSUB 460
200  TC = A(3)
210  PC = A(2)
220  OM = A(4)
230  GOSUB 570
240  S2 = PS
250  PRINT
260  PRINT " X1 P1(KPA) P
      (KPA) Y1"
270  PRINT "=======================
      ======================"
280  FOR I = 1 TO 11
290  X = (1 - 1) * 0.1
```

```
300  P1 = X * S1
310  P1 = P1 / 1000
320  P2 = (1 - X) * S2
330  P2 = P2 / 1000
340  P = P1 + P2
350  Y1 - P1 / P
360  PRINT TAB(2); X; TAB(6); P1;
      TAB(17); P; TAB(29); Y1
370  NEXT I
380  PRINT
390  PRINT "======================
      ====================="
400  PRINT : PRINT
410  INPUT "NEXT COMPOUNDS ? (Y / N
      ) "; A$
420  IF A$ = "Y" THEN GOTO 70
430  INPUT "NEXT TEMPERATURE ? (
      Y / N) "; A$
440  IF A$ = "Y" THEN GOTO 50
450  END
460  REM :READING OF SPECIFIED
      RECORD OF PHDAT FILE INTO
      ARRAY A (13)
470  REM : THE SEQUENTIAL NUMBER OF
      THE COMPOUND IN THE CALLING
      PROGRAM HAS TO BE "IS"
480  REM :ARRAY A HAS TO BE DECLA
      RED IN THE CALLING PROGRAM
490  D$ = CHR$ (4)
500  PRINT D$; "OPEN PHDAT"
510  PRINT D$; "POSITION PHDAT, R";
      IS - 1
520  PRINT D$; "READ PHDAT"
530  INPUT A(1), A(2), A(3), A(4), A(5),
       A(6) A(7) A(8) A(9) A(10),
      A(11), A(12), A(13), A(14), A(15)
540  PRINT D$; "CLOSE PHDAT"
550  PRINT D$; "IN#0"
560  RETURN
570  TR = T / TC
580  PS = 5.92714 - 6.09648 / TR -
      1.28862 * LOG (TR) + 0.1693
      47 * TR ∧ 6 + OM * (15.2518 -
      15.6875 / TR - 13.4721 * LOG
      (TR) + 0.43577 * TR ∧ 6)
590  PS = EXP (PS) * PC
600  RETURN
```

Example. Run the program for chloroform–1,1-dichloroethane mixture at temperatures 288.2 and 270 K. Repeat the calculation for chloroform–1,2-dichloroethane.

```
]RUN

ENTER TEMPERATURE (K) 288.2

ENTER THE SEQUENTIAL NUMBER
OF FIRST COMPONENT 10

ENTER THE SEQUENTIAL NUMBER
OF SECOND COMPONENT 11
```

X1	P1(KPA)	P(KPA)	Y1
0	0	19.0693866	0
.1	1.6656119	18.8280598	.0884643406
.2	3.33122379	18.5867331	.179225891
.3	4.99683569	18.3454063	.272375308
.4	6.66244758	18.1040795	.368008082
.5	8.32805948	17.8627528	.466224864
.6	9.99367137	17.621426	.567131818
.7	11.6592833	17.3800992	.670841006
.8	13.3248952	17.1387725	.777470801
.9	14.9905071	16.8974457	.887146336
1	16.656119	16.656119	1

```
NEXT COMPOUNDS ? (Y / N) N
NEXT TEMPERATURE ? (Y / N) Y

ENTER TEMPERATURE (K) 270

ENTER THE SEQUENTIAL NUMBER
OF FIRST COMPONENT 10

ENTER THE SEQUENTIAL NUMBER
OF SECOND COMPONENT 11
```

X1	P1(KPA)	P(KPA)	Y1
0	0	7.6857438	0
.1	.668289414	7.58620676	.0880926972
.2	1.33657883	7.48583869	.17854764
.3	2.00486824	7.38547062	.271461136
.4	2.67315766	7.28510256	.366934801
.5	3.34144707	7.18473448	.46507593
.6	4.00973649	7.08436642	.565997896
.7	4.6780259	6.98399835	.669820591
.8	5.34631531	6.88363028	.776670899
.9	6.01460473	6.78326221	.886683212
1	6.68289414	6.68289414	1

```
NEXT COMPOUNDS ? (Y / N) Y
```

ENTER THE SEQUENTIAL NUMBER
OF FIRST COMPONENT 10

ENTER THE SEQUENTIAL NUMBER
OF SECOND COMPONENT 12

X1	P1(KPA)	P(KPA)	Y1
0	0	2.09799865	0
.1	.668298414	2.5564882	.261409152
.2	1.33657883	3.01497775	.443312999
.3	2.00486824	3.4734673	.577195082
.4	2.67315766	3.93195685	.679854271
.5	3.34144707	4.3904464	.761072285
.6	4.00973649	4.84893595	.826931213
.7	4.6780259	5.3074255	.881411506
.8	5.34631531	5.76591505	.927227556
.9	6.01460473	6.22440459	.966293986
1	6.68289414	6.68289414	1

NEXT COMPOUNDS ? (Y / N) N
NEXT TEMPERATURE ? (Y / N) Y

ENTER TEMPERATURE (K) 288.2

ENTER THE SEQUENTIAL NUMBER
OF FIRST COMPONENT 10

ENTER THE SEQUENTIAL NUMBER
OF SECOND COMPONENT 12

X1	P1(KPA)	P(KPA)	Y1
0	0	5.94136129	0
.1	1.6656119	7.01283706	.237508997
.2	3.33122379	8.08431283	.412060229
.3	4.99683569	9.15578859	.545756997
.4	6.66244758	10.2272644	.651439852
.5	8.32805948	11.2987401	.737078593
.6	9.99367137	12.3702159	.807881727
.7	11.6592833	13.4416917	.867397019
.8	13.3248952	14.5131674	.91812454
.9	14.9905071	15.5846432	.961876822
1	16.656119	16.656119	1

NEXT COMPOUNDS ? (Y / N) N
NEXT TEMPERATURE ? (Y / N) N

Remark. Explain the results.

Mixture of Real Gases — Ideal Liquid Mixture

For mixtures made up of gaseous components with real behavior, the chemical potential of any species is given as

$$\mu_j[T, P, y_j] = \mu_j[T, P_u] + \mathbf{R}T \ln \frac{\bar{f}_j}{P_u}$$

where $\bar{f}_j$ denotes the partial fugacity of component j in the mixture.

On the other hand, the ideal liquid solution, being in equilibrium with the vapor phase, has chemical potential for component j

$$\mu_j^{(1)} = \mu_j^\circ + \mathbf{R}T \ln x_j$$

According to the equilibrium criterion, the chemical potentials must be identical in the coexisting phases:

$$\mu_j[T, P, y_j] = \mu_j^{(1)}$$

Hence

$$\mu_j[T, P_u] + \mathbf{R}T \ln \frac{\bar{f}_j}{P_u} = \mu_j^\circ + \mathbf{R}T \ln x_j$$

The above relationship holds for all values of x, and consequently also for $x_j = 1$, that is, for a pure component. Accordingly, the pressure in the vapor phase is identical to the equilibrium vapor pressure, and the fugacity is equal to the corresponding f_j°:

$$\mu_j^\circ = \mu_j[T, P_u] + \mathbf{R}T \ln \frac{f_j^\circ}{P_u}$$

After substituting back into the former equation,

$$\mu_j[T, P_u] + \mathbf{R}T \ln \frac{\bar{f}_j}{P_u} = \mu_j[T, P_u] + \mathbf{R}T \ln \frac{f_j^\circ}{P_u} + \mathbf{R}T \ln x_j$$

Carrying out the reductions, we obtain

$$\bar{f}_j = f_j^\circ x_j$$

This means that the vapor-phase fugacity of the component in question is proportional to its liquid-phase mole fraction, and the equilibrium fugacity of this component is the proportionality factor. This is called the Lewis–Randall law, and can be considered as a more general form of Raoult's law. The calculation of the equilibrium fugacity f_j° from the vapor pressure is not difficult; see the second section in Chapter 10, dealing with the fugacity of a one-component gas.

On the other hand, the partial fugacity $\bar{f}_j$ of component j in the gaseous phase is proportional to its mole fraction y_j:

$$\bar{f}_j = y_j f_j[T, P]$$

where the proportionality factor $f_j[T, P]$ represents the fugacity of the pure component j at the temperature T and pressure P of the system, in the gaseous state.

By using the above two relationships the partial fugacity $\bar{f}_j$ can be eliminated:

$$y_j = \frac{x_j f_j^\circ[T]}{f_j[T, P]}$$

Mixture of Perfect Gases — Nonideal Liquid Mixture

For an ideal-gas solution, the chemical potential of any component is given as

$$\mu_j[T, P, y_j] = \mu_j[T, P_u] + \mathbf{R}T \ln \frac{P}{P_u} + \mathbf{R}T \ln y_j$$

For a real liquid solution, the chemical potential of this same component is

$$\mu_j^{(1)} = \mu_j^\circ + \mathbf{R}T \ln x_j + \mathbf{R}T \ln \gamma_j$$

In equilibrium, the chemical potentials of the component are identical in the coexisting phases, so

$$\mu_j[T, P, y_j] = \mu_j^{(1)}$$

Hence

$$\mu_j[T, P_u] + \mathbf{R}T \ln \frac{P}{P_u} + \mathbf{R}T \ln y_j = \mu_j^\circ + \mathbf{R}T \ln x_j + \mathbf{R}T \ln \gamma_j$$

This relationship holds for all compositions, and consequently also for $x_j = y_j = 1$. In this case, since $P_j = P_j^\circ$ and $\gamma_j = 1$,

$$\mu_j^\circ = \mu_j[T, P_u] + \mathbf{R}T \ln \frac{P_j^\circ}{P_u}$$

Let us substitute this into the foregoing equation:

$$\mu_j[T, P_u] + \mathbf{R}T \ln \frac{P}{P_u} + \mathbf{R}T \ln y_j = \mu_j[T, P_u] + \mathbf{R}T \ln \frac{P_j^\circ}{P_u} + \mathbf{R}T \ln x_j \gamma_j$$

After proper reduction and also using Dalton's law, we obtain

$$P_j = y_j P = P_j^\circ x_j \gamma_j$$

This is called the modified Raoult's law. The activity coefficient in it is computable from experimental data, since x_j and y_j can be determined analytically and the vapor pressure P_j° is separately measurable. So γ_j is given as a function of the composition. The γ_j may also be calculated independently on the basis of mixture models, as discussed in the preceding chapter.

Algorithm and Program: Calculation of Equilibrium Composition with Margules Equation

Write a program for the calculation of the total vapor pressure, partial pressures, and vapor-phase compositions of the components in a binary mixture as a function of liquid composition. The actual temperature is known. Calculate the liquid activity coefficients with the Margules equation. The vapor phase is an ideal gas.

Algorithm.

1. Read the actual temperature.
2. Read the sequential numbers of the mixture components.
3. Call the property-datafile reader routine.
4. Calculate the pure-component vapor pressures at the actual temperature.
5. Read the constants A_{12} and A_{21} of the Margules equation.
6. Calculate the activity coefficients of the components with the Margules equation. (See program VLEMARGULES in Chapter 10.)
7. Calculate the partial pressures of the components.
8. Calculate the total pressure.
9. Calculate the vapor-phase composition.
10. Print the results.

Program DISTIDMARGULES. A possible realization:

```
]PR#0
]LOAD DISTIDMARGULES
]LIST

 10  REM :CALCULATION OF VLE COMPO
      SITIONS AND PARTIAL AND TOTAL
      PRESSURE
 20  REM :SUPPOSING THE GAS TO BE
      IDEAL
 30  REM :THE LIQUID ACTIVITY COEF
      FICIENT IS CALCULATED BY MAR
      GULES EQUATION
 40  PRINT : PRINT
```

```
 50 DIM A(15)
 60 PRINT
 70 INPUT "ENTER TEMPERATURE (K)
     "; T
 80 PRINT
 90 PRINT "ENTER THE SUBSEQUENTIAL
     NUMBER"
100 INPUT "OF FIRST COMPONENT
     "; IS
110 GOSUB 610
120 TC = A(3)
130 PC = A(2)
140 OM = A(4)
150 GOSUB 720
160 S1 = PS
170 PRINT
180 PRINT "ENTER THE SEQUENTIAL
     NUMBER"
190 INPUT "OF SECOND COMPONENT
     "; IS
200 GOSUB 610
210 TC = A(3)
220 PC = A(2)
230 OM = A(4)
240 GOSUB 720
250 S2 = PS
260 PRINT
270 INPUT "ENTER MARGULES CONSTA
     NT A12 "; A1
280 PRINT
290 INPUT "ENTER MARGULES CONSTA
     NT A21 "; A2
300 PRINT
310 PRINT "PRESSURES ARE IN KILO
     PASCAL !!!"
320 PRINT
330 PRINT " X1 P1 P2
     P Y1"
340 PRINT "=====================
     ===================="
350 FOR I = 1 TO 11
360 X = (I - 1) * 0.1
370 GC = EXP ((1 - X) ∧ 2 * (A1 +
     2 * (A2 - A1) * X))
380 P1 = X * S1 * GC
390 P1 = P1 / 1000
400 P% = P1 * 100
410 P1 = P% / 100
420 GC = EXP (X ∧ 2 * (A2 + 2 *
```

```
        (A1 - A2) * (1 - X)))
430  P2 = (1 - X) * S2 * GC
440  P2 = P2 / 1000
450  P% = P2 * 100
460  P2 = P% / 100
470  P = P1 + P2
480  Y1 = P1 / P
490  Y% = Y1 * 1000
500  Y1 = Y% / 1000
510  PRINT TAB(1); X; TAB(7); P1;
      TAB(16); P2; TAB(26); P; TAB(
      34); Y1
520  NEXT I
530  PRINT
540  PRINT "========================
      ====================="
550  PRINT : PRINT
560  INPUT "NEXT COMPOUNDS ? (Y / N
      ) "; A$
570  IF A$ = "Y" THEN GOTO 80
580  INPUT "NEXT TEMPERATURE ? (
      Y / N) "; A$
590  IF A$ = "Y" THEN GOTO 60
600  END
610  REM :READING OF SPECIFIED
      RECORD OF PHDAT
      FILE INTO ARRAY
      A(13)
620  REM :THE SEQUENTIAL NUMBER OF
      THE COMPOUND IN THE CALLING
      PROGRAM HAS TO BE "IS"
630  REM :ARRAY A HAS TO BE DECLA
      RED IN THE CALLING PROGRAM
640  D$ = CHR$ (4)
650  PRINT D$; "OPEN PHDAT"
660  PRINT D$; "POSITION PHDAT, R";
      IS - 1
670  PRINT D$; "READ PHDAT"
680  INPUT A(1), A(2), A(3), A(4), A(5),
      A(6) A(7) A(8) A(9) A(10),
      A(11), A(12), A(13), A(14), A(15)
690  PRINT D$; "CLOSE PHDAT"
700  PRINT D$; "IN#0"
710  RETURN
720  TR = T / TC
730  PS = 5.92714 - 6.09648 / TR -
      1.28862 * LOG (TR) + 0.1693
      47 * TR ∧ 6 + OM * (15.2518 -
      15.6875 / TR - 13.4721 * LOG
```

```
        (TR) + 0.43577 * TR ∧ 6)
740  PS = EXP (PS) * PC
750  RETURN
```

Example. Run the program for a chloroform–1,2-dichloroethane mixture at temperature 288.2 K. Repeat the calculation with a chloroform–*n*-heptane mixture at 288.2 K.

```
]RUN

ENTER TEMPERATURE (K) 288.2

ENTER THE SEQUENTIAL NUMBER
OF FIRST COMPONENT 10

ENTER THE SEQUENTIAL NUMBER
OF SECOND COMPONENT 12

ENTER MARGULES CONSTANT A12 0.3412

ENTER MARGULES CONSTANT A21 0.08948

PRESSURES ARE IN KILOPASCAL !!!
```

X1	P1	P2	P	Y1
0	0	5.94	5.94	0
.1	2.1	5.37	7.47	.281
.2	3.88	4.84	8.72	.444
.3	5.48	4.32	9.8	.559
.4	7	3.79	10.79	.648
.5	8.51	3.23	11.74	.724
.6	10.05	2.63	12.68	.792
.7	11.64	2	13.64	.853
.8	13.29	1.34	14.63	.908
.9	14.97	.66	15.63	.957
1	16.65	0	16.65	1

```
NEXT COMPOUNDS ? (Y / N) Y

ENTER THE SEQUENTIAL NUMBER
OF FIRST COMPONENT 10

ENTER THE SEQUENTIAL NUMBER
OF SECOND COMPONENT 7

ENTER MARGULES CONSTANT A12 0.3531

ENTER MARGULES CONSTANT A21 0.716

PRESSURES ARE IN KILOPASCAL !!!
```

X1	P1	P2	P	Y1
0	0	3.48	3.48	0
.1	2.35	3.13	5.48	.428
.2	4.58	2.8	7.38	.62
.3	6.6	2.48	9.08	.726
.4	8.39	2.18	10.57	.793
.5	9.96	1.9	11.86	.839
.6	11.33	1.62	12.95	.874
.7	12.59	1.33	13.92	.904
.8	13.83	1	14.83	.932
.9	15.14	.58	15.72	.963
1	16.65	0	16.65	1

NEXT COMPOUNDS ? (Y / N) N
NEXT TEMPERATURE ? (Y / N) N

Remark. Explain the results.

AZEOTROPE MIXTURES

The modified Raoult's law for binary mixtures applies:

$$y_i P = P_1^\circ x_1 \gamma_1$$

$$y_2 P = P_2^\circ x_2 \gamma_2$$

Let us divide the one equation by the other:

$$\frac{y_1}{y_2} = \frac{P_1^\circ \gamma_1}{P_2^\circ \gamma_2} \cdot \frac{x_1}{x_2}$$

and introduce the following notation:

$$\frac{P_1^\circ \gamma_1}{P_2^\circ \gamma_2} = \alpha^\circ \frac{\gamma_1}{\gamma_2} \equiv \alpha$$

This quantity is usually called the relative volatility or separation factor. Since γ_1 and γ_2 depend on the composition, α is also composition-dependent. It may happen that at certain composition

$$\frac{\gamma_2}{\gamma_1} = \frac{P_1^\circ}{P_2^\circ}$$

From this, however, it follows that the separation factor at a certain unique concentration become equal to $\alpha = 1$, and so

$$\frac{y_1}{y_2} = \frac{x_1}{x_2}$$

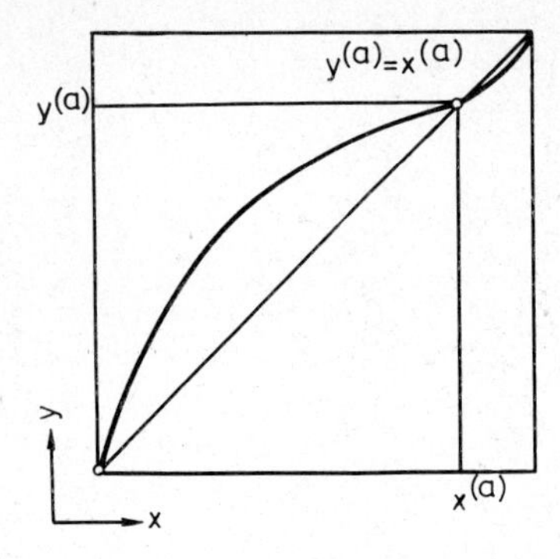

FIGURE 12.5. Azeotrope point in (x, y) plot.

This may occur only if the composition of the vapor and liquid phases is identical. A mixture having such a composition is called an *azeotrope*. At this composition, the equilibrium curve intersects the diagonal of the (x, y) diagram (Figure 12.5).

Let us take the logarithm of the defining equation for the relative volatility:

$$\ln \alpha = \ln \alpha^\circ + \ln \frac{\gamma_1}{\gamma_2}$$

In the second section of Chapter 11 it was indicated that if the excess free enthalpy is symmetrical in the concentration,

$$\ln \frac{\gamma_1}{\gamma_2} = A_0(1 - 2x_1)$$

Thus

$$\ln \alpha = \ln \alpha^\circ + A_0(1 - 2x_1)$$

The mixture is an azeotrope if $\alpha = 1$, $\ln \alpha = 0$. Consequently, for the azeotrope one may write

$$A_0\left(2x_1^{(a)} - 1\right) = \ln \alpha^\circ$$

The mole fraction the azeotrope given as

$$x_1^{(a)} = \frac{1}{2} + \frac{\ln \alpha^\circ}{2A_0}$$

So, with knowledge of A_0 and α° on the right-hand side, the azeotropic composition can be predicted. Since the azeotropic concentration has a meaning only for the region $1 > x_1^{(a)} > 0$, if the $x_1^{(a)}$ value obtained with the above formula is greater than 1 or less than 0, it means that the mixture is nonazeotropic.

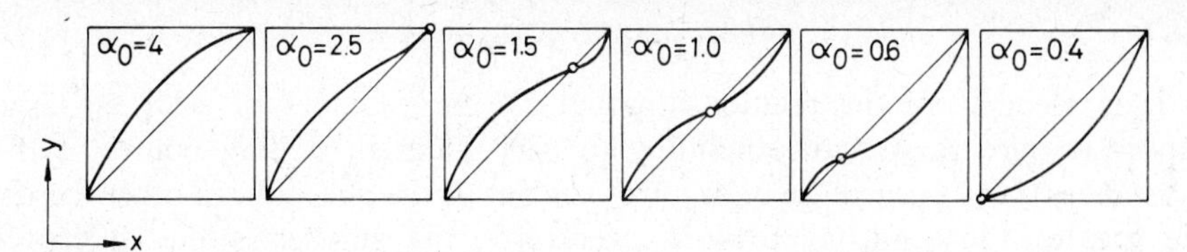

FIGURE 12.6. Effect of relative volatility on equilibrium at constant excess free enthalpy.

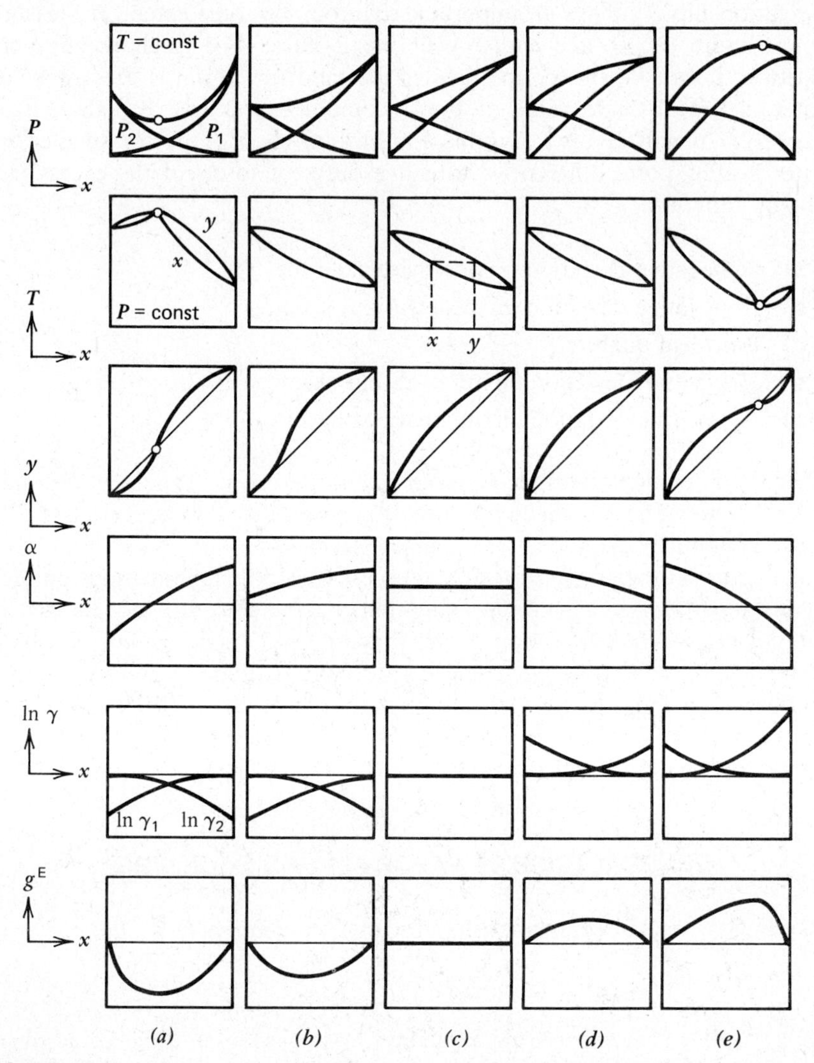

FIGURE 12.7. Effect of excess free enthalpy on equilibrium properties at given boiling-point difference: (*a*) a negative-deviation mixture with azeotrope, (*b*) a negative-deviation mixture without azeotrope, (*c*) an ideal mixture, (*d*) a positive-deviation mixture without azeotrope, (*e*) a positive-deviation mixture with azeotrope.

It is clear from the relationship that for large values of α° (i.e., large vapor-pressure ratios, corresponding to very different boiling points of the substances), any occurrence of azeotropy is out of the question. In other words, the greater the tendency towards azeotropy, the smaller is the difference between the boiling points of the components. In Figure 12.6 we show how the (y, x) equilibrium curves for different α° values look, at $A_0 = 0.4$. In the relationship A_0 plays an important role too. To some extent, it measures the maximum excess free enthalpy of the mixture. For given boiling-point departure, it can be stated: the greater the tendency towards azeotropy, the greater is the absolute value of the maximum excess free enthalpy. From Figure 12.7 it will be seen how the equilibrium curves look for smaller and greater boiling-point differences, with five different forms of the excess free-enthalpy function.

(*a*) A negative-deviation mixture with azetrope.

(*b*) A negative-deviation mixture without azeotrope.

(*c*) An ideal mixture.

(*d*) A positive-deviation mixture without azeotrope.

(*e*) A positive-deviation mixture with azeotrope.

The figure demonstrates the typology of the azeotropy. Typology itself does not say anything about whether an azeotrope exists at all in a given case. This question cannot be answered on the macro level.

As mentioned before, a symmetrical excess-free-enthalpy function can often be considered as a sufficient approximation; nevertheless, it is not the general case. Generally, the calculation of the functions γ_1 and γ_2 cannot be carried out by means of thermodynamical considerations alone. However, a single experimental vapor–liquid equilibrium datum may be sufficient for a good estimation of the functions in question, provided the appropriate model for the activity coefficient is applied.

GENERAL CASE OF VAPOR–LIQUID EQUILIBRIA

According to the known relationship, for the ith component we have

$$\bar{f}_i = \bar{f}_i^{(1)}$$

where $\bar{f}_i$ and $\bar{f}_i^{(1)}$ are the partial fugacities in the gaseous and the liquid phase, respectively. In order to calculate the phase equilibrium, the fugacities have to be calculated.

1. For the fugacity we previously derived

$$\ln \bar{f}_i = \frac{1}{\mathbf{R}T} \int_v^\infty \left(\left(\frac{\partial P}{\partial n_i} \right)_{T, v, n_j} - \frac{\mathbf{R}T}{v} \right) dv - \ln \frac{v}{n_i \mathbf{R}T}$$

where n_i = quantity of the component i,
v = molar volume.

If this basic relationship is to be used for both phases, then an equation of state must be known that adequately describes the behavior of both phases.

2. There is, however, another approach: The calculations for the vapor and liquid phases can be carried out in a different way, applying the following equations:

$$\bar{f}_i = \nu_i y_i P$$

$$\bar{f}_i^{(1)} = \gamma_i x_i f_i^{\circ}$$

where x, y = mole fractions,
ν = fugacity coefficient computed from an equation of state,
γ = activity coefficient based on a liquid-mixture model.

First of all, a decision must be made on which of the two approaches to choose. The advantages and disadvantages of the two approaches are summarized as follows:

Equation-of-state approach:

Advantages	Disadvantages
1. No standard state	1. No equation of state valid for a really wide range of application
2. No need for experimental equilibrium data	2. Sensitive to mixing rules
3. Corresponding-state theory can be used	3. Hard to use for polar and large-size molecules

Activity-coefficient approach:

Advantages	Disadvantages
1. Simple liquid models often satisfactory	1. No easy way to estimate molar volume
2. Temperature affects mainly f_i°, not γ_i	2. Fails to work in critical region
3. Suitable for molecules of large size, too	3. May be clumsy for supercritical components

CALCULATION OF VAPOR–LIQUID EQUILIBRIUM USING AN EQUATION OF STATE

Two modified versions of the Redlich–Kwong equation of state will be demonstrated here. These properly describe the vapor–liquid equilibria of saturated nonpolar species.

Calculation of Fugacity and Activity Coefficient with the Redlich–Kwong–Soave Equation of State

Experiments have shown that the original version of the Redlich–Kwong equation of state is not very satisfactory for vapor–liquid equilibrium calculations. Therefore Soave modified the original form by replacing the constant a of the equation with a temperature function involving also Pitzer's omega factor. Thus, the modified equation of state looks like

$$P = \frac{\mathbf{R}T}{v - b} - \frac{a[T]}{v(v + b)}$$

Based on fundamental considerations, it is required that a be given at the critical point by

$$a_c = 0.42747 \frac{R^2 T_c^2}{P_c}$$

Thus, at any temperature

$$a = a_c \alpha[T]$$

where $\alpha[T]$ represents a correction factor that is a function only of the temperature with omega factor as a parameter. How do we find this function?

The equation of state, after rearranging, can be written for z:

$$z^3 - z^2 + z(A - B - B^2) - AB = 0$$

where

$$A = \frac{a[T]P}{\mathbf{R}^2 T^2}$$

$$B = \frac{bP}{\mathbf{R}T}$$

This equation is valid for the equilibrium vapor pressure too, and the solutions for $z, z^{(1)}$ are the compressibility factors of the saturated vapor and liquid, respectively. Moreover it is known that

$$\ln \frac{f}{P} = \int_0^P \left(\frac{v}{\mathbf{R}T} - \frac{1}{P} \right) dP$$

Using the RKS equation of state, for the fugacity coefficient we have

$$\ln \frac{f}{P} = (z - 1) - \ln(z - B) - \frac{A}{B} \ln\left(\frac{z + B}{z} \right)$$

At equilibrium, the fugacity coefficient of the saturated vapor is equal to that of the saturated liquid:

$$\ln \frac{f}{P^\circ} = \ln \frac{f^{(1)}}{P^\circ}$$

$$= (z - 1) - \ln(z - B) - \frac{A}{B} \ln\left(\frac{z + B}{z} \right)$$

$$= (z^{(1)} - 1) - \ln(z^{(1)} - B) - \frac{A}{B} \ln\left(\frac{z^{(1)} + B}{z^{(1)}} \right)$$

This means on the other hand, that at any pressure P°, corresponding to a given temperature T, there is a single value for A which will make the difference between the left- and right-hand sides vanish. The resulting values can effectively be linearized (Figure 12.8) as

$$\sqrt{\alpha_i} = 1 + m_1\left(1 - \sqrt{T_{ri}}\right)$$

where α_i = correction factor to be determined,
m_i = substance-specific parameter for component i.

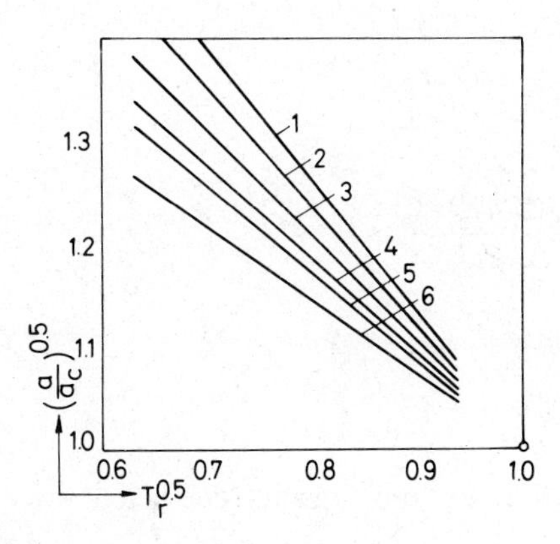

FIGURE 12.8. Redlich–Kwong–Soave equation-of-state correction factor and component-specific parameter: 1, decane; 2, octane; 3, toluene; 4, cyclopentane; 5, propane; 6, ethene.

The third parameter in the three-parameter equation of state is the omega factor. The parameters ω_i and m_i are functionally related, and the dependence is nonlinear. Assuming a quadratic relationship, a parameter estimation results in

$$m_i = 0.480 + 1.57\omega_i - 0.17\omega_i^2$$

That is how the situation looks for one-component systems. If there are more than one component in the system, the question may arise how to use the Redlich–Kwong–Soave equation for the calculation of the fugacity coefficients of the two equilibrium phases. For the fugacity coefficient the following thermodynamical relationship was derived in Chapter 10:

$$\ln \frac{\bar{f}_i}{Px_i} = \int_{\infty}^{v} \left(\frac{1}{v} - \frac{1}{\mathbf{R}T} \left(\frac{\partial P}{\partial n_i} \right)_{T,P,n_j} \right) dv - \ln z$$

After substituting the equation of state and the following mixing rules are used:

$$a = \sum_i \left(x_i a_i^{0.5} \right)^2$$

$$b = \sum x_i b_i$$

we have

$$\ln \frac{\bar{f}_i}{Px_i} = \frac{b_i}{b}(z-1) - \ln(z-B) - \frac{A}{B}\left(2\frac{a_i^{0.5}}{a^{0.5}} - \frac{b_i}{b} \right) \ln\left(1 + \frac{B}{z} \right)$$

So, if the equation of state is to be applied to equilibrium calculations, then the $a[T]$ value at temperature T can be obtained from a knowledge of ω_i, as follows:

$$\alpha_i[T_r] = \left(1 + m_i \left(1 - \sqrt{T_{ri}} \right) \right)^2$$

$$a_i[T_r] = \alpha_i[T_r] \frac{0.42747 R^2 T_{ci}^2}{P_{ci}}$$

$$b_i = 0.08664 \frac{RT_{ci}}{P_{ci}}$$

For the parameters involved in the equation for the fugacity coefficient we obtain

$$\frac{a_i^{0.5}}{a^{0.5}} = \frac{\alpha_i^{0.5} T_{ci}/P_{ci}^{0.5}}{\sum_i \left(x_i \alpha_i^{0.5} T_{ci}/P_{ci}^{0.5} \right)}$$

$$\frac{b_i}{b} = \frac{T_{ci}/P_{ci}}{\sum_i (x_i T_{ci}/P_{ci})}$$

$$A = 0.42747 \frac{P}{T^2} \left(\sum_i \frac{x_i \alpha_i^{0.5} T_{ci}}{P_{ci}^{0.5}} \right)^2$$

$$B = 0.08664 \frac{P}{T} \sum_i \frac{x_i T_{ci}}{P_{ci}}$$

If the values of A and B obtained by the above formula are substituted into the equation of the compressibility factor, we have

$$z^3 - z^2 + z(A - B - B^2) - AB = 0$$

Solving for both the vapor and the liquid phase, the necessary values for the fugacity coefficient are obtained.

Algorithm and Program: Distribution Coefficient

Write a program for the calculation of the distribution coefficient by the Redlich–Kwong–Soave equation of state for binary mixture. The distribution coefficient (or equilibrium ratio) is defined as

$$K_i \equiv \frac{\gamma_i P_i^\circ}{\nu_i P}$$

Algorithm.

1. Read the sequential numbers of the mixture components.
2. Read the mole fraction of the binary mixture.
3. Read the temperature and pressure.
4. Call the property-datafile reader routine.
5. Calculate α_i for both components.
6. Calculate the coefficients A and B of the equation of state.
7. Solve the equation of state by direct iteration for the liquid and the vapor phase.
8. Calculate $a_i^{0.5}/a^{0.5}$ for both components.

9. Calculate b_i/b for both components.
10. Calculate the activity coefficients γ_i using the liquid compressibility factor.
11. Calculate fugacity coefficients ν_i using the vapor compressibility factor.
12. Calculate the pure-component vapor pressures P_i°.
13. Calculate

$$K_i = \frac{\gamma_i P_i^\circ}{\nu_i P}$$

14. Print the results.

Program DISTRIBUTION COEFFRKS. A possible realization:

```
]PR#0
]LOAD DISTRIBUTION COEFFRKS,D2
]LIST

 10  REM :CALCULATION OF DISTRIBUT
      ION COEFFICIENT BY REDLICH-
      KWONG-SOAVE EQUATION OF STATE
 20  DIM A(15)
 30  PRINT : PRINT
 40  INPUT "ENTER SEQ. NO. OF
      FIRST COMPONENT "; C1
 50  PRINT
 60  INPUT "ENTER SEQ. NO. OF SECOND
      COMPONENT "; C2
 70  PRINT
 80  PRINT "ENTER MOLEFRACTION OF
      FIRST COMPONENT "
 90  INPUT "(BRUTTO FOR THE MIXTURE)
      "; X1
100  PRINT
110  INPUT "ENTER PRESSURE (PA) "
      ; P
120  PRINT
130  INPUT "ENTER TEMPERATURE (K)
      "; T
140  IS = C1
150  GOSUB 560
160  P1 = A(2):T1 = A(3):01 = A(4)
      :S1 = A(5)
170  IS = C2
180  GOSUB 560
190  P2 = A(2):T2 = A(3):O2 = A(4)
      :S2 = A(5)
```

```
200  L1 = 1 + (0.48 + 1.574 * O1 -
      0.176 * O1 * O1) * (1 - SQR
      (T / T1))
210  L2 = 1 + (0.48 + 1.574 * O2 -
      0.176 * O2 * O2) * (1 - SQR
      (T / T2))
220  X2 = 1 - X1
230  A = 0.42747 * P * (X1 * T1 *
      SQR (L1) / SQR (P1) + X2 *
      T2 * SQR (A2) / SQR (P2)) ∧
      2 / T / T
240  B = 0.08664 * P * (X1 * T1 /
      P1 + X2 * T2 / P2) / T
250  GOSUB 710
260  R1 = SQR (L1) * T1 / SQR (P
      1) / (X1 * SQR (L1) * T1 /
      SQR (P1) + X2 * SQR (L2) *
      T2 / SQR (P2))
270  R2 = T1 / P1 / (X1 * T1 / P1 +
      X2 * T2 / P2)
280  G1 = R2 * (ZL - 1) - LOG (ZL
      - B) - A * (2 * R1 - R2) *
      LOG (1 + B / ZL)
290  N1 = R2 * (ZV - 1) - LOG (ZV
      - B) - A * (2 * R1 - R2) *
      LOG (1 + B / ZV)
300  R1 = SQR (L2) * T2 / SQR (P
      2) / (X1 * SQR (L1) * T1 /
      SQR (P1) + X2 * SQR (L2) *
      T2 / SQR (P2))
310  R2 = T2 / P2 / (X1 * T1 / P1 +
      X2 * T2 / P2)
320  G2 = R2 * (ZL - 1) - LOG (ZL
      - B) - A * (2 * R1 - R2) *
      LOG (1 + B / ZL)
330  N2 = R2 * (ZV - 1) - LOG (ZV
      - B) - A * (2 * R1 - R2) *
      LOG (1 + B / ZV)
340  PC = P1:TC = T1:OM = O1:TB =
      S1
350  GOSUB 660
360  R1 = PS
370  PC = P2:TC = T2:OM = O2:TB =
      S2
380  GOSUB 660
390   R2 = PS
400  K1 = G1 * R1 / P / N1
410  K2 = G2 * R2 / P / N2
420  PRINT : PRINT
```

```
430  PRINT "DISTRIBUTION COEFFICI
      ENT OF THE"
440  PRINT "FIRST COMPOUND "; K1
450  PRINT
460  PRINT "DISTRIBUTION COEFFICI
      ENT OF THE"
470  PRINT "SECOND COMPOUND "; K2
480  PRINT : PRINT
490  INPUT "NEXT TEMPERATURE ? (Y
      /N) "; A$
500  IF A$ = "Y" THEN GOTO 130
510  INPUT "NEXT PRESSURE ? (Y / N)
      "; A$
520  IF A$ = "Y" THEN GOTO 110
530  INPUT "NEXT COMPOSITION ? (Y
      /N) "; A$
540  IF A$ = "Y" THEN GOTO 30
550  END
560  REM :READING OF SPECIFIED
      RECORD OF PHDAT FILE INTO
      ARRAY A(15)
570  REM :ARRAY A HAS TO BE DECLA
      RED IN THE CALLING PROGRAM
580  D$ = CHR$ (4)
590  PRINT D$; "OPEN PHDAT, D1"
600  PRINT D$; "POSITION PHDAT, R";
      IS - 1
610  PRINT D$; "READ PHDAT"
620  INPUT A(1), A(2), A(3), A(4), A(5),
      A(6), A(7), A(8), A(9), A(10),
      A(11), A(12), A(13), A(14), A(15)
630  PRINT D$; "CLOSE PHDAT"
640  PRINT D$; "IN#0"
650  RETURN
660  REM :CALCULATION OF VAPOR
      PRESSURE OF PURE COMPONENT
670  TR = T / TC
680  PS = 5.92714 - 6.09648 / TR -
      1.28862 * LOG (TR) + 0.1693
      47 * TR ∧ 6 + OM * (15.2518 -
      15.6875 / TR - 13.4721 * LOG
      (TR) + 0.43577 * TR ∧ 6)
690  PS = EXP (PS) * PC
700  RETURN
710  REM :SOLVING THE REDLICH-KWONG-
      SOAVE EQUATION OF STATE FOR
      LIQUID AND VAPOR COMPRES
      SIBILITY FACTOR
720  ZL = - 1.
```

```
730  II = 0
740  Z2 - ZL - (ZL ∧ 3 - ZL ∧ 2 +
      ZL * (A - B - B ∧ 2) - A * B
      ) / (3 * ZL ∧ 2 - 2 * ZL + A
      - B - B ∧ 2)
750  IF ABS (ZL - Z2) < 1E - 5 THEN
      GOTO 800
760  II = II + 1
770  ZL = Z2
780  IF II > 50 THEN PRINT "ITER
      ATION FOR LIQUID PHASE NOT
      CONVERGENT": GOTO 800
790  GOTO 740
800  ZV = 1
810  II = 0
820  Z2 = ZV - (ZV ∧ 3 - ZV ∧ 2 +
      ZV * (A - B - B ∧ 2) - A * B
      ) / (3 * ZV ∧ 2 - 2 * ZV + A
      - B - B ∧ 2)
830  IF ABS (ZV - Z2) < 1E - 5 THEN
      GOTO 880
840  II = II + 1
850  ZV = Z2
860  IF II > 50 THEN PRINT "ITER
      ATION FOR VAPOR PHASE NOT
      CONVERGENT“: GOTO 880
870  GOTO 820
880  RETURN
```

Example. Run the program at temperature 338.7 K and pressure 1,723,690 Pa for an *n*-butane–*n*-ethane mixture. The brutto composition of the mixture is at mole fraction 0.5.

```
]RUN

ENTER SEQ. NO. OF FIRST COMPONENT 2

ENTER SEQ. NO. OF SECOND COMPONENT 4

ENTER MOLE FRACTION OF FIRST COMPONENT
(BRUTTO FOR THE MIXTURE) 0.5

ENTER PRESSURE (PA) 1723690

ENTER TEMPERATURE (K) 338.7

DISTRIBUTION COEFFICIENT OF THE
FIRST COMPOUND 5.49146199

DISTRIBUTION COEFFICIENT OF THE
SECOND COMPOUND .427986743
```

```
NEXT TEMPERATURE ? (Y / N) N
NEXT PRESSURE ? (Y / N) Y
ENTER PRESSURE (PA) 1780000
ENTER TEMPERATURE (K) 338.7
ITERATION FOR LIQUID PHASE NOT CONVERGENT

DISTRIBUTION COEFFICIENT OF THE
FIRST COMPOUND 5.31790386

DISTRIBUTION COEFFICIENT OF THE
SECOND COMPOUND .414438458

NEXT TEMPERATURE ? (Y / N)
NEXT PRESSURE ? (Y / N) N
NEXT COMPOSITION ? (Y / N) N
```

Peng – Robinson Equation of State

This equation of state was discussed among the two-parameter cubic equations in Chapter 3:

$$P = \frac{\mathbf{R}T}{v - b} - \frac{a[T]}{v(v + b) + b(v - b)}$$

where

$$a[T] = 0.45724\frac{\mathbf{R}^2 T_c^2}{P_c}\left(1 + m\left(1 - T_r^{0.5}\right)\right)^2$$

$$m = 0.37464 + 1.54226\omega + 0.26992\omega^2$$

For mixtures is can be applied with the following mixing rules:

$$a = \sum_i \sum_j x_i x_j (a_i a_j)^{0.5}(1 - K_{ij})$$

$$b = \sum_i x_i b_i$$

where K_{ij} is a parameter to be adjusted for the i–j binary solution.

Let us substitute the equation of state into the generalized relationship derived for the fugacity coefficient:

$$\ln\frac{\bar{f}_i}{Px_i} = \frac{b_i}{b}(z - 1) - \ln(z - B) - \frac{A}{B}\left(\frac{2\sum x_j a_{ji}}{a} - \frac{b_i}{b}\right)\ln\left(\frac{z + 2.414B}{z - 0.414B}\right)$$

where

$$A = \frac{aP}{\mathbf{R}^2T^2}$$

$$B = \frac{bP}{\mathbf{R}T}$$

Here, a and b are the parameters calculated by the above mixing rule, and z is the compressibility factor obtained from the Peng–Robinson equation (for vapor and liquid phase):

$$z^3 - (1 - B)z^2 + (A + 3B^2 - 2B)z - (AB - B^2 - B^3) = 0$$

CALCULATION OF VAPOR-LIQUID EQUILIBRIUM BASED ON SEPARATE MODELS FOR THE FUGACITY AND ACTIVITY COEFFICIENTS

In the foregoing it was shown how the nonideality of the two equilibrium phases is taken into account by means of their common constitutive equation, using a suitable equation of state. Now we pass on to the other approach, characterized by different models of the two phases.

At equilibrium, for each species,

$$\bar{f}_i = \bar{f}_i^{(1)}$$

where $\bar{f}_i$ = partial fugacity of the component i in the vapor phase,
$\bar{f}_i^{(1)}$ = partial fugacity of the component i in the liquid phase.

For the vapor phase, on the other hand,

$$\bar{f}_i = \nu_i y_i P$$

where ν_i = fugacity coefficient of component i in the vapor phase,
P = total pressure, and
y_i = mole fraction in the vapor phase.

For the liquid solution,

$$\bar{f}_i^{(1)} = \gamma_i x_i f_i^{\circ}$$

where γ_i = activity coefficient of component i is the liquid phase,
x_i = the liquid mole fraction of component i,
f_i° = reference-state fugacity of component i.

The last quantity can be given as

$$f_i^\circ = P_i^\circ \nu_i^\circ \exp\left(\frac{1}{\mathbf{R}T} \int_{P_i^\circ}^{P} v_i^{(1)}\, dP \right)$$

where ν_i° = fugacity coefficient of species i in the pure state, at equilibrium vapor pressure P_i°,

$v_i^{(1)}$ = molar volume of species i in the liquid phase,

P_i° = equilibrium vapor pressure of pure component i at the system temperature.

The quantity $(1/\mathbf{R}T) \int_{P_i^\circ}^{P} v_i\, dP$ represents the work done by the intermolecular forces on pure substance i while its pressure is changing from the equilibrium pressure P_i° to the system pressure P. This is called the Poynting correction. It may be neglected at low pressures.

In calculating vapor–liquid equilibria, the equilibrium ratio plays an important role:

$$K_i \equiv \frac{y_i}{x_i}$$

or

$$K_i \equiv \frac{\gamma_i P_i^\circ \nu_i^\circ}{\nu_i P} \exp\left(\frac{1}{\mathbf{R}T} \int_{P_i^\circ}^{P} v_i^{(1)}\, dP \right)$$

Assuming that the molar volume $v_i^{(1)}$ remains constant over the range of integration, we can write

$$K_i = \frac{\gamma_i P_i^\circ \nu_i^\circ}{\nu_i P} \exp\left(\frac{v_i (P - P_i^\circ)}{\mathbf{R}T} \right)$$

This expression suggests that the temperature dependence of K_i is due to the temperature dependence of the vapor pressure. The dependence of γ_i on the temperature is not great.

The calculation of the vapor–liquid equilibria requires the subsequent determination of P_i°, ν_i, and γ_i. Among these parameters, the calculation of vapor pressure of component i needs no further consideration. The fugacity coefficient can be calculated from the virial equation, from the Redlich–Kwong–Soave or Peng–Robinson equation of state. For the latter two cases the procedure is identical with that demonstrated in the previous section, the only difference being that the result associated with the liquid phase is ignored. Hence, we will show only the use of the virial equation.

The calculation of the activity coefficients for liquid components was described in Chapter 11 on mixtures.

Calculation of the Fugacity Coefficient with the Virial Equation

This problem usually arises when the system pressure is low. For the fugacity coefficient of component 1, according to a previously derived relationship,

$$\ln \nu_1 = \int_0^P (z_1 - 1)\frac{dP}{P}$$

where z_1 is the compressibility factor of component 1 in the mixture of given composition, at temperature T and,

$$z_1 = \left(\frac{\partial(nz)}{\partial n_1}\right)_{T,P,n_2}$$

Let us express the compressibility factor of the mixture by the virial equation

$$z = 1 + \frac{BP}{\mathbf{R}T}$$

So for the compressibility factor of component 1 we have

$$z_1 = 1 + \frac{P}{\mathbf{R}T}\left(\frac{\partial(nB)}{\partial n_1}\right)_{T,P,n_2}$$

Let us substitute this into the fugacity-coefficient relationship for component 1:

$$\ln \nu_1 = \frac{1}{\mathbf{R}T}\int_0^P \left(\frac{\partial(nB)}{\partial n_1}\right)_{T,P,n_2} dP$$

The second virial coefficient for the mixture can be written as

$$B = \sum_i \sum_j (y_i y_j B_{ij})$$

where

$$B_{ij} = B_{ji}$$

In the case of binary mixtures, this leads to

$$B = y_1 Y_1 B_{11} + y_1 y_2 B_{12} + y_2 y_1 B_{21} + y_{22} B_{22}$$

$$= y_1^2 B_{11} + 2 y_1 y_2 B_{12} + y_2^2 B_{22}$$

On the other hand, we can write

$$B = y_1(1 - y_2)B_{11} + 2y_1y_2B_{12} + y_2(1 - y_1)B_{22}$$

$$= y_1B_{11} - y_1y_2B_{11} + 2y_1y_2B_{12} + y_2B_{22} - y_1y_2B_{12}$$

$$= y_1B_{11} + y_2B_{22} + y_1y_2\delta_{12}$$

where

$$\delta_{12} = 2B_{12} - B_{11} - B_{22}$$

Since

$$y_1 = \frac{n_1}{n}$$

we can write

$$nB = n_1B_{11} + n_2B_{22} + \frac{n_1n_2}{n}\delta_{12}$$

By differentiating we get

$$\left(\frac{\partial(nB)}{\partial n_1}\right)_{T,P,n_2} = B_{11} + \left(\frac{1}{n} - \frac{n_1}{n_2}\right)n_2\delta_{12}$$

$$= B_{11} + \left(1 - \frac{n_1}{n}\right)\frac{n_2}{n}\delta_{12}$$

$$= B_{11} + (1 - y_1)y_2\delta_{12}$$

$$= B_{11} + y_2^2\delta_{12}$$

Finally, substituting this into the fugacity-coefficient relationship, we have

$$\ln \nu_1 = \frac{P}{\mathbf{R}T}(B_{11} + y_2^2\delta_{12})$$

$$\ln \nu_2 = \frac{P}{\mathbf{R}T}(B_{22} + y_1^2\delta_{12})$$

Here, for the calculation of δ_{12}, knowledge of B_{12} is needed. According to the basic formula,

$$B_{ij} = \frac{\mathbf{R}T_{cij}}{P_{cij}}\left(B_r^{(0)} + \omega_{ij}B_r^{(1)}\right)$$

where

$$\omega_{ij} = \tfrac{1}{2}(\omega_i + \omega_j)$$

and the pseudocritical values are determined by the well-known mixing rules.

The calculation of vapor–liquid equilibrium is not finished by finding the compressibility factor of the vapor mixture and of the liquid mixture. Two mixing rules are used during the calculation, supplying the estimated values of the mole fractions to be calculated. Iteration will therefore be needed for the calculation; it will be dealt with in the next section.

Algorithm and Program: Calculation of Equilibrium Composition with van Laar and Virial Equations

Write a program for the calculation of the total vapor pressure, partial pressures, vapor-phase composition, and relative volatility in a binary mixture as a function of liquid composition. The actual temperature is known. Calculate the liquid activity coefficient with the van Laar equation, and the vapor fugacity coefficient with the virial equation.

Algorithm.

1. Read the temperature.
2. Read the sequential numbers of the mixture components.
3. Call the property-datafile reader routine.
4. Calculate the pure-component vapor pressures.
5. Read the coefficients A_{12} and A_{21} of the van Laar equation.
6. Calculate the activity coefficients of the components with the van Laar equation (see program VLEVANLAAR in Chapter 10).
7. Calculate the partial *fugacities* of the components:

$$f_i = x_i \gamma_i P_i^\circ$$

8. Calculate the fugacity coefficient, total pressure, and vapor composition from the fugacities, given that

$$f_i = y_i \nu_i P$$

The three factors of this product have to be found; as initial guesses use

$$P_i = f_i \qquad P = P_1 + P_2 \qquad y_i = \frac{P_i}{P}$$

(a) Calculate the pseudo-one-component properties of the mixture with the mixture rules (see Chapter 8).

(b) Calculate the reduced second virial coefficient for the pure components and for the mixture (for the pseudo-one-component model).

(c) Calculate δ_{12}:

$$\delta_{12} = 2B_{\text{mix}} - B_1 - B_2$$

where B is the reduced second virial coefficient.

(d) Calculate the fugacity coefficients derived from the virial equation

$$\ln \nu_i = \frac{P}{\mathbf{R}T}\left(B_i + y_i^2\delta_{12}\right)$$

(e) Calculate the next estimate of the vapor-phase compositions:

$$y_i = \frac{f_i}{P\upsilon_i}$$

(f) Calculate the next estimate of the total pressure:

$$P \leftarrow P(y_1 + y_2)$$

(g) Stop the iteration and go to step 9 if

$$1 - (y_1 + y_2) \leq 10^{-4}$$

(h) Reestimate the normal vapor compositions:

$$y_i \leftarrow \frac{y_i}{y_1 + y_2}$$

(i) Go to step (a).

9. Calculate $\alpha = y_2/y_1$, and print the results.

Program DISTVIRVANLAAR. A possible realization:

```
]PR#0
]LOAD DISTVIRVANLAAR
]LIST

  10  REM :CALCULATION OF VLE COMPO
       SITIONS AND PARTIAL AND TOTAL
       PRESSURE
  20  REM :CALCULATE THE VAPOR
       FUGACITY COEFF. BY VIRIAL EQUA
       TION OF STATE
  30  REM :THE LIQUID ACTIVITY COE
       FFICIENT IS CALCULATED BY VAN
       LAAR EQUATION
```

```
 40  PRINT : PRINT
 50  DIM A(15)
 60  PRINT
 70  INPUT "ENTER TEMPERATURE (K)
      "; T
 80  PRINT
 90  QJ = 0
100  PRINT "ENTER THE SEQUENTIAL
      NUMBER"
110  INPUT "OF FIRST COMPONENT
      "; IS
120  C1 = IS
130  GOSUB 1170
140  TC = A(3)
150  PC = A(2)
160  OM = A(4)
170  GOSUB 770
180  S1 = PS
190  PRINT
200  PRINT "ENTER THE SEQUENTIAL
      NUMBER"
210  INPUT "OF SECOND COMPONENT
      "; IS
220  C2 = IS
230  GOSUB 1170
240  TC = A(3)
250  PC = A(2)
260  OM = A(4)
270  GOSUB 770
280  S2 = PS
290  PRINT
300  INPUT "ENTER VAN LAAR CONSTANT
      A12 "; A1
310  PRINT
320  INPUT "ENTER VAN LAAR CONSTANT
      A21 "; A2
330  PRINT
340  PRINT "PRESSURES ARE IN KILO
      PASCAL !!!"
350  PRINT
360  PRINT "X1 Y1(ID) Y1(VI
      R) P ALPHA"
370  PRINT "==========
      ====================="
380  FOR I = 1 TO 11
390  X = (I - 1) * 0.1
400  GC = EXP (A1 * (1 - X) ∧ 2 /
      (A1 / A2 * X + (1 - X)) ∧ 2)
410  P3 = X * S1 * GC
```

```
415  P8 = S1 * GC
420  GC = EXP (A2 * X ∧ 2 / (A2 /
      A1 * (1 − X) + X) ∧ 2)
430  P4 = (1 − X) * S2 * GC
435  P9 = S2 * GC
440  P = P3 + P4
450  Y1 = P3 / P
460  I1 = Y1
470  Y2 = 1 − Y1
480  E1 = Y1
490  E2 = Y2
500  GOSUB 940
510  Y1 = P3 / P / N1
520  Y2 = P4 / P / N2
525  S = Y1 + Y2
530  Y1 = Y1 / (Y1 + Y2)
540  Y2 = Y2 / (Y1 + Y2)
550  IF ABS (S − 1) < 1E − 4 THEN
      GOTO 590
560  P = P * S
580  GOTO 500
590  Y% = Y1 * 1000
600  Y1 = Y% / 1000
620  AL = P8 / P9
630  A% = AL * 1000
640  AL = A% / 1000
650  P = P / 1000:P% = P * 1000:P =
      P% / 1000
660  I% = I1 * 1000:I1 = I% / 1000
670  PRINT TAB(1); X; TAB(7); I1;
      TAB(16); Y1) TAB(26); P; TAB(
      34); AL
680  NEXT I
690  PRINT
700  PRINT "==============
      ===================="
710  PRINT : PRINT
720  INPUT "NEXT COMPOUNDS ? (Y / N
      ) "; A$
730  IF A$ = "Y" THEN GOTO 80
740  INPUT "NEXT TEMPERATURE ? (
      Y / N) "; A$
750  IF A$ = "Y" THEN GOTO 60
760  END
770  TR = T / TC
780  PS = 5.92714 − 6.09648 / TR −
     1.28862 * LOG (TR) + 0.1693
      47 * TR ∧ 6 + OM * (15.2518 −
     15.6875 / TR − 13.4721 * LOG
      (TR) + 0.43577 * TR ∧ 6)
```

```
 790  PS = EXP (PS) * PC
 800  RETURN
 810  REM :CALCULATION OF SECOND
       VIRIAL COEFF. FOR THE TWO PURE
       COMPONENTS AND FOR THE MIXTURE
 820  GOSUB 1000
 830  R = 8.3144
 840  TC = T1 :PC = P1: OM = O1
 850  GOSUB 1270
 860  B6 = R * TC / PC * B
 870  TC = T2:PC = P2:OM = O2
 880  GOSUB 1270
 890  B7 = R * TC / PC * B
 900  TC = TP:PC = PP:OM = OP
 910  GOSUB 1270
 920  BB = R * TC / PC * B
 930  RETURN
 940  REM :CALCULATION OF FUGACITY
       COEFFICIENTS FOR BINARY MIX
       TURE WITH VIRIAL EQN. OF STATE
 950  GOSUB 810
 960  DE = 2 * BB - B6 - B7
 970  N1 = EXP (P / R / T * (B6 +
       (1 - Y1) ∧ 2 * DE))
 980  N2 = EXP (P / R / T * (B7 +
      Y1 ∧ 2 * DE))
 990  RETURN
1000  REM :CALCULATION OF PSEUDO-
       ONE-COMPONENT PROPERTIES FROM
       COMPOSITION AND PURE COMPO
       NENT DATA FOR BINARY MIXTURE
1010  REM :USING LEE-KESLER MIXING
       RULES
1020  REM :COMPOSITION IS IN 'Y1',
       SEQUENTIAL NUMBERS OF COMPO
       UNDS IN MIXTURE ARE IN 'C1'
       AND IN 'C2'
1030  IF QJ = 1 THEN GOTO 1110
1040  QJ = 1
1050  IS = C1
1060  GOSUB 1170
1070  T1 = A(3):P1 = A(2):V1 = A(1
       4):O1 = A(4)
1080  IS = C2
1090  GOSUB 1170
1100  T2 = A(3):P2 = A(2):V2 = A(1
       4):O2 = A(4)
1110  VP = 1 / 8 * (Y1 * (1 - Y1) *
       (V1 ∧ (1 / 3) + V2 ∧ (1 / 3)
       ) ∧ 3 + Y1 * Y1 * (2 * V1 ∧
```

```
     (1 / 3)) ∧ 3 + (1 − Y1) * Y1
     * (V2 ∧ (1 / 3) + V1 ∧ (1 /
     3)) ∧ 3 + (1 − Y1) ∧ 2 * (2 *
     V2 ∧ (1 / 3)) ∧ 3)
1120 TP = 1 / 8 / VP * (Y1 * (1 −
     Y1) * (V1 ∧ (1 / 3) + V2 ∧ (
     1 / 3)) ∧ 3 * SQR (T1 * T2)
     + Y1 * Y1 * (2 * V1 ∧ (1 /
     3)) ∧ 3 * T1 + (1 − Y1) * Y1
     * (V2 ∧ (1 / 3) + V1 ∧ (1 /
     3)) ∧ 3 * SQR (T2 * T1) + (
     1 − Y1) ∧ 2 * (2 * V2 ∧ (1 /
     3)) ∧ 3 * T2)
1130 OP = (Y1 * O1 + (1 − Y1) * O
     2)
1140 R = 8.3144
1150 PP = (0.2905 - 0.085 * OP) *
     P * TP / VP
1160 RETURN
1170 REM :READING OF SPECIFIED
     RECORD OF PHDAT FILE INTO
     ARRAY A(15)
1180 REM :ARRAY A HAS TO BE DECL
     ARED IN THE CALLING PROGRAM
1190 D$ = CHR$ (4)
1200 PRINT D$; "OPEN PHDAT"
1210 PRINT D$; "POSITION PHDAT, R";
     IS − 1
1220 PRINT D$; "READ PHDAT"
1230 INPUT A(1), A(2), A(3), A(4), A(5),
     A(6), A(7), A(8), A(9), A(10),
     A(11), A(12), A(13), A(14), A(15)
1240 PRINT D$; "CLOSE PHDAT"
1250 PRINT D$; "IN#0"
1260 RETURN
1270 RESTORE
1280 READ N1,N2,N3,N4
1290 READ R1,R2,R3,R4
1300 DATA 0.1181193,0.265728,0.
     15479,0.0030323
1310 DATA 0.2026579,0.331511,0.
     027655,0.203488
1320 MR = 0.3978
1330 B1 = N1 + OM * (R1 − N1) / MR
1340 B2 = N2 + OM * (R2 − N2) / MR
1350 b4 = N3 + OM ° (R3 − N3) / MR
1360 B4 = N4 + OM * (R4 − N4) / MR
1370 TR = T / TC
1380 PR = P / PC
```

```
1390  B = B1 - B2 / TR - B3 / TR /
      TR - B4 / TR / TR / TR
1400  BA = TR / PR
1410  RETURN
```

Example. Run the program for a chloroform–n-heptane mixture at temperature 288.2 K. Repeat the calculation for a chloroform–1,2-dichloroethane mixture at the same temperature.

```
]RUN

ENTER TEMPERATURE (K) 288.2

ENTER THE SEQUENTIAL NUMBER
OF FIRST COMPONENT 10

ENTER THE SEQUENTIAL NUMBER
OF SECOND COMPONENT 7

ENTER VAN LAAR CONSTANT A12 0.3531

ENTER VAN LAAR CONSTANT A21 0.7116

PRESSURES ARE IN KILOPASCAL !!!
```

X1	Y1(ID)	Y1(VIR)	P	ALFA
0	0	0	3.496	6.808
.1	.421	.42	5.454	6.555
.2	.61	.609	7.252	6.27
.3	.718	.717	8.892	5.949
.4	.788	.787	10.375	5.587
.5	.838	.837	11.705	5.178
.6	.876	.875	12.893	4.719
.7	.907	.907	13.956	4.204
.8	.935	.935	14.925	3.633
.9	.964	.964	15.848	3.009
1	1	1	16.789	2.347

```
NEXT COMPOUNDS ? (Y / N) Y

ENTER THE SEQUENTIAL NUMBER
OF FIRST COMPONENT 10

ENTER THE SEQUENTIAL NUMBER
OF SECOND COMPONENT 12

ENTER VAN LAAR CONSTANT A12 0.3412

ENTER VAN LAAR CONSTANT A21 0.08948

PRESSURES ARE IN KILOPASCAL !!!
```

X1	Y1(ID)	Y1(VIR)	P	ALFA
	0	0	5.963	3.943
.1	.267	.267	7.392	3.291
.2	.428	.428	8.539	3.001
.3	.549	.549	9.603	2.845
.4	.647	.646	10.638	2.75
.5	.728	.728	11.662	2.689
.6	.798	.798	12.685	2.647
.7	.859	.859	13.707	2.616
.8	.912	.911	14.732	2.594
.9	.958	.958	15.759	2.577
1	1	1	16.789	2.563

NEXT COMPOUNDS ? (Y / N) N
NEXT TEMPERATURE ? (Y / N) N

Remark. For comparison of the results of the virial and ideal-gas vapor-phase calculations, we have run both compositions. Explain the results.

Algorithm and Program: Calculation of Equilibrium Composition with Regular-Solution Theory and Virial Equation

Write a program for the calculation of the total vapor pressure, partial pressures, vapor-phase composition, and relative volatility in a binary mixture as a function of liquid composition. The actual temperature is given. Calculate the liquid activity coefficient using regular-solution theory and using the vapor fugacity coefficient in the virial equation.

Algorithm.

1. Read the temperature.
2. Read the sequential numbers of the mixture components.
3. Call the property-datafile reader routine.
4. Calculate the pure-component vapor pressures.
5. Calculate the enthalpy of vaporization of the components at the actual temperature.
6. Calculate the pure-component liquid molar volumes.
7. Calculate the solubility parameters and volume fraction of components in the mixture.
8. Calculate the partial fugacities of the components:

$$f_i = x_i \gamma_i P_i^\circ$$

9. Calculate the fugacity coefficients, total pressure, and vapor composition from the fugacities, given that

$$f_i = y_i \nu_i P$$

The three factors of this product have to be found; as initial guesses use

$$P_i = f_i \qquad P = P_1 + P_2 \qquad y_i = \frac{P_i}{P}$$

(a) Calculate the pseudo-one-component properties of the mixture with mixture rules (see Chapter 8).

(b) Calculate the reduced second virial coefficient for the pure components and for the mixture (for the pseudo-one-component model).

(c) Calculate δ_{12}

$$\delta_{12} = 2B_{\text{mix}} - B_1 - B_2$$

where B is the reduced second virial coefficient.

(d) Calculate fugacity coefficients derived from the virial equation

$$\ln \nu_i = \frac{P}{RT}\left(B_i + y_i^2 \delta_{12}\right)$$

(e) Calculate the next estimate of the vapor-phase compositions:

$$y_i = \frac{f_i}{P v_i}$$

(f) Calculate the next estimation of the total pressure:

$$P \leftarrow P(y_1 + y_2)$$

(g) Stop the iteration and go to step 10 if

$$1 - (y_1 + y_2) \le 10^{-4}$$

(h) Reestimate the normal vapor compositions:

$$y_i \leftarrow \frac{y_i}{y_1 + y_2}$$

(i) Go to step (a).

10. Calculate $\alpha = y_2/y_1$, and print the results.

Program DISTVIRREGULAR. A possible realization:

```
]PR#0
]LOAD DISTVIRREGULAR
```

```
]LIST

 10  REM :CALCULATION OF VLE COMPO
      SITIONS AND PARTIAL AND TOTAL
      PRESSURE
 20  REM :CALCULATE THE VAPOR FUGACITY
      COEFF. BY VIRIAL EQUATION
      OF STATE
 30  REM THE LIQUID ACTIVITY COEFF.
      IS CALCULATED BY REGULAR
      SOLUTION THEORY
 40  PRINT : PRINT
 50  DIM A(15)
 60  P = 101325
 70  R = 8.3144
 80  PRINT
 90  INPUT "ENTER TEMPERATURE (K)
      "; T
100  PRINT
110  QJ = 0
120  PRINT "ENTER THE SEQUENTIAL
      NUMBER"
130  INPUT "OF FIRST COMPONENT
      "; IS
140  C1 = IS
150  GOSUB 1330
160  TC = A(3)
170  PC = A(2)
180  OM = A(4)
190  TB = A(5)
200  DH = A(6)
210  GOSUB 930
220  GOSUB 1650
230  GOSUB 1580
240  Q1 = V
250  H1 = HV
260  S1 = PS
270  PRINT
280  PRINT "ENTER THE SEQUENTIAL
      NUMBER"
290  INPUT "OF SECOND COMPONENT
      "; IS
300  C2 = IS
310  GOSUB 1330
320  TC = A(3)
330  PC = A(2)
340  OM = A(4)
350  TB = A(5)
360  DH = A(6)
```

```
370  GOSUB 930
380  GOSUB 1650
390  GOSUB 1580
400  Q2 = V
410  H2 = HV
420  S2 = PS
430  D1 = SQR ((H1 - R * T) / Q1)
440  D2 = SQR ((H2 - R * T) / Q2)
450  PRINT
460  PRINT "PRESSURES ARE IN KILO
      PASCAL !!!"
470  PRINT
480  PRINT " X1 Y1 P(VIR)
      P(ID) ALPHA"
490  PRINT "=================
      ======================="
500  FOR I = 1 TO 11
510  X = (I - 1) * 0.1
520  F2 = (1 - X) * Q2 / (X * Q1 +
      (1 - X) * Q2)
530  GC = EXP (Q1 * F2 * (D1 - D2
      ) ∧ 2 / R / T)
540  P3 = X * S1 * GC
545  P8 = S1 * GC
550  F1 = X * Q1 / (X * Q1 + (1 -
      X) * Q2)
560 GC = EXP (Q2 * F1 * (D1 - D2
      ) ∧ 2 / R / T)
570  P4 = (1 - X) * S2 * GC
575  P9 = S2 * GC
580  P = P3 + P4
590  Y1 = P3 / P
600  I1 = Y1
610  Y2 = 1 - Y1
620  I2 = P
630  I2 = I2 / 1000:P% = I2 * 100:
      I2 = P% / 100
640  E1 = Y1
650  E2 = Y2
660  GOSUB 1100
670  Y1 = P3 / P / N1
680  Y2 = P4 / P / N2
690  S = Y1 + Y2
700  Y1 = Y1 / (Y1 + Y2)
710  Y2 = Y2 / (Y1 + Y2)
720  IF ABS (S - 1) < 1E - 4 THEN
      GOTO 750
730  P = P * S
740  GOTO 660
```

```
750  Y% = Y1 * 1000
760  Y1 = Y% / 1000
780  AL = P8 / P9
790  A% = AL * 1000
800  AL = A% / 1000
810  P = P / 1000:P% = P * 100:P =
      P% / 100
820  I% = I1 * 1000:I1 = I% / 1000
830  PRINT TAB(1); X; TAB(7); Y1;
      TAB(16); P; TAB(26); I2; TAB(
      34); AL
840  NEXT I
850  PRINT
860  PRINT "=================
      ======================"
870  PRINT : PRINT
880  INPUT "NEXT COMPOUNDS ? (Y / N
      ) "; A$
890  IF A$ = "Y" THEN GOTO 100
900  INPUT "NEXT TEMPERATURE ? (
      Y / N) "; A$
910  IF A$ = "Y" THEN GOTO 80
920  END
930  TR = T / TC
940  PS = 5.92714 - 6.09648 / TR -
      1.28862 * LOG (TR) + 0.1693
      47 * TR ∧ 6 + OM * (15.2518 -
      15.6875 / TR - 13.4721 * LOG
      (TR) + 0.43577 * TR ∧ 6)
950  PS = EXP (PS) * PC
960  RETURN
970  REM :CALCULATION OF SECOND
      VIRIAL COEFF. FOR THE TWO PURE
      COMPONENTS AND FOR THE MIXTURE
980  GOSUB 1160
990  R = 8.3144
1000 TC = T1 :PC = P1: OM = O1
1010 GOSUB 1430
1020 B6 = R * TC / PC * B
1030 TC = T2:PC = P2:OM = 02
1040 GOSUB 1430
1050 B7 = R * TC / PC * B
1060 TC = TP:PC = PP:OM = OP
1070 GOSUB 1430
1080 BB = R * TC / PC * B
1090 RETURN
1100 REM :CALCULATION OF FUGACITY
      COEFFICIENTS FOR BINARY MI
      XTURE WITH VIRIAL EQN. OF STATE
```

```
1110  GOSUB 970
1120  DE = 2 * BB - B6 - B7
1130  N1 = EXP (P / R / T * (B6 +
       (1 - Y1) ∧ 2 * DE))
1140  N2 = EXP (P / R / T * (B7 +
       Y1 ∧ 2 * DE))
1150  RETURN
1160  REM :CALCULATION OF PSEUDO-
       ONE-COMPONENT PROPERTIES FROM
       COMPOSITION AND PURE COMPO
       NENT DATA FOR BINARY MIXTURE
1170  REM :USING LEE-KESLER MIXING
       RULES
1180  REM :COMPOSITION IS IN 'Y1',
       SEQUENTIAL NUMBERS OF COMPO
       UNDS IN MIXTURE ARE IN 'C1'
       AND IN 'C2'
1190  IF QJ = 1 THEN GOTO 1270
1200  QJ = 1
1210  IS = C1
1220  GOSUB 1330
1230  T1 = A(3):P1 = A(2):V1 = A(1
       4):O1 = A(4)
1240  IS = C2
1250  GOSUB 1330
1260  T2 = A(3):P2 = A(2):V2 = A(1
       4):O2 = A(4)
1270  VP = 1 / 8 * (Y1 * (1 - Y1) *
       (V1 ∧ (1 / 3) + V2 ∧ (1 / 3)
       ) ∧ 3 + Y1 * Y1 * (2 * V1 ∧
       (1 / 3)) ∧ 3 + (1 - Y1) * Y1
      * (V2 ∧ (1 / 3) + V1 ∧ (1 /
       3)) ∧ 3 + (1 - Y1) ∧ 2 * (2 *
      V2 ∧ (1 / 3)) ∧ 3)
1280  TP = 1 / 8 / VP * (Y1 * (1 -
       Y1) * (V1 ∧ (1 / 3) + V2 ∧ (
       1 / 3)) ∧ 3 * SQR (T1 * T2)
       + Y1 * Y1 * (2 * V1 ∧ (1 /
       3)) ∧ 3 * T1 + (1 - Y1) * Y1
      * (V2 ∧ (1 / 3) + V1 ∧ (1 /
       3)) ∧ 3 * SQR (T2 * T1) + (
       1 - Y1) ∧ 2 * (2 * V2 ∧ (1 /
       3)) ∧ 3 * T2)
1290  OP = (Y1 * O1 + (1 - Y1) * O
      2)
1300  R = 8.3144
1310  PP = (0.2905 - 0.085 * OP) *
       R * TP / VP
1320  RETURN
```

```
1330  REM :READING OF SPECIFIED
       RECORD OF PHDAT FILE INTO
       ARRAY A(15)
1340  REM :ARRAY A HAS TO BE DECL
       ARED IN THE CALLING PROGRAM
1350  D$ = CHR$ (4)
1360  PRINT D$; "OPEN PHDAT"
1370  PRINT D$; "POSITION PHDAT, R";
       IS - 1
1380  PRINT D$; "READ PHDAT"
1390  INPUT A(1), A(2), A(3), A(4), A(5),
       A(6), A(7), A(8), A(9), A(10),
       A(11), A(12), A(13), A(14), A(15)
1400  PRINT D$; "CLOSE PHDAT"
1410  PRINT D$; "IN#0"
1420  RETURN
1430  RESTORE
1440  READ N1,N2,N3,N4
1450  READ R1,R2,R3,R4
1460  DATA 0.1181193,0.265728,0.
       15479,0.0030323
1470  DATA 0.2026579,0.331511,0.
      027655,0.203488
1480  MR = 0.3978
1490  B1 = N1 + OM * (R1 - N1) / MR
1500  B2 = N2 + OM * (R2 - N2) / MR
1510  B3 = N3 + OM * (R3 - N3) / MR
1520  B4 = N4 + OM * (R4 - N4) / MR
1530  TR = T / TC
1540  PR = P / PC
1550  B = B1 - B2 / TR - B3 / TR /
       TR - B4 / TR / TR / TR
1560  BA = TR / PR
1570  RETURN
1580  DEF FN Z0(PR) = (.46407 -
       0.73221 * TR + 0.45256 * TR ^
       2) * PR - (0.00871 - 0.02939
       * TR + 0.02775 * TR ^ 2) *
       PR ^ 2
1590  DEF FN Z1(PR) = - (0.0267
       6 + 0.28376 * TR - 0.2834 *
       TR ^ 2) * PR - (0.10209 - 0.
       32504 * TR + 0.25376 * TR ^
       2) * PR ^ 2 + (0.00919 - 0.0
       3016 * TR - 0.02485 * TR ^ 2
       ) * PR ^ 3
1600  TR = T / TC
1610  PR = P / PC
1620  Z = FN Z0(PR) + OM * FN Z1
       (PR)
```

```
1630  V = Z * 8.3144 * T / P
1640  RETURN
1650  REM :CALCULATION OF ENTHALPY
       OF VAPORIZATION BY WATSON EQN.
1660  HV = DH * ((T - TC) / (TB -
       TC)) ∧ 0.38
1670  RETURN
```

Example. Run the program for a benzene–2,4-dimethylpentene mixture at temperature 309 K.

```
]RUN

ENTER TEMPERATURE (K) 309

ENTER THE SEQUENTIAL NUMBER
OF FIRST COMPONENT 22

ENTER THE SEQUENTIAL NUMBER
OF SECOND COMPONENT 23

PRESSURES ARE IN KILOPASCAL !!!
```

X1	Y1	P(VIR)	P(ID)	ALPHA
0	0	20.79	20.47	8.294
.1	.466	36.28	35.6	8.006
.2	.653	51.24	50.22	7.702
.3	.755	65.69	64.26	7.382
.4	.821	79.56	77.63	7.044
.5	.867	92.74	90.24	6.686
.6	.903	105.09	101.99	6.309
.7	.931	116.45	112.72	5.911
.8	.956	126.62	122.26	5.49
.9	.978	135.35	130.41	5.047
1	1	142.33	136.88	4.581

```
NEXT COMPOUNDS ? (Y / N) N
NEXT TEMPERATURE ? (Y / N) N
```

Remark. Explain the results.

Algorithm and Program: Calculation of Equilibrium Composition with Wilson and Virial Equations

Write a program for the calculation of the total vapor pressure, partial pressures, vapor-phase composition, and relative volatility in a binary mixture as a function of liquid composition. The actual temperature is given. Calculate the liquid activity coefficient with the Wilson equation and the vapor fugacity coefficient with the virial equation.

Algorithm.

1. Read the temperature.
2. Read the sequential numbers of the mixture components.
3. Call the property-datafile reader routine.
4. Calculate pure-component vapor pressures.
5. Read the coefficients A_{12} and A_{21} of the Wilson equation.
6. Calculate the activity coefficients of the components with the Wilson equation (see program VLEWILSON in Chapter 10).
7. Calculate the partial *fugacities* of the components:

$$f_i = x_i \gamma_i P_i^\circ$$

8. Calculate the fugacity coefficient, total pressure, and vapor composition from fugacities, given that

$$f_i = y_i \nu_i P$$

The three factors of this product have to be found; as initial guesses use

$$P_i = f_i \qquad P = P_1 + P_2 \qquad y_i = \frac{P_i}{P}$$

(a) Calculate the pseudo-one-component properties of the mixture with the mixture rules (see Chapter 8).

(b) Calculate the reduced second virial coefficient for the pure components and for the mixture (for the pseudo-one-component model).

(c) Calculate δ_{12}:

$$\delta_{12} = 2B_{\text{mix}} - B_1 - B_2$$

where B is the reduced second virial coefficient.

(d) Calculate the fugacity coefficients derived from the virial equation

$$\ln \nu_i = \frac{P}{\mathbf{RT}}\left(B_i + y_i^2 \delta_{12}\right)$$

(e) Calculate the next estimate of the vapor-phase compositions:

$$y_i = \frac{f_i}{P v_i}$$

(f) Calculate the next estimate of the total pressure:

$$P \leftarrow P(y_1 + y_2)$$

(g) Stop the iteration and go to step 9 if

$$1 - (y_1 + y_2) \leq 10^{-4}$$

(h) Reestimate the normal vapor compositions:

$$y_i \leftarrow \frac{y_i}{y_1 + y_2}$$

(i) Go to step (a).

9. Calculate $\alpha = y_2/y_1$, and print the results.

Program DISTVIRWILSON. A possible realization:

```
]LOAD DISTVIRWILSON
]LIST

 10  REM ;CALCULATION OF VLE COMPO
      SITIONS AND PARTIAL AND TOTAL
      PRESSURE
 20  REM :CALCULATE THE VAPOR FU
      GACITY COEFF. BY VIRIAL EQUA
      TION OF STATE
 30  REM :THE LIQUID ACTIVITY COEFF.
      IS CALCULATED BY WILSON EQ
      UATION
 40  PRINT : PRINT
 50  DIM A(15)
 60  PRINT
 70  INPUT "ENTER TEMPERATURE (K)
      "; T
 80  PRINT
 90  QJ = 0
100  PRINT "ENTER THE SEQUENTIAL
      NUMBER"
110  INPUT "OF FIRST COMPONENT
      "; IS
120  C1 = IS
130  GOSUB 1210
140  TC = A(3)
150  PC = A(2)
160  OM = A(4)
170  GOSUB 810
180  S1 = PS
```

```
190  PRINT
200  PRINT "ENTER THE SEQUENTIAL
      NUMBER"
210  INPUT "OF SECOND COMPONENT
      "; IS
220  C2 = IS
230  GOSUB 1210
240  TC = A(3)
250  PC = A(2)
260  OM = A(4)
270  GOSUB 810
280  S2 = PS
290  PRINT
300  INPUT "ENTER WILSON CONSTANT
      A12 "; A1
310  PRINT
320  INPUT "ENTER WILSON CONSTANT
      A21 "; A2
340  PRINT
350  PRINT
360  PRINT "PRESSURES ARE IN KILO
      PASCAL !!!"
390  PRINT
400  PRINT "X1 Y1(ID) Y1(VI
      R) P ALPHA"
410  PRINT "=======================
      ======================="
420  FOR I = 1 TO 11
430  X = (I - 1) * 0.1
440  GC = EXP ( - LOG (X + (1 -
     X) * A1) + (1 - X) * (A1 / (
      X + A1 * (1 - X)) - A2 / ((1
      - X) + A2 * X)))
450  P3 = X * S1 * GC
455  P8 = S1 * GC
460  GC = EXP ( - LOG (1 - X + X
      * A2) - X * (A1 / (X + A1 *
      (1 - X)) - A2 / ((1 - X) + A
      2 * X)))
470  P4 = (1 - X) * S2 * GC
475  P9 = S2 * GC
480  P = P3 + P4
490  Y1 = P3 / P
500  I1 = Y1
510  Y2 = 1 - Y1
520  E1 = Y1
530  E2 = Y2
540  GOSUB 980
550  Y1 = P3 / P / N1
```

```
560  Y2 = P4 / P / N2
570  S = Y1 + Y2
580  Y1 = Y1 / (Y1 + Y2)
590  Y2 = Y2 / (Y1 + Y2)
600  IF ABS (S - 1) < 1E - 4 THEN
      GOTO 630
610  P = P * S
620  GOTO 540
630  Y% = Y1 * 1000
640  Y1 = Y% / 1000
660  AL = P8 / P9
670  A% = AL * 1000
680  AL = A% / 1000
690  P = P / 1000:P% = P * 1000:P =
      P% / 1000
700  I% = I1 * 1000: I1 = I% / 1000
710  PRINT TAB(1); X; TAB(7); I1;
      TAB(16); Y1; TAB(26); P; TAB(
      34); AL
720  NEXT I
730  PRINT
740  PRINT "=======================
      ======================="
750  PRINT : PRINT
760  INPUT "NEXT COMPOUNDS ? (Y / N
      ) "; A$
770  IF A$ = "Y" THEN GOTO 80
780  INPUT "NEXT TEMPERATURE ? (
      Y / N) "; A$
790  IF A$ = "Y" THEN GOTO 60
800  END
810  TR = T / TC
820  PS = 5.92714 - 6.09648 / TR -
      1.28862 * LOG (TR) + 0.1693
      47 * TR ∧ 6 + OM * (15.2518 -
      15.6875 / TR - 13.4721 * LOG
      (TR) + 0.43577 * TR ∧ 6)
830  PS = EXP (PS) * PC
840  RETURN
850  REM :CALCULATION OF SECOND
      VIRIAL COEFF. FOR THE TWO PURE
      COMPONENTS AND FOR THE MIXTURE
860  GOSUB 1040
870  R = 8.3144
880  TC = T1:PC = P1:OM = O1
890  GOSUB 1310
900  B6 = R * TC / PC * B
910  TC = T2:PC = P2:OM = O2
920  GOSUB 1310
```

```
 930  B7 = R* TC / PC * B
 940  TC = TP:PC = PP:OM = OP
 950  GOSUB 1310
 960  BB = R * TC / PC * B
 970  RETURN
 980  REM :CALCULATION OF FUGACITY
       COEFFICIENTS FOR BINARY MIX
       TURE WITH VIRIAL EQN. OF STATE
 990  GOSUB 850
1000  DE = 2 * BB - B6 - B7
1010  N1 = EXP (P / R / T * (B6 +
       (1 - Y1) ∧ 2 * DE))
1020  N2 = EXP (P / R / T * (B7 +
       Y1 ∧ 2 * DE))
1030  RETURN
1040  REM :CALCULATION OF PSEUDO-
       ONE-COMPONENT PROPERTIES FROM
       COMPOSITION AND PURE COMPO
       NENT DATA FOR BINARY MIXTURE
1050  REM :USING LEE-KESLER MIXING
       RULES
1060  REM :COMPOSITION IS IN 'Y1',
       SEQUENTIAL NUMBERS OF COMPO
       UNDS IN MIXTURE ARE IN 'C1'
       AND IN 'C2'
1070  IF QJ = 1 THEN GOTO 1150
1080  QJ = 1
1090  IS = C1
1100  GOSUB 1210
1110  T1 = A(3):P1 = A(2):V1 = A(1
       4):O1 = A(4)
1120  IS = C2
1130  GOSUB 1210
1140  T2 = A(3):P2 = A(2):V2 = A(1
       4):O2 = A(4)
1150  VP = 1 / 8 * (Y1 * (1 - Y1) *
       (V1 ∧ (1 / 3) + V2 ∧ (1 / 3)
       ) ∧ 3 + Y1 * Y1 * (2 * V1 ∧
      (1 / 3)) ∧ 3 + (1 - Y1) * Y1
       * (V2 ∧ (1 / 3) + V1 ∧ (1 /
       3)) ∧ 3 + (1 - Y1) ∧ 2 * (2 *
       V2 ∧ (1 / 3)) ∧ 3)
1160  TP = 1 / 8 / VP * (Y1 * (1 -
       Y1) * (V1 ∧ (1 / 3) + V2 ∧ (
       1 / 3)) ∧ 3 * SQR (T1 * T2)
       + Y1 * Y1 * (2 * V1 ∧ (1 /
       3)) ∧ 3 * T1 + (1 - Y1) * Y1
       * (V2 ∧ (1 / 3) + V1 ∧ (1 /
       3)) ∧ 3 * SQR (T2 * T1) + (
       1 - Y1) ∧ 2 * (2 * V2 ∧ (1 /
```

```
      3)) ∧ 3 * T2)
1170  OP  =  (Y1 * O1 + (1 - Y1) * O
      2)
1180  R  =  8.3144
1190  PP  =  (0.2905 - 0.085 * OP) *
      R * TP / VP
1200  RETURN
1210  REM :READING OF SPECIFIED
      RECORD OF PHDAT FILE INTO
      ARRAY A(15)
1220  REM :ARRAY A HRS TO BE DECL
      ARED IN THE CALLING PROGRAM
1230  D$ =  CHR$ (4)
1240  PRINT D$; "OPEN PHDAT"
1250  PRINT D$; "POSITION PHDAT, R";
      IS - 1
1260  PRINT D$; "READ PHDAT"
1270  INPUT A(1), A(2), A(3), A(4), A(5),
      A(6), A(7), A(8), A(9), A(10),
      A(11), A(12), A(13), A(14), A(15)
1280  PRINT D$; "CLOSE PHDAT"
1290  PRINT D$; "IN#0"
1300  RETURN
1310  RESTORE
1320  READ N1,N2,N3,N4
1330  READ R1,R2,R3,R4
1340  DATA 0.1181193,0.265728,0.
      15479,0.0030323
1350  DATA 0.2026579,0.331511,0.
      027655,0.203488
1360  MR  =  0.3978
1370  B1 = N1 + OM * (R1 - N1) / MR
1380  B2  =  N2 + OM * (R2 - N2) / MR
1390  B3  =  N3 + OM * (R3) / MR
1400  B4  =  N4 + OM * (R4 - N4) / MR
1410  TR  =  T / TC
1420  PR  =  P / PC
1430  B  =  B1 - B2 / TR - B3 / TR /
      TR - B4 / TR / TR / TR
1440  BA  =  TR / PR
1450  RETURN
```

Example. Run the program for a methanol–chloroform mixture at temperature 288.2 K, with $A_{12} = 0.119362$, $A_{21} = 0.881787$. Repeat the calculation for an acetic acid–water mixture at temperature 288.2 K, with $A_{12} = 0.2022$, $A_{21} = 1.3433$.

```
]RUN
ENTER TEMPERATURE (K) 288.2
```

ENTER THE SEQUENTIAL NUMBER
OF FIRST COMPONENT 13

ENTER THE SEQUENTIAL NUMBER
OF SECOND COMPONENT 10

ENTER WILSON CONSTANT A12 0.119362

ENTER WILSON CONSTANT A21 0.881787

PRESSURES ARE IN KILOPASCAL !!!

X1	Y1(ID)	Y1(VIR)	P	ALPHA
0	0	0	16.789	3.542
.1	.126	.126	18.084	1.303
.2	.158	.158	18.079	.753
.3	.181	.181	17.848	.517
.4	.205	.205	17.42	.387
.5	.234	.234	16.74	.306
.6	.273	.273	15.736	.25
.7	.329	.329	14.315	.21
.8	.418	.418	12.364	.18
.9	.584	.584	9.741	.156
1	1	1	6.275	.137

NEXT COMPOUNDS ? (Y / N) Y

ENTER THE SEQUENTIAL NUMBER
OF FIRST COMPONENT 16

ENTER THE SEQUENTIAL NUMBER
OF SECOND COMPONENT 15

ENTER WILSON CONSTANT A12 0.2022

ENTER WILSON CONSTANT A21 1.3433

PRESSURES ARE IN KILOPASCAL !!!

X1	Y1(ID)	Y1(VIR)	P	ALPHA
0	0	0	1.446	1.635
.1	.096	.096	1.475	.955
.2	.146	.146	1.458	.685
.3	.189	.189	1.421	.544
.4	.234	.234	1.369	.46
.5	.287	.287	1.299	.404
.6	.353	.353	1.211	.364
.7	.439	.439	1.105	.335
.8	.556	.556	.98	.313
.9	.727	.727	.836	.296
1	1	1	.674	.281

NEXT COMPOUNDS ? (Y / N) N
NEXT TEMPERATURE ? (Y / N) N

Remark. The results show that calculating the vapor fugacity coefficient with the virial equation of state is not justified in most cases.

Algorithm and Program: Calculation of Equilibrium Composition with NRTL and Virial Equation

Write a program for the calculation of the total vapor pressure, partial pressures, vapor-phase composition, and relative volatility in a binary mixture as a function of liquid composition. The actual temperature is given. Calculate the liquid activity coefficient with the NRTL equation, and the vapor fugacity coefficient with the virial equation.

Algorithm.

1. Read the temperature.
2. Read the sequential numbers of the mixture components.
3. Call the property-datafile reader routine.
4. Calculate the pure-component vapor pressures.
5. Read the coefficients A_{12}, A_{21}, and α of the NRTL equation.
6. Calculate the activity coefficients of the components with the NRTL equation (see program VLENRTL in Chapter 10).
7. Calculate the partial *fugacities* of the components:

$$f_i = x_i \gamma_i P_i^\circ$$

8. Calculate the fugacity coefficient, total pressure, and vapor composition from the fugacities, given that

$$f_i = y_i v_i P$$

The three factors of this product have to be found; as initial guess use

$$P_i = f_i \qquad P = P_1 + P_2 \qquad y_i = \frac{P_i}{P}$$

(a) Calculate the pseudo-one-component properties of the mixture with the mixture rules (see Chapter 8).

(b) Calculate the reduced second virial coefficient for the pure components and for the mixture (for the pseudo-one-component model).

(c) Calculate δ_{12}:

$$\delta_{12} = 2B_{\text{mix}} - B_1 - B_2$$

where B is the reduced second virial coefficient.

(d) Calculate the fugacity coefficients derived from the virial equation

$$\ln \nu_i = \frac{P}{RT}\left(B_i + y_i^2\delta_{12}\right)$$

(e) Calculate the next estimate of the vapor-phase compositions:

$$y_i = \frac{f_i}{P\upsilon_i}$$

(f) Calculate the next estimate of the total pressure:

$$P \leftarrow P(y_1 + y_2)$$

(g) Stop the iteration and go to step 9 if

$$1 - (y_1 + y_2) \leq 10^{-4}$$

(h) Reestimate the normal vapor compositions:

$$y_i \leftarrow \frac{y_i}{y_1 + y_2}$$

(i) Go to step (a).

9. Calculate $\alpha = y_2/y_1$, and print the results.

Program DISTVIRNRTL. A possible realization:

```
]PR#0
]LOAD DISTVIRNRTL
]LIST

 10  REM :CALCULATION OF VLE COMPO
      SITIONS AND PARTIAL AND TOTAL
      PRESSURE
 20  REM :CALCULATE THE VAPOR FU
      GACITY COEFF. BY VIRIAL EQUA
      TION OF STATE
 30  REM :THE LIQUID ACTIVITY COEFF.
      IS CALCULATED BY NRTL EQ
      UATION
 40  PRINT : PRINT
 50  DIM A(15)
 60  PRINT
 70  INPUT "ENTER TEMPERATURE (K)
      "; T
 80  PRINT
 90  QJ = 0
```

```
100  PRINT "ENTER THE SEQUENTIAL
      NUMBER"
110  INPUT "OF FIRST COMPONENT
      "; IS
120  C1 = IS
130  GOSUB 1210
140  TC = A(3)
150  PC = A(2)
160  OM = A(4)
170  GOSUB 810
180  S1 = PS
190  PRINT
200  PRINT "ENTER THE SEQUENTIAL
      NUMBER"
210  INPUT "OF SECOND COMPONENT
      "; IS
     C2 = IS
230  GOSUB 1210
240  TC = A(3)
250  PC = A(2)
260  OM = A(4)
270  GOSUB 810
280  S2 = PS
290  PRINT
300  INPUT "ENTER NRTL CONSTANT
      A12 "; A1
310  PRINT
320  INPUT "ENTER NRTL CONSTANT
      A21 "; A2
330  INPUT "ENTER NRTL CONSTANT
      ALPHA "; AL
340  PRINT
350  PRINT
360  PRINT "PRESSURES ARE IN KILO
      PASCAL !!!"
370  T6 = LOG (A1) / - AL
380  T7 = LOG (A2) / - AL
390  PRINT
400  PRINT " X1 Y1(ID) Y1(VI
      R) P ALPHA"
410  PRINT "======================
      ======================"
420  FOR I = 1 TO 11
430  X = (I - 1) * 0.1
440  GC = EXP ((1 - X) ∧ 2 * (T7 *
      A2 ∧ 2 / (X + (1 - X) * A2) ∧
      2 + T6 * A1 ∧ 2 / ((1 - X) +
      X * A1) ∧ 2))
450  P3 = X * S1 * GC
```

```
455  P8 = S1 * GC
460  GC = EXP (X ∧ 2 * (T6 * A1 ∧
      2 / ((1 − X) + X * A1) ∧ 2 +
      T7 * A2 ∧ 2 / (X + (1 − X) *
      A2) ∧ 2))
470  P4 = (1 − X) * S2 * GC
475  P9 = S2 * GC
480  P = P3 + P4
490  Y1 = P3 / P
500  I1 = Y1
510  Y2 = 1 − Y1
520  E1 = Y1
530  E2 = Y2
540  GOSUB 980
550  Y1 = P3 / P / N1
560  Y2 = P4 / P / N2
570  S = Y1 + Y2
580  Y1 = Y1 / (Y1 + Y2)
590  Y2 = Y2 / (Y1 + Y2)
600  IF ABS (S − 1) < 1E − 4 THEN
      GOTO 630
610  P = P * S
620  GOTO 540
630  Y% = Y1 * 1000
640  Y1 = Y% / 1000
660  AL = P8 / P9
670  A% = AL * 1000
680  AL = A% / 1000
690  P = P / 1000:P% = P * 1000:P =
      P% / 1000
700  I% = I1 * 1000: I1 = I% / 1000
710  PRINT TAB (1); X; TAB(7); I1;
      TAB(16); Y1; TAB(26); P; TAB(
      34);AL
720  NEXT I
730  PRINT
740  PRINT "======================
      ======================"
750  PRINT : PRINT
760  INPUT "NEXT COMPOUNDS ? (Y / N
      ) "; A$
770  IF A$ = "Y" THEN GOTO 80
780  INPUT "NEXT TEMPERATURE ?
      (Y / N) " A$
790  IF A$ = "Y" THEN GOTO 60
800  END
810  TR = T / TC
820  PS = 5.92714 − 6.09648 / TR −
      1.28862 * LOG (TR) + 0.1693
```

```
     47 * TR ∧ 6 + OM * (15.2518 -
     15.6875 / TR - 13.4721 * LOG
     (TR) + 0.43577 * TR ∧ 6)
830  PS = EXP (PS) * PC
840  RETURN
850  REM :CALCULATION OF SECOND
     VIRIAL COEFF. FOR THE TWO PURE
     COMPONENTS AND FOR THE MIXTURE
860  GOSUB 1040
870  R = 8.3144
880  TC = T1:PC = P1:OM = O1
890  GOSUB 1310
900  B6 = R * TC / PC * B
910  TC = T2:PC = P2:OM = O2
920  GOSUB 1310
930  B7 = R * TC / PC * B
940  TC = TP:PC = PP :OM = OP
950  GOSUB 1310
960  BB = R * TC / PC * B
970  RETURN
980  REM :CALCULATION OF FUGACITY
     COEFFICIENTS FOR BINARY MIX
     TURE WITH VIRIAL EQN. OF STATE
990  GOSUB 850
1000 DE = 2 * BB - B6 - B7
1010 N1 = EXP (P / R / T * (B6 +
     (1 - Y1) ∧ 2 * DE))
1020 N2 = EXP (P / R / T * (B7 +
     Y1 ∧ 2 * DE))
1030 RETURN
1040 REM :CALCULATION OF PSEUDO-
     ONE-COMPONENT PROPERTIES FROM
     COMPOSITION AND PURE COMPO
     NENT DATA FOR BINARY MIXTURE
1050 REM :USING LEE-KESLER MIXING
     RULES
1060 REM :COMPOSITION IS IN 'Y1',
     SEQUENTIAL NUMBERS OF COMPO
     UNDS IN MIXTURE ARE IN 'C1'
     AND IN 'C2'
1070 IF QJ = 1 THEN GOTO 1150
1080 QJ = 1
1090 IS = C1
1100 GOSUB 1210
1110 T1 = A(3):P1 = A(2):V1 = A(1
     4):O1 = A(4)
1120 IS = C2
1130 GOSUB 1210
1140 T2 = A(3):P2 = A(2):V2 = A(1
```

```
        4):O2 = A(4)
1150  VP = 1 / 8 * (Y1 * (1 - Y1) *
        (V1 ∧ (1 / 3) + V2 ∧ (1 / 3)
        ) ∧ 3 + Y1 * Y1 * (2 * V1 ∧
        (1 / 3)) ∧ 3 + (1 - Y1) * Y1
        * (V2 ∧ (1 / 3) + V1 ∧ (1 /
        3)) ∧ 3 + (1 - Y1) ∧ 2 * (2 *
        V2 ∧ (1 / 3)) ∧ 3)
1160  TP = 1 / 8 / VP * (Y1 * (1 -
        Y1) * (V1 ∧ (1 / 3) + V2 ∧ (
        1 / 3)) ∧ 3 * SQR (T1 * T2)
        + Y1 * Y1 * (2 * V1 ∧ (1 /
        3)) ∧ 3 * SQR (T2 * T1) + (
        1 - Y1) ∧ 2 * (2 * V2 ∧ (1 /
        3)) ∧ 3 * T2)
1170  OP = (Y1 * O1 + (1 - Y1) * O
        2)
1180  R = 8.3144
1190  PP = (0.2905 - 0.085 * OP) *
        R * TP / VP
1200  RETURN
1210  REM :READING OF SPECIFIED
        RECORD OF PHDAT FILE INTO
        ARRAY A(15)
1220  REM :ARRAY A HAS TO BE DECL
        ARED IN THE CALLING PROGRAM
1230  D$ = CHR$ (4)
1240  PRINT D$; "OPEN PHDAT"
1250  PRINT D$; "POSITION PHDAT, R";
        IS - 1
1260  PRINT D$; "READ PHDAT"
1270  INPUT A(1), A(2), A(3), A(4), A(5),
        A(6), A(7), A(8), A(9), A(10),
        A(11), A(12), A(13), A(14), A(15)
1280  PRINT D$; "CLOSE PHDAT"
1290  PRINT D$; "IN#0"
1300  RETURN
1310  RESTORE
1320  READ N1,N2,N3,N4
1330  READ R1,R2,R3,R4
1340  DATA 0.1181193,0.265728,0.
        15479,0.0030323
1350  DATA 0.2026579,0.331511,0.
        027655,0.203488
1360  MR = 0.3978
1370  B1 = N1 + OM * (R1 - N1) / MR
1380  B2 = N2 + OM * (R2 - N2) / MR
```

```
1390  B3  =  N3 + OM * (R3 - N3) / MR
1400  B4  =  N4 + OM * (R4 - N4) / MR
1410  TR  =  T / TC
1420  PR  =  P / PC
1430  B  =  B1 - B2 / TR - B3 / TR /
        TR - B4 / TR / TR / TR
1440  BA  =  TR / PR
1450  RETURN
```

Example. Run the program for an acetic acid–water mixture at temperature 288.2 K. The coefficients of the NRTL equation are

$$A_{12} = 1.115279 \qquad A_{21} = 0.499985 \qquad \alpha = 0.44506$$

```
]RUN
ENTER TEMPERATURE (K) 288.2

ENTER THE SEQUENTIAL NUMBER
OF FIRST COMPONENT 16

ENTER THE SEQUENTIAL NUMBER
OF SECOND COMPONENT 15

ENTER NRTL CONSTANT A12 1.115279
ENTER NRTL CONSTANT A21 0.499985
ENTER NRTL CONSTANT ALPHA 0.44506

PRESSURES ARE IN KILOPASCAL !!!
```

X1	Y1(ID)	Y1(VIR)	P	ALPHA
0	0	0	1.446	1.631
.1	.102	.102	1.464	1.028
.2	.157	.157	1.417	.748
.3	.204	.204	1.348	.601
.4	.256	.256	1.267	.516
.5	.317	.318	1.177	.466
.6	.395	.395	1.081	.435
.7	.493	.493	.981	.417
.8	.619	.619	.878	.407
.9	.783	.784	.775	.403
1	1	1	.674	.403

```
NEXT COMPOUNDS ? (Y / N) N
NEXT TEMPERATURE ? (Y / N) N
```

Remark. Explain the results.

SPECIAL CASES OF THE VAPOR–LIQUID EQUILIBRIUM CALCULATION

The calculation of vapor–liquid equilibrium is tantamount to the solution of the following equation set:

$$y_i \nu_i P = x_i \gamma_i f_i^{\circ} \qquad i = 1, 2, \ldots, k$$

where

$$\nu_i = f_1[T, P, y_1, y_2, \ldots, y_k]$$

$$\gamma_i = f_2[T, P, x_1, x_2, \ldots, x_k]$$

$$f_i^{\circ} = f_3[T]$$

In this equation set, there are $2k$ variables but only k equations. The solution then is unambiguously defined only if k variables are fixed. However, the k variables to be fixed, in accordance with the degrees of freedom, may be selected in many ways:

1. The pressure and liquid composition are specified; the vapor composition and the bubble-point temperature are to be searched for.

2. The system temperature and liquid composition are specified; the vapor composition and the bubble-point pressure are to be searched for.

3. The system pressure and vapor composition are specified; the dew-point temperature and liquid composition are to be searched for.

4. The system temperature and vapor composition are specified; the dew-point pressure and liquid composition are to be searched for.

5. This case is somewhat different from the other four: the system pressure, system temperature, and feed composition are specified; the vapor–liquid equilibrium ratio and the compositions of both phases are to be searched for. Thus, there are $2k + 1$ variables and also $2k + 1$ equations. That is to say, for each component we have a material balance:

$$F_i = x_i L + y_i(1 - L)$$

where F_i and L denote the mole fraction of component i in the feed and the proportion of the liquid phase, respectively. Furthermore, for each component the equilibrium condition is valid:

$$y_i = K_i x_i$$

So far we have got $2k$ equations. An additional equation is provided by the constraint related to the sum of the mole fractions:

$$\sum_{i}^{k} x_i = 1$$

Combining the first two equations gives

$$x_i = \frac{F_i}{L + K_i(1 - L)}$$

and using the mole-fraction constraint,

$$\sum_{i=1}^{k} \left(\frac{F_i}{L + K_i(1 - L)} \right) - 1 = 0$$

where L is the only variable to be searched for, lying in the range $0 < L < 1$. This one-variable nonlinear equation has to be solved. Knowing the liquid-to-feed ratio, the liquid mole fractions can be obtained by substituting L into the former equation, and the vapor mole fractions are given by the relationship $y_i = K_i x_i$.

These are the special cases of the vapor–liquid equilibrium calculation. For each case k vapor–liquid equilibrium ratios need to be calculated. It is seen that there are many options for this purpose. The question is which combination of the options to use for a given problem. Due to the structure of the system of nonlinear equations, the vapor–liquid equilibrium calculation requires iterations for all the five cases presented. The programming of such an iterative procedure is not hard work in itself; nevertheless, fast convergence can often be achieved only by subtle tricks.

MULTIPHASE EQUILIBRIUM

Several methods for estimating vapor–liquid equilibrium have been discussed. A common feature of these methods is the strict equilibrium criterion, that is, at equilibrium the chemical potentials of a given component should be identical in all the phases. The manner of application of this equilibrium criterion depends on the explicit forms of the chemical potential in the liquid and vapor phases.

As mentioned before, there is another, equivalent criterion for equilibrium. It reads: the free enthalpy of any closed multicomponent system consisting of equilibrium phases has a minimum value (at constant T and P).

So far, the former, chemical-potential-based approach has been widely used in both the theoretical and the practical spheres. Currently, however, the approach based on the minimization of the free enthalpy is coming to be preferred. This is due to the rapid development of numerical methods, to the new versions of nonlinear programming techniques, and of course, to the new facilities provided by computers.

The procedure based on free-enthalpy minimization doesn't relieve us of the need to an adequate explicit form of the chemical potential for a given case. For a multicomponent, multiphase system the free energy can be written as

$$G = \sum_i \sum_j N_{ij}\mu_{ij} \qquad i = 1, 2, 3 \quad j = 1, 2, \ldots, k$$

where N_{ij} = amount of component j in phase i,
μ_{ij} = chemical potential of component j in phase i,

and where the index $i = 1$ relates to the only possible gas phase, whereas $i = 2, 3$ relate to the two possible liquid phases.

The minimum of the free enthalpy of any system at equilibrium must also satisfy two constraints. The first one is

$$\sum_i N_{ij} = b_j$$

where b_j is the total amount of component j in the closed system. This restriction, precluding the possibility of chemical reactions, means that the amount of the component should remain unchanged in closed systems. The second restriction is that none of the variables may have a negative value:

$$N_{ij} \geq 0$$

For a gaseous phase, assuming an ideal mixture, the explicit form of the chemical potential involved in the free-enthalpy function is as follows:

$$\mu_j = \mu_j[T, P_u] + \mathbf{R}T \ln \frac{y_j P}{P_u} \qquad i = 1$$

Similarly, for the liquid phase

$$\mu_{ij} = \mu_{ij}^{\circ} + \mathbf{R}T \ln \frac{P_j^{\circ} x_{ij} \gamma_{ij}}{P_u} \qquad i = 2, 3$$

where μ_j = chemical potential of component j in the hypothetical ideal-gas state at system temperature and pressure,
P_j° = equilibrium vapor pressure of component j,
γ_{ij} = activity coefficient of component j in phase i.

The activity coefficient is calculable by any appropriate model. However, it should be noted that this model must also take the immiscibilities into account.

For such cases, there is no doubt about the priority of use of the NRTL or UNIQUAC equations.

For the minimization of the free energy any mathematical method can be used. Nevertheless, in order to reduce the size of the mathematical problem, care should be taken to give a realistic description of the problem, that is, the mathematical model should consist of equations for the really existing phases only. If such preliminary considerations cannot be made, then the solution of the maximum-size problem will dispose of the nonexistent phases.

Moreover, let us remember the following. If component j does not appear in phase i, then its chemical potential will take on an infinite negative value because $\ln 0 \rightarrow \infty$. This has no physical meaning. To avoid such awkward situations, the variable is given a lower limit but one that does not upset the component balance. Another possible solution is to replace the function $\ln N_{ij}$, at a given lower limit, by a function $f[N_{ij}]$ that approaches some other value than minus infinity in the vicinity of zero.

In the case of a binary mixture, there is no need for such a large apparatus; we can assume that:

1. The gasous phase forms an ideal mixture.
2. The behavior of the liquid phase is adequately described by the NRTL equation.

Making use of these two assumptions, the excess free enthalpies for both phases can be calculated separately:

$$\frac{g^{\mathrm{E}}}{\mathbf{R}T} = x_1 \ln \frac{\bar{f}_1}{P_{\mathrm{u}}} + x_2 \ln \frac{\bar{f}_2}{P_{\mathrm{u}}}$$

The partial fugacities of component i in the liquid and gaseous phases are

$$\bar{f}_i^{(1)} = \gamma_i x_i P_i^{\circ}$$

The activity coefficient is calculable by any appropriate model. However, it should be noted that this model must also take the immiscibilities into account.

where γ_i = activity coefficient of component i calculated by the NRTL equation,

P_i° = equilibrium vapor pressure of component i,

P = pressure.

Consequently the reference state is chosen so that the pure-species fugacity is unity at system pressure. Thus, the procedure consists in comparing the excess free enthalpy of the vapor and liquid at identical compositions. Obviously, that phase will exist, the excess free enthalpy of which is more negative.

If the excess-free-enthalpy function of the liquid phase has two inflection points, then the conclusion can be drawn that there may exist two (immiscible) liquid phases in equilibrium.

Algorithm and Program: Unbounded Mixing

Write a program for the graphical presentation of the endpoints of the unbounded-mixing domain for two pure liquids.

Algorithm.

1. Read the temperature.
2. Read the vapor pressures of the pure components.
3. Read the NRTL parameters of the two-component mixture.
4. Calculate the parameters A_{12} and A_{21}.
5. Calculate the partial fugacities of the two components as a function of the composition.
6. Calculate the free enthalpy of mixing.
7. Plot the results.

Program MIXA,D2. A possible realization:

```
]LOAD MIXA,D2
]LIST

 10  REM :CALCULATION OF MOLAR FREE
      ENTHALPY OF MIXING AND DETER
      MINATION OF UNBOUNDED MIXING
      DOMAIN
 20  PRINT : PRINT
 30  INPUT "ENTER TEMPERATURE (K)
      "; T
 40  PRINT
 50  PRINT "ENTER VAPOR PRESSURE"
 60  INPUT "OF THE FIRST COMPONENT
      (PA) "; S1
 70  PRINT
 80  PRINT "ENTER VAPOR PRESSURE"
 90  INPUT "OF THE SECOND COMPONEN
      T (PA) "; S2
100  PRINT
110  PRINT
120  INPUT "ENTER 'TAU12' (NRTL EQN.)
      "; T1
130  INPUT "ENTER 'TAU21'
      "; T2
```

```
140  INPUT "ENTER 'ALPHA' (NRTL EQN.)
      "; AL
150  PRINT : PRINT
160  A1 = EXP ( - AL * T1)
170  A2 = EXP ( - AL * T2)
180  PRINT "NRTL CONSTANT A12 = ";
      A1
190  PRINT "NRTL CONSTANT A21 = ";
      A2
200  PRINT : PRINT
210  INPUT "PRESS 'RETURN' TO BEGIN
      GRAPHICAL REPRESENTATION
      "; A$
220  HPLOT 0,0
230  HGR
235  HCOLOR = 7
240  HPLOT 10,0 TO 10,150
250  HPLOT 10,0 TO 200,0
260  HPLOT 200,150 TO 200,0
270  HPLOT 10,0
280  FOR X = 0 TO 0.5 STEP 0.005
290  IF X = 0 THEN GOTO 330
300  GC = EXP ((1 - X) ∧ 2 * (T2 *
      A2 ∧ 2 / (X + (1 - X) * A2) ∧
      2 + T1 * A1 ∧ 2 / ((1 - X) +
      X * A1) ∧ 2))
310  F1 = GC * X * S1 / 101325
320  IF X = 1 THEN GOTO 350
330  GC = EXP (X ∧ 2 * (T1 * A1 ∧
      2 / ((1 - X) + X * A1) ∧ 2 +
      T2 * A2 ∧ 2 / (X + (1 - X) *
      A2) ∧ 2))
340  F2 = GC * (1 - X) * S2 / 1013
      25
350  IF X = 0 THEN GM = 1 + LOG
      (F2): GOTO 380
360  IF X = 1 THEN GM = LOG (F1)
      + 1: GOTO 270
370  GM = X * LOG (F1) + (1 - X) *
      LOG (F2)
380  XX = 10 + 380 * X
390  Y = - 700 * GM - 670
400  IF Y < 0 THEN Y = 0
410  HPLOT TO XX,Y
420  NEXT X
430  INPUT "PRESS 'RETURN' TO FINISH";
      A$
440  TEXT
450  END
```

Example. Run the program for an ethyl acetate–water system at temperature 343.7 K. The NRTL parameters are

$$\tau_{12} = 0.896425 \qquad \tau_{21} = 2.739735 \qquad \alpha = 0.3$$

The vapor pressure of ethyl acetate is 81,501 Pa, and of water is 31,984 Pa at the above temperature.

```
]RUN

ENTER TEMPERATURE (K) 343.7

ENTER VAPOR PRESSURE
OF THE FIRST COMPONENT (PA) 81501

ENTER VAPOR PRESSURE
OF THE SECOND COMPONENT (PA) 31984

ENTER 'TAU12' (NRTL EQN.) 0.896425
ENTER 'TAU21' 2.739735
ENTER 'ALPHA' (NRTL EQN.) 0.3

NRTL CONSTANT A12 =  .764198658
NRTL CONSTANT A21 =  .439586617

PRESS 'RETURN' TO BEGIN GRAPHICAL REPRESENTATION
PRESS 'RETURN' TO FINISH
```

Remark. Watch the screen and explain the result.

Algorithm and Program: Existence of Vapor – Liquid Equilibrium

Write a program for the determination of whether vapor–liquid equilibrium exists or not.

Algorithm.

1. Read the temperature.
2. Read the vapor pressures of the pure components.
3. Read the NRTL parameters of the two-component mixture.
4. Calculate the constants A_{12} and A_{21}.
5. Calculate the partial fugacities of the components as a function of the composition.
6. Calculate and plot the free enthalpy of mixing for the liquid phase.
7. Calculate and plot the free enthalpy of mixing for the vapor phase with the equation

$$\frac{\Delta g}{\mathbf{R}T} = x_1 \ln x_1 + x_2 \ln x_2$$

Program MIXB,D2. A possible realization:

```
]LOAD MIXB,D2
]LIST

 10  REM :INVESTIGATION OF THE EXI
      STENCE OF VAPOR-LIQUID EQUI
      LIBRIUM SUPPOSING THE GAS
      PHASE TO BE IDEAL
 20  PRINT : PRINT
 30  INPUT "ENTER TEMPERATURE (K)
      "; T
 40  PRINT
 50  PRINT "ENTER VAPOR PRESS
      URE"
 60  INPUT "OF THE FIRST COMPONENT
      (PA) "; S1
 70  PRINT
 80  PRINT "ENTER VAPOR PRESS
      URE"
 90  INPUT "OF THE SECOND COMPONENT
      (PA) "; S2
100  PRINT
110  PRINT
120  INPUT "ENTER 'TAU12' (NRTL EQN.)
      "; T1
130  INPUT "ENTER 'TAU21'
      "; T2
140  INPUT "ENTER 'ALPHA' (NRTL EQN.)
      "; AL
150  PRINT : PRINT
160  A1 = EXP ( - AL * T1)
170  A2 = EXP ( - AL * T2)
180  PRINT "NRTL CONSTANT A12 = ";
      A1
190  PRINT "NRTL CONSTANT A21 = ";
      A2
200  PRINT : PRINT
210  INPUT "PRESS 'RETURN' TO BEGIN
      GRAPHICAL REPRESENTATION
      "; A$
220  HPLOT 0,0
230  HGR
235  HCOLOR = 7
240  HPLOT 10,0 TO 10,150
250  HPLOT 10,0 TO 200,0
260  HPLOT 200,150 TO 200,0
270  HPLOT 10,0
280  FOR X = 0 TO 1 STEP 0.05
```

```
290  IF X = 0 THEN GOTO 330
300  GC = EXP ((1 − X) ∧ 2 * (T2 *
      A2 ∧ 2 / (X + (1 − X) * A2) ∧
      2 + T1 * A1 ∧ 2 / ((1 − X) +
      X * A1) ∧ 2))
310  F1 = GC * X * S1 / 101325
320  IF X = 1 THEN GOTO 350
330  GC = EXP (X ∧ 2 * (T1 * A1 ∧
      2 / ((1 − X) + X * A1) ∧ 2 +
      T2 * A2 ∧ 2 / (X + (1 − X) *
      A2) ∧ 2))
340  F2 = GC * (1 − X) * S2 / 1013
      25
350  IF X = 0 THEN GM = 1 + LOG
      (F2); GOTO 380
360  IF X = 1 THEN GM = LOG (F1)
      + 1; GOTO 380
370  GM = X * LOG (F1) + (1 − X) *
      LOG (F2)
380  XX = 10 + 200 * X
390  Y = − 100 * GM
400  IF Y < 0 THEN Y = 0
405  IF X = 0 THEN HPLOT XX,Y
410  HPLOT TO XX,Y
420  NEXT X
450  HPLOT 10,0
500  FOR X = 0 TO 1 STEP 0.05
510  IF X = 0 THEN GOTO 560
530  F1 = X
540  IF X = 1 THEN GOTO 570
560  F2 = 1 − X
570  IF X = 0 THEN GM = 1 + LOG
      (F2): GOTO 600
580  IF X = 1 THEN GM = LOG (F1)
      + 1: GOTO 600
590  GM = X * LOG (F1) + (1 − X) *
      LOG (F2)
600  XX = 10 + 200 * X
610  Y = − 100 * GM
620  IF Y < 0 THEN Y = 0
625  IF X = 0 THEN HPLOT XX,Y
630  HPLOT TO XX,Y
640  NEXT X
650  INPUT "PRESS 'RETURN' TO FINISH";
      A$
660  TEXT
670  END
```

Example. Run the program for the ethyl acetate–water system at temperature 343.7 K. The NRTL parameters are

$$\tau_{12} = 0.896425 \qquad \tau_{21} = 2.739735 \qquad \alpha = 0.3$$

The vapor pressures are 81,501 and 31,984 Pa.

Rerun the program to calculate the bounds of the two-phase region for the methanol–ethyl acetate system. The temperature is 343.7 K. The NRTL constants are

$$\tau_{12} = 0.82712 \qquad \tau_{21} = 0.18919 \qquad \alpha = 0.1$$

The vapor pressures of the pure components are 128,198 and 81,501 Pa respectively.

```
]RUN

ENTER TEMPERATURE (K) 343.7

ENTER VAPOR PRESSURE
OF THE FIRST COMPONENT (PA) 81501

ENTER VAPOR PRESSURE
OF THE SECOND COMPONENT (PA) 31984

ENTER 'TAU12' (NRTL EQN.) 0.896425
ENTER 'TAU21' 2.739735
ENTER 'ALPHA' (NRTL EQN. - 0.3

NRTL CONSTANT A12 = .764198658
NRTL CONSTANT A21 = .439586617

PRESS 'RETURN' TO BEGIN GRAPHICAL REPRESENTATION
PRESS 'RETURN' TO FINISH

]RUN

ENTER TEMPERATURE (K) 343.7

ENTER VAPOR PRESSURE
OF THE FIRST COMPONENT (PA) 128198

ENTER VAPOR PRESSURE
OF THE SECOND COMPONENT (PA) 81501

ENTER 'TAU12' (NRTL EQN.) 0.82712
ENTER 'TAU21' 0.18919
ENTER 'ALPHA' (NRTL EQN.) 0.1

NRTL CONSTANT A12 = .920616247
NRTL CONSTANT A21 = .981258841

PRESS 'RETURN' TO BEGIN GRAPHICAL REPRESENTATION
PRESS 'RETURN' TO FINISH
```

Remark. Watch the monitor and explain the curves.

13

EQUILIBRIUM OF CHEMICAL REACTIONS

The molar free enthalpy of the n-butane–isobutane mixture were calculated as a function of the composition.

$$g = x_1 g_1^\circ + x_2 g_2^\circ + \mathbf{R}T(x_1 \ln x_1 + x_2 \ln x_2)$$

The result is depicted in Figure 13.1.

Let us choose such a mixture that is in state A at a controlled constant temperature and pressure, and make this mixture—say, by means of an appropriate catalyst—capable of reaction. What will be the consequence of this? The mixture decreases in free enthalpy until it arrives at the state C corresponding to the minimum. This spontaneous decrease in free enthalpy can be attained only by changing the composition, that is, by chemical reaction. The reaction

$$n\text{-butane} \rightarrow \text{isobutane}$$

stops at point C, meaning a chemical equilibrium due to the minimum of the free enthalpy for the closed, isothermal, isobaric system. The same considerations may be repeated but starting from the point B. In this case, the spontaneous change is also directed to the state C and goes on until the reaction mixture arrives there, but its direction is reversed:

$$\text{isobutane} \rightarrow n\text{-butane}$$

From thermodynamical viewpoint, however, a single reaction is taken to include both directions: we write

$$n\text{-butane} \rightleftarrows \text{isobutane},$$

and the problem is to find the equilibrium position of this reaction.

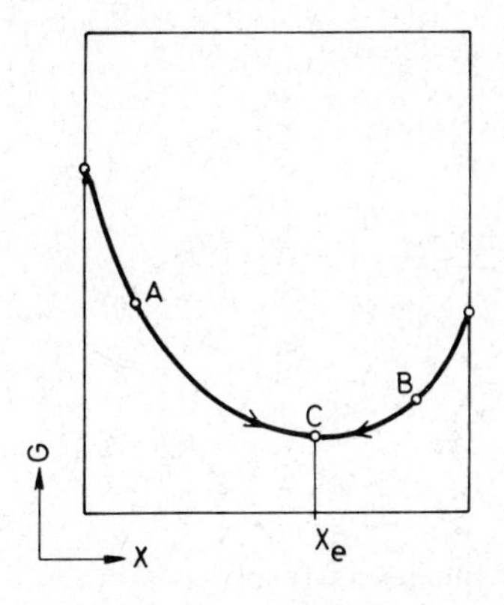

FIGURE 13.1. Isomerization of *n*-butane.

The foregoing considerations could be followed easily because the system in question was a binary mixture with a simple reaction of isomerization type based on intramolecular rearrangement. There are, of course, far more complex cases.

STOICHIOMETRY

The mass of a closed system remains constant even if a chemical reaction is taking place in it. This statement is treated by most books as an important conservation principle, in itself, but that is misleading. The true basic principle of stoichiometry is as follows: For chemical reactions taking place in a closed system, the amount of any *chemical constituent* remains unchanged.

The well-known formula of sulfuric acid, H_2SO_4, means that one mole of sulfuric acid contains two moles of hydrogen, one mole of sulfur, and four moles of oxygen *as constituents*. So in the formula H_2SO_4, the symbols H, S, and O refer to the constituents and the numbers 2,1,4 constitute a vector of stoichiometric indices. In what follows this vector will be denoted by $\mathbf{m}$.

Let us consider a closed system including k components that are composed of e consituents. Let N_j designate the amount of component j, and a_{ij} the amount of constituent i in one mole of component j; in other words, a_{ij} is the stoichiometric index. The matrix $\mathbf{A} = [a_{ij}]$ is conventionally called the *atom matrix*. Hence, the conservation law for the constituents can be written for any closed system as follows:

$$a_{11}\,dN_1 + a_{12}\,dN_2 + \cdots + a_{1k}\,dN_k = 0$$

$$a_{21}\,dN_1 + a_{22}\,dN_2 + \cdots + a_{2k}\,dN_k = 0$$

$$\vdots$$

$$a_{e1}\,dN_1 + a_{e2}\,dN_2 + \cdots + a_{ek}\,dN_k = 0$$

Integrating between any initial and final state, we have

$$a_{11}N_1 + a_{12}N_2 + \cdots + a_{1k}N_k = a_{11}N_1^\circ + a_{12}N_2^\circ + \cdots + a_{1k}N_k^\circ$$

$$a_{21}N_1 + a_{22}N_2 + \cdots + a_{2k}N_k = a_{21}N_1^\circ + a_{22}N_2^\circ + \cdots + a_{2k}N_k^\circ$$

$$a_{e1}N_1 + a_{e2}N_2 + \cdots + a_{ek}N_k = a_{e1}N_1^\circ + a_{e2}N_2^\circ + \cdots + a_{ek}N_k^\circ$$

or in matrix notation,

$$\mathbf{AN} = \mathbf{AN}^\circ$$

Let the matrix-vector product $\mathbf{AN}^\circ$ be designated please set from:

$$\mathbf{AN}^\circ = \mathbf{b}$$

Each element of the vector $\mathbf{b}$ is identical with the corresponding conserved quantity of a constituent involved in the given system. The system in question has k components composed of e constituents, where $e \leq k$. Let the rank of the atomic matrix be $\rho[\mathbf{A}]$. It is obvious that $\rho[\mathbf{A}] \leq e$. Moreover, let the atom matrix be arranged so that the first $r = \rho[\mathbf{A}]$ column vectors are independent. Then the column vectors $r + 1, r + 2, \ldots, k$ are dependent.

On the other hand, the atom matrix $\mathbf{A}$ can be decomposed into two submatrices: $\mathbf{B}$, made up of linearly independent vectors, and $\mathbf{C}$, made up of vectors that are linearly dependent on the ones in $\mathbf{B}$.

Let us denote by $\nu_{j,r+1}$ those scalar coefficients by means of which the vector $\mathbf{m}_{r+1}$ can be expressed as a linear combination of the r independent vectors. So we can write

$$\mathbf{m}_{r+1} = \nu_{1,r+1}\mathbf{m}_1 + \nu_{2,r+1}\mathbf{m}_2 + \cdots + \nu_{r,r+1}\mathbf{m}_r$$

This relation can be interpreted also in the following manner. If the vector $\mathbf{m}_j$ is replaced by the molecular formula of component j, then we arrive at a chemical-reaction equation which means that one mole of component $r + 1$ can be produced from r independent components reacting with each other to the following extent:

$\nu_{1,r+1}$ moles from component 1

$\nu_{2,r+1}$ moles from component 2

$\vdots$

$\nu_{r,r+1}$ moles from component r

The term "reacting" has expressly stoichiometric meaning and has nothing to do with any reaction kinetics.

The scalar multipliers ν_{ij} in the stoichiometric equations are called stoichiometric coefficients in chemical terminology. They are concerned with that reaction in which component $r+1$ takes part. Components $r+2$, $r+3, \ldots,$ and k can be produced in similar manner. Hence

$$\begin{aligned}
\nu_{1,r+1}\mathbf{m}_1 + \nu_{2,r+1}\mathbf{m}_2 + \cdots + \nu_{r,r+1}\mathbf{m}_r &= \mathbf{m}_{r+1} \\
\nu_{1,r+2}\mathbf{m}_1 + \nu_{2,r+2}\mathbf{m}_2 + \cdots + \nu_{r,r+2}\mathbf{m}_r &= \qquad \mathbf{m}_{r+2} \\
&\vdots \\
\nu_{1,k}\mathbf{m}_1 + \nu_{2,k}\mathbf{m}_2 + \cdots + \nu_{r,k}\mathbf{m}_r &= \qquad\qquad \mathbf{m}_k
\end{aligned}$$

Since each reaction serves for the formation of one and only one new component, $k - r$ different reactions have to occur in the system in order to produce all the other components. More than $k - r$ reactions lead to redundancy, and less would be insufficient.

As a matter of fact, for a given system with components capable of reaction, the input of r components also is sufficient to get all k components as the result of the reactions. This is why the r components are usually called independent components and the reactions resulting in further components are called independent reactions. Now, let us arrange the stoichiometric coefficients of the $k - r$ reactions in the reaction matrix $\mathbf{R} = [\nu_{ji}]$, where ν_{ji} is the stoichiometric coefficient of component j in the ith reaction. Then the rank of the reaction matrix is

$$p = \rho[\mathbf{R}] = k - r$$

Among the arbitrary number of written reactions, only $p = k - r$ are independent, and the rank of the reaction matrix should be $\rho[\mathbf{R}] = p$. For checking, the following relation may be useful to keep in mind:

$$k = \rho[\mathbf{A}] + \rho[\mathbf{R}]$$

The dependent reactions can be expressed as linear combinations of the independent ones. Thus, if the following two reactions are independent:

$$\sum_j \nu_{j1} m_j = 0$$

$$\sum_j \nu_{j2} m_j = 0$$

then the following one is not independent:

$$\sum_j \nu_{j3} m_j = \sum_j \left(\lambda_1 \nu_{j1} + \lambda_2 \nu_{j2}\right) m_j = 0$$

if λ_1 and λ_2 are not equal to zero.

The rank $\rho[\mathbf{A}]$ of the matrix **A** often happens to be identical with the number e of constituents in the closed system.

Algorithm and Program: Determination of Stoichiometrically Independent Reactions from Atom Matrix

Write a program for the determination of independent reactions by the transformation of atom matrix. The number of independent reactions is equal to the number of components minus the rank of the atom matrix; the stoichiometric coefficients of the independent reactions are the elements of the inverse matrix. The rows of the atom matrix correspond to the constituents (elements) in the system, and the columns to the compounds.

Algorithm.

1. Read the nonzero elements of the atom matrix.
2. Choose the best pivot element.
3. Transform the matrix by changing the basis of unit vectors.
4. Repeat steps 2 and 3 until nonzero pivot elements exist.
5. Compress the resulting matrix.
6. Print of the results.

Program RANK. A possible realization:

```
]LOAD RANK,D2
]LIST

  10  REM :CALCULATION OF THE RANK
       OF AN ATOM MATRIX
  20  PRINT : PRINT
  30  INPUT "ENTER THE NUMBER OF
       ELEMENTS "; IP
  40  PRINT
  50  INPUT "ENTER THE NUMBER OF
       COMPOUNDS "; K
  60  PRINT
  70  PRINT "ENTER THE NONZERO ELE
       MENTS OF THE ATOM MATRIX"
  80  PRINT "GIVE FIRST THE ROW POS
       ITION"
  90  PRINT "THEN THE COLUMN POSITION
       AND VALUE"
 100  PRINT "OF NONZERO ELEMENT"
 120  PRINT ENTER '999' FOR ROW PO
       SITION AT THE END"
 130  DIM A(IP,K),B(IP,K),RO(IP),I
       S(IP)
```

```
140  PRINT : PRINT
150  INPUT "ENTER ROW POSITION "; I
160  IF I = 999 THEN GOTO 200
170  INPUT "ENTER COLUMN POSITION
      "; J
180  INPUT "ENTER VALUE "; A(I,J)
190  GOTO 140
200  FOR I = 1 TO IP
210  FOR J = 1 TO K
220  B(I,J) = A(I,J)
230  NEXT J
240  NEXT I
250  IR = 0
260  I = 1
270  J = 1
280  IF B(I,J) < > 0. THEN GOTO
      820
290  I = I + 1
300  IF I < = IP THEN GOTO 280
310  I = 1
320  J = J + 1
330  IF J < = K THEN GOTO 280
340  PRINT : PRINT
350  PRINT "NO PIVOT ELEMENT OTHER
      THAN "ZERO"
360  STOP
370  IS(IR) = I
380  RO(IR) = J
390  JJ = J + 1
400  IF JJ > K THEN GOTO 650
410  FOR J1 = JJ TO K
420  FOR I1 = 1 TO IP
430  BB = B(I1,J1)
440  IF I1 = I THEN GOTO 620
450  BD = ABS (B(I,J1) * B(I1,J))
460  BA = ABS (B(I,J1) / B(I,J))
470  DA = ABS (B(I1,J) / B(I,J))
480  A1 = ABS (1 / B(I,J))
490  B1 = ABS (B(I,J1))
500  D1 = ABS (B(I1,J))
510  IF BD < A1 THEN BD = A1
520  IF BA < D1 THEN BA = D1
530  IF DA < B1 THEN DA = B1
540  IF BD < BA AND DA > = BD THEN
      GOTO 600
550  IF DA > = (BA) THEN GOTO 5
      80
560  B(I1,J1) = BB - (B(I1,J) / B(
      I,J)) * B(I,J1)
570  GOTO 610
```

```
 580  B(I1,J1) = BB = (B(I,J1) / B(
       I,J)) * B(I1,J)
 590  GOTO 610
 600  B(I1,J1) = BB - (B(I,J1) * B(
       I1,J)) / B(I,J)
 610  IF ABS (B(I1,J1)) < = 1E -
       9 * ABS (BB) THEN B(I1,J1) =
       0
 620  NEXT I1
 630  B(I,J1) = B(I,J1) / B(I,J)
 640  NEXT J1
 650  FOR I1 = 1 TO IP
 660  B(I1,J) = 0
 670  NEXT I1
 680  B(I,J) = 1
 690  J = J + 1
 700  IF J > K THEN GOTO 980
 710  I = 1
 720  IF B(I,J) < > 0 THEN GOTO
       770
 730  I = I + 1
 740  IF I - IP < 0 THEN GOTO 720
 750  IF I - IP = 0 THEN GOTO 720
 760  IF I - IP > 0 THEN GOTO 690
 770  KO = 0
 780  FOR L = 1 TO IR
 790  IF I = IS(L) THEN KO = 1
 800  NEXT L
 810  IF KO = 1 THEN GOTO 730
 820  IR = IR + 1
 830  IF J = K THEN GOTO 370
 840  IF I = IP THEN GOTO 370
 850  GG = ABS (B(I,J))
 860  IL = I
 870  IF IL > IP THEN GOTO 370
 880  UJ = ABS (B(IL,J))
 890  IF UJ = 0 THEN GOTO 960
 900  FOR L = 1 TO IR
 910  IF IL = IS(L) THEN GOTO 960
 920  NEXT L
 930  IF UJ > = GG THEN GOTO 960
 940  GG = UJ
 950  I = IL
 960  IL = IL + 1
 970  GOTO 870
 980  R1 = IR - 1
 990  IF R1 < = 0 THEN GOTO 1150
1000  FOR I = 1 TO R1
1010  I1 = I + 1
```

```
1020  FOR J = I1 TO IR
1030  KK = IS(I)
1040  IF IS(J) > = KK THEN GOTO
       1140
1050  IS(I) = IS(J)
1060  IS(J) = KK
1070  SI = IS(I)
1080  SJ = IS(J)
1090  FOR L = 1 TO K
1100  BS = B(SI,L)
1110  B(SI,L) = B(SJ,L)
1120  B(SJ,L) = BS
1130  NEXT L
1140  NEXT J
1150  NEXT I
1160  FOR I = 1 TO IR
1170  SS = IS(I)
1180  FOR J = 1 TO K
1190  B(I,J) = B(SS,J)
1200  NEXT J
1210  NEXT I
1220  IF IR = IP THEN GOTO 1290
1230  RR = IR + 1
1240  FOR I = RR TO IP
1250  FOR J = 1 TO K
1260  B(I,J) = 0
1270  NEXT J
1280  NEXT I
1290  PRINT : PRINT
1300  PRINT "THE RANK OF THE ATOM
       MATRIX IS "; IR
1310  PRINT
1320  PRINT "THE INVERSE MATRIX :
       "
1330  PRINT : PRINT
1340  PRINT TAB(7);
1350  FOR J = 1 TO K
1360  PRINT J; SPC(4);
1370  NEXT J
1380  PRINT
1390  PRINT "--------------------
       -----------------": PRINT TAB(
       4); "I"
1400  FOR I = 1 TO IR
1410  PRINT TAB(1); RO(I); TAB(
       4); "I"; SPC(2);
1420  FOR J = 1 TO K
1430  B% = B(I,J) * 100.
1440  B(I,J) = B% / 100.
```

```
1450  PRINT TAB(J* 5 + 2); B(I,J);
1460  NEXT J
1470  PRINT
1480  NEXT I
1490  END
```

Example. Run the program for a system of the following compounds:

$$CH_4, \quad H_2O, \quad CO_1, \quad H_2, \quad CO_2$$

In this system the constituents (elements) are C, H, O. The atom matrix is

	CH_4	H_2O	CO	H_2	CO_2
C	1		1		1
H	4	2		2	
O		1	1		2

```
]RUN

ENTER THE NUMBER OF ELEMENTS 3

ENTER THE NUMBER OF COMPOUNDS 5

ENTER THE NONZERO ELEMENTS OF THE ATOM MATRIX
GIVE FIRST THE ROW POSITION
THEN THE COLUMN POSITION AND VALUE
OF NONZERO ELEMENT

ENTER '999' FOR ROW POSITION AT THE END

ENTER ROW POSITION 1
ENTER COLUMN POSITION 1
ENTER VALUE 1

ENTER ROW POSITION 1
ENTER COLUMN POSITION 3
ENTER VALUE 1

ENTER ROW POSITION 1
ENTER COLUMN POSITION 5
ENTER VALUE 1

ENTER ROW POSITION 2
ENTER COLUMN POSITION 1
ENTER VALUE 4

ENTER ROW POSITION 2
ENTER COLUMN POSITION 2
ENTER VALUE 2
```

```
ENTER ROW POSITION 2
ENTER COLUMN POSITION 4
ENTER VALUE 2

ENTER ROW POSITION 3
ENTER COLUMN POSITION 2
ENTER VALUE 1

ENTER ROW POSITION 3
ENTER COLUMN POSITION 3
ENTER VALUE 1

ENTER ROW POSITION 3
ENTER COLUMN POSITION 5
ENTER VALUE 2

ENTER ROW POSITION 999

THE RANK OF THE ATOM MATRIX IS 3

THE INVERSE MATRIX :
```

	1	2	3	4	5
1	1	0	0	.33	−.34
2	0	1	0	.33	.66
3	0	0	1	−.34	1.33

Remark. The number of components in the system is 5, and the rank of the atom matrix is 3, so the number of independent reactions is 5 − 3 = 2. These reactions are

$$H_2 \rightleftarrows 0.33CH_4 + 0.33H_2O - 0.33CO$$

$$CO_2 \rightleftarrows -0.33CH_4 + 0.66H_2O + 1.33CO$$

These reactions are independent and stoichiometricaly correct.

Here we do not deal with the details of the matrix transformation algorithm; these can be found in linear-algebra textbooks.

Example. Run the program **RANK** for a system of the following components:

$$H_2SO_4, NaOH, Na_2SO_4, H_2O$$

The atom matrix is

	Na_2SO_4	NaOH	H_2SO_4	H_2O
S	1		1	
Na	2	1		
O	4	1	4	1
H		1	2	2

```
]RUN
ENTER THE NUMBER OF ELEMENTS 4

ENTER THE NUMBER OF COMPOUNDS 4
ENTER THE NONZERO ELEMENTS OF THE ATOM MATRIX
GIVE FIRST THE ROW POSITION
THEN THE COLUMN POSITION AND VALUE
OF NONZERO ELEMENT

ENTER '999' FOR ROW POSITION AT THE END

ENTER ROW POSITION 1
ENTER COLUMN POSITION 1
ENTER VALUE 1

ENTER ROW POSITION 2
ENTER COLUMN POSITION 1
ENTER VALUE 2

ENTER ROW POSITION 3
ENTER COLUMN POSITION 1
ENTER VALUE 4

ENTER ROW POSITION 2
ENTER COLUMN POSITION 2
ENTER VALUE 1

ENTER ROW POSITION 3
ENTER COLUMN POSITION 2
ENTER VALUE 1

ENTER ROW POSITION 4
ENTER COLUMN POSITION 2
ENTER VALUE 1

ENTER ROW POSITION 1
ENTER COLUMN POSITION 3
ENTER VALUE 1

ENTER ROW POSITION 3
ENTER COLUMN POSITION 3
ENTER VALUE 4
```

```
ENTER ROW POSITION 4
ENTER COLUMN POSITION 3
ENTER VALUE 2

ENTER ROW POSITION 3
ENTER COLUMN POSITION 3
ENTER VALUE 4

ENTER ROW POSITION 3
ENTER COLUMN POSITION 4
ENTER VALUE 1

ENTER ROW POSITION 4
ENTER COLUMN POSITION 4
ENTER VALUE 2

ENTER ROW POSITION 999

THE RANK OF THE ATOM MATRIX IS 3

THE INVERSE MATRIX
```

	1	2	3	4
1	1	0	0	−.5
2	0	1	0	1
3	0	0	1	.5

Remark. The number of components in the system is 4, and the rank of the atom matrix is 3, so the number of the independent reactions is $4 - 3 = 1$. This reaction is

$$0.5\ H_2SO_4 + NaOH = 0.5\ Na_2SO_4 + H_2O$$

Algorithm and Program: Determination of the Independent Reactions from a Reaction Matrix

Write a program for the determination of the stoichiometrically independent reactions from a given system of existing or supposed reactions. Formulate the reaction matrix of the system. Rewrite the reaction equations with zero on the left. The rows of the reaction matrix correspond to the components, and the columns to the reactions. The number of independent reactions is equal to the rank of the reaction matrix.

Algorithm.

1. Read the nonzero elements of the reaction matrix.
2. Choose best pivot element.
3. Transform of the matrix by changing the basis.

4. Repeat steps 2 and 3 until nonzero pivot elements exist.
5. Compress the resulting matrix.
6. Print the results.

Program RANKREACTIONMATRIX. A possible realization

```
]LOAD RANKREACTIONMATRIX,D2
]LIST

 10  REM :CALCULATION OF THE RANK
      OF A REACTION MATRIX
 20  PRINT : PRINT
 30  INPUT "ENTER THE NUMBER OF
      COMPOUNDS "; IP
 40  PRINT
 50  INPUT "ENTER THE NUMBER OF
      REACTIONS "; K
 60  PRINT
 70  PRINT "ENTER THE NONZERO ELE
      MENTS OF THE REACTION MATRIX"
 80  PRINT "GIVE FIRST THE ROW
      POSITION"
 90  PRINT "THEN THE COLUMN POSITION
      AND VALUE"
100  PRINT "OF NONZERO ELEMENT"
110  PRINT
120  PRINT "ENTER '999' FOR ROW
      POSITION AT THE END"
130  DIM A(IP,K),B(IP,K),RO(IP),
      IS(IP)
140  PRINT : PRINT
150  INPUT "ENTER ROW POSITION "; I
160  IF I = 999 THEN GOTO 200
170  INPUT "ENTER COLUMN POSITION
      "; J
180  INPUT "ENTER VALUE "; A(I,J)
190  GOTO 140
200  FOR I = 1 TO IP
210  FOR J = 1 TO K
220  B(I,J) = A(I,J)
230  NEXT J
240  NEXT I
250  IR = 0
260  I = 1
270  J = 1
280  IF B(I,J) < > 0. THEN GOTO
      820
290  I = I + 1
```

```
300  IF I < = IP THEN GOTO 280
310  I = 1
320  J = J + 1
330  IF J ( = K THEN GOTO 280
340  PRINT : PRINT
350  PRINT "NO PIVOT ELEMENT OTHER
      R THAN ZERO"
360  STOP
370  IS(IR) = 1
380  RO(IR) = J
390  JJ = J + 1
400  IF JJ > K THEN GOTO 650
410  FOR J1 = JJ TO K
420  FOR I1 = 1 TO IP
430  BS = B (I1,J1)
440  IF I1 = I THEN GOTO 620
450  BD = ABS (B(I,J1) * B(I1,J))
460  BA = ABS (B(I,J1) / B(I,J))
470  DA = ABS (B(I1,J) / B(I,J))
480  A1 = ABS (1 / B(I,J))
490  B1 = ABS (B(I,J1))
500  D1 = ABS (B(I1,J))
510  IF BD < A1 THEN BD = A1
520  IF BA < D1 THEN BA = D1
530  IF DA < B1 THEN DA = B1
540  IF BD < BA AND DA > = BD THEN
      GOTO 600
550  IF DA > = (BA) THEN GOTO 5
      80
560  B(I1,J1) = BB - (B(I1,J) / B(
      I,J)) * B(I,J1)
570  GOTO 610
580  B(I1,J1) = BB - (B(I,J1) / B(
      I,J)) * B(I1,J)
590  GOTO 610
600  B(I1,J1) = BB - (B(I,J1) * B(
      I1,J)) / B(I,J)
610  IF ABS (B(I1,J1)) < = 1E -
      9 * ABS (BB) THEN B(I1,J1) =
      0
620  NEXT I1
630  B(I,J1) = B(I,J1) / B(I,J)
640  NEXT J1
650  FOR I1 = 1 TO IP
660  B(I1,J) = 0
670  NEXT I1
680  B(I,J) = 1
690  J = J + 1
700  IF J > K THEN GOT0 980
```

```
 710 I = 1
 720 IF B(I,J) < > 0 THEN GOTO
      770
 730 I = I + 1
 740 IF I - IP < 0 THEN GOTO 720
 750 IF I - IP = 0 THEN GOTO 720
 760 IF I - IP > 0 THEN GOTO 690
 770 KO = 0
 780 FOR L = 1 TO IR
 790 IF I = IS(L) THEN KO = 1
 800 NEXT L
 810 IF KO = 1 THEN GOTO 730
 820 IR = IR + 1
 830 IF J = K THEN GOTO 370
 840 IF I = IP THEN GOTO 370
 850 GG = ABS (B(I,J))
 860 IL = 1
 870 IF IL > IP THEN GOTO 370
 880 UJ = ABS (B(IL,J))
 890 IF UJ = 0 THEN GOTO 960
 900 FOR L = 1 TO IR
 910 IF IL = IS(L) THEN GOTO 960
 920 NEXT L
 930 IF UJ > = GG THEN GOTO 960
 940 GG = UJ
 950 I = IL
 960 IL = IL + 1
 970 GOTO 870
 980 R1 = IR - 1
 990 IF R1 < = 0 THEN GOTO 1150
1000 FOR I = 1 TO R1
1010 I1 = I + 1
1020 FOR J = I1 TO IR
1030 KK = IS(I)
1040 IF IS(J) > = KK THEN GOTO
      1140
1050 IS (I) = IS(J)
1060 IS(J) = KK
1070 SI = IS(I)
1080 SJ = IS(J)
1090 FOR L = 1 TO K
1100 BS = B(SI,L)
1110 B(SI,L) = B(SJ,L)
1120 B(SJ,L) = BS
1130 NEXT L
1140 NEXT J
1150 NEXT I
1160 FOR I = 1 TO IR
1170 SS = IS(I)
```

```
1180  FOR J = 1 TO K
1190  B(I,J) = B(SS,J)
1200  NEXT J
1210  NEXT I
1220  IF IR = IP THEN GOTO 1290
1230  RR = IR + 1
1240  FOR I = RR TO IP
1250  FOR J = 1 TO K
1260  B(I,J) = 0
1270  NEXT J
1280  NEXT I
1290  PRINT : PRINT
1300  PRINT "THE RANK OF THE REAC
       TION MATRIX IS "; IR
1310  PRINT
1320  PRINT "THE INVERSE MATRIX :"
1330  PRINT : PRINT
1340  PRINT TAB(7);
1350  FOR J = 1 TO K
1360  PRINT J; SPC(4);
1370  NEXT J
1380  PRINT
1390  PRINT "--------------------
       -----------------": PRINT TAB(
       4); "I"
1400  FOR I = 1 TO IR
1410  PRINT TAB(1);RO(I); TAB(
       4);"I"; SPC(2);
1420  FOR J = 1 TO K
1430  B% = B(I,J) * 100.
1440  B(I,J) = B% / 100.
1450  PRINT TAB(J * 5 + 2);B(I,
       J);
1460  NEXT J
1470  PRINT
1480  NEXT I
1490  END
```

Example. Find the independent reactions in the following system:

$$CH_4 + H_2O \rightleftarrows CO + 3H_2$$

$$CO + H_2O \rightleftarrows CO_2 + H_2$$

$$CH_4 + 2H_2O \rightleftarrows CO_2 + 4H_2$$

Equate the reaction to zero:

$$0 = -CH_4 - H_2 + CO + 3H_2 \quad (1)$$

$$0 = -CO - H_2O + CO_2 + H_2 \quad (2)$$

$$0 = -CH_4 - 2H_2O + CO_2 + 4H_2O \quad (3)$$

The reaction matrix:

Reaction: Component	(1)	(2)	(3)
1 CH_4	−1		−1
2 H_2O	−1	−1	−2
3 CO	1	−1	
4 H_2	3	1	4
5 CO_2		1	1

```
]RUN

ENTER THE NUMBER OF COMPOUNDS 5

ENTER THE NUMBER OF REACTIONS 3

ENTER THE NONZERO ELEMENTS OF THE REACTION MATRIX
GIVE FIRST THE ROW POSITION
THEN THE COLUMN POSITION AND VALUE
OF NONZERO ELEMENT

ENTER '999' FOR ROW POSITION AT THE END

ENTER ROW POSITION 1
ENTER COLUMN POSITION 1
ENTER VALUE -1

ENTER ROW POSITION 1
ENER COLUMN POSITION 3
ENTER VALUE -1

ENTER ROW POSITION 2
ENTER COLUMN POSITION 1
ENTER VALUE -1

ENTER ROW POSITION 2
ENTER COLUMN POSITION 2
ENTER VALUE -1

ENTER ROW POSITION 2
ENTER COLUMN POSITION 3
ENTER VALUE -1
```

```
ENTER ROW POSITION 2
ENTER COLUMN POSITION 3
ENTER VALUE -2

ENTER ROW POSITION 3
ENTER COLUMN POSITION 1
ENTER VALUE 1

ENTER ROW POSITION 3
ENTER COLUMN POSITION 2
ENTER VALUE -1

ENTER ROW POSITION 4
ENTER COLUMN POSITION 1
ENTER VALUE 3

ENTER ROW POSITION 4
ENTER COLUMN POSITION 2
ENTER VALUE 1

ENTER ROW POSITION 4
ENTER COLUMN POSITION 3
ENTER VALUE 4

ENTER ROW POSITION 5
ENTER COLUMN POSITION 2
ENTER VALUE 1

ENTER ROW POSITION 5
ENTER COLUMN POSITION 3
ENTER VALUE 1

ENTER ROW POSITION 999

THE RANK OF THE REACTION MATRIX IS 2

THE INVERSE MATRIX :
```

	1	2	3
1	1	0	1
2	0	1	1

Remark. The results show that the rank of the reaction matrix is equal to 2, the independent reactions are the first and the second, and the third reaction can be derived from the independent reactions:

$$(3) = 1 \times (1) + 1 \times (2)$$

Extent of Chemical Reaction

In the previous subsection it has already been indicated for any closed system that $r = \rho[\mathbf{A}]$ independent components reacting in $p = \rho[\mathbf{R}]$ independent reactions make up the rest of the components produced. For instance, let the rank of the reaction matrix be $\rho[\mathbf{R}] = 1$. This means that only a single reaction can occur in the system. Using the previous notation this can be written

$$\sum \nu_j m_j = 0$$

Let us denote the initial and actual quantities of component j by N_j° and N_j respectively. Thus the actual amount of the component is given by

$$N_j = N_j^\circ + \nu_j X$$

where X represents the extent of the reaction. It is an extensive variable whose unit is the mole.

If the rank of the reaction matrix is $\rho[\mathbf{R}] > 1$, that is, more than one independent reaction may take place, then X_j can be defined independently for every reaction:

$$\frac{N_j - N_j^\circ}{\nu_{ji}} = X_i \qquad i = 1, 2, \ldots, p$$

So the actual amount of the component j is

$$N_j = N_j^\circ + \sum_{i=1} \nu_{ji} X_i$$

or in matrix notation

$$\mathbf{N} = \mathbf{N}^\circ + \mathbf{RX}$$

CHEMICAL EQUILIBRIUM

In previous chapters devoted to equilibrium we have already indicated that the criteria for equilibrium in any system can be formulated in two different ways:

1. Any body is in equilibrium of its relevant thermodynamic potential takes on an extremum. For instance, for isothermal, isobaric equilibrium, the free enthalpy of the body exhibits a minimum. This theorem does not involve any restriction on the type of equilibrium. Consequently, it is valid also for chemical equilibrium.

2. In the second formulation, intensive variables are used for the description of the equilibrium. It was shown that equilibrium means the identity of the relevant properties in the coexisting phases, provided no chemical reaction takes place.

How do matters stand in the case of chemical reactions? If only a single reaction occurs in a homogeneous system, under isothermal, isobaric conditions, the amount of component j can be given as

$$N_j = N_j^\circ + \nu_j X$$

At equilibrium, the free enthalpy of the system takes on an extremum:

$$dG = \sum_j \mu_j \, dN_j$$

$$= \sum_j \mu_j \nu_j \, dX = 0$$

From this it follows that at virtual displacement dX

$$\sum \mu_j \nu_j = 0$$

Accordingly, for the change of free enthalpy in an isothermal, isobaric system where two reactions take place, we have

$$dG = \sum_j \mu_j \, dN_j$$

$$= \sum_j \mu_j \nu_{j1} \, dX_1 + \sum_j \mu_j \nu_{j2} \, dX_2$$

On the other hand, since X_1 and X_2 are independent of each other and at equilibrium $dG = 0$, it follows that the equilibrium has two criteria:

$$\sum_j \mu_j \nu_{j1} = 0$$

$$\sum_j \mu_j \nu_{j2} = 0$$

Obviously, for every additional reaction an additional condition of similar form will appear. In practice, this leads to a nonlinear equation system with the equilibrium values of X_1 and X_2 to be solved for.

The nonlinearity is due to the composition dependence of the chemical potential. For a practical procedure, two questions may arise, namely:

1. How to formulate the chemical potential explicitly.
2. How to formulate the nonlinear equation set describing the equilibrium criteria.

Algorithm and Program: Reaction Equilibrium at Given Temperature and Pressure

Write a program for the determination of the equilibrium concentration of a system where only one chemical reaction takes place. The chemical potentials of the pure components at the given temperature and pressure are known.

Algorithm.

1. Read the stoichiometric coefficients of the reaction.
2. Read the initial mole numbers.
3. Read the chemical potentials of the pure compounds at the existing temperature and pressure.
4. Determine the maximal possible conversion from the initial mole numbers and the stoichiometric coefficients.
5. Read the temperature.
6. Calculate the chemical potential of the mixture at different conversions.
7. Choose the equilibrium conversion (where the chemical potential is closest to minimum).
8. Print the result.

Program REACTIONEQUILIBRIUM. A possible realization:

```
]LOAD REACTIONEQUILIBRIUM,D2
]LIST

 10  REM :CALCULATION OF EQUILIBRIUM
      CONCENTRATIONS FOR A GIVEN
      REACTION AND DEFINED INITIAL
      CONCENTRATIONS
 20  PRINT : PRINT
 30  PRINT "ENTER THE NUMBER OF
      COMPOUNDS"
 40  R = 8.3144
 50  INPUT "IN THE REACTION "; N
 60  PRINT
 70  DIM NO(N),C(N),NF(N),MU(N),X(
      N),N(N)
 80  PRINT "ORDER THE REACTION EQU
      ATION TO 0 "
 90  PRINT "THE STOICHIOMETRIC COEF
      FICIENTS ARE"
100  PRINT "CONSIDERED OF ABOVE
      FORM"
110  PRINT
120  FOR I = 1 TO N
```

```
130  PRINT "ENTER STOICH. COEFF.
      OF "; I;-TH COMPOUND"
140  INPUT " "; C(I)
150  NEXT I
160  PRINT : PRINT
170  FOR I = 1 TO N
180  PRINT "ENTER INITIAL MOLE NUMB
      ER OF "; I;"-TH COMPOUND
190  INPUT " "; NO(I)
200  NEXT I
210  PRINT : PRINT
220  FOR I = 1 TO N
230  PRINT "ENTER CHEM. POTENTIAL
      OF "; I; "-TH COMPOUND;"
240  INPUT "(J / MOL) "; MU(I)
250  NEXT I
260  A = 1
270  FOR I = 1 TO N
280  IF C(I) < 0 THEN A = A * N0(
      I)
290  NEXT I
300  B = 1
310  FOR I = 1 TO N
320  IF C(I) > THEN B = B * N0(
      I)
330  NEXT I
340  IF A = 0 AND B = 0 THEN PRINT
      "IMPOSSIBLE TO MODIFY INITIAL
      CONCENTRATION; GIVE NEW DATA
      ": GOTO 80
350  PRINT
360  INPUT "ENTER TEMPERATURE (K)
      "; T
370  MA = 1000
380  IF A = 0 THEN GOTO 430
390  FOR I = 1 TO N
400  IF C(I) < 0 THEN X = - N0(I
      ) / C(I)
410  IF X < (MA) THEN MA = X:IA = I
420  NEXT I
430  MB = - 1000
440  IF B = 0 THEN GOTO 490
450  FOR I = 1 TO N
460  IF C(I) > 0 THEN X = - N0(I
      ) / C(I)
470  IF X > MB THEN MB = X:IB = I
480  NEXT I
490  IF MA < ABS (MB) THEN XM =
      MA
```

```
500  IF MA > ABS (MB) THEN XM =
      MB
510  G = 100000
520  ST = XM / 21
530  RX = 10000
540  FOR XE = ST TO 20 * ST STEP
      ST
550  SN = 0
560  FOR I = 1 TO N
570  N(I) = N0(I) + C(I) * XE
580  SN = SN + N(I)
590  NEXT I
600  PX = 0
610  FOR I = 1 TO N
620  X(I) = N(I) / SN
630  PX = PX + C(I) * (MU(I) + R *
      T * LOG (X(I)))
640  NEXT I
650  IF ABS (PX) < RX THEN XR =
      XE:RX = ABS (PX)
660  NEXT XE
670  FOR I = 1 TO N
680  NF(I) = N0(I) + C(I) * XR
690  NEXT I
700  PRINT : PRINT
710  PRINT "CONVERSION OF EQUILIB
      RIUM: "; XR
720  PRINT : PRINT
730  PRINT : PRINT
740  PRINT "THE MOLE NUMBERS IN THE
      EQUILIBRIUM ARE:"
750  PRINT : PRINT
760  PRINT "NO. OF COMPOUND
      NUMBER OF MOLES"
770  PRINT "--------------------
      -----------------"
780  PRINT
790  FOR I = 1 TO N
800  PRINT TAB(6); I; TAB(23); N
      F(I)
810  NEXT I
820  PRINT : PRINT
850  END
```

Example. Run the program for a system where the initial molenumbers are

$$N^{\circ}_{H_2O} = 10 \qquad N^{\circ}_{CH_4} = 1 \qquad N^{\circ}_{H_2} = N^{\circ}_{CO} = 0$$

and the reaction is

$$CH_4 + H_2O \rightleftarrows CO + 3H_2$$

The temperature is 600 K; the molar free enthalpy of the pure components at 600 K are

$$\mu^\circ_{CH_4} = -173{,}335 \text{ J/mole}$$

$$\mu^\circ_{H_2O} = -346{,}376 \text{ J/mole}$$

$$\mu^\circ_{CO} = -227{,}474 \text{ J/mole}$$

$$\mu^\circ_{H_2} = -73{,}359 \text{ J/mole}$$

```
]RUN

ENTER THE NUMBER OF COMPOUNDS
IN THE REACTION 4

ORDER THE REACTION EQUATION TO 0
THE STOICHIOMETRIC COEFFICIENTS ARE
CONSIDERED OF ABOVE FORM

ENTER STOICH. COEFF OF 1-TH COMPOUND 1
ENTER STOICH. COEFF OF 2-TH COMPOUND 1
ENTER STOICH. COEFF OF 3-TH COMPOUND -1
ENTER STOICH. COEFF OF 4-TH COMPOUND -3

ENTER INITIAL MOLE NUMBER OF 1-TH COMPOUND 1
ENTER INITIAL MOLE NUMBER OF 2-TH COMPOUND 10
ENTER INITIAL MOLE NUMBER OF 3-TH COMPOUND 0
ENTER INITIAL MOLE NUMBER OF 4-TH COMPOUND 0

ENTER CHEM. POTENTIAL OF 1-TH COMPOUND
(J / MOL) -173335
ENTER CHEM. POTENTIAL OF 2-TH COMPOUND
(J / MOL) -346376
ENTER CHEM. POTENTIAL OF 3-TH COMPOUND
(J / MOL) -227474
ENTER CHEM. POTENTIAL OF 4-TH COMPOUND
(J / MOL) -73359

ENTER TEMPERATURE (K) 600

CONVERSION IN EQUILIBRIUM: -.0952380953
```

THE MOLE NUMBERS IN THE EQUILIBRIUM ARE:

COMPOUND NO.	NUMBER OF MOLES
1	.904761905
2	9.90476191
3	.0952380953
4	.285714286

ENTHALPY AND FREE ENTHALPY OF REACTION

For any reaction mixture with a single isothermal, isobaric reaction, the actual enthalpy is

$$H = \sum_j N_j \bar{h}_j = \sum_j \bar{h}_j (N_j^\circ + \nu_j X)$$

where $\bar{h}_j$ denotes the partial molar enthalpy of component j.

Differentiate this with respect to the extent of reaction, X

$$\left(\frac{\partial H}{\partial X}\right)_{P,T} = \sum_j h_j \nu_j$$

The derivative $(dH/dX)_{P,T}$ is called the *enthalpy of reaction*. In accordance with the sign convention, it has a positive value for endothermic, and a negative value for exothermic reactions. In the literature, this quantity is referred to also by the symbol ΔH

$$\left(\frac{\partial H}{\partial X}\right)_{P,T} = \Delta H$$

The enthalpy of reaction is usually standardized. The standard enthalpy of reaction can be written as

$$\Delta H^\circ = \sum_j h_j^\circ \nu_j$$

where h_j° represents the standard enthalpy of pure component j at the temperature of the reaction and at atmospheric pressure, in a hypothetically ideal-gas state. From a variety of possible conventions we will use the one introduced in Chapter 8 on enthalpy. This is to say, the molar enthalpy of any pure component at standard pressure, in a hypothetical ideal-gas state, at temperature T can be given as

$$h_j{}^\circ[T] = h_0^\circ + a_1 T + a_2 T^2 + a_3 T^3 + a_4 T^4 + a_5 T^5$$

where h_0° is an integration constant representing the value of $h_j{}^\circ[T]$ extrapolated to temperature 0 K. The enthalpy of elements, according to this convention, is zero at 0 K.

Proceeding in the same way for the standard free enthalpy of reaction, we may write

$$\Delta G^\circ = \sum_j \nu_j g_i{}^\circ$$

$$= \sum \nu_j \mu_j{}^\circ$$

where $g_j^\circ = \mu_j{}^\circ$ designates the molar free enthalpy or chemical potential of the pure substance. Thus

$$g_j^\circ[T] = h_0^\circ + a_1(1 - \ln T) - a_2 T^2 - \tfrac{1}{2} a_3 T^3 - \tfrac{1}{3} a_4 T^4 - \tfrac{1}{4} a_5 T^5 - I_s$$

where I_s is the integration constant.

Define the reactions for the formation of components in such a way that 1 mole of component j is formed from its constituents. In this sense we may refer to the enthalpy of formation ($\Delta H_f{}^\circ$) and the free enthalpy of formation ($\Delta G_f{}^\circ$). The respective values are tabulated in the literature.

Algorithm and Program: Standard Enthalpy of Chemical Reaction

Write a program for the calculation of the standard enthalpy ("heat of enthalpy") of a chemical reaction.

Algorithm.

1. Read the stoichiometric coefficients of the reaction.
2. Read the sequential numbers of the components in the physical-property datafile.
3. Read the temperature.
4. Read the coefficients of the enthalpy polynomials of the pure compounds from the physical-property datafile.
5. Calculate the enthalpies of the pure components at the given temperature.
6. Form a linear combination of the pure-component enthalpies with stoichiometric coefficients.
7. Print the result.

Program REACTIONENTHALPY. A possible realization:

```
]LOAD REACTIONENTHALPY,D2
]LIST

 10  REM CALCULATION OF REACTION
      ENTHALPY FOR A GIVEN REACTION
 20  PRINT : PRINT
 30  PRINT "ENTER THE NUMBERS OF
      COMPOUNDS"
 40  INPUT "IN THE REACTION "; N
 50  PRINT
 60  DIM A(15),C(N),NO(N),H(N)
 70  PRINT "ORDER THE REACTION EQU
      ATION TO 0 "
 80  PRINT "THE STOICHIOMETRIC COEF
      FICIENTS ARE"
 90  PRINT "CONSIDERED OF ABOVE
      FORM"
100  PRINT
110  FOR I = 1 TO N
120  PRINT "ENTER STOICH. COEFF.
      OF "; I; "-TH COMPOUND"
130  INPUT " "; C(I)
140  NEXT I
150  PRINT : PRINT
160  FOR I = 1 TO N
170  PRINT "ENTER SEQUENTIAL NUMBER
      OF "; I; "-TH COMPOUND"
180  INPUT " "; NO(I)
190  NEXT I
200  PRINT : PRINT
210  INPUT "ENTER TEMPERATURE (K)
      "; T
220  PRINT : PRINT
230  FOR I = 1 TO N
240  IS = NO(I)
250  GOSUB 420
260  H(I) = ((((A(12) * T + A(11))
      * T + A(10)) * T + A(9)) *
      T + A(8)) * T + A(7)
270  NEXT I
280  DH = 0
290  FOR I = 1 TO N
300  DH = DH + C(I) * H(I)
310  NEXT I
320  PRINT : PRINT
330  PRINT "THE STANDARD ENTHALPY OF THE"
340  PRINT "REACTION AT "; T;" K I
      S "; DH
```

```
350  PRINT : PRINT
360  INPUT "NEXT TEMPERATURE ? (Y
      /N) "; A$
370  IF A$ = "Y" THEN GOTO 210
380  PRINT : PRINT
390  INPUT "NEXT REACTION ? (Y / N)
      NO. OF COMPOUNDS MUST NOT
      CHANGE "; A$
400  IF A$ = "Y" THEN GOTO 70
410  END
420  REM :READING OF SPECIFIED
      RECORD OF PHDAT
      FILE INTO ARRAY
      A (13)
430  REM : THE SEQUENTIAL NUMBER
      OF THE COMPOUND IN THE CALLING
      PROGRAM HAS TO BE "IS"
440  REM :ARRAY A HAS TO BE DECLA
      RED IN THE CALLING PROGRAM
450  D$ =  CHR$ (4)
460  PRINT D$; "OPEN PHDAT, D1"
470  PRINT D$; "POSITION PHDAT, R";
      IS - 1
480  PRINT D$; "READ PHDAT"
490  INPUT A(1), A(2), A(3), A(4), A(5),
      A(6), A(7), A(8), A(9), A(10),
      A(11), A(12), A(13), A(14), A(15)
500  PRINT D$; "CLOSE PHDAT"
510  PRINT D$; "IN#0"
520   RETURN
```

Example. Run the program for the following reactions at temperature 600 K:

$$CH_4 + H_2O \rightleftarrows CO + 3H_2$$

$$CO + H_2O \rightleftarrows CO_2 + H_2$$

$$CH_4 + 2H_2O \rightleftarrows CO_2 + 4H_2$$

```
]RUN

ENTER THE NUMBER OF COMPOUNDS
IN THE REACTION 4

ORDER THE REACTION EQUATION TO 0
THE STOICHIOMETRIC COEFFICIENTS ARE
CONSIDERED OF ABOVE FORM

ENTER STOICH. COEFF OF 1-TH COMPOUND -1
ENTER STOICH. COEFF OF 2-TH COMPOUND -1
ENTER STOICH. COEFF OF 3-TH COMPOUND 1
ENTER STOICH. COEFF OF 4-TH COMPOUND 3
```

```
ENTER SEQUENTIAL NUMBER OF 1-TH COMPOUND 1
ENTER SEQUENTIAL NUMBER OF 2-TH COMPOUND 15
ENTER SEQUENTIAL NUMBER OF 3-TH COMPOUND 18
ENTER SEQUENTIAL NUMBER OF 4-TH COMPOUND 17

ENTER TEMPERATURE (K) 600

THE STANDARD ENTHALPY OF THE
REACTION AT 600 K IS 217834.071

NEXT TEMPERATURE ? (Y / N) N
NEXT REATION ? (Y / N) NO. OF COMPOUNDS MUST NOT CHANGE Y
ORDER THE REACTION EQUATION TO 0
THE STOICHIOMETRIC COEFFICIENTS ARE
CONSIDERED OF ABOVE FORM

ENTER STOICH. COEFF OF 1-TH COMPOUND -1
ENTER STOICH. COEFF OF 2-TH COMPOUND -1
ENTER STOICH. COEFF OF 3-TH COMPOUND 1
ENTER STOICH. COEFF OF 4-TH COMPOUND 1

ENTER SEQUENTIAL NUMBER OF 1-TH COMPOUND 18
ENTER SEQUENTIAL NUMBER OF 2-TH COMPOUND 15
ENTER SEQUENTIAL NUMBER OF 3-TH COMPOUND 19
ENTER SEQUENTIAL NUMBER OF 4-TH COMPOUND 17

ENTER TEMPERATURE (K) 600

THE STANDARD ENTHALPY OF THE
REACTION AT 600 K IS -39019.6144

NEXT TEMPERATURE ? (Y / N) N
NEXT REACTION ? (Y / N) NO. OF COMPOUNDS MUST NOT CHANGE Y
ORDER THE REACTION EQUATION TO 0
THE STOICHIOMETRIC COEFFICIENTS ARE
CONSIDERED OF ABOVE FORM

ENTER STOICH. COEFF OF 1-TH COMPOUND -1
ENTER STOICH. COEFF OF 2-TH COMPOUND -2
ENTER STOICH. COEFF OF 3-TH COMPOUND 1
ENTER STOICH. COEFF OF 4-TH COMPOUND 4

ENTER SEQUENTIAL NUMBER OF 1-TH COMPOUND 1
ENTER SEQUENTIAL NUMBER OF 2-TH COMPOUND 15
ENTER SEQUENTIAL NUMBER OF 3-TH COMPOUND 19
ENTER SEQUENTIAL NUMBER OF 4-TH COMPOUND 17

ENTER TEMPERATURE (K) 600

THE STANDARD ENTHALPY OF THE
REACTION AT 600 K IS 178814.456

NEXT TEMPERATURE ? (Y / N) N
NEXT REACTION ? (Y / N) NO. OF COMPOUNDS MUST NOT CHANGE N
```

Remark. Explain the results.

Algorithm and Program: Free-Energy Change of System During Chemical Reaction

Write a program for the calculation of the reaction free enthalpy at given temperature and pressure.

Algorithm.

1. Read the stoichiometric coefficients of the reaction.
2. Read the sequential numbers of the components in the physical-property datafile.
3. Read the temperature and pressure.
4. Read the coefficients of the enthalpy polynomials and entropy equations of the pure components from the physical-property datafile.
5. Calculate the enthalpies and entropies of the pure components.
6. Calculate the pure-component free energies.
7. Linear combination of the pure-component free energies with the stoichiometric coefficients.
8. Print the results.

Program REACTIONFREENERGY. A possible realization:

```
]LOAD REACTIONFREENERGY,D2
]LIST

 10  REM :CALCULATION OF REACTION
      FREE ENERGY FOR A GIVEN REAC
      TION
 20  PRINT : PRINT
 30  PRINT "ENTER THE NUMBER OF CO
      MPOUNDS"
 40  INPUT "IN THE REACTION "; N
 50  PRINT
 60  R = 8.3144
 70  PU = 101325
 80  DIM A(15),C(N),NO(N),G(N)
 90  PRINT "ORDER THE REACTION EQU
      ATION TO 0 "
100  PRINT "THE STOICHIOMETRIC COE
      FFICIENTS ARE"
110  PRINT "CONSIDERED OF ABOVE
      FORM"
120  PRINT
130  FOR I = 1 TO N
140  PRINT "ENTER STOICH. COEFF.
      OF"; I; "-TH COMPOUND"
150  INPUT " "; C(I)
```

```
160  NEXT I
170  PRINT : PRINT
180  FOR I = 1 TO N
190  PRINT "ENTER SEQUENTIAL NUMBER
      OF "; I; "-TH COMPOUND"
200  INPUT " "; NO(I)
210  NEXT I
220  PRINT : PRINT
230  INPUT "ENTER PRESSURE (PA) "
      ;P
240  PRINT : PRINT
250  INPUT "ENTER TEMPERATURE (K)
      "; T
260  PRINT : PRINT
270  FOR I = 1 TO N
280  IS = NO(I)
290  GOSUB 510
300  G(I) = ((((A(12) * T + A(11))
      * T + A(10)) * T + A(9)) *
      T + A(8)) * T + A(7)
310  S0 = A(13) + A(8) * LOG (T) +
      (((5 / 4 * A(12)) * T + 4 / 3
      * A(11)) * T + 3 / 2 * A(10
      )) * T + 2 * A(9)) * T - R *
      LOG (P / PU)
320  G(I) = G(I) - T * S0
330  NEXT I
340  DG = 0
350  FOR I = 1 TO N
360  DG = DG + C(I) * G(I)
370  NEXT I
380  PRINT : PRINT
390  PRINT "THE STANDARD FREE ENERGY
      OF THE"
400  PRINT "REACTION AT "; T; " K I
      S "; DG
410  PRINT : PRINT
420  INPUT "NEXT TEMPERATURE ? (Y
      /N) "; A$
430  IF A$ = "Y" THEN GOTO 250
440  PRINT : PRINT
450  INPUT "NEXT PRESSURE ? (Y / N)
      "; A$
460  IF A$ = "Y" THEN GOTO 230
470  PRINT : PRINT
480  INPUT "NEXT REACTION ? (Y / N)
      NO. OF COMPOUNDS MUST NOT
      CHANGE "; A$
490  IF A$ = "Y" THEN GOTO 90
```

```
500  END
510  REM :READING OF SPECIFIED
      RECORD OF PHDAT
      FILED INTO ARRAY
      A (13)
520  REM :THE SEQUENTIAL NUMBER
      OF THE COMPOUND IN THE CALLING
      PROGRAM HAS TO BE "IS"
530  REM :ARRAY A HAS TO BE DECLA
      RED IN THE CALLING PROGRAM
540  D$ = CHR$ (4)
550  PRINT D$; "OPEN PHDAT, D1"
560  PRINT D$; "POSITION PHDAT, R";
      IS - 1
570  PRINT D$; "READ PHDAT"
580  INPUT A(1), A(2), A(3), A(4), A(5),
      A(6), A(7), A(8), A(9), A(10),
      A(11), A(12), A(13), A(14), A(15)
590  PRINT D$; "CLOSE PHDAT"
600  PRINT D$; "IN#0"
610  RETURN
```

Example. Run the program for the following reactions at temperature 600 K and pressure 101,325 Pa:

$$CH_4 + H_2O \rightleftarrows CO + 3H_2$$

$$CO + H_2O \rightleftarrows CO_2 + H_2$$

$$CH_4 + 2H_2O \rightleftarrows CO_2 + 4H_2$$

```
]RUN

ENTER THE NUMBER OF COMPOUNDS
IN THE REACTION 4

ORDER THE REACTION EQUATION TO 0
THE STOICHIOMETRIC COEFFICIENTS ARE
CONSIDERED OF ABOVE FORM

ENTER STOICH. COEFF OF 1-TH COMPOUND -1
ENTER STOICH. COEFF OF 2-TH COMPOUND -1
ENTER STOICH. COEFF OF 3-TH COMPOUND 1
ENTER STOICH. COEFF OF 4-TH COMPOUND 3

ENTER SEQUENTIAL NUMBER OF 1-TH COMPOUND 1
ENTER SEQUENTIAL NUMBER OF 2-TH COMPOUND 15
ENTER SEQUENTIAL NUMBER OF 3-TH COMPOUND 18
ENTER SEQUENTIAL NUMBER OF 4-TH COMPOUND 17
```

```
ENTER PRESSURE (PA) 101325

ENTER TEMPERATURE (K) 600

THE STANDARD FREE ENERGY OF THE
REACTION AT 600 K IS 72160.4148

NEXT TEMPERATURE ? (Y / N) N

NEXT PRESSURE ? (Y / N) N

NEXT REACTION ? (Y / N) NO. OF COMPOUNDS MUST NOT CHANGE Y
ORDER THE REACTION EQUATION TO 0
THE STOICHIOMETRIC COEFFICIENTS ARE
CONSIDERED OF ABOVE FORM

ENTER STOICH. COEFF OF 1-TH COMPOUND -1
ENTER STOICH. COEFF OF 2-TH COMPOUND -1
ENTER STOICH. COEFF OF 3-TH COMPOUND 1
ENTER STOICH. COEFF OF 4-TH COMPOUND 1

ENTER SEQUENTIAL NUMBER OF 1-TH COMPOUND 18
ENTER SEQUENTIAL NUMBER OF 2-TH COMPOUND 15
ENTER SEQUENTIAL NUMBER OF 3-TH COMPOUND 19
ENTER SEQUENTIAL NUMBER OF 4-TH COMPOUND 17

ENTER PRESSURE (PA) 101325

ENTER TEMPERATURE (K) 600

THE STANDARD FREE ENERGY OF THE
REACTION AT 600 K IS -16644.4952

NEXT TEMPERATURE ? (Y / N) N

NEXT PRESSURE ? (Y / N) N

NEXT REACTION ? (Y / N) NO. OF COMPOUNDS MUST NOT CHANGE Y
ORDER THE REACTION EQUATION TO 0
THE STOICHIOMETRIC COEFFICIENTS ARE
CONSIDERED OF ABOVE FORM

ENTER STOICH. COEFF OF 1-TH COMPOUND -1
ENTER STOICH. COEFF OF 2-TH COMPOUND -2
ENTER STOICH. COEFF OF 3-TH COMPOUND 1
ENTER STOICH. COEFF OF 4-TH COMPOUND 4

ENTER SEQUENTIAL NUMBER OF 1-TH COMPOUND 1
ENTER SEQUENTIAL NUMBER OF 2-TH COMPOUND 15
ENTER SEQUENTIAL NUMBER OF 3-TH COMPOUND 19
ENTER SEQUENTIAL NUMBER OF 4-TH COMPOUND 17

ENTER PRESSURE (PA) 101325

ENTER TEMPERATURE (K) 600

THE STANDARD FREE ENERGY OF THE
```

```
REACTION AT 600 K IS 55515.9197

NEXT TEMPERATURE ? (Y / N) N

NEXT PRESSURE ? (Y / N) N

NEXT REACTION ? (Y / N) NO. OF COMPOUNDS MUST NOT CHANGE N
```

Remark. Explain the results.

DISPLACEMENT OF EQUILIBRIUM

The question may arise how the equilibrium composition changes if a perturbation of temperature or pressure leads to a new equilibrium state. To tackle the problem, let us set out from the free enthalpy. It should be born in our mind that at equilibrium the free enthalpy of the reaction mixture exhibits an extremum, that is,

$$\left(\frac{\partial G}{\partial X}\right)_{P,T} = \sum_j \nu_j \mu_j = 0$$

On the other hand, according to the minimum principle of the free enthalpy we can write

$$\frac{\partial}{\partial X}\left(\frac{\partial G}{\partial X}\right)_{P,T} = \frac{\partial\left(\sum_j \nu_j \mu_j\right)}{\partial X} = \frac{\partial^2 G}{\partial X^2} > 0$$

We are interested in how the equilibrium extent of reaction changes with the temperature. That is to say, the sign of the partial derivative

$$\left(\frac{\partial X}{\partial T}\right)_{P,\Sigma \nu_j \mu_j}$$

is critical. From the chain-rule,

$$\left(\frac{\partial X}{\partial T}\right)_{(P,\Sigma \nu_j \mu_j)} = -\frac{\partial\left(\sum \nu_j \mu_j\right)}{\partial T} \bigg/ \frac{\partial\left(\sum \nu_j \mu_j\right)}{\partial X}$$

The numerator however, can be related to the enthalpy of reaction:

$$\left(\frac{\partial\left(\sum \nu_j \mu_j\right)}{\partial T}\right)_{P,X} = -\frac{1}{T}\frac{\partial H}{\partial X}$$

Substituting this relation into the above formula, we have

$$\left(\frac{\partial X}{\partial T}\right)_{P,\Sigma \nu_j \mu_j} = \frac{1}{T}\left(\frac{\partial\left(\sum \nu_j \mu_j\right)}{\partial X}\right)^{-1}\frac{\partial H}{\partial X}$$

The factor $1/T$ of the right-hand side is always positive because of the positivity of the temperature. The second factor is just the second derivative of the free enthalpy of the reaction mixture; consequently, it is always positive at equilibrium. Hence, the sign of the partial derivative of interest is determined by the third factor on the right-hand side. So, if dH/dX has a positive value (i.e., the reaction is endothermic), the sign of the right-hand side is positive, as well. In other words, if the reaction is endothermic and the temperature is increasing, then the reaction is going forward to achieve a new equilibrium state. For an exothermic reaction, however, the equilibrium conversion decreases as the temperature increases.

Similar considerations can be made when investigating the displacement of equilibrium caused by a pressure change at constant temperature. In this case we are interested in the sign of the following partial derivate:

$$\left(\frac{\partial X}{\partial P}\right)_{T,\Sigma\nu_j\mu_j}$$

Applying the chain rule, this can be written

$$\left(\frac{\partial X}{\partial P}\right)_{T,\Sigma\nu_j\mu_j} = -\left(\frac{\partial\left(\sum\nu_j\mu_j\right)}{\partial P}\right)_{T,X} \Bigg/ \left(\frac{\partial\left(\sum\nu_j\mu_j\right)}{\partial X}\right)_{T,P}$$

The denominator of the right-hand side has already been shown to be always positive. What about the numerator?

Since

$$\left(\frac{\partial G}{\partial X}\right)_{P,T} = \sum\nu_j\mu_j$$

and on the other hand

$$\left(\frac{\partial G}{\partial P}\right)_T = V$$

the numerator can be written as

$$\left(\frac{\partial\left(\sum\nu_j\mu_j\right)}{\partial P}\right)_{T,X} = \frac{\partial^2 G}{\partial P\,\partial X} = \frac{\partial^2 G}{\partial X\,\partial P} = \left(\frac{\partial V}{\partial X}\right)_T$$

and so

$$\left(\frac{\partial X}{\partial P}\right)_{T,\Sigma\nu_j\mu_j} = -\frac{\left(\dfrac{\partial V}{\partial X}\right)_T}{\left(\dfrac{\partial\left(\sum\nu_j\mu_j\right)}{\partial X}\right)_{T,P}}$$

The partial derivative $(\partial V/\partial X)_T$ expresses the change of volume due to the reaction.

If the reaction takes place without change in volume, the pressure change does not displace the equilibrium. For any increase in pressure, the equilibrium extent of reaction increases if the reaction leads to a decrease in volume, and decreases if it leads to an increase in volume.

Algorithm and Program: Reaction Equilibrium at Arbitrary Temperature and Pressure

Write a program for the determination of the equilibrium concentration of a one-chemical-reaction system. Calculate the chemical potential from the molar enthalpy and entropy.

Algorithm.

1. Read the stoichiometric coefficients of the reaction.
2. Read the initial mole numbers.
3. Read the sequential numbers of the components in the datafile.
4. Determine the maximal possible conversion from the initial mole numbers and the stoichiometric coefficients.
5. Read the temperature and pressure.
6. Read the coefficients of the enthalpy polynomials and entropy equations of the pure compounds from the physical-property datafile.
7. Calculate the enthalpies and entropies of the pure compounds.
8. Calculate the chemical potentials of the pure compounds at the given temperature and pressure.
9. Tabulate the values of the mixture chemical potential as a function of the conversion, and find the equilibrium conversion (where the function value is closest to minimum).
10. Print the results.

Program REACTIONEQUILIBRIUMDISC. A possible realization:

```
]PR#0
]LOAD REACTIONEQUILIBRIUMDISC,D2
]LIST

  10  REM :CALCULATION OF EQUILIBRI
       UM CONCENTRATIONS FOR A GIVEN
       REACTION AND DEFINED INITIAL
       CONCENTRATIONS
  20  PRINT : PRINT
```

```
 30  PRINT "ENTER THE NUMBER OF
      COMPOUNDS"
 40  R = 8.3144
 50  INPUT "IN THE REACTION "; N
 60  PRINT
 70  DIN N0(N),C(N),NF(N),MU(N),X(
      N),N(N),A(15),NO(N)
 80  PRINT "ORDER THE REACTION EQU
      ATION TO 0 "
 90  PRINT "THE STOICHIOMETRIC COEF
      FICIENTS ARE"
100  PRINT "CONSIDERED OF ABOVE
      FORM"
110  PRINT
120  FOR I = 1 TO N
130  PRINT "ENTER STOICH. COEFF. OF
      "; I; "-TH COMPOUND"
140  INPUT " "; C(I)
150  NEXT I
160  PRINT : PRINT
170  FOR I = 1 TO N
180  PRINT "ENTER INITIAL MOLE NUMBER
      OF "; I;"-TH COMPOUND"
190  INPUT " "; N0(I)
200  NEXT I
210  PRINT : PRINT
220  FOR I = 1 TO N
230  PRINT "ENTER SEQUENTIAL NO.
      OF "; I; "-TH COMPOUND"
240  INPUT " "; NO(I)
250  NEXT I
260  A = 1
270  FOR I = 1 TO N
280  IF C(I) < 0 THEN A = A * N0(I)
290  NEXT I
300  B = 1
310  FOR I = 1 TO N
320  IF C(I) > 0 THEN B = B * N0(I)
330  NEXT I
340  IF A = 0 AND B = 0 THEN PRINT
      "IMPOSSIBLE TO MODIFY INITIAL
      CONCENTRATION, GIVE NEW DATA
      ": GOTO 80
350  PRINT
360  INPUT "ENTER PRESSURE (PA)"
      ; P
370  PRINT
380  INPUT "ENTER TEMPERATURE (K)"
      T;
```

```
390  FOR I = 1 TO N
400  IS = NO(I)
410  GOSUB 990
420  MU(I) = G0
430  NEXT I
440  MA = 1000
450  IF A = 0 THEN GOTO 500
460  FOR I = 1 TO N
470  IF C(I) < 0 THEN X = - N0(I
      ) / C(I)
480  IF X < (MA) THEN MA = X:IA = I
490  NEXT I
500  MB = - 1000
510  IF B = 0 THEN GOTO 560
520  FOR I = 1 TO N
530  IF C(I) > THEN X = - N0(I
      ) / C(I)
540  IF X > MB THEN MB = X:IB = I
550  NEXT I
560  IF MA < ABS (MB) THEN XM =
      MA
570  IF MA > ABS (MB) THEN XM =
      MB
580  G = 100000
590  ST = XM / 41
600  RX = 1000000
610  FOR XE = ST TO 40 * ST STEP
      ST
620  SN = 0
630  FOR I = 1 TO N
640  N(I) = N0(I) + C(I) * XE
650  SN = SN + N(I)
660  NEXT I
670  PX = 0
680  FOR I = 1 TO N
690  X(I) = N(I) / SN
700  PX = PX + C(I) * (MU(I) + R *
      T * LOG (X(I)))
710  NEXT I
720  IF ABS (PX) < RX THEN XR =
      XE:RX = ABS (PX)
730  NEXT XE
740  FOR I = 1 TO N
750  NF (I) = N0(I) + C(I) * XR
760  NEXT I
770  PRINT : PRINT
780  PRINT "CONVERSION IN EQUILIB
      RIUM: "; XR
790  PRINT : PRINT
```

```
800  PRINT : PRINT
810  PRINT "THE MOLE NUMBERS IN THE
      EQUILIBRIUM ARE:"
820  PRINT : PRINT
830  PRINT "COMPOUND NO.
      NUMBER OF MOLES"
840  PRINT "---------------------
      ------------------"
850  PRINT
860  FOR I = 1 TO N
870  PRINT TAB(6); I; TAB(23); N
      F(I)
880  NEXT I
890  PRINT : PRINT
900  INPUT "NEXT TEMPERATURE ? (Y
      /N) "; A$
910  IF A$ = "Y" THEN GOTO 380
920PRINT : PRINT
930  INPUT "NEXT PRESSURE ? (Y / N)
      "; A$
940  IF A$ = "Y" THEN GOTO 350
950  PRINT : PRINT
960  INPUT "NEXT INITIAL MOLE NUMBERS
      ? (Y / N) "; A$
970  IF A$ = "Y" THEN GOTO 160
980  END
990  GOSUB 1060
1000 PU = 101325
1010 R = 8.3144
1020 S0 = A(13) + A(8) * LOG (T)
      + (((5 / 4 * A(12) * T + 4 /
      3 * A(11)) * T + 3 / 2 * A(1
      0)) * T + 2 * A(9)) * T - R *
      LOG (P / PU)
1030 H0 = ((((A(12) * T + A (11)) *
      T + A(10)) * T + A(9)) * T +
      A(8)) * T + A(7)
1040 G0 = H0 - T * S0
1050 RETURN
1060 REM :READING OF SPECIFIED
      RECORD OF PHDATFILE INTO
      ARRAY A(13)
1070 REM :THE SEQUENTIAL NUMBER
      OF THE COMPOUND IN THE CALLING
     NG PROGRAM HAS TO BE "IS"
1080 REM :ARRAY A HAS TO BE DECL
      ARED IN THE CALLING PROGRAM
1090 D$ = CHR$ (4)
1100 PRINT D$; "OPEN PHDAT, D1"
```

```
1110  PRINT D$; "POSITION PHDAT, R"
       ; IS - 1
1120  PRINT D$; "READ PHDAT"
1130  INPUT A(1), A(2), A(3), A(4), A(5),
       A(6), A(7), A(8), A(9), A(10),
       A(11), A(12), A(13), A(14), A(5)
1140  PRINT D$; "CLOSE PHDAT"
1150  PRINT D$;"IN#0"
1160  RETURN
```

Example. Run the program for the reaction

$$CH_4 + H_2O \rightleftarrows CO + 3H_2$$

at the temperature 600 and 1000 K, and the pressures 101,325 and 1,013,250 Pa. The initial mole numbers are

$$N^{\circ}_{H_2O} = 10, \qquad N^{\circ}_{CH_4} = 1, \qquad N^{\circ}_{H_2} = N^{\circ}_{CO} = 0$$

$$N^{\circ}_{H_2O} = 10, \qquad N^{\circ}_{CH_4} = 5, \qquad N^{\circ}_{H_2} = N^{\circ}_{CO} = 0$$

Repeat the run for the reaction

$$CO + H_2O = CO_2 + H_2$$

```
]RUN

ENTER THE NUMBER OF COMPOUNDS
IN THE REACTION 4

ORDER THE REACTION EQUATION TO 0
THE STOICHIOMETRIC COEFFICIENTS ARE
CONSIDERED OF ABOVE FORM

ENTER STOICH. COEFF OF 1-TH COMPOUND 1
ENTER STOICH. COEFF OF 2-TH COMPOUND 1
ENTER STOICH. COEFF OF 3-TH COMPOUND -1
ENTER STOICH. COEFF OF 4-TH COMPOUND -3

ENTER INITIAL MOLE NUMBER OF 1-TH COMPOUND 1
ENTER INITIAL MOLE NUMBER OF 2-TH COMPOUND 10
ENTER INITIAL MOLE NUMBER OF 3-TH COMPOUND 0
ENTER INITIAL MOLE NUMBER OF 4-TH COMPOUND 0

ENTER SEQUENTIAL NO. OF 1-TH COMPOUND 1
ENTER SEQUENTIAL NO. OF 2-TH COMPOUND 15
ENTER SEQUENTIAL NO. OF 3-TH COMPOUND 18
ENTER SEQUENTIAL NO. OF 4-TH COMPOUND 17

ENTER PRESSURE (PA) 101325
```

ENTER TEMPERATURE (K) 600

CONVERSION IN EQUILIBRIUM: −.0731707317

THE MOLE NUMBERS IN THE EQUILIBRIUM ARE:

COMPOUND NO.	NUMBER OF MOLES
1	.926829268
2	9.92682927
3	.0731707317
4	.219512195

NEXT TEMPERATURE ? (Y / N) Y
ENTER TEMPERATURE (K) 1000

CONVERSION IN EQUILIBRIUM: −.951219513

THE MOLE NUMBERS IN THE EQUILIBRIUM ARE:

COMPOUND NO.	NUMBER OF MOLES
1	.0487804872
2	9.04878049
3	.951219513
4	2.85365854

NEXT TEMPERATURE ? (Y / N) N

ENTER PRESSURE (PA) 1013250

NEXT TEMPERATURE (K) 600

CONVERSION IN EQUILIBRIUM: −.0243902439

THE MOLE NUMBERS OF THE EQUILIBRIUM ARE:

COMPOUND NO.	NUMBER OF MOLES
1	.975609756
2	9.97560976
3	.0243902439
4	.0731707317

NEXT TEMPERATURE ? (Y / N) N

NEXT PRESSURE ? (Y / N) N

NEXT INITIAL MOLE NUMBERS ? (Y / N) Y

ENTER INITIAL MOLE NUMBER OF 1-TH COMPOUND 5
ENTER INITIAL MOLE NUMBER OF 2-TH COMPOUND 10
ENTER INITIAL MOLE NUMBER OF 3-TH COMPOUND 0
ENTER INITIAL MOLE NUMBER OF 4-TH COMPOUND 0

ENTER SEQUENTIAL NO. OF 1-TH COMPOUND 1
ENTER SEQUENTIAL NO. OF 2-TH COMPOUND 15
ENTER SEQUENTIAL NO. OF 3-TH COMPOUND 18
ENTER SEQUENTIAL NO. OF 4-TH COMPOUND 17

ENTER PRESSURE (PA) 101325

ENTER TEMPERATURE (K) 600

CONVERSION IN EQUILIBRIUM: −.12195122

THE MOLE NUMBERS IN THE EQUILIBRIUM ARE:

COMPOUND NO.	NUMBER OF MOLES
1	4.87804878
2	9.87804878
3	.12195122
4	.365853659

NEXT TEMPERATURE ? (Y / N) N

NEXT PRESSURE ? (Y / N) N

NEXT INITIAL MOLE NUMBERS ? (Y / N) N

]RUN

ENTER THE NUMBER OF COMPOUNDS
IN THE REACTION 4

ORDER THE REACTION EQUATION TO 0
THE STOICHIOMETRIC COEFFICIENTS ARE
CONSIDERED OF ABOVE FORM

ENTER STOICH. COEFF OF 1-TH COMPOUND 1
ENTER STOICH. COEFF OF 2-TH COMPOUND 1
ENTER STOICH. COEFF OF 3-TH COMPOUND −1
ENTER STOICH. COEFF OF 4-TH COMPOUND −1

ENTER INITIAL MOLE NUMBER OF 1-TH COMPOUND 1
ENTER INITIAL MOLE NUMBER OF 2-TH COMPOUND 10
ENTER INITIAL MOLE NUMBER OF 3-TH COMPOUND 0
ENTER INITIAL MOLE NUMBER OF 4-TH COMPOUND 0

ENTER SEQUENTIAL NO. OF 1-TH COMPOUND 18
ENTER SEQUENTIAL NO. OF 2-TH COMPOUND 15
ENTER SEQUENTIAL NO. OF 3-TH COMPOUND 19
ENTER SEQUENTIAL NO. OF 4-TH COMPOUND 17

ENTER PRESSURE (PA) 101325

ENTER TEMPERATURE (K) 600

CONVERSION IN EQUILIBRIUM: −.951219513

THE MOLE NUMBERS IN THE EQUILIBRIUM ARE:

COMPOUND NO.	NUMBER OF MOLES
1	.0487804872
2	9.04878049
3	.951219513
4	.951219513

NEXT TEMPERATURE ? (Y / N) Y
ENTER TEMPERATURE (K) 1000

CONVERSION IN EQUILIBRIUM: −.926829269

THE MOLE NUMBERS IN EQUILIBRIUM ARE:

COMPOUND NO.	NUMBER OF MOLES
1	.0731707311
2	9.07317073
3	.926829269
4	.926829269

NEXT TEMPERATURE ? (Y / N) N

NEXT PRESSURE ? (Y / N) Y

ENTER PRESSURE (PA) 1013250

ENTER TEMPERATURE (K) 1000

CONVERSION IN EQUILIBRIUM: −.926829269

THE MOLE NUMBERS IN THE EQUILIBRIUM ARE:

COMPOUND NO.	NUMBER OF MOLES
1	.0731707311
2	9.07317073
3	.926829269
4	.926829269

NEXT TEMPERATURE ? (Y / N) N

NEXT PRESSURE ? (Y / N) N

NEXT INITIAL MOLE NUMBERS ? (Y / N) Y

ENTER INITIAL MOLE NUMBERS OF 1-TH COMPOUND 1
ENTER INITIAL MOLE NUMBERS OF 2-TH COMPOUND 5
ENTER INITIAL MOLE NUMBERS OF 3-TH COMPOUND 0
ENTER INITIAL MOLE NUMBERS OF 4-TH COMPOUND 0

ENTER SEQUENTIAL NO. OF 1-TH COMPOUND 18
ENTER SEQUENTIAL NO. OF 2-TH COMPOUND 15

```
ENTER SEQUENTIAL NO. OF 3-TH COMPOUND 19
ENTER SEQUENTIAL NO. OF 4-TH COMPOUND 17

ENTER PRESSURE (PA) 101325

ENTER TEMPERATURE (K) 1000

CONVERSION IN EQUILIBRIUM: -.878048781

THE MOLE NUMBERS IN THE EQUILIBRIUM ARE:
```

COMPOUND NO.	NUMBER OF MOLES
1	.121951219
2	4.12195122
3	.878048781
4	.878048781

```
NEXT TEMPERATURE ? (Y / N) N

NEXT PRESSURE ? (Y / N) N

NEXT INITIAL MOLE NUMBERS ? (Y / N) N
```

Remarks. The first reaction (as usual) is endothermic, and as the reaction proceeds, the pressure increases at constant volume. As can be seen, the model describes this behavior well. The higher the temperature, the higher the H_2 equilibrium concentration. The higher the pressure, the lower the H_2 equilibrium concentration. The bigger the initial mole number of the diminishing component, the bigger the equilibrium conversion. The second chemical reaction is exothermic; that is why at higher temperature the equilibrium conversion is lower. As the reaction proceeds, the pressure does not change, so the equilibrium conversion is independent of the pressure.

STANDARD FREE ENTHALPY OF REACTION AND REACTION CONSTANT

The relationships so for discussed give theoretical information on the equilibrium of chemical reactions. Quantitative results can be gained if the explicit form of the chemical potential

$$\mu_j = \mu\left[T, P, N_j\right]$$

is substituted into the equation expressing the criterion for chemical equilibrium:

$$\sum_j \nu_j \mu_j = 0$$

For the chemical potential, from our previous knowledge, we have

$$\mu_j = \mu_j^\circ + \mathbf{R}T \ln a_j$$

where μ_j° depends only on the temperature and represents the chemical potential of component j in the standard state, and a_j is the activity of component j in the equilibrium mixture. Let us substitute this explicit form of the chemical potential into the equation of equilibrium:

$$\sum_j \nu_j \left(\mu_j^\circ + \mathbf{R}T \ln a_j \right) = 0$$

Then

$$\sum_j \nu_j \mu_j^\circ = -\mathbf{R}T \sum \ln a_j^{\nu_j}$$

$$= -\mathbf{R}T \ln \prod_j a_j^{\nu_j}$$

The product is called, by definition, the equilibrium constant:

$$K \equiv \prod_j a_j^{\nu_j}$$

Making use of this notation, we may write

$$\sum_j \nu_j \mu_j^\circ = -\mathbf{R}T \ln K$$

This relationship shows that the equilibrium constant, defined in terms of the activities, is a function of the temperature alone, because the left-hand side involves only the standard chemical potentials, depending only on the temperature.

The left-hand side of the equation, defining the equilibrium constant, may also be expressed as

$$\sum_j \nu_j \mu_j^\circ = \Delta G^\circ$$

Here ΔG° is standard free enthalpy of the reaction. This is nothing but the arithmetic sum of the standard molar free energies of the forming and reacting components, weighted by the corresponding stoichiometric coefficients.

In accordance with our earlier considerations, the relationship derived in the foregoing can be generalized also for p independent reactions in the system:

$$\sum_j \nu_{ji} \mu_j^\circ = -\mathbf{R}T \ln K_i \qquad j = 1, 2, \ldots, k \quad i = 1, 2, \ldots, p$$

We will return to this relationship later on.

Algorithm and Program: Calculation of Equilibrium Constant

Write a program for the calculation of the equilibrium constant.

Algorithm.

1. Read the stoichiometric coefficients of the reaction.
2. Read the sequential numbers of the components in the physical-property datafile.
3. Read the temperature and pressure.
4. Read the coefficients of the enthalpy polynomials and entropy equations of the pure components from the physical-property datafile.
5. Calculate the enthalpies and entropies of the pure components.
6. Calculate the pure-component free energies.
7. Calculate the reaction free energy by the linear combination of the pure-component free energies with stoichiometric coefficients.
8. Calculate the equilibrium constant from the equation

$$-RT \ln K = \Delta G^\circ$$

9. Print the results.

Program EQUILIBRIUMCONSTANT. A possible realization:

```
]LOAD EQUILIBRIUMCONSTANT,D2
]LIST

 10  REM : CALCULATION OF EQUILIBRIUM
      CONSTANT FOR A GIVEN REACTION
 20  PRINT : PRINT
 30  PRINT "ENTER THE NUMBER OF
      COMPOUNDS"
 40  INPUT "IN THE REACTION "; N
 50  PRINT
 60  R = 8.3144
 70  PU = 101325
 80  DIM A(15),C(N),NO(N),G(N)
 90  PRINT "ORDER THE REACTION EQU
      ATION TO 0 "
100  PRINT "THE STOICHIOMETRIC COEF
      FICIENTS ARE"
110  PRINT "CONSIDERED OF ABOVE
      FORM"
120  PRINT
130  FOR I = 1 TO N
140  PRINT "ENTER STOICH. COEFF.
      OF "; I; "-TH COMPOUND"
```

```
150  INPUT " "; C(I)
160  NEXT I
170  PRINT : PRINT
180  FOR I = 1 TO N
190  PRINT " ENTER SEQUENTIAL NUMBER
      OF "; I; "-TH COMPOUND
200  INPUT " "; NO(I)
210  NEXT I
220  PRINT : PRINT
230  INPUT "ENTER PRESSURE (PA) "
      ;P
240  PRINT : PRINT
250  INPUT "ENTER TEMPERATURE (K)
      "; T
260  PRINT : PRINT
270  FOR I = 1 TO N
280  IS = NO(I)
290  GOSUB 530
300  G(I) = ((((A(12) * T + A(11))
      * T + A(10)) * T + A(9)) *
      T + A(8)) * T + A(7)
310  S0 = A(13) + A(8) * LOG (T) +
      (((5 / 4 * A(12) * T + 4 / 3
      * A(11)) * T + 3 / 2 * A(10
      )) * T + 2 * A(9)) * T - R *
      LOG (P / PU)
320  G(I) = G(I) - T * S0
330  NEXT I
340  DG = 0
350  FOR I = 1 TO N
360  DG = DG + C(I) * G(I)
370  NEXT I
380  K = DG / ( - R * T)
390  K = EXP (K)
400  PRINT : PRINT
410  PRINT :THE EQUILIBRIUM CONST
      ANT OF THE"
420  PRINT "REACTION AT "; T; " K I
      S "; K
430  PRINT : PRINT
440  INPUT "NEXT TEMPERATURE ? (Y
      /N) "; A$
450  IF A$ = "Y" THEN GOTO 250
460  PRINT : PRINT
470  INPUT "NEXT TEMPERATURE ? (Y / N)
      "; A$
480  IF A$ = "Y" THEN GOTO 230
490  PRINT : PRINT
500  INPUT "NEXT REACTION ? (Y / N)
```

```
         NO. OF COMPOUNDS MUST NOT
         CHANGE "; A$
510  IF A$ = "Y" THEN GOTO 90
520  END
530  REM :READING OF SPECIFIED
         RECORD OF PHDAT FILE INTO
         ARRAY A(13)
540  REM :THE SEQUENTIAL NUMBER OF
         THE COMPOUND IN THE CALLING
         PROGRAM HAS TO BE "IS"
550  REM :ARRAY A HAS TO BE DECLA
         RED IN THE CALLING PROGRAM
560  D$ = CHR$ (4)
570  PRINT D$; "OPEN PHDAT, D1"
580  PRINT D$; "POSITION PHDAT, R";
         IS - 1
590  PRINT D$; "READ PHDAT"
600  INPUT A(1), A(2), A(3), A(4), A(5),
         A(6), A(7), A(8), A(9), A(10),
         A(11), A(12), A(13), A(14), A(15)
610  PRINT D$; "CLOSE PHDAT"
620  PRINT D$; "IN#0"
630  RETURN
```

Example. Run the program for the calculation of the equilibrium constants of the following reactions at temperature 600 K and pressure 101,325 Pa:

$$CH_4 + H_2O \rightleftarrows CO + 3H_2$$

$$CO + H_2O \rightleftarrows CO_2 + H_2$$

```
]RUN

ENTER THE NUMBER OF COMPOUNDS
IN THE REACTION 4

ORDER THE REACTION EQUATION TO 0
THE STOICHIOMETRIC COEFFICIENTS ARE
CONSIDERED OF ABOVE FORM

ENTER STOICH. COEFF OF 1-TH COMPOUND -1
ENTER STOICH. COEFF OF 2-TH COMPOUND -1
ENTER STOICH. COEFF OF 3-TH COMPOUND 1
ENTER STOICH. COEFF OF 4-TH COMPOUND 3

ENTER SEQUENTIAL NUMBER OF 1-TH COMPOUND 1
ENTER SEQUENTIAL NUMBER OF 2-TH COMPOUND 15
ENTER SEQUENTIAL NUMBER OF 3-TH COMPOUND 18
ENTER SEQUENTIAL NUMBER OF 4-TH COMPOUND 17
```

```
ENTER PRESSURE (PA) 101325

ENTER TEMPERATURE (K) 600

THE EQUILIBRIUM CONSTANT OF THE
REACTION AT 600 K IS 5.22339897E - 07

NEXT TEMPERATURE ? (Y / N) Y
ENTER TEMPERATURE (K) 700

THE EQUILIBRIUM CONSTANT OF THE
REACTION AT 700 K IS 2.77577853E - 04

NEXT TEMPERATURE ? (Y / N) N

NEXT PRESSURE ? (Y / N) Y
ENTER PRESSURE (PA) 1013250

ENTER TEMPERATURE (K) 700

THE EQUILIBRIUM CONSTANT OF THE
REACTION AT 700 K IS 2.77577857E - 06

NEXT TEMPERATURE ? (Y / N) N

NEXT PRESSURE ? (Y / N) N

NEXT REACTION ? (Y / N) NO. OF COMPOUNDS MUST NOT CHANGE Y
ORDER THE REACTION EQUATION TO 0
THE STOICHIOMETRIC COEFFICIENTS ARE
CONSIDERED OF ABOVE FORM

ENTER STOICH. COEFF OF 1-TH COMPOUND -1
ENTER STOICH. COEFF OF 2-TH COMPOUND -1
ENTER STOICH. COEFF OF 3-TH COMPOUND 1
ENTER STOICH. COEFF OF 4-TH COMPOUND 1

ENTER SEQUENTIAL NUMBER OF 1-TH COMPOUND 18
ENTER SEQUENTIAL NUMBER OF 2-TH COMPOUND 15
ENTER SEQUENTIAL NUMBER OF 3-TH COMPOUND 19
ENTER SEQUENTIAL NUMBER OF 4-TH COMPOUND 17

ENTER PRESSURE (PA) 101325

ENTER TEMPERATURE (K) 600

THE EQUILIBRIUM CONSTANT OF THE
REACTION AT 600 K IS 28.1199566

NEXT TEMPERATURE ? (Y / N) Y
ENTER TEMPERATURE (K) 700

THE EQUILIBRIUM CONSTANT OF THE
REACTION AT 700 K IS 9.32323025

NEXT TEMPERATURE ? (Y / N) N
```

NEXT PRESSURE ? (Y / N) N

NEXT REACTION ? (Y / N) NO. OF COMPOUNDS MUST NOT CHANGE N

Remark. Explain the results.

Additivity of Stoichiometric Reactions. Hess's Law

As indicated before, the stoichiometric equations of two or more linearly independent reactions can be combined linearly as to give an additional, linearly dependent stoichiometric equation:

$$\sum_j \nu_{j1} m_j = 0$$

$$\sum_j \nu_{j2} m_j = 0$$

$$\sum_j \nu_{j3} m_j = \sum_j (\lambda_1 \nu_{j1} + \lambda_2 \nu_{j2}) m_j$$

where λ_1 and λ_2 are constants. From this, it follows that any of the homogeneous linear functions associated with the reaction (e.g. enthalpy of reaction, free enthalpy of reaction) can be derived in similar way. For instance, in the case of the enthalpy of reaction,

$$\Delta H_3^\circ = \lambda_1 \Delta H_1^\circ + \lambda_2 \Delta H_2^\circ$$

This is called Hess's law.

For the standard free enthalpy of reaction, we may write

$$\Delta G_3^\circ = \lambda_1 \Delta G_1^\circ + \lambda_2 \Delta G_2^\circ$$

Consequently, the logarithms of the constants of reactions are additive; therefore

$$\ln K_3 = \lambda_1 \ln K_1 + \lambda_2 \ln K_2$$

In its physical meaning, Hess's law associated with the enthalpy of reaction is just a special form of the energy conservation principle applied to chemical reactions. Hess's law is of great importance mainly for the computation of the calorimetrically unmeasurable enthalpy of reaction. This computation is based on additivity.

CALCULATION OF EQUILIBRIUM COMPOSITION

The criterion for chemical equilibrium may be formulated in two different ways. That is to say, an isothermal, isobaric reacting system is at equilibrium:

1. If the free enthalpy of the system takes on a minimum.
2. If for each stoichiometrically independent reaction, the mass-action quotient is equal to the equilibrium constant of the reaction:

$$\prod_j a_j^{\nu_j} = \prod_j \left(a_j^{(e)}\right)^{\nu_j} \equiv K$$

These two formulations of the equilibrium criterion make two kinds of calculation approaches possible:

1. The first formulation leads to a nonlinear-programming task, with a variety of solution techniques,
2. The second formulation leads to a nonlinear equation set, which can also be solved in many different ways.

We will show some calculation methods for each case.

Calculation of Equilibrium Composition with Known Equilibrium Constant

In order to calculate the composition of the reaction mixture from a knowledge of the equilibrium constant at initial composition N_j°, the equation

$$K = \prod_j a_j^{\nu_j}$$

derived before, must be properly transformed. First of all, let us substitute the explicit form of the activities:

$$a_j = \frac{\bar{f}_j}{f_j^{\circ}} = \frac{\bar{\phi}_j y_j P}{f_j^{\circ}}$$

where $\bar{f}_j$ = fugacity of component j at equilibrium,
$\bar{\phi}_j$ = fugacity coefficient of component j at equilibrium (this notation
is used instead of the Greek ν, which is reserved for the stoichiometric coefficients in this chapter),
f_j° = fugacity of component j at system temperature in the standard state,
y_j = mole fraction of component j at equilibrium,
P = total pressure in the mixture.

After substitution we have

$$K = \prod_j \left(\frac{\phi_j y_j P}{f_j^\circ} \right)^{\nu_j} = \left(\frac{P}{f_j^\circ} \right)^{\Sigma\nu_j} \prod_j (\bar{\phi}_j y_j)^{\nu_j}$$

where $\Sigma\nu_j$ is the stoichiometric change in the number of moles, and (P/f_j°) can be factored out because the standard fugacity does not depend on the actual composition. Assuming furthermore, that the reaction mixture constitutes an ideal mixture of real gaseous components, we may write

$$\bar{\phi}_j = \phi_j$$

where ϕ_j represents the fugacity coefficient of pure component j at the temperature and pressure of the mixture.

Let us define the following notation:

$$K_\phi = \prod_j \phi_j^{\nu_j}$$

This number is usually called the fugacity quotient on the analogy of the mass-action quotient. The fugacity quotient on the analogy of the mass-action quotient. The fugacity quotients in accordance with the Lewis-Randall rule, is independent of the composition. After bringing the constants over to the left-hand side, we can write

$$A = K_\phi^{-1} \left(\frac{P}{f_j^0} \right)^{-\Sigma\nu_j} K = \prod_j (y_j)^{\nu_j}$$

The amount of component j in the equilibrium mixture is

$$N_j = N_j^\circ + \nu_j X$$

Thus the mole fraction of component j is

$$y_j = \frac{N_j}{\sum N_j}$$

Using this notation, we can write

$$A = \frac{1}{\left(\sum_j N_j \right)^{\Sigma\nu_j}} \prod_j \left(N_j^\circ + \nu_j X_e \right)^{\nu_j}$$

where X_e means the solution of the nonlinear equation representing the extent of reaction in the equilibrium mixture.

Algorithm and Program: Calculation of Equilibrium Concentration from Equilibrium Constant

Write a program for the calculation of the equilibrium constant and thence the equilibrium concentration.

Algorithm.

1. Read the stoichiometric coefficients of the reaction.
2. Read the sequential numbers of the components in the physical-property datafile.
3. Read the temperature and pressure.
4. Read the parameters of the enthalpy and entropy equations from the physical-property datafile.
5. Calculate the enthalpies, entropies, and free energies of the pure components.
6. Calculate the reaction free enthalpy.
7. Calculate the equilibrium constant.
8. Read the initial mole numbers.
9. Determine the maximal possible conversion from the initial mole numbers and the stoichiometric coefficients.
10. Tabulate the mass-action quotient the equilibrium constant as a function of the conversion and compare with the equilibrium constant.

Program EQUILIBRIUMCONCENTRATIONFROMK,D2: A possible realization:

```
]PR#0
]LOAD EQUILIBRIUMCONCENTRATIONFROMK,D2
]LIST

 10  REM :CALCULATION OF EQUILIBRI
      UM CONCENTRATION FROM EQUILI
      BRIUM CONSTANT
 20  PRINT : PRINT
 30  PRINT "ENTER THE NUMBER OF
      COMPOUNDS"
 40  INPUT "IN THE REACTION "; N
 50  PRINT
 60  R = 8.3144
 70  PU = 101325
 80  DIM A(15),C(N),NO(N),G(N)
```

```
 90  DIM NF(N),MU(N),X(N),N(N),NO(N)
100  PRINT "ORDER THE REACTION
      EQUATION TO 0 "
110  PRINT " THE STOICHIOMETRIC COEF
      FICIENTS ARE"
120  PRINT "CONSIDERED OF ABOVE
      FORM"
130  PRINT
140  FOR I = 1 TO N
150  PRINT "ENTER STOICH. COEFF.
      "; I; "-TH COMPOUND"
160  INPUT " "; C(I)
170  NEXT I
180  PRINT : PRINT
190  FOR I = 1 TO N
200  PRINT "ENTER SEQUENTIAL NUMBER
      OF "; I; "-TH COMPOUND"
210  INPUT " "; NO(I)
220  NEXT I
230  PRINT : PRINT
240  INPUT "ENTER PRESSURE (PA) "
      ; P
250  PRINT : PRINT
260  INPUT "ENTER TEMPERATURE (K)
      "; T
270  PRINT : PRINT
280  FOR I = 1 TO N
290  IS = NO(I)
300  GOSUB 450
310  G(I) = ((((A(12) * T + A(11))
      * T + A(10)) * T + A(9)) *
      T + A(8)) * T + A(7)
320  S0 = A(13) + A(8) * LOG (T) +
      (((5 / 4 * A(12)) * T + 4 / 3
      * A(11)) * T + 3 / 2 * A(10
      )) * T + 2 * A(9)) * T - R *
      LOG (P / PU)
330  G(I) = G(I) - T * S0
340  NEXT I
350  DG = 0
360  FOR I = 1 TO N
370  DG = DG + C(I) * G(I)
380  NEXT I
390  K = DG / ( - R * T)
400  K = EXP (K)
410  PRINT : PRINT
420  PRINT "THE EQUILIBRIUM CONSTANT
      OF THE"
430  PRINT "REACTION AT "; T; " K I
```

```
     S "; K
440  PRINT : PRINT
445  GOTO 560
450  REM :READING OF SPECIFIED
      RECORD OF PHDAT FILE INTO ARRAY
      A(15)
460  REM :THE SEQUENTIAL NUMBER OF
      THE COMPOUND IN THE CALLING
      PROGRAM HAS TO BE "IS"
470  REM :ARRAY A HAS TO BE DECLA
      RED IN THE CALLING PROGRAM
480  D$ = CHR$ (4)
490  PRINT D$; "OPEN PHDAT, D1"
500  PRINT D$; "POSITION PHDAT, R";
      IS - 1
510  PRINT D$; "READ PHDAT"
520  INPUT A(1), A(2), A(3), A(4), A(5),
      A(6), A(7), A(8), A(9), A(10),
      A(11), A(12), A(13), A(14), A(15)
530  PRINT D$; "CLOSE PHDAT"
540  PRINT D$; "IN#0"
550  RETURN
560  REM :CALCULATION OF THE EQUI
      LIBRIUM CONCENTRATIONS FROM
      THE EQUILIBRIUM CONSTANT. THE
      INITIAL MOLE NUMBERERS HAVE TO
      BE KNOWN
570  FOR I = 1 TO N
580  PRINT "ENTER INITIAL MOLE NUMBER
      OF "; I; "-TH COMPOUND"
590  INPUT " "; N0(I)
600  NEXT I
610  PRINT : PRINT
620  A = 1
630  FOR I = 1 TO N
640  IF C(I) < 0 THEN A = A * N0(
      I)
650  NEXT I
660  B = 1
670  FOR I = 1 TO N
680  IF C(I) > 0 THEN B = B * N0(
      I)
690  NEXT I
700  IF A = 0 AND B = 0 THEN PRINT
      "IMPOSSIBLE TO MODIFY INITIAL
      CONCENTRATION, GIVE NEW DATA
      ": GOTO 710
710  MA = 1000
720  IS A = 0 THEN GOTO 770
```

```
 730  FOR I = 1 TO N
 740  IF C(I) < 0 THEN X = - N0(I
       ) / (C(I)
 750  IF X < (MA) THEN MA = X: IA =
       I
 760  NEXT I
 770  MB = - 1000
 780  IF B = 0 THEN GOTO 830
 790  FOR I = 1 TO N
 800  IF C(I) > 0 THEN X = - N0(I
       ) / C(I)
 810  IF X > MB THEN MB = X: IB = I
 820  NEXT I
 830  IF MA < ABS (MB) THEN XM =
       MA
 840  IF MA > ABS (MB) THEM XM =
       MA
 850  G = 100000
 860  ST = XM / 41
 870  RX = 1E30
 880  SC = 0
 890  FOR I = 1 TO N
 900  SC = SC + C(I)
 910  NEXT I
 920  FOR XE = ST TO 40 * ST STEP
       ST
 930  SN = 0
 940  FOR I = 1 TO N
 950  N(I) = N0(I) + C(I) * XE
 960  SN = SN + N(I)
 970  NEXT I
 980  PX = 1
 990  FOR I = 1 TO N
1000  X(I) = N(I) / SN
1010  PX = PX * X(I) ∧ C(I)
1020  NEXT I
1040  PX = PX * (P / PU) ∧ SC
1050  IF ABS (PX - K) < RX THEN
       XR = XE:RX = ABS (PX - K)
1060  NEXT XE
1070  FOR I = 1 TO N
1080  NF(I) = N0(I) + C(I) * XR
1090  NEXT I
1100  PRINT : PRINT
1110  PRINT "CONVERSION IN EQUILI
       BRIUM: "; XR
1020  PRINT : PRINT
1030  PRINT : PRINT
1140  PRINT "THE MOLE NUMBERS IN
```

```
         THE EQUILIBRIUM ARE:"
1150  PRINT : PRINT
1160  PRINT "COMPOUND NO.
         NUMBER OF MOLES"
1170  PRINT "-------------------
         -------------------"
1180  PRINT
1190  FOR I = 1 TO N
1200  PRINT TAB(6); I; TAB(23);
         NF(I)
1210  NEXT I
1220  PRINT : PRINT
1230  PRINT : PRINT
1240  INPUT "NEXT INITIAL MOLE NUMBERS
         ? (Y / N) "; A$
1250  IF A$ = "Y" THEN GOTO 560
1260  INPUT "NEXT TEMPERATURE ? (
         Y / N) "; A$
1270  IF A$ = "Y" THEN GOTO 250
1280  INPUT "NEXT PRESSURE ? (Y / N
         ) "; A$
1290  IF A$ = "Y" THEN GOTO 230
1300  END
```

Example. Run the program for the following reaction:

$$CH_4 + H_2O \rightleftarrows CO + 3H_2$$

The temperature is 600 K, the pressure is 101325 Pa, and the initial concentrations are

$$N^\circ_{H_2O} = 10, \qquad N^\circ_{CH_4} = 1 \qquad N^\circ_{H_2} = N^\circ_{CO} = 0$$

```
]RUN

ENTER THE NUMBER OF COMPOUNDS
IN THE REACTION 4

ORDER THE REACTION EQUATION TO 0
THE STOICHIOMETRIC COEFFICIENTS ARE
CONSIDERED OF ABOVE FORM

ENTER STOICH. COEFF OF 1-TH COMPOUND 1
ENTER STOICH. COEFF OF 2-TH COMPOUND 1
ENTER STOICH. COEFF OF 3-TH COMPOUND -1
ENTER STOICH. COEFF OF 4-TH COMPOUND -3

ENTER SEQUENTIAL NUMBER OF 1-TH COMPOUND 1
ENTER SEQUENTIAL NUMBER OF 2-TH COMPOUND 15
```

ENTER SEQUENTIAL NUMBER OF 3-TH COMPOUND 18
ENTER SEQUENTIAL NUMBER OF 4-TH COMPOUND 17

ENTER PRESSURE (PA) 101325

ENTER TEMPERATURE (K) 600

THE EQUILIBRIUM CONSTANT OF THE
REACTION AT 600 K IS 1914462.22

ENTER INITIAL MOLE NUMBER OF 1-TH COMPOUND 1
ENTER INITIAL MOLE NUMBER OF 2-TH COMPOUND 10
ENTER INITIAL MOLE NUMBER OF 3-TH COMPOUND 0
ENTER INITIAL MOLE NUMBER OF 4-TH COMPOUND 0

CONVERSION IN EQUILIBRIUM: −.0731707317

THE MOLE NUMBERS IN THE EQUILIBRIUM ARE:

COMPOUND NO.	NUMBER OF MOLES
1	.926829268
2	9.92682927
3	.0731707317
4	.219512195

NEXT INITIAL MOLE NUMBERS ? (Y / N) Y
ENTER INITIAL MOLE NUMBER OF 1-TH COMPOUND 5
ENTER INITIAL MOLE NUMBER OF 2-TH COMPOUND 10
ENTER INITIAL MOLE NUMBER OF 3-TH COMPOUND 0
ENTER INITIAL MOLE NUMBER OF 4-TH COMPOUND 0

CONVERSION IN EQUILIBRIUM: −.12195122

THE MOLE NUMBERS IN THE EQUILIBRIUM ARE:

COMPOUND NO.	NUMBER OF MOLES
1	4.87804878
2	9.87804878
3	.12195122
4	.365853659

NEXT INITIAL MOLE NUMBERS ? (Y / N) N
NEXT TEMPERATURE ? (Y / N) Y

ENTER TEMPERATURE (K) 1000

THE EQUILIBRIUM CONSTANT OF THE
REACTION AT 1000 K IS .0367460493

ENTER INITIAL MOLE NUMBER OF 1-TH COMPOUND 1
ENTER INITIAL MOLE NUMBER OF 2-TH COMPOUND 10

```
ENTER INITIAL MOLE NUMBER OF 3-TH COMPOUND 0
ENTER INITIAL MOLE NUMBER OF 4-TH COMPOUND 0

CONVERSION IN EQUILIBRIUM: -.951219513

THE MOLE NUMBERS IN THE EQUILIBRIUM ARE:
```

COMPOUND NO.	NUMBER OF MOLES
1	.0487804872
2	9.04878049
3	.951219513
4	2.85365854

```
NEXT INITIAL MOLE NUMBERS ? (Y / N) N
NEXT TEMPERATURE ? (Y / N) N
NEXT PRESSURE ? (Y / N) Y

ENTER PRESSURE (PA) 1013250

ENTER TEMPERATURE (K) 600

THE EQUILIBRIUM CONSTANT OF THE
REACTION AT 600 K IS 191446221

ENTER INITIAL MOLE NUMBER OF 1-TH COMPOUND 1
ENTER INITIAL MOLE NUMBER OF 2-TH COMPOUND 10
ENTER INITIAL MOLE NUMBER OF 3-TH COMPOUND 0
ENTER INITIAL MOLE NUMBER OF 4-TH COMPOUND 0
CONVERSION IN EQUILIBRIUM: -.0243902439
THE MOLE NUMBERS IN THE EQUILIBRIUM ARE:
```

COMPOUND NO.	NUMBER OF MOLES
1	.975609756
2	9.97560976
3	.0243902439
4	.0731707317

```
NEXT INITIAL MOLE NUMBERS ? (Y / N) N
NEXT TEMPERATURE ? (Y / N) N
NEXT PRESSURE ? (Y / N) N
```

Remark. Explain the results.

Calculation of Complex Chemical Equilibrium

If, in any k-component mixture, the number of independent reactions is p, the criterion for the complex chemical equilibrium can be given by a generalization of the criterion derived for a single reaction:

$$A_i = \frac{1}{\left(\sum N_j\right)^{\sum \nu_{ji}}} \prod_j \left(N_j^\circ + \nu_{ji} X_i\right)^{\nu_{ji}} \qquad j = 1, 2, \ldots, k \quad i = 1, 2, \ldots p$$

This constitutes a system of p nonlinear equations in the variables $X_1, X_2, \ldots, X_p$.

The solution of the above system of equations can be attained by the mathematical simulation of the following "chemical" experiment. Let the mixture of N_j° initial composition be fed to a reactor in which, due to its selective catalyst packing, only the reaction $i = 1$ takes place; moreover, equilibrium is reached under the given conditions. The equilibrium mixture, as regards the first reaction, will then be fed into the second reactor, in which only the reaction $i = 2$ occurs, also up to equilibrium under the given conditions. This change, however, upsets the equilibrium from the point of view of the first reaction.

Let us proceed in a similar manner for each reaction in turn. The mixture leaving the last reactor will be at equilibrium only with regard to the $i = p$ reaction; nevertheless, the reaction mixture as a whole has got nearer to the equilibrium in the state space than it was in the initial state. Consider the final state reached in the cycle as an initial state for a subsequent cycle; let the reaction mixture be returned to the first reactor for a second cycle. After this, the reaction mixture will again be closer to the final equilibrium state. After a sufficient number of cycles, no change will be observed from the final state of the previous cycle. This means that the reaction mixture is at equilibrium for all the reactions. The essence of this "chemical" experiment rests in the recycling, the mathematical simulation of which leads to the solution of the system of equations.

Mathematically, the following computational method is applied. Instead of a simultaneous solution, the equations are solved sequentially and the result of a sequential calculation is taken as the initialization of the subsequent cycle. This is simply direct iteration.

Algorithm and Program: Calculation of Equilibrium Concentration from Equilibrium Constants for Multireaction Systems

Write a program for the calculation of equilibrium constants, and from these constants calculate the equilibrium concentration for a multireaction system.

Algorithm.

1. Read the number of reactions.
2. Read the numbers of components in the reactions.
3. Read the stoichiometric coefficients of the reactions.
4. Read the sequential numbers of the components in the reactions.
5. Read the pressure and temperature.

6. Calculate the equilibrium constants of the reactions (see program EQUILIBRIUMCONSTANT).
7. Read the initial mole numbers.
8. Calculate the equilibrium concentration from the initial mole numbers, supposing only one reaction in the system (program EQUILIBRIUM-CONCENTRATIONFROMK).
9. Calculate the equilibrium concentration from the result of the previous step as initial concentration, supposing only the second reaction in the system.
10. Perform step 9 for the remaining reactions.
11. Continue steps 9–10 until the change in mole numbers is less than 3×10^{-2}.
12. Print the results.

Program MULTIREACTIONEQUILIBRIUM. A possible realization:

```
]LOAD MULTIREACTIONEQUILIBRIUM
]LIST

 10  REM :CALCULATION OF EQUILIBRIUM
      CONCENTRATION FROM EQUILIBRIUM
      CONSTANT
 20  PRINT : PRINT
 30  INPUT "ENTER THE NUMBER OF
      REACTIONS "; IR
 40  PRINT : PRINT
 50  DIM NN(IR)
 60  NM = 0
 70  FOR I = 1 TO IR
 80  PRINT "ENTER THE NUMBER OF
      COMPOUNDS"
 90  PRINT "IN THE "; I; "-TH REACTION ";
100  INPUT NN(I)
110  IF NN(I) > NM THEN NM = NN(I)
120  PRINT
130  NEXT I
140  DIM C(NM,IR)
150  R = 8.3144
160  PU = 101325
170  DIM K(IR)
180  PRINT "ORDER THE REACTION
      EQUATION TO 0 "
190  PRINT "THE STOICHIOMETRIC COEF
      FICIENTS ARE"
200  PRINT "CONSIDERED OF ABOVE
      FORM"
210  PRINT
```

```
220  FOR J = 1 TO IR
230  PRINT : PRINT
240  PRINT "STOICHIOMETRIC COEFFI
      CIENTS OF "; J;"-TH REACTION"
250  FOR I = 1 TO NN(J)
260  PRINT "ENTER STOICH. COEFF.
      "; I; "-TH COMPOUND"
270  INPUT " "; C(I,J)
280  NEXT I
290  NEXT J
300  DIM NO(NM,IR)
310  PRINT : PRINT
320  FOR J = 1 TO IR
330  PRINT : PRINT
340  PRINT "SEQUENTIAL NUMBERS IN
      "; J;"-TH REACTION"
350  FOR I = 1 TO NN(J)
360  PRINT "ENTER SEQUENTIAL NUMBER
      OF "; I; "-TH COMPOUND"
370  INPUT " "; NO(I,J)
380  NEXT I
390  NEXT J
400  NK(1) = NO(1,1)
410  NO(1,1) = 1
420  N = 1
430  FOR J = 1 TO IR
440  FOR I = 1 TO NN(J)
450  FOR II = 1 TO N
460  IF NO(I,J) < > NK(II) AND I
      I = N THEN GOTO 500
470  IF NO(I,J) = NK(II) THEN NO(
      I,J) = II: GOTO 530
480  NEXT II
490  GOTO 530
500  N = N + 1
510  NK(N) = NO(I,J)
520  NO(I,J) = N
530  NEXT I
540  NEXT J
550  PRINT : PRINT
560  DIM AA(15,N),G(N),NF(N),MU(N),
      X(N),N(N),N0(N),A(15)
570  FOR I = 1 TO N
580  IS = NK(I)
590  GOSUB 900
600  FOR J = 1 TO 15
610  AA(J,I) = A(J)
620  NEXT J
630  NEXT I
```

```
640  INPUT "ENTER PRESSURE (PA) "
      ;P
650  PRINT : PRINT
660  INPUT "ENTER TEMPERATURE (K)
      "; T
670  PRINT : PRINT
680  FOR I = 1 TO N
690  G(I) = ((((AA(12,I) * T + AA(
      11,I)) * T + AA(10,I)) * T +
     AA(9,I)) * T + AA(8,I)) * T +
     AA(7,I)
700  S0 = AA(13,I) + AA(8,I) * LOG
      (T) + (((5 / 4 * AA(12,I) *
      T + 4 / 3 * AA(11,I)) * T +
      3 / 2 * AA(10,I)) * T + 2 *
     AA(9,I)) * T - R * LOG (P /
      PU)
710  G(I) = G(I) - T * S0
720  NEXT I
730  FOR L = 1 TO IR
740  DG = 0
750  FOR I = 1 TO NN(L)
760  LL = NO(I,L)
770  DG = DG + C(I,L) * G(LL)
780  NEXT I
790  K(L) = DG / ( - R * T)
800  K(L) = EXP (K(L))
810  NEXT L
820  PRINT : PRINT
830  PRINT "THE EQUILIBRIUM CONST
      ANTS OF THE"
840  PRINT "REACTIONS AT "; T;" K
      ARE"
850  FOR I = 1 TO IR
860  PRINT " "; K(I)
870  NEXT I
880  PRINT : PRINT
890  GOTO 1010
900  REM :READING OF SPECIFIED
      RECORD OF PHDAT FILE INTO ARRAY
      A(15)
910  REM :THE SEQUENTIAL NUMBER
      OF THE COMPOUND IN THE CALLING
      PROGRAM HAS TO BE "IS"
920  REM :ARRAY A HAS TO BE DECLA
      RED IN THE CALLING PROGRAM
930  D$ = CHR$ (4)
940  PRINT D$; "OPEN PHDAT, D1"
950  PRINT D$; "POSITION PHDAT, R";
      IS - 1
```

```
 960  PRINT D$; "READ PHDAT"
 970  INPUT A(1), A(2), A(3), A(4), A(5),
       A(6), A(7), A(8), A(9), A(10),
       A(11), A(12), A(13), A(14), A(15)
 980  PRINT D$; "CLOSE PHDAT"
 990  PRINT D$; "IN#0"
1000  RETURN
1010  REM :CALCULATION OF THE EQU
       ILIBRIUM CONCENTRATIONS FROM
       THE EQUILIBRIUM CONSTANT. THE
       INITIAL MOLE NUMBERS HAVE TO
       BE KNOWN
1020  FOR I = 1 TO N
1030  PRINT "ENTER INITIAL MOLE NUMBER
       OF "; NK(I); " COMPOUND"
1040  INPUT " "; N0(I)
1050  NEXT I
1060  PRINT : PRINT
1070  S3 = 0
1080  S1 = 0
1090  S2 = S1
1100  FOR J = 1 TO IR
1110  A = 1
1120  FOR I = 1 TO NN(J)
1130  IF C(I,J) < 0 THEN A = A *
       N0(NO(I,J))
1140  NEXT I
1150  B = 1
1160  FOR I = 1 TO NN(J)
1170  IF C(I,J) > 0 THEN B = B *
       N0(NO(I,J))
1180  NEXT I
1190  IF A = 0 AND B = 0 THEN S3 =
       S3 + 1: GOTO 1630
1200  NEXT J
1210  S2 = S1
1220  FOR J = 1 TO IR
1230  MA = 1000
1240  IF A = 0 THEN GOTO 1290
1250  FOR I = 1 TO NN(J)
1260  IF C(I,J) < 0 THEN X = - N
       0(NO(I,J)) / C(I,J)
1270  IF X < (MA) THEN MA = X: IA = I
1280  NEXT I
1290  MB = - 1000
1300  IF B = 0 THEN GOTO 1350
1310  FOR I = 1 TO NN(J)
1320  IF C(I,J) ≤ 0 THEN X = - N
       0(NO(I,J)) / C(I,J)
1330  IF X > MB THEN MB = X: IB = I
```

```
1340  NEXT I
1350  IF MA <  ABS (MB) THEN XM  =
       MA
1360  IF MA >  ABS (MB) THEN XM  =
       MB
1370  G  =  100000
1380  ST  =  XM / 11
1390  RX  =  1E30
1400  SC  =  0
1410  FOR I  =  1 TO NN(J)
1420  SC  =  SC + C(I,J)
1430  NEXT I
1440  FOR XE  =  ST TO 10 * ST STEP
       ST
1450  SN  =  0
1460  FOR I  =  1 TO NN(J)
1470  N(NO(I,J))  =  N0(NO(I,J)) + C
       (I,J) * XE
1480  SN  =  SN + N(NO(I,J))
1490  NEXT I
1500  PX  =  1
1510  FOR I  =  1 TO NN(J)
1520  X(NO(I,J))  =  N(NO(I,J)) / SN
1530  PX  =  PX * X(NO(I,J)) ∧ C(I,J
       )
1540  NEXT I
1550  PX  =  PX * (P / PU) ∧ SC
1560  IF ABS (PX - K(J)) <  RX THEN
       XR  =  XE:RX  =  ABS (PX - K(J)
       )
1570  NEXT XE
1580  FOR I  =  1 TO NN(J)
1590  LL  =  NO(I,J)
1600  NF(LL)  =  N0(LL) + C(I,J) * X
       R
1610  N0(LL)  =  NF(LL)
1620  NEXT I
1630  IF S3  =  IR THEN PRINT "IMP
       OSSIBLE TO MODIFY INITIAL CO
       NCENTRATION, GIVE NEW DATA":
       GOTO 1020
1640  NEXT J
1650  S1  =  0
1660  FOR I  =  1 TO N
1670  S1  =  S1 + NF(I)
1680  NEXT I
1690  PRINT S1,NF(4)
1700  IF ABS (S1 - S2) >  5E - 2 THEN
       GOTO 1090
1710  PRINT : PRINT
```

```
1720 PRINT : PRINT
1730 PRINT "THE MOLE NUMBERS IN T
      HE EQUILIBRIUM ARE:"
1740 PRINT : PRINT
1750 PRINT "COMPOUND NO.
      NUMBER OF MOLES"
1760 PRINT "= = = = = = = = = = = = = = = = = = = =
      = = = = = = = = = = = = = = = = = = = = "
1770 PRINT
1780 FOR I = 1 TO N
1790 PRINT TAB(6);NK(I); TAB(
      23);NF(I)
1800 NEXT I
1810 PRINT : PRINT
1820 PRINT : PRINT
1830 INPUT "NEXT INITIAL MOLE NUMB
      ERS ? (Y / N) "; A$
1840 IF A$ = "Y" THEN GOTO 1020
1850 INPUT "NEXT TEMPERATURE ? (
      Y / N) "; A$
1860 IF A$ = "Y" THEN GOTO 650
1870 INPUT "NEXT PRESSURE ? (Y / N
      ) "; A$
1880 IF A$ = "Y" THEN GOTO 640
1890 END
]PR#0
```

Example. Run the program for the following system:

$$CH_4 + H_2O \rightleftarrows CO + 3H_2$$

$$CO + H_2O \rightleftarrows CO_2 + H_2$$

The temperature is 600 K, the pressure 101,325 Pa, and the initial mole numbers are

$$N^{\circ}_{CH_4} = 1 \quad N^{\circ}_{H_2O} = 10 \quad N^{\circ}_{CO} = 0.1 \quad N^{\circ}_{H_2} = N^{\circ}_{CO_2} = 0$$

Continue with initial mole numbers

$$N^{\circ}_{CH_4} = 1 \quad N^{\circ}_{H_2O} = 1 \quad N^{\circ}_{CO} = 0.1 \quad N^{\circ}_{H_2} = N^{\circ}_{CO_2} = 0$$

```
]RUN

ENTER THE NUMBER OF REACTIONS 2

ENTER THE NUMBER OF COMPOUNDS
IN THE 1-TH REACTION ?4
```

```
ENTER THE NUMBER OF COMPOUNDS
IN THE 2-TH REACTION ?4

ORDER THE REACTION EQUATION TO 0
THE STOICHIOMETRIC COEFFICIENTS ARE
CONSIDERED OF ABOVE FORM

STOICHIOMETRIC COEFFICIENTS OF 1-TH REACTION
ENTER -1 STOICH. COEFF. OF 1-TH COMPOUND
ENTER -1 STOICH. COEFF. OF 2-TH COMPOUND
ENTER 1 STOICH. COEFF. OF 3-TH COMPOUND
ENTER 3 STOICH. COEFF. OF 4-TH COMPOUND

STOICHIOMETRIC COEFFICIENTS OF 2-TH REACTION
ENTER -1 STOICH. COEFF. OF 1 - TH COMPOUND
ENTER -1 STOICH. COEFF. OF 2-TH COMPOUND
ENTER 1 STOICH. COEFF. OF 3-TH COMPOUND
ENTER 1 STOICH. COEFF. OF 4-TH COMPOUND

SEQUENTIAL NUMBERS IN 1-TH REACTION
ENTER SEQUENTIAL NUMBER OF 1-TH COMPOUND 1
ENTER SEQUENTIAL NUMBER OF 2-TH COMPOUND 15
ENTER SEQUENTIAL NUMBER OF 3-TH COMPOUND 18
ENTER SEQUENTIAL NUMBER OF 4-TH COMPOUND 17

SEQUENTIAL NUMBERS IN 2-TH REACTION
ENTER SEQUENTIAL NUMBER OF 1-TH COMPOUND 18
ENTER SEQUENTIAL NUMBER OF 2-TH COMPOUND 15
ENTER SEQUENTIAL NUMBER OF 3-TH COMPOUND 19
ENTER SEQUENTIAL NUMBER OF 4-TH COMPOUND 17

ENTER PRESSURE (PA) 101325

ENTER TEMPERATURE (K) 600

THE EQUILIBRIUM CONSTANTS OF THE
REACTIONS AT 600 K ARE
 5.22339897E - 07
 28.1199566

ENTER INITIAL MOLE NUMBER OF 1 COMPOUND 1
ENTER INITIAL MOLE NUMBER OF 15 COMPOUND 10
ENTER INITIAL MOLE NUMBER OF 18 COMPOUND 1
ENTER INITIAL MOLE NUMBER OF 17 COMPOUND 0
ENTER INITIAL MOLE NUMBER OF 19 COMPOUND 0

THE MOLE NUMBERS IN THE EQUILIBRIUM ARE:
```

COMPOUND NO.	NUMBER OF MOLES
1	.239392049
15	7.51364771
18	.0348636143
17	4.00756819
19	1.72574434

```
NEXT INITIAL MOLE NUMBERS ? (Y / N) Y
ENTER INITIAL MOLE NUMBER OF 1 COMPOUND 1
ENTER INITIAL MOLE NUMBER OF 15 COMPOUND 1
ENTER INITIAL MOLE NUMBER OF 18 COMPOUND 0.1
ENTER INITIAL MOLE NUMBER OF 17 COMPOUND 0
ENTER INITIAL MOLE NUMBER OF 19 COMPOUND 0

THE MOLE NUMBERS IN THE EQUILIBRIUM ARE:
```

COMPOUND NO.	NUMBER OF MOLES
1	.891735537
15	.689782119
18	6.31104433E − 03
17	.526746807
19	.201953418

```
NEXT INITIAL MOLE NUMBERS ? (Y / N) N
NEXT TEMPERATURE ? (Y / N) N
NEXT PRESSURE ? (Y / N) N
```

Calculation of Chemical Equilibrium by Minimizing the Free Enthalpy

In calculating the equilibrium composition we have so far assumed knowledge of the constant of equilibrium. Now, we go over to the case based on the minimization of the free enthalpy of the mixture. One of the possible techniques for this is the calculation of a constrained minimum.

In a homogeneous phase, for the free enthalpy of any k-component mixture we have

$$G[\mathbf{N}] = \sum_j N_j \left(g_j^\circ + \mathbf{R}T \ln \gamma_j y_j + \mathbf{R}T \ln \frac{P}{P_u} \right)$$

where $\mathbf{N} = [N_1, N_2, \ldots, N_k]$ = vector of the amounts of components,
g_j° = standard free enthalpy of pure component j at temperature T and pressure P_u
P = total pressure of the mixture.

The determination of the equilibrium amounts to finding the vector $\mathbf{N}$ made up of nonnegative elements which minimizes the function $G[\mathbf{N}]$ so that the following stoichiometric constraints are also satisfied:

$$\sum_j a_{ij} N_j = b_i$$

where a_{ij} = stoichiometric index of constituent i in component j,
e = number of constituents,
b_i = quantity of constituent i in the closed system.

The actual state of an isothermal isobaric reaction mixture is represented by a point in a space with $k+1$ dimensions, and every change is described by displacement of this point. However, this point cannot move without restriction in the space; the constraints $N_j \geq 0$ and $\Sigma a_{ij} N_j = b_i$ determine the possible states, that is, the state surface, on which we are to find the minimum.

Thus, a constrained extremum problem must be solved; it can be handled with the Lagrange-multiplier method. The zero-arranged form of the constraint,

$$\sum_j a_{ij} N_j - b_i = 0 \qquad i = 1, 2, \ldots, e \qquad j = 1, 2, \ldots, k$$

holds for each constituent. Each such expression is multiplied by its own factor λ_i and added to the function G to be minimized. Thus we will have the following Lagrange function:

$$\mathscr{L} = G + \sum_i \lambda_i \left(\sum_j a_{ij} N_j - b_i \right)$$

Now we have only to look for the unconstrained extremum of this transformed function.

The extremum is expected to be at that point where the partial derivatives of the Lagrange function with respect to all the variables equal zero. The variables are of two types, namely the amounts of components (N_j) and the multipliers λ_i. Hence, the partial derivatives are

$$\left(\frac{\partial \mathscr{L}}{\partial N_j} \right)_{T, P, {}_{\llcorner}N_j} = 0$$

$$\left(\frac{\partial \mathscr{L}}{\partial \lambda_i} \right)_{T, P, N_j} = 0$$

where ${}_{\llcorner}N_j$ means N_i for $i \neq j$. The second set of equations just give back the constraints. The first set are as follows:

$$\left(\frac{\partial \mathscr{L}}{\partial N_j} \right)_{T, P, {}_{\llcorner}N_j} = \left(\frac{\partial G}{\partial N_j} \right)_{T, P, {}_{\llcorner}N_j} + \sum_i \lambda_i a_{ij} = 0$$

The first term in the middle represents, by definition, the chemical potential:

$$\left(\frac{\partial G}{\partial N_j} \right)_{T, P, N_j} = \mu_j$$

At the constrained extremum (i.e. at equilibrium), the following system of equations is therefore valid:

$$\mu_j + \sum_i \lambda_i a_{ij} = 0$$

For homogeneous reactions, the following substitution can be carried out:

$$\mu_j = \Delta g_j^\circ + \mathbf{R}T \ln \frac{\bar{f}_j}{P_u}$$

where Δg_j° = standard free enthalpy of formation of component j,
$\bar{f}_j$ = fugacity of component j in the mixture.

The latter can be expressed in accordance with the foregoing as

$$\bar{f}_j = y_j \bar{\phi}_j P$$

At last, the system of nonlinear equations to be solved becomes

$$\Delta g_j^\circ + \mathbf{R}T \ln(y_j \phi_j P) + \sum_i \lambda_i a_{ij} = 0 \qquad j = 1, 2, \ldots, k$$

$$\sum_j a_{ij} N_j = b_i \qquad i = 1, 2, \ldots, e$$

The system consists of $k + e$ equations in the variables N_j and λ_i. The variables $y_1, y_2, \ldots, y_k$ can be computed by means of their defining equation:

$$y_j = \frac{N_j}{\sum N_j}$$

If, at the given temperature and pressure, the reaction system can form two (or three) phases, then the situation becomes more complicated. As shown at the end of the previous chapter, the free enthalpy of a multiphase, multicomponent system is

$$G = \sum_j \sum_\iota N_j^{(\iota)} \mu_j^{(\iota)} \qquad \iota = 1, 2, 3 \quad j = 1, 2, \ldots, k$$

This has a minimum in the equilibrium state of the system. It is a minimum that fulfills the following conservation-type restriction:

$$\sum_j \sum_\iota a_{ij} N_j^{(\iota)} = b_i$$

where a_{ij} = stoichiometric index of constituent i in component j,
$N_j^{(\iota)}$ = amount of component j in phase ι,
b_i = total amount of constituent i in the closed system.

At chemical equilibrium, one type of constraint on the minimization of the free enthalpy is the conservation of the amount of each constituent involved in a closed, reactive system. Moreover, the nonnegativity constraint for the variables N_j holds also in the case of chemical equilibrium:

$$N_j^{(\iota)} \geq 0$$

The calculation of the equilibrium of such multicomponent, multiphase reaction systems requires a nonlinear optimum search as well.

APPENDIX: DATAFILE

The elements of a record are:

(1) Molecular mass, kg/mole.
(2) Critical pressure, Pa.
(3) Critical temperature, K.
(4) Acentric factor.
(5) Normal boiling point, K.
(6) Enthalpy of vaporization at normal boiling point, J/mole.
(7)–(12) Six constants of the ideal-gas enthalpy polynomial, J/mole, J/mole K, J/mole K^2, J/mole K^3, J/mole K^4, J/mole K^5.
(13) Constant I of the ideal-gas entropy equation, J/mole K.
(14) Critical molar volume, m^3/mole.

A Substance Index at the back of the book gives an alphabetical listing of all substances in this appendix.

1 **Methane**

(1)	1.60430E − 02	(2)	4.60016E + 06	(3)	190.560	(4)	4.06115E − 03
(5)	111.660	(6)	8185.19	(7)	− 67142.5	(8)	37.9158
(9)	− 3.41914E − 02	(10)	9.07813E − 05	(11)	− 5.97243E − 08	(12)	1.38036E − 11
(13)	− 19.6134	(14)	9.85000E − 05				

2 **Ethane**

(1)	3.00700E − 02	(2)	4.89400E + 06	(3)	305.460	(4)	9.40157E − 02
(5)	184.960	(6)	14725.0	(7)	− 69206.5	(8)	34.3598
(9)	− 9.72846E − 03	(10)	1.27521E − 04	(11)	− 1.01987E − 07	(12)	2.65088E − 11
(13)	25.9113	(14)	1.46300E − 04				

3 **Propane**

(1)	4.40960E − 02	(2)	4.24552E + 06	(3)	369.860	(4)	.148716
(5)	231.150	(6)	18786.2	(7)	− 81692.6	(8)	33.1620
(9)	2.20716E − 02	(10)	1.50048E − 04	(11)	− 1.34233E − 07	(12)	3.66751E − 11
(13)	52.4052	(14)	2.03000E − 04				

4 ***n*-Butane**

(1)	5.81240E − 02	(2)	3.79970E + 06	(3)	425.200	(4)	.196726
(5)	272.660	(6)	22407.8	(7)	− 95167.7	(8)	.496135
(9)	.190375	(10)	− 6.44646E − 05	(11)	1.02617E − 08	(12)	− 3.71725E − 13
(13)	202.241	(14)	2.55007E − 04				

5 ***n*-Pentane**

(1)	7.21509E − 02	(2)	3.37410E + 06	(3)	469.800	(4)	.248393
(5)	309.230	(6)	25790.7	(7)	− 109449.	(8)	− .843798
(9)	.239141	(10)	− 8.44019E − 05	(11)	1.43958E − 08	(12)	− 6.24804E − 13
(13)	222.130	(14)	3.11020E − 04				

6 ***n*-Hexane**

(1)	8.61779E − 02	(2)	3.02960E + 06	(3)	507.900	(4)	.296379
(5)	341.900	(6)	28872.2	(7)	− 122996.	(8)	− 8.32813
(9)	.299432	(10)	− 1.13795E − 04	(11)	2.17377E − 08	(12)	− 1.15993E − 12
(13)	272.440	(14)	3.68011E − 04				

7 ***n*-Heptane**

(1)	.100206	(2)	2.73580E + 06	(3)	540.200	(4)	.348687
(5)	371.590	(6)	31715.0	(7)	− 137089.	(8)	− 9.70349
(9)	.347907	(10)	− 1.33232E − 04	(11)	2.56146E − 08	(12)	− 1.37942E − 12
(13)	293.272	(14)	4.26012E − 04				

8 ***n*-Octane**

(1)	.114233	(2)	2.49260E + 06	(3)	569.400	(4)	.394001
(5)	398.830	(6)	34390.4	(7)	− 151646.	(8)	− 10.7076
(9)	.395522	(10)	− 1.51863E − 04	(11)	2.92008E − 08	(12)	− 1.57327E − 12
(13)	312.017	(14)	4.90022E − 04				

9 *n*-**Nonane**

(1)	.128260	(2)	2.27980E + 06	(3)	595.000	(4)	.440465
(5)	423.950	(6)	36927.6	(7)	− 166173.	(8)	− 11.6210
(9)	.442929	(10)	− 1.70356E − 04	(11)	3.27551E − 08	(12)	− 1.76437E − 12
(13)	330.347	(14)	5.43032E − 04				

10 *n*-**Decane**

(1)	.142287	(2)	2.10760E + 06	(3)	619.000	(4)	.479366
(5)	447.280	(6)	39314.1	(7)	− 180487.	(8)	− 14.1916
(9)	.494205	(10)	− 1.92484E − 04	(11)	3.76217E − 08	(12)	− 2.06708E − 12
(13)	356.250	(14)	6.02009E − 04				

11 *n*-**Undecane**

(1)	.156314	(2)	1.94540E + 06	(3)	640.000	(4)	.521348
(5)	469.040	(6)	41533.1	(7)	− 194950.	(8)	− 15.5946
(9)	.542465	(10)	− 2.11572E − 04	(11)	4.13542E − 08	(12)	− 2.27415E − 12
(13)	377.033	(14)	6.60051E − 04				

12 *n*-**Dodecane**

(1)	.170341	(2)	1.81370E + 06	(3)	659.000	(4)	.567165
(5)	489.440	(6)	43710.2	(7)	− 209971.	(8)	− 13.2014
(9)	.582317	(10)	− 2.22763E − 04	(11)	4.21463E − 08	(12)	− 2.21957E − 12
(13)	380.054	(14)	7.18031E − 04				

13 ***n*-Hexadecane**

(1)	.226449	(2)	1.41850E + 06	(3)	725.000	(4)	.581241
(5)	560.000	(6)	51497.6	(7)	− 267462.	(8)	− 21.6270
(9)	.782876	(10)	− 3.07058E − 04	(11)	6.01835E − 08	(12)	− 3.32117E − 12
(13)	475.399	(14)	9.50065E − 04				

14 **2-Methylpropane**

(1)	5.81240E − 02	(2)	3.64770E + 06	(3)	408.100	(4)	.180685
(5)	261.430	(6)	21306.6	(7)	− 103955.	(8)	8.91025
(9)	.153057	(10)	4.22436E − 06	(11)	− 4.22830E − 08	(12)	1.37523E − 11
(13)	153.678	(14)	2.63015E − 04				

15 **2-Methylbutane**

(1)	7.21509E − 02	(2)	3.33360E + 06	(3)	461.000	(4)	.215465
(5)	301.000	(6)	24459.3	(7)	− 115757.	(8)	− 9.51092
(9)	.255341	(10)	− 9.61347E − 05	(11)	1.81321E − 08	(12)	− 9.34432E − 13
(13)	258.043	(14)	3.08017E − 04				

16 **2,2-Dimethylpropane**

(1)	7.21509E − 02	(2)	3.20190E + 06	(3)	433.800	(4)	.194894
(5)	282.650	(6)	22767.8	(7)	− 128876.	(8)	1.31989
(9)	.220878	(10)	− 2.70417E − 05	(11)	− 3.82860E − 08	(12)	1.48477E − 11
(13)	172.323	(14)	3.03004E − 04				

17 **2-Methylpentane**

(1)	8.61779E − 02	(2)	3.02960E + 06	(3)	497.900	(4)	.275661
(5)	333.430	(6)	27812.9	(7)	− 124478.	(8)	− 52.1864
(9)	.423766	(10)	− 2.59846E − 04	(11)	9.17362E − 08	(12)	− 9.30477E − 12
(13)	457.650	(14)	3.67023E − 04				

18 **3-Methylpentane**

(1)	8.61779E − 02	(2)	3.12080E + 06	(3)	504.700	(4)	.268627
(5)	336.440	(6)	28097.6	(7)	− 126601.	(8)	− 14.6056
(9)	.316784	(10)	− 1.34186E − 04	(11)	3.04736E − 08	(12)	− 1.52601E − 12
(13)	291.705	(14)	3.67013E − 04				

19 **2,2-Dimethylbutane**

(1)	8.61779E − 02	(2)	3.11070E + 06	(3)	489.400	(4)	.229045
(5)	322.900	(6)	26322.4	(7)	− 136766.	(8)	− 43.0902
(9)	.389783	(10)	− 2.08017E − 04	(11)	6.14241E − 08	(12)	− 5.09082E − 12
(13)	398.271	(14)	3.59018E − 04				

20 **2,3-Dimethylbutane**

(1)	8.61779E − 02	(2)	3.13090E + 06	(3)	500.300	(4)	.243007
(5)	331.150	(6)	27293.7	(7)	− 134005.	(8)	2.23376
(9)	.243117	(10)	1.60497E − 06	(11)	− 7.79305E − 08	(12)	2.99328E − 11
(13)	210.627	(14)	3.58016E − 04				

21 **3-Methylhexane**

(1)	.100198	(2)	2.84723E + 06	(3)	535.500	(4)	.324532
(5)	365.000	(6)	30806.5	(7)	− 144911.	(8)	16.2204
(9)	.257092	(10)	1.99066E − 05	(11)	− 9.63185E − 08	(12)	3.54125E − 11
(13)	179.177	(14)	4.18000E − 04				

22 **Cyclopropane**

(1)	4.20800E − 02	(2)	5.53741E + 06	(3)	398.060	(4)	.125183
(5)	240.360	(6)	20096.6	(7)	76107.4	(8)	− 20.4069
(9)	.142041	(10)	− 2.06430E − 05	(11)	− 3.30536E − 08	(12)	1.57400E − 11
(13)	272.965	(14)	1.62800E − 04				

23 **Cyclopentane**

(1)	7.01399E − 02	(2)	4.50795E + 06	(3)	511.660	(4)	.192347
(5)	322.420	(6)	27314.7	(7)	− 35294.3	(8)	− 51.2280
(9)	.257741	(10)	− 7.51394E − 05	(11)	− 4.43299E − 09	(12)	5.77160E − 12
(13)	441.691	(14)	2.59800E − 04				

24 **Methylcyclopentane**

(1)	8.41630E − 02	(2)	3.78960E + 06	(3)	532.800	(4)	.228406
(5)	344.970	(6)	28955.9	(7)	− 58881.9	(8)	− 57.5776
(9)	.336923	(10)	− 1.41455E − 04	(11)	3.00983E − 08	(12)	− 1.83982E − 12
(13)	485.556	(14)	3.19011E − 04				

25 Ethylcyclopentane

(1)	9.81899E − 02	(2)	3.39440E + 06	(3)	569.500	(4)	.268073
(5)	376.630	(6)	32301.2	(7)	− 72428.4	(8)	− 62.6355
(9)	.390398	(10)	− 1.64012E − 04	(11)	3.51683E − 08	(12)	− 2.15551E − 12
(13)	525.002	(14)	3.75007E − 04				

26 *n*-Propylcyclopentane

(1)	.112217	(2)	2.99300E + 06	(3)	581.430	(4)	.434446
(5)	404.110	(6)	34130.8	(7)	− 87509.4	(8)	− 63.1832
(9)	.436901	(10)	− 1.81489E − 04	(11)	3.82888E − 08	(12)	− 2.30556E − 12
(13)	542.118	(14)	4.13648E − 04				

27 Cyclohexane

(1)	8.41600E − 02	(2)	4.07327E + 06	(3)	553.460	(4)	.210222
(5)	353.900	(6)	29977.5	(7)	− 74667.9	(8)	− 52.7681
(9)	.289848	(10)	− 4.41991E − 05	(11)	− 3.67868E − 08	(12)	1.40234E − 11
(13)	433.627	(14)	3.08300E − 04				

28 Methylcyclohexane

(1)	9.81899E − 02	(2)	3.47750E + 06	(3)	572.300	(4)	.233633
(5)	374.100	(6)	31149.8	(7)	− 99554.2	(8)	− 69.1828
(9)	.402661	(10)	− 1.50006E − 04	(11)	1.79976E − 08	(12)	2.61372E − 12
(13)	517.561	(14)	3.44009E − 04				

29 Ethylcyclohexane

(1)	.112217	(2)	3.09730E + 06	(3)	582.000	(4)	.454395
(5)	404.950	(6)	34331.8	(7)	− 114333.	(8)	− 39.8912
(9)	.361577	(10)	− 4.30668E − 05	(11)	− 6.07070E − 08	(12)	2.24559E − 11
(13)	402.510	(14)	3.97611E − 04				

30 *n*-Propylcyclohexane

(1)	.126244	(2)	2.77500E + 06	(3)	609.270	(4)	.478480
(5)	429.880	(6)	36090.2	(7)	− 124168.	(8)	− 89.5777
(9)	.557933	(10)	− 2.58420E − 04	(11)	6.09359E − 08	(12)	− 4.11202E − 12
(13)	631.347	(14)	4.61665E − 04				

31 Ethene (ethylene)

(1)	2.80520E − 02	(2)	5.14731E + 06	(3)	283.060	(4)	8.64535E − 02
(5)	169.460	(6)	13552.7	(7)	64404.1	(8)	2.43249
(9)	8.14324E − 02	(10)	− 3.14586E − 05	(11)	6.31949E − 09	(12)	− 3.60827E − 13
(13)	161.166	(14)	1.32600E − 04				

32 Propene (propylene)

(1)	4.20780E − 02	(2)	4.65082E + 06	(3)	365.060	(4)	.139813
(5)	225.460	(6)	18430.3	(7)	38014.3	(8)	7.89944
(9)	.102764	(10)	− 1.74579E − 05	(11)	− 8.59762E − 09	(12)	3.49276E − 12
(13)	163.490	(14)	1.81600E − 04				

33 **1-Butene**

(1)	5.61040E − 02	(2)	4.02260E + 06	(3)	419.360	(4)	.188228
(5)	266.860	(6)	21930.5	(7)	25036.5	(8)	− 4.34736
(9)	.180153	(10)	− 7.15400E − 05	(11)	1.46797E − 08	(12)	− 8.62095E − 13
(13)	232.332	(14)	2.40400E − 04				

34 **1-Pentene**

(1)	7.01350E − 02	(2)	4.05300E + 06	(3)	474.000	(4)	.207900
(5)	303.130	(6)	25212.9	(7)	9647.45	(8)	− 2.01725
(9)	.222412	(10)	− 8.63622E − 05	(11)	1.71796E − 08	(12)	− 9.73201E − 13
(13)	236.114	(14)	2.58651E − 04				

35 **1-Hexene**

(1)	8.41630E − 02	(2)	3.25250E + 06	(3)	516.660	(4)	.200198
(5)	336.650	(6)	28302.8	(7)	− 5257.09	(8)	− 1.50089
(9)	.266019	(10)	− 1.00648E − 04	(11)	1.90040E − 08	(12)	− 1.00006E − 12
(13)	247.842	(14)	3.62412E − 04				

36 **1-Heptene**

(1)	9.81899E − 02	(2)	3.34740E + 06	(3)	541.140	(4)	.363139
(5)	366.800	(6)	31107.9	(7)	− 19700.9	(8)	− 3.20749
(9)	.314991	(10)	− 1.20464E − 04	(11)	2.29929E − 08	(12)	− 1.22793E − 12
(13)	269.971	(14)	3.51349E − 04				

37 **1-Octene**

(1)	.112217	(2)	3.15970E + 06	(3)	571.000	(4)	.427203
(5)	394.440	(6)	33787.5	(7)	− 34032.6	(8)	− 6.05144
(9)	.366916	(10)	− 1.43346E − 04	(11)	2.82204E − 08	(12)	− 1.57010E − 12
(13)	297.165	(14)	3.85547E − 04				

38 **1-Nonene**

(1)	.126244	(2)	2.97090E + 06	(3)	598.600	(4)	.479695
(5)	420.030	(6)	36341.4	(7)	− 48435.8	(8)	− 8.12844
(9)	.417202	(10)	− 1.64690E − 04	(11)	3.28459E − 08	(12)	− 1.84943E − 12
(13)	321.209	(14)	4.23335E − 04				

39 **1-Decene**

(1)	.140271	(2)	2.67510E + 06	(3)	624.230	(4)	.502047
(5)	443.730	(6)	38686.0	(7)	− 63222.5	(8)	− 7.74212
(9)	.461599	(10)	− 1.80292E − 04	(11)	3.53053E − 08	(12)	− 1.94820E − 12
(13)	333.645	(14)	4.87366E − 04				

40 **1-Undecene**

(1)	.154298	(2)	2.43090E + 06	(3)	648.100	(4)	.519028
(5)	465.830	(6)	40905.0	(7)	− 77481.7	(8)	− 10.7644
(9)	.513897	(10)	− 2.03463E − 04	(11)	4.06540E − 08	(12)	− 2.30338E − 12
(13)	361.739	(14)	5.54395E − 04				

41 1-Dodecene

(1)	.168325	(2)	2.22920E + 06	(3)	670.440	(4)	.532148
(5)	486.520	(6)	42998.4	(7)	− 92409.0	(8)	− 9.21447
(9)	.555795	(10)	− 1.93989E − 04	(11)	4.22977E − 08	(12)	− 2.32852E − 12
(13)	368.729	(14)	6.23396E − 04				

42 1-Hexadecene

(1)	.224433	(2)	1.67190E + 06	(3)	558.000	(4)	.432923
(5)	505.940	(6)	50450.9	(7)	− 149881.	(8)	− 18.8614
(9)	.759199	(10)	− 3.03878E − 04	(11)	6.14332E − 08	(12)	− 3.52149E − 12
(13)	469.835	(14)	7.13439E − 04				

43 *cis*-2-Butene

(1)	5.61080E − 02	(2)	4.15430E + 06	(3)	428.000	(4)	.255122
(5)	276.880	(6)	23362.3	(7)	20275.6	(8)	− 10.0469
(9)	.170483	(10)	− 5.20497E − 05	(11)	1.84139E − 09	(12)	2.16595E − 12
(13)	263.596	(14)	2.36010E − 04				

44 *trans*-2-Butene

(1)	5.61080E − 02	(2)	4.15430E + 06	(3)	428.000	(4)	.219547
(5)	274.040	(6)	22772.0	(7)	12076.8	(8)	8.69393
(9)	.150068	(10)	− 4.26298E − 05	(11)	2.16972E − 09	(12)	1.09589E − 12
(13)	163.351	(14)	2.36010E − 04				

45 2-**Methylpropene**

(1)	5.61040E − 02	(2)	4.00234E + 06	(3)	417.860	(4)	.191885
(5)	266.260	(6)	22131.4	(7)	6055.08	(8)	7.74891
(9)	.159836	(10)	− 5.57752E − 05	(11)	9.55022E − 09	(12)	− 4.30838E − 13
(13)	161.520	(14)	2.38700E − 04				

46 *cis*-2-**Pentene**

(1)	7.01350E − 02	(2)	4.09350E + 06	(3)	475.560	(4)	.279095
(5)	310.100	(6)	26125.6	(7)	5913.91	(8)	− 17.5571
(9)	.239287	(10)	− 9.68497E − 05	(11)	2.10074E − 08	(12)	− 1.60075E − 12
(13)	316.085	(14)	2.58770E − 04				

47 *trans*-2-**Pentene**

(1)	7.01350E − 02	(2)	4.09350E + 06	(3)	475.560	(4)	.272084
(5)	309.510	(6)	26125.6	(7)	− 1338.30	(8)	.951005
(9)	.211173	(10)	− 7.47170E − 05	(11)	1.20295E − 08	(12)	− 1.88938E − 13
(13)	218.891	(14)	2.59156E − 04				

48 2-**Methyl**-1-**Butene**

(1)	7.01350E − 02	(2)	3.79650E + 06	(3)	473.660	(4)	.202809
(5)	304.320	(6)	29056.4	(7)	− 5789.18	(8)	− 5.08994
(9)	.239486	(10)	− 1.09568E − 04	(11)	3.09766E − 08	(12)	− 4.05191E − 12
(13)	241.994	(14)	2.84227E − 04				

49 3-Methyl-1-Butene

(1)	7.01350E − 02	(2)	3.81630 + 06	(3)	425.500	(4)	.492679
(5)	293.220	(6)	24074.1	(7)	− 851.128	(8)	3.43958
(9)	.240561	(10)	− 1.20208E − 04	(11)	3.90679E − 08	(12)	− 5.96308E − 12
(13)	185.411	(14)	2.32867E − 04				

50 2-Methyl-2-Butene

(1)	7.01350E − 02	(2)	3.44500E + 06	(3)	470.000	(4)	.284359
(5)	311.730	(6)	26335.0	(7)	− 11125.1	(8)	− 1.81045
(9)	.207404	(10)	− 6.74287E − 05	(11)	7.79213E − 09	(12)	6.68273E − 13
(13)	234.198	(14)	3.03547E − 04				

51 Allene (propadiene)

(1)	4.00650E − 02	(2)	5.35000E + 06	(3)	393.000	(4)	.131750
(5)	238.660	(6)	18631.3	(7)	202150.	(8)	5.65704
(9)	.112106	(10)	− 5.77163E − 05	(11)	1.82300E − 08	(12)	− 2.52542E − 12
(13)	152.156	(14)	1.70829E − 04				

52 1,2-Butadiene

(1)	5.40920E − 02	(2)	4.53060E + 06	(3)	444.160	(4)	.241986
(5)	284.010	(6)	24283.4	(7)	178120.	(8)	8.95372
(9)	.141767	(10)	− 5.54440E − 05	(11)	1.10927E − 08	(12)	− 6.29983E − 13
(13)	164.762	(14)	2.20568E − 04				

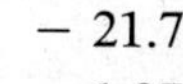

53 **1,3-Butadiene**

(1)	5.40900E − 02	(2)	4.32658E + 06	(3)	425.160	(4)	.191350
(5)	268.750	(6)	22495.7	(7)	129925.	(8)	− 22.7691
(9)	.230248	(10)	− 1.55711E − 04	(11)	6.37594E − 08	(12)	− 1.12579E − 11
(13)	290.230	(14)	2.20800E − 04				

54 2-**Methyl**-1,3-**Butadiene**

(1)	6.81300E − 02	(2)	3.73889E + 06	(3)	483.360	(4)	.161876
(5)	307.230	(6)	26083.8	(7)	100692.	(8)	− 21.7646
(9)	.282184	(10)	− 1.84084E − 04	(11)	7.37112E − 08	(12)	− 1.27862E − 11
(13)	293.712	(14)	2.66000E − 04				

55 **Cyclopentene**

(1)	6.81199E − 02	(2)	4.54950E + 06	(3)	504.160	(4)	.192270
(5)	317.400	(6)	27004.9	(7)	63323.5	(8)	− 17.0812
(9)	.151741	(10)	3.07431E − 05	(11)	− 6.45854E − 08	(12)	1.94064E − 11
(13)	294.748	(14)	2.53101E − 04				

56 **Cyclohexene**

(1)	8.21469E − 02	(2)	4.23540E + 06	(3)	559.160	(4)	.208083
(5)	356.140	(6)	30479.9	(7)	33505.7	(8)	− 64.8076
(9)	.360057	(10)	− 1.91943E − 04	(11)	6.02056E − 08	(12)	− 8.25627E − 12
(13)	489.099	(14)	3.00213E − 04				

57 **Ethyne (Acetylene)**

(1)	2.60380E − 02	(2)	6.24160E + 06	(3)	309.000	(4)	.195260
(5)	189.160	(6)	16956.5	(7)	227987.	(8)	14.6732
(9)	7.15095E − 02	(10)	− 5.98663E − 05	(11)	2.95722E − 08	(12)	− 5.97064E − 12
(13)	81.7326	(14)	1.13005E − 04				

58 **Propyne**

(1)	4.00650E − 02	(2)	5.35000E + 06	(3)	401.000	(4)	.207705
(5)	249.940	(6)	22148.2	(7)	194077.	(8)	13.4761
(9)	9.77702E − 02	(10)	− 4.77841E − 05	(11)	1.56346E − 08	(12)	− 2.46548E − 12
(13)	119.089	(14)	1.70318E − 04				

59 1-**Butyne**

(1)	5.40920E − 02	(2)	4.73880E + 06	(3)	463.700	(4)	9.67736E − 02
(5)	281.230	(6)	24995.2	(7)	180214.	(8)	12.0717
(9)	.141003	(10)	− 5.89074E − 05	(11)	1.38961E − 08	(12)	− 1.26873E − 12
(13)	145.555	(14)	2.30161E − 04				

60 2-**Butyne**

(1)	5.40920E − 02	(2)	5.12040E + 06	(3)	488.700	(4)	.151588
(5)	300.150	(6)	26669.9	(7)	161404.	(8)	19.2631
(9)	.107581	(10)	− 1.87735E − 05	(11)	− 6.60854E − 09	(12)	2.59264E − 12
(13)	112.334	(14)	2.20049E − 04				

61 Benzene

(1)	7.81100E − 02	(2)	4.89400E + 06	(3)	562.100	(4)	.210503
(5)	353.260	(6)	30785.5	(7)	107077.	(8)	− 40.0895
(9)	.253602	(10)	− 1.20521E − 04	(11)	2.84953E − 08	(12)	− 1.93768E − 12
(13)	361.913	(14)	2.56980E − 04				

62 Toluene

(1)	9.21000E − 02	(2)	4.16446E + 06	(3)	591.960	(4)	.255292
(5)	383.760	(6)	33201.3	(7)	80095.3	(8)	− 38.9802
(9)	.293121	(10)	− 1.32530E − 04	(11)	3.00631E − 08	(12)	− 1.95827E − 12
(13)	384.054	(14)	3.16000E − 04				

63 Ethylbenzene

(1)	.106200	(2)	3.67810E + 06	(3)	617.160	(4)	.308144
(5)	409.360	(6)	35587.8	(7)	65755.9	(8)	− 41.6071
(9)	.351188	(10)	− 1.62158E − 04	(11)	3.77902E − 08	(12)	− 2.53389E − 12
(13)	409.112	(14)	3.74000E − 04				

64 *n*-Propylbenzene

(1)	.120196	(2)	3.24240E + 06	(3)	638.800	(4)	.344798
(5)	432.380	(6)	38267.4	(7)	50154.2	(8)	− 50.2459
(9)	.422579	(10)	− 2.06827E − 04	(11)	5.18534E − 08	(12)	− 3.66626E − 12
(13)	461.411	(14)	4.30023E − 04				

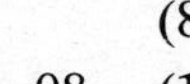

65 Isopropylbenzene

(1)	.120200	(2)	3.19174E + 06	(3)	633.160	(4)	.297822
(5)	425.560	(6)	37555.6	(7)	45786.6	(8)	− 42.6350
(9)	.400646	(10)	1.81358E − 04	(11)	4.10640E − 08	(12)	− 2.69022E − 12
(13)	416.078	(14)	4.34000E − 04				

66 1-Methyl-3-Ethylbenzene

(1)	.120190	(2)	3.10000E + 06	(3)	636.200	(4)	.366618
(5)	434.500	(6)	38560.4	(7)	37105.5	(8)	− 18.5671
(9)	.329495	(10)	− 9.17955E − 05	(11)	− 1.29996E − 08	(12)	9.97030E − 12
(13)	326.648	(14)	4.35114E − 04				

67 *o*-Xylene

(1)	.106160	(2)	3.77132E + 06	(3)	630.960	(4)	.300738
(5)	417.560	(6)	36843.8	(7)	49906.6	(8)	− 6.64074
(9)	.267245	(10)	− 6.96687E − 05	(11)	− 1.19210E − 08	(12)	7.95307E − 12
(13)	241.371	(14)	3.69000E − 04				

68 *m*-Xylene

(1)	.106160	(2)	3.54131E + 06	(3)	616.960	(4)	.321082
(5)	412.260	(6)	36425.2	(7)	52035.0	(8)	− 30.6060
(9)	.319422	(10)	− 1.33091E − 04	(11)	2.74544E − 08	(12)	− 1.60999E − 12
(13)	359.014	(14)	3.76000E − 04				

69 ***p*-Xylene**

(1)	.106160	(2)	3.56461E + 06	(3)	617.160	(4)	.310167
(5)	411.260	(6)	36090.2	(7)	50889.3	(8)	− 13.3659
(9)	.263894	(10)	− 5.67547E − 05	(11)	− 2.12726E − 08	(12)	1.01338E − 11
(13)	279.884	(14)	3.79000E − 04				

70 **Isopropenylbenzene**

(1)	.118180	(2)	3.31000E + 06	(3)	664.000	(4)	.254289
(5)	438.540	(6)	38309.2	(7)	145871.	(8)	− 28.9754
(9)	.362032	(10)	− 1.76279E − 04	(11)	4.94000E − 08	(12)	− 5.95123E − 12
(13)	355.031	(14)	4.26983E − 04				

71 **Styrene**

(1)	.104144	(2)	3.80982E + 06	(3)	642.160	(4)	.297383
(5)	418.160	(6)	36807.8	(7)	175835.	(8)	− 43.2255
(9)	.355947	(10)	− 2.01741E − 04	(11)	6.76566E − 08	(12)	− 9.93361E − 12
(13)	403.943	(14)	3.69700E − 04				

72 ***m*-Methylstyrene**

(1)	.118170	(2)	3.39440E + 06	(3)	656.000	(4)	.344198
(5)	422.160	(6)	40360.8	(7)	148047.	(8)	− 28.1735
(9)	.359376	(10)	− 1.72568E − 04	(11)	4.70551E − 08	(12)	− 5.40915E − 12
(13)	357.490	(14)	4.04923E − 04				

73 Naphthalene

(1)	.128170	(2)	4.01146E + 06	(3)	748.180	(4)	.304639
(5)	491.160	(6)	43291.5	(7)	181669.	(8)	− 52.3736
(9)	.371136	(10)	− 1.35190E − 04	(11)	− 8.56020E − 09	(12)	1.56258E − 11
(13)	431.201	(14)	4.10140E − 04				

74 Oxygen

(1)	3.20000E − 02	(2)	5.04599E + 06	(3)	154.660	(4)	1.80445E − 02
(5)	90.1600	(6)	6820.30	(7)	− 73.0293	(8)	30.4592
(9)	− 8.99091E − 03	(10)	2.09544E − 05	(11)	− 1.44652E − 08	(12)	3.47863E − 12
(13)	34.5342	(14)	7.33700E − 05				

75 Nitrogen

(1)	2.80100E − 02	(2)	3.40047E + 06	(3)	126.360	(4)	3.57003E − 02
(5)	77.3600	(6)	5593.56	(7)	− 44.8778	(8)	29.7294
(9)	− 3.06926E − 03	(10)	4.76331E − 06	(11)	− 1.16921E − 09	(12)	− 1.01366E − 13
(13)	23.3357	(14)	8.90500E − 05				

76 Nitrous oxide

(1)	4.40100E − 02	(2)	7.28527E + 06	(3)	309.660	(4)	.144822
(5)	183.660	(6)	16558.8	(7)	86077.9	(8)	20.3751
(9)	4.01464E − 02	(10)	− 2.49584E − 05	(11)	9.03380E − 09	(12)	− 1.41456E − 12
(13)	83.0131	(14)	9.65000E − 05				

77 **Nitric oxide**

(1)	3.00100E − 02	(2)	6.52533E + 06	(3)	180.060	(4)	.623028
(5)	121.360	(6)	13816.4	(7)	89835.7	(8)	33.5412
(9)	− 1.29661E − 02	(10)	1.92639E − 05	(11)	− 1.03129E − 08	(12)	1.98607E − 12
(13)	25.2117	(14)	5.77000E − 05				

78 **Nitrogen dioxide**

(1)	4.60100E − 02	(2)	1.01325E + 07	(3)	431.160	(4)	.827737
(5)	294.360	(6)	19070.9	(7)	36460.3	(8)	27.8782
(9)	1.50600E − 02	(10)	3.70847E − 06	(11)	− 6.21673E − 09	(12)	1.61779E − 12
(13)	71.9926	(14)	8.22000E − 05				

79 **Dinitrogen tetroxide**

(1)	9.20110E − 02	(2)	1.00310E + 07	(3)	431.160	(4)	.933004
(5)	299.150	(6)	38141.7	(7)	19131.8	(8)	24.6786
(9)	.117691	(10)	− 7.67353E − 05	(11)	2.75413E − 08	(12)	− 4.15728E − 12
(13)	103.031	(14)	7.71574E − 05				

80 **Hydrogen**

(1)	2.01600E − 03	(2)	1.296E96 + 06	(3)	32.9600	(4)	− .224469
(5)	20.3600	(6)	1092.75	(7)	57.7671	(8)	26.9900
(9)	5.96394E − 03	(10)	− 8.02023E − 06	(11)	5.36261E − 09	(12)	− 1.22897E − 12
(13)	− 25.7821	(14)	6.41400E − 05				

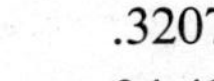
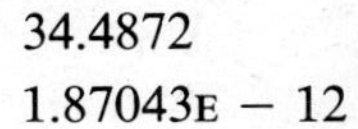

81 Hydroxyl radical

(1)	1.70070E − 02	(2)	.0	(3)	.0	(4)	.0
(5)	.0	(6)	.0	(7)	41668.7	(8)	33.1765
(9)	− 8.13278E − 03	(10)	7.56222E − 06	(11)	− 2.68275E − 09	(12)	3.68462E − 13
(13)	− 1.36204	(14)	.0				

82 Water

(1)	1.80200E − 02	(2)	2.21192E + 07	(3)	647.360	(4)	.320795
(5)	373.160	(6)	40670.6	(7)	− 239133.	(8)	34.4872
(9)	− 7.12689E − 03	(10)	1.57800E − 05	(11)	− 8.91593E − 09	(12)	1.87043E − 12
(13)	− 5.36330	(14)	5.63000E − 05				

83 Carbon

(1)	1.20110E − 02	(2)	.0	(3)	.0	(4)	.0
(5)	3927.00	(6)	.0	(7)	114.907	(8)	− 2.39955
(9)	1.84310E − 02	(10)	3.21120E − 06	(11)	− 9.74131E − 09	(12)	3.27324E − 12
(13)	8.28338	(14)	.0				

84 Carbon monoxide

(1)	2.80100E − 02	(2)	3.50585E + 06	(3)	133.060	(4)	3.24908E − 02
(5)	81.6600	(6)	6045.74	(7)	− 114004.	(8)	30.0646
(9)	− 4.83330E − 03	(10)	8.46034E − 06	(11)	− 3.84994E − 09	(12)	5.60878E − 13
(13)	28.5013	(14)	9.31000E − 05				

85 **Carbon dioxide**

(1)	4.40100E − 02	(2)	7.38659E + 06	(3)	304.260	(4)	.397400
(5)	194.660	(6)	17165.9	(7)	− 392981.	(8)	21.0725
(9)	3.35218E − 02	(10)	− 1.58073E − 05	(11)	3.72724E − 09	(12)	− 2.54010E − 13
(13)	75.6681	(14)	9.40000E − 05				

86 **Carbon suboxide**

(1)	6.80320E − 02	(2)	5.87720E + 06	(3)	485.260	(4)	3.14940E − 02
(5)	280.150	(6)	22138.8	(7)	− 96515.4	(8)	27.3341
(9)	9.02966E − 02	(10)	− 6.46094E − 05	(11)	2.65871E − 08	(12)	− 4.59392E − 12
(13)	74.3291	(14)	1.97984E − 04				

87 **Air**

(1)	2.89670E − 02	(2)	3.89090E + 06	(3)	132.450	(4)	3.57368E − 02
(5)	80.0000	(6)	5941.07	(7)	− 8329.30	(8)	26.6020
(9)	4.78960E − 03	(10)	− 1.09470E − 06	(11)	1.05630E − 10	(12)	.0
(13)	− 137.702	(14)	8.50025E − 05				

88 **Sulfur**

(1)	6.41280E − 02	(2)	1.17540E + 07	(3)	1313.00	(4)	5.87472E − 02
(5)	717.760	(6)	10467.0	(7)	− 582.366	(8)	9.96985
(9)	2.85627E − 02	(10)	− 1.78621E − 05	(11)	9.83753E − 09	(12)	− 1.99504E − 12
(13)	− 39.8212	(14)	2.65351E − 04				

89 **Sulfur dioxide**

(1)	6.40600E − 02	(2)	7.91348E + 06	(3)	430.760	(4)	.242046
(5)	263.160	(6)	24932.4	(7)	− 294452.	(8)	29.5550
(9)	1.59356E − 02	(10)	7.73965E − 06	(11)	− 1.20857E − 08	(12)	3.63786E − 12
(13)	69.6109	(14)	1.22000E − 04				

90 **Sulfur trioxide**

(1)	8.00599E − 02	(2)	8.39984E + 06	(3)	491.460	(4)	.486613
(5)	317.960	(6)	40678.9	(7)	− 389422.	(8)	18.6347
(9)	7.18768E − 02	(10)	− 4.62873E − 05	(11)	1.59896E − 08	(12)	− 2.28858E − 12
(13)	113.457	(14)	1.26100E − 04				

91 **Hydrogen sulfide**

(1)	3.40800E − 02	(2)	9.00780E + 06	(3)	373.600	(4)	9.60316E − 02
(5)	212.960	(6)	18673.1	(7)	− 17549.8	(8)	34.0204
(9)	− 6.27754E − 03	(10)	1.89737E − 05	(11)	− 1.08217E − 08	(12)	2.16835E − 12
(13)	13.4369	(14)	9.77038E − 05				

92 **Carbon disulfide**

(1)	7.61390E − 02	(2)	7.90330E + 06	(3)	552.000	(4)	.102247
(5)	319.350	(6)	26753.6	(7)	116177.	(8)	23.4691
(9)	5.20732E − 02	(10)	− 4.13994E − 05	(11)	1.76126E − 08	(12)	− 3.04730E − 12
(13)	78.1261	(14)	1.70010E − 04				

93 **Carbon oxysulfide**

(1)	6.00750E − 02	(2)	6.18080E + 06	(3)	378.160	(4)	− .385254
(5)	135.150	(6)	18518.2	(7)	− 138080.	(8)	18.9856
(9)	5.25354E − 02	(10)	− 4.01215E − 05	(11)	1.70032E − 08	(12)	− 2.96099E − 12
(13)	96.9260	(14)	1.71432E − 04				

94 **Fluorine**

(1)	3.80000E − 02	(2)	5.37023E + 06	(3)	144.160	(4)	6.58921E − 02
(5)	85.0600	(6)	6531.41	(7)	403.861	(8)	24.3732
(9)	1.58205E − 02	(10)	− 9.87194E − 06	(11)	3.23126E − 09	(12)	− 4.18936E − 13
(13)	55.7578	(14)	6.60000E − 05				

95 **Chlorine**

(1)	7.09100E − 02	(2)	7.83242E + 06	(3)	417.160	(4)	7.59783E − 02
(5)	239.100	(6)	20423.2	(7)	187.803	(8)	25.1954
(9)	2.20184E − 02	(10)	− 2.03584E − 05	(11)	9.47557E − 09	(12)	− 1.72680E − 12
(13)	68.8413	(14)	1.24500E − 04				

96 **Bromine**

(1)	.159800	(2)	1.03352E + 07	(3)	588.160	(4)	.118573
(5)	331.940	(6)	30186.8	(7)	.0	(8)	34.0720
(9)	4.74480E − 03	(10)	− 2.82660E − 06	(11)	6.72550E − 10	(12)	.0
(13)	48.9041	(14)	1.27000E − 04				

97 Iodine

(1)	.253800	(2)	1.17537E + 07	(3)	819.160	(4)	.229000
(5)	457.560	(6)	41868.0	(7)	.0	(8)	35.7720
(9)	2.49510E − 03	(10)	− 1.37900E − 06	(11)	3.20570E − 10	(12)	.0
(13)	55.6190	(14)	1.55000E − 04				

98 Hydrogen fluoride

(1)	2.00000E − 02	(2)	6.48480E + 06	(3)	461.160	(4)	8.41920E − 02
(5)	292.660	(6)	6698.88	(7)	− 272774.	(8)	29.1289
(9)	3.77755E − 04	(10)	− 1.35115E − 06	(11)	1.61131E − 09	(12)	− 4.19106E − 13
(13)	7.73786	(14)	6.90000E − 05				

99 Hydrogen chloride

(1)	3.64600E − 02	(2)	8.30865E + 06	(3)	324.660	(4)	.143201
(5)	188.130	(6)	16161.1	(7)	− 92310.3	(8)	30.0355
(9)	− 3.19557E − 03	(10)	4.04648E − 06	(11)	− 1.10765E − 09	(12)	5.13685E − 14
(13)	17.2090	(14)	8.70000E − 05				

100 Hydrogen bromide

(1)	8.09200E − 02	(2)	8.54170E + 06	(3)	363.160	(4)	7.64930E − 02
(5)	206.360	(6)	17626.4	(7)	− 28696.2	(8)	30.4322
(9)	− 4.99362E − 03	(10)	7.24097E − 06	(11)	− 3.01194E − 09	(12)	4.37678E − 13
(13)	27.4651	(14)	1.10000E − 04				

101 **Hydrogen iodide**

(1)	.127900	(2)	8.30865E + 06	(3)	424.160	(4)	4.03171E − 02
(5)	237.660	(6)	19778.4	(7)	28405.6	(8)	30.6965
(9)	− 6.55800E − 03	(10)	1.06424E − 05	(11)	− 5.23783E − 09	(12)	9.18515E − 13
(13)	34.4083	(14)	1.35000E − 04				

102 **Phosgene**

(1)	9.89169E − 02	(2)	5.67420E + 06	(3)	455.000	(4)	.199616
(5)	281.350	(6)	24409.0	(7)	− 219042.	(8)	20.5087
(9)	9.65605E − 02	(10)	− 9.56650E − 05	(11)	5.35547E − 08	(12)	− 1.26896E − 11
(13)	120.537	(14)	1.90006E − 04				

103 **Cyanogen**

(1)	5.20360E − 02	(2)	5.97820E + 06	(3)	400.000	(4)	.280991
(5)	252.450	(6)	24183.0	(7)	306838.	(8)	29.1239
(9)	6.41550E − 02	(10)	− 5.01290E − 05	(11)	2.24023E − 08	(12)	− 4.11168E − 12
(13)	43.3940	(14)	1.48647E − 04				

104 **Hydrogen cyanide**

(1)	2.70260E − 02	(2)	5.39050E + 06	(3)	456.700	(4)	.395852
(5)	299.650	(6)	25233.8	(7)	136195.	(8)	20.7361
(9)	3.39807E − 02	(10)	− 2.32336E − 05	(11)	1.01060E − 08	(12)	− 1.85002E − 12
(13)	66.2223	(14)	1.39004E − 04				

105 Cyanogen chloride

(1)	6.14710E − 02	(2)	7.49700E + 06	(3)	495.780	(4)	8.50238E − 02
(5)	286.160	(6)	26376.8	(7)	137189.	(8)	25.1767
(9)	4.76877E − 02	(10)	− 3.94971E − 05	(11)	1.79929E − 08	(12)	− 3.29543E − 12
(13)	69.1855	(14)	1.55878E − 04				

106 Cyanogen bromide

(1)	.105927	(2)	8.55630E + 06	(3)	580.290	(4)	.110834
(5)	334.450	(6)	46096.7	(7)	192549.	(8)	28.2524
(9)	4.51887E − 02	(10)	− 3.89715E − 05	(11)	1.83802E − 08	(12)	− 3.45677E − 12
(13)	65.1073	(14)	1.58508E − 04				

107 Cyanogen iodide

(1)	.152922	(2)	5.86830E + 06	(3)	719.780	(4)	2.40090E − 02
(5)	414.160	(6)	58531.5	(7)	22504.	(8)	32.2604
(9)	3.88544E − 02	(10)	− 3.42275E − 05	(11)	1.63458E − 08	(12)	− 3.10676E − 12
(13)	54.4658	(14)	2.94826E − 04				

108 Acetonitrile

(1)	4.10530E − 02	(2)	5.11387E + 06	(3)	547.860	(4)	.402606
(5)	354.800	(6)	31401.0	(7)	95577.9	(8)	25.0389
(9)	4.27503E − 02	(10)	1.46317E − 05	(11)	− 2.33294E − 08	(12)	7.45617E − 12
(13)	74.3023	(14)	2.29187E − 04				

109 **Acrylonitrile**

(1)	5.30640E − 02	(2)	4.45830E + 06	(3)	523.500	(4)	.476788
(5)	350.460	(6)	32657.0	(7)	193583.	(8)	10.2904
(9)	.111894	(10)	− 5.49403E − 05	(11)	1.38613E − 08	(12)	− 7.54997E − 13
(13)	155.632	(14)	2.45829E − 04				

110 **Ethylene oxide (oxirane)**

(1)	4.40500E − 02	(2)	7.19104E + 06	(3)	468.960	(4)	.203827
(5)	283.710	(6)	25623.2	(7)	− 38551.4	(8)	17.2474
(9)	2.83409E − 02	(10)	8.45087E − 05	(11)	− 8.33134E − 08	(12)	2.40697E − 11
(13)	119.363	(14)	1.40300E − 04				

111 **Ketene**

(1)	4.20380E − 02	(2)	5.00000E + 06	(3)	325.600	(4)	.440282
(5)	217.150	(6)	20640.9	(7)	− 55244.8	(8)	11.0106
(9)	8.03971E − 02	(10)	− 4.96166E − 05	(11)	1.96352E − 08	(12)	− 3.52417E − 12
(13)	134.105	(14)	1.46187E − 04				

112 **Acetone**

(1)	5.80810E − 02	(2)	4.72170E + 06	(3)	508.700	(4)	.299415
(5)	329.350	(6)	29140.1	(7)	− 194810.	(8)	18.3041
(9)	9.91952E − 02	(10)	1.00074E − 05	(11)	− 3.55753E − 08	(12)	1.19047E − 11
(13)	132.469	(14)	2.13013E − 04				

113 **Ammonia**

(1)	1.70320E − 02	(2)	1.12775E + 07	(3)	405.560	(4)	.242196
(5)	239.730	(6)	23366.5	(7)	− 38968.8	(8)	33.9375
(9)	− 9.06695E − 03	(10)	3.50930E − 05	(11)	− 2.27854E − 08	(12)	5.19718E − 12
(13)	.865205	(14)	7.24000E − 05				

114 **Hydrazine**

(1)	3.20480E − 02	(2)	1.46921E + 07	(3)	653.160	(4)	.309927
(5)	386.660	(6)	44798.8	(7)	111752.	(8)	5.26528
(9)	.111671	(10)	− 8.48402E − 05	(11)	3.92980E − 08	(12)	− 7.56051E − 12
(13)	153.801	(14)	1.01100E − 04				

115 **Methylamine**

(1)	3.10580E − 02	(2)	7.45750E + 06	(3)	430.100	(4)	.281352
(5)	266.450	(6)	26000.0	(7)	− 9381.90	(8)	15.2577
(9)	6.73182E − 02	(10)	− 1.24035E − 05	(11)	− 2.96552E − 09	(12)	1.28573E − 12
(13)	116.486	(14)	1.27888E − 04				

116 **Dimethylamine**

(1)	4.50850E − 02	(2)	5.30940E + 06	(3)	437.700	(4)	.293327
(5)	280.050	(6)	27758.5	(7)	1075.49	(8)	3.46605
(9)	.124618	(10)	− 3.13613E − 05	(11)	− 2.15651E − 09	(12)	2.00324E − 12
(13)	183.571	(14)	1.82530E − 04				

117 **Trimethylamine**

(1) 5.91120E − 02 (2) 4.07330E + 06 (3) 433.300 (4) .197008
(5) 276.050 (6) 24116.0 (7) − 9956.15 (8) − 6.22128
(9) .193305 (10) − 6.71419E − 05 (11) 7.13259E − 09 (12) 1.23720E − 12
(13) 217.886 (14) 2.54005E − 04

118 **Methyl chloride**

(1) 5.04900E − 02 (2) 6.67732E + 06 (3) 416.260 (4) .148080
(5) 249.360 (6) 21436.4 (7) − 77904.6 (8) 27.1942
(9) 6.83199E − 03 (10) 5.35238E − 05 (11) − 4.66441E − 08 (12) 1.27739E − 11
(13) 69.7965 (14) 1.39100E − 04

119 **Dichloromethane**

(1) 8.49299E − 02 (2) 6.13016E + 06 (3) 514.160 (4) .199125
(5) 312.960 (6) 28009.7 (7) − 87833.2 (8) 21.0299
(9) 5.69482E − 02 (10) − 1.10415E − 05 (11) − 9.10412E − 09 (12) 4.26999E − 12
(13) 118.358 (14) 1.93000E − 04

120 **Chloroform**

(1) .119400 (2) 5.47155E + 06 (3) 536.560 (4) .211184
(5) 334.460 (6) 29726.3 (7) − 97809.8 (8) 20.0731
(9) .108978 (10) − 8.81267E − 05 (11) 4.00510E − 08 (12) − 7.65608E − 12
(13) 126.806 (14) 2.40000E − 04

121 **Carbon tetrachloride**

(1)	.153800	(2)	4.55963E + 06	(3)	556.360	(4)	.189001
(5)	349.860	(6)	30019.4	(7)	– 94016.9	(8)	22.3247
(9)	.166207	(10)	– 1.80576E – 04	(11)	1.01971E – 07	(12)	– 2.31645E – 11
(13)	104.341	(14)	2.75000E – 04				

122 1-**Chloroethane**

(1)	6.45150E – 02	(2)	5.26890E + 06	(3)	460.360	(4)	.187409
(5)	285.450	(6)	24702.1	(7)	– 85234.4	(8)	6.06359
(9)	.108674	(10)	– 3.01008E – 05	(11)	– 4.68273E – 09	(12)	3.64016E – 12
(13)	179.797	(14)	1.99775E – 04				

123 1,1-**Dichloroethane**

(1)	9.89600E – 02	(2)	5.06620E + 06	(3)	523.160	(4)	.234670
(5)	330.450	(6)	28721.4	(7)	– 153851.	(8)	11.9805
(9)	.137819	(10)	– 7.64077E – 05	(11)	2.43439E – 08	(12)	– 3.16941E – 12
(13)	163.466	(14)	2.32762E – 04				

124 1,2-**Dichloroethane**

(1)	9.89600E – 02	(2)	5.37020E + 06	(3)	561.000	(4)	.274134
(5)	356.650	(6)	32029.0	(7)	– 155183.	(8)	29.6713
(9)	.101758	(10)	– 4.51556E – 05	(11)	1.19351E – 08	(12)	– 1.41668E – 12
(13)	85.0595	(14)	2.20008E – 04				

125 1,1,1-**Trichloroethane**

(1)	.133405	(2)	4.44040E + 06	(3)	550.160	(4)	.0
(5)	.0	(6)	33326.9	(7)	− 249.313	(8)	22.7421
(9)	.167191	(10)	− 1.31149E − 04	(11)	6.16812E − 08	(12)	− 1.24888E − 11
(13)	108.929	(14)	3.19138E − 04				

126 1,1,2-**Trichloroethane**

(1)	.133405	(2)	4.50040E + 06	(3)	562.570	(4)	.545986
(5)	386.650	(6)	33326.9	(7)	1387.60	(8)	12.7253
(9)	.177016	(10)	− 1.30165E − 04	(11)	5.58075E − 08	(12)	− 1.02441E − 11
(13)	174.778	(14)	2.56612E − 04				

127 1,1,1,2-**Tetrachloroethane**

(1)	.167850	(2)	4.06780E + 06	(3)	581.000	(4)	.561932
(5)	403.660	(6)	36843.8	(7)	105.012	(8)	23.6384
(9)	.188541	(10)	− 1.49605E − 04	(11)	6.80465E − 08	(12)	− 1.31571E − 11
(13)	126.875	(14)	2.92015E − 04				

128 1,1,2,2-**Tetrachloroethane**

(1)	.167850	(2)	3.98250E + 06	(3)	642.160	(4)	.264942
(5)	418.160	(6)	38518.6	(7)	− 131935.	(8)	20.6902
(9)	.189048	(10)	− 1.45683E − 04	(11)	6.40499E − 08	(12)	− 1.19707E − 11
(13)	149.623	(14)	3.60637E − 04				

129 **Pentachloroethane**

(1)	.202295	(2)	3.74390E + 06	(3)	615.00	(4)	.629454
(5)	435.150	(6)	40612.0	(7)	− 147532.	(8)	30.3181
(9)	.218225	(10)	− 1.94593E − 04	(11)	9.52283E − 08	(12)	− 1.93700E − 11
(13)	100.677	(14)	3.29154E − 04				

130 **Hexachloroethane**

(1)	.236740	(2)	3.45760E + 06	(3)	639.150	(4)	.663962
(5)	.457550	(6)	51079.0	(7)	− 134129.	(8)	44.4464
(9)	.240894	(10)	− 2.38209E − 04	(11)	1.24244E − 07	(12)	− 2.64348E − 11
(13)	27.5764	(14)	3.66715E − 04				

131 **Chloroethylene**

(1)	6.24990E − 02	(2)	5.37020E + 06	(3)	429.700	(4)	− .271878
(5)	261.150	(6)	22320.9	(7)	45076.6	(8)	6.04499
(9)	.100543	(10)	− 5.06345E − 05	(11)	1.16307E − 08	(12)	4.34963E − 14
(13)	176.075	(14)	2.12558E − 04				

132 1,1-**Dichloroethylene**

(1)	9.69440E − 02	(2)	5.80610E + 06	(3)	480.060	(4)	.358010
(5)	310.150	(6)	29727.9	(7)	8805.23	(8)	6.92786
(9)	.145287	(10)	− 1.21466E − 04	(11)	6.05354E − 08	(12)	− 1.30342E − 11
(13)	176.374	(14)	1.79356E − 04				

133 *Cis*-1,2-**Dichloroethylene**

(1)	9.69440E − 02	(2)	5.86670E + 06	(3)	544.160	(4)	.181618
(5)	333.450	(6)	29119.3	(7)	9992.44	(8)	5.91790
(9)	.138986	(10)	− 1.06475E − 04	(11)	4.78255E − 08	(12)	− 9.18912E − 12
(13)	185.762	(14)	2.12387E − 04				

134 *trans*-1,2-**Dichloroethylene**

(1)	9.69440E − 02	(2)	5.52220E + 06	(3)	516.460	(4)	.213223
(5)	321.550	(6)	28366.7	(7)	10887.3	(8)	15.5833
(9)	.114879	(10)	− 7.58396E − 05	(11)	2.83438E − 08	(12)	− 4.26580E − 12
(13)	142.000	(14)	2.12129E − 04				

135 **Trichloroethylene**

(1)	.131389	(2)	5.01560E + 06	(3)	571.000	(4)	.410193
(5)	360.350	(6)	31401.0	(7)	− 2000.66	(8)	20.4638
(9)	.147797	(10)	− 1.30468E − 04	(11)	6.64464E − 08	(12)	− 1.45167E − 11
(13)	135.528	(14)	2.43358E − 04				

136 **Tetrachloroethylene**

(1)	.165834	(2)	3.60310E + 06	(3)	620.260	(4)	.150042
(5)	393.950	(6)	34743.3	(7)	− 12780.0	(8)	34.6651
(9)	.155976	(10)	− 1.49721E − 04	(11)	7.84391E − 08	(12)	− 1.70201E − 11
(13)	70.4267	(14)	3.98472E − 04				

137 Trichlorofluoromethane

(1)	.137369	(2)	4.37720E + 06	(3)	471.160	(4)	.182351
(5)	296.760	(6)	24785.9	(7)	− 285435.	(8)	17.9686
(9)	.157962	(10)	− 1.59858E − 04	(11)	8.58883E − 08	(12)	− 1.88359E − 11
(13)	131.766	(14)	2.46651E − 04				

138 Dichlorodifluoromethane

(1)	.120914	(2)	4.01250E + 06	(3)	384.660	(4)	.177945
(5)	243.950	(6)	19979.4	(7)	− 487100.	(8)	13.4172
(9)	.149660	(10)	− 1.39240E − 04	(11)	7.00550E − 08	(12)	− 1.46105E − 11
(13)	151.660	(14)	2.19990E − 04				

139 Chlorotrifluoromethane

(1)	.104459	(2)	3.95170E + 06	(3)	301.960	(4)	.182511
(5)	192.060	(6)	15516.3	(7)	− 702181.	(8)	9.78192
(9)	.139086	(10)	− 1.16180E − 04	(11)	5.29337E − 08	(12)	− 1.01129E − 11
(13)	160.515	(14)	1.75096E − 04				

140 Carbon tetrafluoride

(1)	8.80049E − 02	(2)	3.72880E + 06	(3)	228.160	(4)	− .395706
(5)	145.200	(6)	11974.2	(7)	− 926112.	(8)	6.72578
(9)	.125684	(10)	− 8.90788E − 05	(11)	3.31975E − 08	(12)	− 4.96993E − 12
(13)	159.001	(14)	1.71091E − 04				

141 **Formaldehyde**

(1)	3.00260E − 02	(2)	6.82720E + 06	(3)	410.160	(4)	.232282
(5)	252.150	(6)	23027.4	(7)	− 112080.	(8)	34.7764
(9)	− 2.19612E − 02	(10)	6.99069E − 05	(11)	− 5.15896E − 08	(12)	1.35518E − 11
(13)	26.2250	(14)	1.35366E − 04				

142 **Acetaldehyde**

(1)	4.40540E − 02	(2)	5.57140E + 06	(3)	461.000	(4)	.288607
(5)	293.350	(6)	25748.8	(7)	− 154937.	(8)	25.3297
(9)	3.52672E − 02	(10)	4.99176E − 05	(11)	− 5.33836E − 08	(12)	1.61630E − 11
(13)	94.1467	(14)	1.83412E − 04				

143 **Propionaldehyde**

(1)	5.80810E − 02	(2)	4.67630E + 06	(3)	493.160	(4)	.336277
(5)	322.650	(6)	28302.8	(7)	− 167742.	(8)	19.3629
(9)	.102141	(10)	5.63819E − 06	(11)	− 3.38601E − 08	(12)	1.18669E − 11
(13)	134.063	(14)	2.30607E − 04				

144 ***n*-Butyraldehyde**

(1)	7.21080E − 02	(2)	4.02410E + 06	(3)	521.160	(4)	.377610
(5)	348.850	(6)	31526.6	(7)	− 182146.	(8)	29.6698
(9)	.114861	(10)	4.17416E − 05	(11)	− 7.22069E − 08	(12)	2.41331E − 11
(13)	104.421	(14)	2.79967E − 04				

145 *n*-Pentaldehyde

(1)	8.61350E − 02	(2)	3.44210E + 06	(3)	551.670	(4)	.405199
(5)	376.550	(6)	33661.9	(7)	− 197295.	(8)	32.6380
(9)	.147626	(10)	4.82209E − 05	(11)	− 8.78526E − 08	(12)	2.93653E − 11
(13)	106.473	(14)	3.43934E − 04				

146 *n*-Hexaldehyde

(1)	.100162	(2)	3.10680E + 06	(3)	578.700	(4)	.444475
(5)	401.750	(6)	37013.5	(7)	− 211990.	(8)	31.2151
(9)	.196647	(10)	2.81191E − 05	(11)	− 8.44036E − 08	(12)	2.97426E − 11
(13)	126.895	(14)	3.95386E − 04				

147 *n*-Heptaldehyde

(1)	.114189	(2)	2.86670E + 06	(3)	605.000	(4)	.481607
(5)	425.950	(6)	39509.1	(7)	− 226991.	(8)	34.7412
(9)	.226752	(10)	4.16198E − 05	(11)	− 1.08010E − 07	(12)	3.80736E − 11
(13)	126.795	(14)	4.43238E − 04				

148 *n*-Octaldehyde

(1)	.128216	(2)	2.58920E + 06	(3)	622.300	(4)	.483297
(5)	442.060	(6)	40422.6	(7)	− 241480.	(8)	32.2633
(9)	.281195	(10)	9.30067E − 06	(11)	− 9.17387E − 08	(12)	3.34687E − 11
(13)	151.220	(14)	5.04975E − 04				

149 ***n*-Nonaldehyde**

(1)	.142243	(2)	2.36240E + 06	(3)	637.400	(4)	.478410
(5)	455.950	(6)	41054.3	(7)	− 257194.	(8)	43.5831
(9)	.281325	(10)	7.60551E − 05	(11)	− 1.60041E − 07	(12)	5.61315E − 11
(13)	118.925	(14)	5.68006E − 04				

150 ***n*-Decaldehyde**

(1)	.156270	(2)	2.17040E + 06	(3)	664.500	(4)	.503994
(5)	481.000	(6)	43299.2	(7)	− 271087.	(8)	35.4085
(9)	.356073	(10)	1.00688E − 05	(11)	− 1.17703E − 07	(12)	4.3888E − 11
(13)	167.343	(14)	6.39958E − 04				

151 **Methyl alcohol**

(1)	3.20400E − 02	(2)	8.09587E + 06	(3)	512.560	(4)	.614448
(5)	337.860	(6)	35278.0	(7)	− 190193.	(8)	35.7061
(9)	− 1.89472E − 02	(10)	1.02843E − 04	(11)	− 8.37282E − 08	(12)	2.36231E − 11
(13)	36.7554	(14)	1.17800E − 04				

152 **Ethyl alcohol**

(1)	4.60700E − 02	(2)	6.38348E + 06	(3)	516.260	(4)	.749090
(5)	351.460	(6)	38769.8	(7)	− 216021.	(8)	17.8292
(9)	7.42695E − 02	(10)	3.10839E − 05	(11)	− 5.05473E − 08	(12)	1.70314E − 11
(13)	134.412	(14)	1.66900E − 04				

153 ***n*-Propanol**

(1)	6.00900E − 02	(2)	5.16758E + 06	(3)	536.760	(4)	.614331
(5)	370.360	(6)	41784.3	(7)	− 232463.	(8)	15.2848
(9)	.117850	(10)	3.26349E − 05	(11)	− 7.01413E − 08	(12)	2.52511E − 11
(13)	164.829	(14)	2.18500E − 04				

154 ***n*-Butanol**

(1)	7.41200E − 02	(2)	4.41777E + 06	(3)	562.960	(4)	.661348
(5)	390.860	(6)	43124.0	(7)	− 247005.	(8)	.298103
(9)	.206855	(10)	− 4.39418E − 05	(11)	− 2.74048E − 08	(12)	1.47791E − 11
(13)	243.765	(14)	2.74500E − 04				

155 ***n*-Pentanol**

(1)	8.81510E − 02	(2)	3.70820E + 06	(3)	545.160	(4)	1.10440
(5)	411.250	(6)	44380.1	(7)	− 261894.	(8)	− 8.17162
(9)	.280515	(10)	− 1.04130E − 04	(11)	7.17414E − 09	(12)	5.80345E − 12
(13)	294.442	(14)	2.50213E − 04				

156 ***n*-Hexanol**

(1)	.102178	(2)	3.26030E + 06	(3)	610.100	(4)	.553156
(5)	430.650	(6)	48566.9	(7)	− 277981.	(8)	2.47934
(9)	.285762	(10)	− 5.07780E − 05	(11)	− 4.63579E − 08	(12)	2.26904E − 11
(13)	264.606	(14)	3.84301E − 04				

157 ***n*-Heptanol**

(1)	.116205	(2)	2.87990E + 06	(3)	638.160	(4)	.470239
(5)	448.150	(6)	48148.2	(7)	− 293194.	(8)	7.32932
(9)	.311830	(10)	− 3.27954E − 05	(11)	− 7.20685E − 08	(12)	3.14537E − 11
(13)	259.221	(14)	4.67047E − 04				

158 ***n*-Octanol**

(1)	.130232	(2)	2.62290E + 06	(3)	650.100	(4)	.560679
(5)	467.650	(6)	50660.3	(7)	− 308250.	(8)	10.3528
(9)	.344332	(10)	− 2.27335E − 05	(11)	− 9.39544E − 08	(12)	3.96268E − 11
(13)	260.495	(14)	5.08598E − 04				

159 ***n*-Nonanol**

(1)	.144259	(2)	2.38810E + 06	(3)	670.600	(4)	.567053
(5)	486.650	(6)	45735.4	(7)	− 322596.	(8)	7.16902
(9)	.400278	(10)	− 5.73362E − 05	(11)	− 7.59514E − 08	(12)	3.45498E − 11
(13)	288.669	(14)	5.75518E − 04				

160 ***n*-Decanol**

(1)	.158286	(2)	2.19350E + 06	(3)	691.500	(4)	.575402
(5)	506.050	(6)	50241.6	(7)	− 337832.	(8)	12.2000
(9)	.426431	(10)	− 3.96616E − 05	(11)	− 1.01627E − 07	(12)	4.33892E − 11
(13)	281.822	(14)	6.45055E − 04				

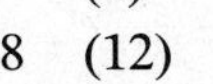
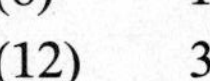

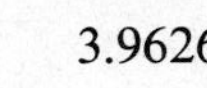
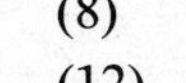

161 **2-Propanol**

(1)	6.00970E − 02	(2)	5.37020E + 06	(3)	508.800	(4)	.715489
(5)	355.650	(6)	39858.3	(7)	− 246685.	(8)	10.7510
(9)	.141714	(10)	− 4.57571E − 06	(11)	− 4.34479E − 08	(12)	1.78642E − 11
(13)	166.790	(14)	2.19010E − 04				

162 **Allyl alcohol**

(1)	5.80810E − 02	(2)	5.47240E + 06	(3)	543.690	(4)	.560152
(5)	369.160	(6)	39983.9	(7)	− 111347.	(8)	.510537
(9)	.151451	(10)	− 5.79624E − 05	(11)	5.63685E − 09	(12)	2.30027E − 12
(13)	222.098	(14)	2.02712E − 04				

163 *tert*-**Butyl alcohol**

(1)	7.41240E − 02	(2)	4.51330E + 06	(3)	508.000	(4)	.440079
(5)	356.050	(6)	39062.8	(7)	− 290454.	(8)	6.70382
(9)	.193842	(10)	− 1.46968E − 05	(11)	− 5.33681E − 08	(12)	2.26849E − 11
(13)	183.499	(14)	2.38639E − 04				

164 **Cyclohexanol**

(1)	.100162	(2)	4.33540E + 06	(3)	611.400	(4)	.685349
(5)	431.860	(6)	45510.5	(7)	− 243732.	(8)	− 13.5670
(9)	.224069	(10)	8.00594E − 05	(11)	− 1.53737E − 07	(12)	5.42954E − 11
(13)	283.802	(14)	2.77305E − 04				

165 **Phenol**

(1)	9.41100E − 02	(2)	6.13016E + 06	(3)	693.160	(4)	.457445
(5)	454.960	(6)	45636.1	(7)	− 73864.4	(8)	− 11.1747
(9)	.194081	(10)	5.97552E − 06	(11)	− 8.46263E − 08	(12)	3.49492E − 11
(13)	267.615	(14)	2.29500E − 04				

166 **Acetic acid**

(1)	6.00530E − 02	(2)	5.78566E + 06	(3)	594.750	(4)	.454000
(5)	391.660	(6)	23697.3	(7)	− 419608.	(8)	13.2461
(9)	9.60823E − 02	(10)	− 4.24413E − 06	(11)	− 3.16518E − 08	(12)	1.35725E − 11
(13)	151.512	(14)	1.71000E − 04				

167 **Acetic anhydride**

(1)	.102090	(2)	4.68022E + 06	(3)	569.150	(4)	.411000
(5)	409.560	(6)	41240.0	(7)	− 547803.	(8)	− 8.30504
(9)	.200492	(10)	− 2.67564E − 05	(11)	− 5.14810E − 08	(12)	2.40099E − 11
(13)	323.175	(14)	2.90000E − 04				

168 **Methanethiol**

(1)	4.81070E − 02	(2)	7.23460E + 06	(3)	469.960	(4)	.147716
(5)	279.110	(6)	24584.9	(7)	− 76070.7	(8)	23.8751
(9)	4.34505E − 02	(10)	8.70391E − 06	(11)	− 1.78740E − 08	(12)	5.81771E − 12
(13)	92.6421	(14)	1.50149E − 04				

169 **Ethanethiol**

(1)	6.21340E − 02	(2)	5.49180E + 06	(3)	499.160	(4)	.157212
(5)	305.160	(6)	26795.5	(7)	− 93387.4	(8)	19.9749
(9)	9.40055E − 02	(10)	− 4.74763E − 06	(11)	− 2.27744E − 08	(12)	9.09906E − 12
(13)	127.820	(14)	2.09710E − 04				

170 ***n*-Propanethiol**

(1)	7.61610E − 02	(2)	4.50880E + 06	(3)	513.000	(4)	.385731
(5)	340.760	(6)	32235.5	(7)	− 109236.	(8)	30.5373
(9)	9.93311E − 02	(10)	3.86527E − 05	(11)	− 5.73915E − 08	(12)	1.79225E − 11
(13)	100.143	(14)	2.45201E − 04				

171 ***n*-Butanethiol**

(1)	9.01880E − 02	(2)	3.83450E + 06	(3)	545.700	(4)	.436782
(5)	371.620	(6)	35100.5	(7)	− 123188.	(8)	25.9498
(9)	.158221	(10)	1.01843E − 05	(11)	− 5.24122E − 08	(12)	1.90376E − 11
(13)	133.388	(14)	3.02320E − 04				

172 ***n*-Pentanethiol**

(1)	.104215	(2)	3.32840E + 06	(3)	594.160	(4)	.329732
(5)	399.800	(6)	35075.7	(7)	− 144337.	(8)	32.2446
(9)	.178578	(10)	3.92472E − 05	(11)	− 8.76607E − 08	(12)	3.07309E − 11
(13)	122.131	(14)	3.91983E − 04				

173 ***n*-Hexanethiol**

(1)	.118242	(2)	2.95070E + 06	(3)	603.600	(4)	.490631
(5)	424.660	(6)	39721.1	(7)	− 154796.	(8)	33.7804
(9)	.216867	(10)	3.69345E − 05	(11)	− 9.74091E − 08	(12)	3.46204E − 11
(13)	130.573	(14)	4.28433E − 04				

174 ***n*-Heptanethiol**

(1)	.132270	(2)	2.65110E + 06	(3)	603.300	(4)	.800971
(5)	449.360	(6)	51079.0	(7)	− 169648.	(8)	35.6664
(9)	.253778	(10)	3.71186E − 05	(11)	− 1.09262E − 07	(12)	3.91770E − 11
(13)	137.157	(14)	4.33474E − 04				

175 ***n*-Octanethiol**

(1)	.146297	(2)	2.40760E + 06	(3)	601.000	(4)	.393815
(5)	422.260	(6)	35525.0	(7)	− 184559.	(8)	37.8542
(9)	.289436	(10)	3.98577E − 05	(11)	− 1.23608E − 07	(12)	4.46450E − 11
(13)	142.507	(14)	5.38798E − 04				

176 ***n*-Nonanethiol**

(1)	.160324	(2)	2.20770E + 06	(3)	677.800	(4)	.545503
(5)	493.360	(6)	45425.8	(7)	− 199453.	(8)	40.1752
(9)	.324674	(10)	4.29903E − 05	(11)	− 1.37866E − 07	(12)	4.99392E − 11
(13)	147.295	(14)	6.33825E − 04				

177 ***n*-Decanethiol**

(1)	.174351	(2)	2.00050E + 06	(3)	700.000	(4)	.544917
(5)	513.750	(6)	46587.8	(7)	− 214348.	(8)	42.1381
(9)	.361242	(10)	4.39755E − 05	(11)	− 1.50630E − 07	(12)	5.48659E − 11
(13)	153.568	(14)	7.22964E − 04				

178 **Dimethyl sulfide**

(1)	6.21340E − 02	(2)	5.53230E + 06	(3)	503.060	(4)	.187112
(5)	310.500	(6)	26991.9	(7)	− 87151.7	(8)	46.2621
(9)	1.19917E − 02	(10)	1.15045E − 04	(11)	− 1.09643E − 07	(12)	3.41327E − 11
(13)	13.8008	(14)	2.13203E − 04				

179 **Methyl ethyl sulfide**

(1)	7.61610E − 02	(2)	4.25570E + 06	(3)	533.160	(4)	.211767
(5)	339.810	(6)	29516.9	(7)	− 80680.6	(8)	27.1918
(9)	.112777	(10)	1.94212E − 05	(11)	− 4.88294E − 08	(12)	1.70111E − 11
(13)	110.192	(14)	2.84576E − 04				

180 **2,2-Dimethylpentane**

(1)	.100198	(2)	2.87763E + 06	(3)	520.800	(4)	.297468
(5)	352.350	(6)	29177.8	(7)	− 158769.	(8)	16.2204
(9)	.257092	(10)	1.99066E − 05	(11)	− 9.63185E − 08	(12)	3.54125E − 11
(13)	147.902	(14)	4.04000E − 04				

181 2,3-**Dimethylpentane**

(1)	.100198	(2)	2.95869E + 06	(3)	537.700	(4)	.298371
(5)	362.930	(6)	30404.5	(7)	− 151861.	(8)	16.2204
(9)	.257092	(10)	1.99066E − 05	(11)	− 9.63185E − 08	(12)	3.54125E − 11
(13)	169.087	(14)	4.05000E − 04				

182 2,4-**Dimethylpentane**

(1)	.100198	(2)	2.77631E + 06	(3)	520.200	(4)	.303833
(5)	353.660	(6)	29516.9	(7)	− 154624.	(8)	16.2204
(9)	.257092	(10)	1.99066E − 05	(11)	− 9.63185E − 08	(12)	3.54125E − 11
(13)	151.670	(14)	4.20000E − 04				

183 3,3-**Dimethylpentane**

(1)	.100198	(2)	3.03975E + 06	(3)	536.000	(4)	.279794
(5)	359.210	(6)	29550.4	(7)	− 154164.	(8)	16.2204
(9)	.257092	(10)	1.99066E − 05	(11)	− 9.63185E − 08	(12)	3.54125E − 11
(13)	154.726	(14)	4.00000E − 04				

184 2-**Methylhexane**

(1)	.100198	(2)	2.75604E + 06	(3)	531.000	(4)	.326749
(5)	363.200	(6)	30685.1	(7)	− 147549.	(8)	16.2204
(9)	.257092	(10)	1.99066E − 05	(11)	− 9.63185E − 08	(12)	3.54125E − 11
(13)	175.032	(14)	4.28000E − 04				

185 **3-Methylhexane**

(1)	.100198	(2)	2.84723E + 06	(3)	535.500	(4)	.324532
(5)	365.000	(6)	30806.5	(7)	− 144911.	(8)	16.2204
(9)	.257092	(10)	1.99066E − 05	(11)	− 9.63185E − 08	(12)	3.54125E − 11
(13)	179.177	(14)	4.18000E − 04				

186 **3-Ethylpentane**

(1)	.100198	(2)	2.89790E + 06	(3)	540.700	(4)	.309253
(5)	366.620	(6)	30973.9	(7)	− 142273.	(8)	16.2204
(9)	.257092	(10)	1.99066E − 05	(11)	− 9.63185E − 08	(12)	3.54125E − 11
(13)	166.533	(14)	4.16000E − 04				

187 2,2-**Dimethylhexane**

(1)	.114224	(2)	2.59392E + 06	(3)	552.000	(4)	.330380
(5)	379.990	(6)	32280.2	(7)	− 171752.	(8)	19.5271
(9)	.288189	(10)	3.14546E − 05	(11)	− 1.18651E − 07	(12)	4.36335E − 11
(13)	148.049	(14)	4.67000E − 04				

188 2,3-**Dimethylhexane**

(1)	.114224	(2)	2.69525E + 06	(3)	566.000	(4)	.336066
(5)	388.760	(6)	33222.3	(7)	− 160950.	(8)	19.5271
(9)	.288189	(10)	3.14546E − 05	(11)	− 1.18651E − 07	(12)	4.36335E − 11
(13)	160.819	(14)	4.61000E − 04				

189 **2,4-Dimethylhexane**

(1)	.114224	(2)	2.61419E + 06	(3)	555.000	(4)	.339639
(5)	382.580	(6)	32615.2	(7)	− 166435.	(8)	19.5271
(9)	.288189	(10)	3.14546E − 05	(11)	− 1.18651E − 07	(12)	4.36335E − 11
(13)	162.494	(14)	4.66000E − 04				

190 **2,5-Dimethylhexane**

(1)	.114224	(2)	2.53313E + 06	(3)	552.000	(4)	.347160
(5)	382.250	(6)	32657.0	(7)	− 169659.	(8)	19.5271
(9)	.288189	(10)	3.14546E − 05	(11)	− 1.18651E − 07	(12)	4.36335E − 11
(13)	155.879	(14)	4.78000E − 04				

191 **3,3-Dimethylhexane**

(1)	.114224	(2)	2.75604E + 06	(3)	564.000	(4)	.319475
(5)	385.120	(6)	32489.6	(7)	− 167147.	(8)	19.5271
(9)	.288189	(10)	3.14546E − 05	(11)	− 1.18651E − 07	(12)	4.36335E − 11
(13)	154.916	(14)	4.50000E − 04				

192 **3,4-Dimethylhexane**

(1)	.114224	(2)	2.77631E + 06	(3)	571.000	(4)	.333107
(5)	390.870	(6)	33293.4	(7)	− 160029.	(8)	19.5271
(9)	.288189	(10)	3.14546E − 05	(11)	− 1.18651E − 07	(12)	4.36335E − 11
(13)	165.173	(14)	4.52000E − 04				

193 2,2,3-**Trimethylpentane**

(1)	.114224	(2)	2.85737E + 06	(3)	567.000	(4)	.288229
(5)	382.990	(6)	32029.0	(7)	− 167147.	(8)	19.5271
(9)	.288189	(10)	3.14546E − 05	(11)	− 1.18651E − 07	(12)	4.36335E − 11
(13)	142.020	(14)	4.37000E − 04				

194 2,3,3-**Trimethylpentane**

(1)	.114224	(2)	2.93843E + 06	(3)	576.000	(4)	.286709
(5)	387.910	(6)	32364.0	(7)	− 163462.	(8)	19.5271
(9)	.288189	(10)	3.14546E − 05	(11)	− 1.18651E − 07	(12)	4.36335E − 11
(13)	148.384	(14)	4.33000E − 04				

195 3-**Ethylhexane**

(1)	.114224	(2)	2.67498E + 06	(3)	567.000	(4)	.358616
(5)	391.680	(6)	33628.4	(7)	− 157894.	(8)	19.5271
(9)	.288189	(10)	3.14546E − 05	(11)	− 1.18651E − 07	(12)	4.36335E − 11
(13)	175.054	(14)	4.66000E − 04				

196 2-**Methylheptane**

(1)	.114224	(2)	2.51286E + 06	(3)	561.000	(4)	.371279
(5)	390.800	(6)	33829.3	(7)	− 162499.	(8)	19.5271
(9)	.288189	(10)	3.14546E − 05	(11)	− 1.18651E − 07	(12)	4.36335E − 11
(13)	172.123	(14)	4.89000E − 04				

197 **3-Methylheptane**

(1)	.114224	(2)	2.59392E + 06	(3)	565.000	(4)	.366655
(5)	392.070	(6)	33913.1	(7)	− 159652.	(8)	19.5271
(9)	.288189	(10)	3.14546E − 05	(11)	− 1.18651E − 07	(12)	4.36335E − 11
(13)	178.446	(14)	4.78000E − 04				

198 **4-Methylheptane**

(1)	.114224	(2)	2.59392E + 06	(3)	563.000	(4)	.368771
(5)	390.860	(6)	33913.1	(7)	− 158252.	(8)	9.71379
(9)	.321579	(10)	− 1.51772E − 05	(11)	− 8.98511E − 08	(12)	3.71379E − 11
(13)	211.464	(14)	4.76000E − 04				

199 **2-Methyloctane**

(1)	.128259	(2)	2.28995E + 06	(3)	586.760	(4)	.421660
(5)	416.410	(6)	36676.4	(7)	− 180951.	(8)	43.5961
(9)	.271376	(10)	1.09384E − 04	(11)	− 1.88244E − 07	(12)	6.48710E − 11
(13)	77.5400	(14)	5.41000E − 04				

200 **3-Methyloctane**

(1)	.128259	(2)	2.34061E + 06	(3)	590.160	(4)	.413760
(5)	417.360	(6)	36802.0	(7)	− 174702.	(8)	21.3866
(9)	.324071	(10)	4.16001E − 05	(11)	− 1.42412E − 07	(12)	5.22456E − 11
(13)	185.978	(14)	5.29000E − 04				

201 **4-Methyloctane**

(1)	.128259	(2)	2.34061E + 06	(3)	587.660	(4)	.413492
(5)	415.570	(6)	36634.5	(7)	− 174702.	(8)	21.3866
(9)	.324071	(10)	4.16001E − 05	(11)	− 1.42412E − 07	(12)	5.22456E − 11
(13)	185.978	(14)	5.23000E − 04				

202 **3-Ethylheptane**

(1)	.128259	(2)	2.40140E + 06	(3)	590.360	(4)	.408356
(5)	416.100	(6)	36760.1	(7)	− 169175.	(8)	6.95506
(9)	.346495	(10)	2.91061E − 05	(11)	− 1.43922E − 07	(12)	5.48598E − 11
(13)	250.745	(14)	5.11000E − 04				

203 **4-Ethylheptane**

(1)	.128259	(2)	2.39127E + 06	(3)	587.960	(4)	.405161
(5)	414.300	(6)	36676.4	(7)	− 169175.	(8)	6.95506
(9)	.346495	(10)	2.91061E − 05	(11)	− 1.43922E − 07	(12)	5.48598E − 11
(13)	250.745	(14)	5.05000E − 04				

204 **2,2-Dimethylheptane**

(1)	.128259	(2)	2.32034E + 06	(3)	577.760	(4)	.376449
(5)	405.840	(6)	34792.3	(7)	− 188890.	(8)	20.8923
(9)	.330986	(10)	5.57103E − 05	(11)	− 1.69425E − 07	(12)	6.38016E − 11
(13)	155.036	(14)	5.26000E − 04				

205 2,3-**Dimethylheptane**

(1)	.128259	(2)	2.40140E + 06	(3)	589.660	(4)	.384528
(5)	413.600	(6)	36132.1	(7)	− 175031.	(8)	7.79306
(9)	.357639	(10)	6.76082E − 07	(11)	− 1.17334E − 07	(12)	4.60623E − 11
(13)	235.022	(14)	5.15000E − 04				

206 2,4-**Dimethylheptane**

(1)	.128259	(2)	2.34061E + 06	(3)	576.860	(4)	.390288
(5)	406.040	(6)	35378.5	(7)	− 176931.	(8)	− 10.2555
(9)	.394900	(10)	− 3.99757E − 05	(11)	− 9.36567E − 08	(12)	4.02413E − 11
(13)	320.275	(14)	5.17000E − 04				

207 2,5-**Dimethylheptane**

(1)	.128259	(2)	2.35074E + 06	(3)	581.160	(4)	.392608
(5)	409.100	(6)	35629.7	(7)	− 176931.	(8)	− 10.2555
(9)	.394900	(10)	− 3.99757E − 05	(11)	− 9.36567E − 08	(12)	4.02413E − 11
(13)	320.275	(14)	5.22000E − 04				

208 2,6-**Dimethylheptane**

(1)	.128259	(2)	2.30008E + 06	(3)	577.960	(4)	.401277
(5)	408.360	(6)	35545.9	(7)	− 182827.	(8)	8.00824
(9)	.357555	(10)	− 1.53125E − 07	(11)	− 1.15620E − 07	(12)	4.52036E − 11
(13)	222.349	(14)	5.35000E − 04				

209 3,3-**Dimethylheptane**

(1)	.128259	(2)	2.43180E + 06	(3)	588.560	(4)	.359143
(5)	410.160	(6)	35336.6	(7)	− 180823.	(8)	6.38989
(9)	.353183	(10)	4.40751E − 05	(11)	− 1.71302E − 07	(12)	6.63027E − 11
(13)	235.148	(14)	5.06000E − 04				

210 3,4-**Dimethylheptane**

(1)	.128259	(2)	2.46220E + 06	(3)	591.960	(4)	.377840
(5)	413.700	(6)	36383.3	(7)	− 169115.	(8)	− 10.6536
(9)	.395677	(10)	− 4.03033E − 05	(11)	− 9.44742E − 08	(12)	4.08362E − 11
(13)	325.036	(14)	5.03000E − 04				

211 3,5-**Dimethylheptane**

(1)	.128259	(2)	2.40140E + 06	(3)	583.260	(4)	.384399
(5)	409.100	(6)	35671.5	(7)	− 169611.	(8)	− 44.7681
(9)	.493680	(10)	− 1.84186E − 04	(11)	1.23475E − 08	(12)	9.40563E − 12
(13)	470.908	(14)	5.10000E − 04				

212 4,4-**Dimethylheptane**

(1)	.128259	(2)	2.43180E + 06	(3)	585.460	(4)	.362583
(5)	408.300	(6)	35378.5	(7)	− 180582.	(8)	2.49899
(9)	.368824	(10)	1.51114E − 05	(11)	− 1.46903E − 07	(12)	5.86868E − 11
(13)	245.287	(14)	5.01000E − 04				

213 **3-Ethyl-2-methylhexane**

(1)	.128259	(2)	2.45207E + 06	(3)	588.260	(4)	.375896
(5)	411.100	(6)	36006.5	(7)	− 169115.	(8)	− 10.6536
(9)	.395677	(10)	− 4.03033E − 05	(11)	− 9.44742E − 08	(12)	4.08362E − 11
(13)	322.147	(14)	4.97000E − 04				

214 **4-Ethyl-2-methylhexane**

(1)	.128259	(2)	2.40140E + 06	(3)	580.160	(4)	.384094
(5)	406.900	(6)	35671.5	(7)	− 171027.	(8)	− 28.4781
(9)	.432125	(10)	− 7.95688E − 05	(11)	− 7.19223E − 08	(12)	3.53641E − 11
(13)	400.683	(14)	5.04000E − 04				

215 **3-Ethyl-3-methylhexane**

(1)	.128259	(2)	2.55339E + 06	(3)	597.560	(4)	.34981
(5)	413.700	(6)	35755.3	(7)	− 172306.	(8)	− 15.6154
(9)	.405687	(10)	− 2.39666E − 05	(11)	− 1.25502E − 07	(12)	5.38905E − 11
(13)	336.791	(14)	4.87000E − 04				

216 **3-Ethyl-4-methylhexane**

(1)	.128259	(2)	2.51286E + 06	(3)	593.860	(4)	.369112
(5)	413.500	(6)	36425.2	(7)	− 163224.	(8)	− 28.7627
(9)	.432469	(10)	− 7.91818E − 05	(11)	− 7.32713E − 08	(12)	3.61062E − 11
(13)	407.865	(14)	4.90000E − 04				

217 2,2,3-**Trimethylhexane**

(1)	.128259	(2)	2.49260E + 06	(3)	588.060	(4)	.335245
(5)	406.750	(6)	34792.3	(7)	− 177862.	(8)	− 14.9620
(9)	.417478	(10)	− 5.34236E − 05	(11)	− 9.81427E − 08	(12)	4.48701E − 11
(13)	312.719	(14)	4.98000E − 04				

218 2,2,4-**Trimethylhexane**

(1)	.128259	(2)	2.38114E + 06	(3)	573.700	(4)	.348840
(5)	399.690	(6)	34038.7	(7)	− 176896.	(8)	− 32.9905
(9)	.454679	(10)	− 9.39875E − 05	(11)	− 7.45387E − 08	(12)	3.90746E − 11
(13)	397.884	(14)	5.07000E − 04				

219 2,2,5-**Trimethylhexane**

(1)	.128259	(2)	2.33048E + 06	(3)	568.060	(4)	.356864
(5)	397.230	(6)	33787.5	(7)	− 190744.	(8)	− 14.5276
(9)	.416589	(10)	− 5.29551E − 05	(11)	− 9.74088E − 08	(12)	4.42944E − 11
(13)	304.912	(14)	5.19000E − 04				

220 2,3,3-**Trimethylhexane**

(1)	.128259	(2)	2.55339E + 06	(3)	596.160	(4)	.328701
(5)	410.830	(6)	35001.6	(7)	− 175476.	(8)	− 14.9720
(9)	.417478	(10)	− 5.34236E − 05	(11)	− 9.81427E − 08	(12)	4.48701E − 11
(13)	316.068	(14)	4.91000E − 04				

221 **2,3,4-Trimethylhexane**

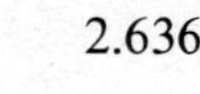

(1)	.128259	(2)	2.52299E + 06	(3)	594.560	(4)	.350663
(5)	412.190	(6)	35713.4	(7)	− 169074.	(8)	− 28.0943
(9)	.444213	(10)	− 1.08597E − 04	(11)	− 4.59175E − 08	(12)	2.70812E − 11
(13)	389.969	(14)	4.94000E − 04				

222 **2,3,5-Trimethylhexane**

(1)	.128259	(2)	2.40140E + 06	(3)	579.260	(4)	.363155
(5)	404.490	(6)	34834.2	(7)	− 176873.	(8)	− 27.8306
(9)	.443914	(10)	− 1.08983E − 04	(11)	− 4.46207E − 08	(12)	2.63660E − 11
(13)	379.990	(14)	5.09000E − 04				

223 **2,4,4-Trimethylhexane**

(1)	.128259	(2)	2.43180E + 06	(3)	581.660	(4)	.341484
(5)	403.800	(6)	34331.8	(7)	− 174510.	(8)	− 32.9905
(9)	.454679	(10)	− 9.39875E − 05	(11)	− 7.45387E − 08	(12)	3.90746E − 11
(13)	401.233	(14)	5.00000E − 04				

224 **3,3,4-Trimethylhexane**

(1)	.128259	(2)	2.62432E + 06	(3)	602.360	(4)	.324011
(5)	413.610	(6)	35169.1	(7)	− 169566.	(8)	− 33.2390
(9)	.454916	(10)	− 9.34174E − 05	(11)	− 7.60492E − 08	(12)	3.98716E − 11
(13)	408.256	(14)	4.84000E − 04				

225 **3,3-Diethylpentane**

(1)	.128259	(2)	2.67498E + 06	(3)	610.060	(4)	.336147
(5)	419.320	(6)	36006.5	(7)	− 164914.	(8)	− 34.1150
(9)	.443887	(10)	− 6.51355E − 05	(11)	− 1.02563E − 07	(12)	4.86606E − 11
(13)	403.461	(14)	4.73000E − 04				

226 **3-Ethyl-2,2-dimethylpentane**

(1)	.128259	(2)	2.55339E + 06	(3)	590.560	(4)	.328800
(5)	406.980	(6)	34834.2	(7)	− 172204.	(8)	− 33.3032
(9)	.461873	(10)	− 1.12802E − 04	(11)	− 5.75342E − 08	(12)	3.37357E − 11
(13)	391.559	(14)	4.86000E − 04				

227 **3-Ethyl-2,3-dimethylpentane**

(1)	.128259	(2)	2.68511E + 06	(3)	606.860	(4)	.345442
(5)	417.850	(6)	35336.6	(7)	− 167179.	(8)	− 33.2390
(9)	.454916	(10)	− 9.34174E − 05	(11)	− 7.60492E − 08	(12)	3.98716E − 11
(13)	402.478	(14)	4.77000E − 04				

228 **3-Ethyl-2,4-dimethylpentane**

(1)	.128259	(2)	2.52299E + 06	(3)	591.260	(4)	.349970
(5)	409.840	(6)	35420.3	(7)	− 169074.	(8)	− 28.0943
(9)	.44213	(10)	− 1.08597E − 04	(11)	− 4.59175E − 08	(12)	2.70812E − 11
(13)	381.302	(14)	4.89000E − 04				

229 2,2,3,3-**Tetramethylpentane**

(1)	.128259	(2)	2.73578E + 06	(3)	610.860	(4)	.279604
(5)	413.420	(6)	35294.7	(7)	− 170527.	(8)	− 47.1117
(9)	.520637	(10)	− 1.90616E − 04	(11)	− 9.19048E − 09	(12)	2.20794E − 11
(13)	430.287	(14)	4.78000E − 04				

230 2,2,3,4-**Tetramethylpentane**

(1)	.128259	(2)	2.56352E + 06	(3)	592.160	(4)	.309956
(5)	406.170	(6)	34289.9	(7)	− 171127.	(8)	− 34.0560
(9)	.472435	(10)	− 1.33365E − 04	(11)	− 3.96916E − 08	(12)	2.79519E − 11
(13)	384.476	(14)	4.900000E − 04				

231 2,2,4,4-**Tetramethylpentane**

(1)	.128259	(2)	2.36087E + 06	(3)	571.360	(4)	.315331
(5)	395.430	(6)	32866.4	(7)	− 175935.	(8)	− 37.2304
(9)	.476255	(10)	− 1.06767E − 04	(11)	− 7.84534E − 08	(12)	4.31763E − 11
(13)	376.606	(14)	5.04000E − 04				

232 2,3,3,4-**Tetramethylpentane**

(1)	.128259	(2)	2.69525E + 06	(3)	609.060	(4)	.298645
(5)	414.700	(6)	34959.8	(7)	− 170524.	(8)	− 32.4893
(9)	.466352	(10)	− 1.22310E − 04	(11)	− 4.91167E − 08	(12)	3.09764E − 11
(13)	375.578	(14)	4.81000E − 04				

SUBSTANCE INDEX FOR APPENDIX

The numbers refer the reader to the Appendix listing of parameters for the substance in question.

INDEX